Satellite Remote Sensing for Ocean and Coastal Environment Monitoring

Satellite Remote Sensing for Ocean and Coastal Environment Monitoring

Guest Editors

Haidong Pan
Daosheng Wang
Jungang Yang

MDPI

Basel • Beijing • Wuhan • Barcelona • Belgrade • Novi Sad • Cluj • Manchester

Guest Editors
Haidong Pan
First Institute of Oceanography
Ministry of Natural Resources
Qingdao
China

Daosheng Wang
College of Marine Science and Technology
China University of Geosciences
Wuhan
China

Jungang Yang
First Institute of Oceanography
Ministry of Natural Resources
Qingdao
China

Editorial Office
MDPI AG
Grosspeteranlage 5
4052 Basel, Switzerland

This is a reprint of the Special Issue, published open access by the journal *Remote Sensing* (ISSN 2072-4292), freely accessible at: https://www.mdpi.com/journal/remotesensing/special_issues/R64SGA4481.

For citation purposes, cite each article independently as indicated on the article page online and as indicated below:

Lastname, A.A.; Lastname, B.B. Article Title. *Journal Name* **Year**, *Volume Number*, Page Range.

ISBN 978-3-7258-6159-0 (Hbk)
ISBN 978-3-7258-6160-6 (PDF)
https://doi.org/10.3390/books978-3-7258-6160-6

Contents

About the Editors

Haidong Pan

Haidong Pan is a tidal scientist at the First Institute of Oceanography, Ministry of Natural Resources. He holds a PhD in Physical Oceanography from the Ocean University of China. His research interests lie in tidal dynamics, particularly tidal evolution under human activities and river flow from multi-source data, including tide gauges and satellite altimeters. He developed the S_TIDE tidal analysis toolbox, which has been widely used by the tidal research community to analyze tidal levels and tidal currents worldwide. His findings are mainly published in SCI journals including *JGR: Oceans, Ocean Modelling, Continental Shelf Research, Estuarine, Coastal and Shelf Science*, and *Remote Sensing*. He is the Principal Investigator (PI) of a National Natural Science Foundation of China (NSFC) project (42206022). He also serves as a Topical Advisory Panel Member for the SCI-indexed journal *Remote Sensing*.

Daosheng Wang

Daosheng Wang is an associate professor at the China University of Geosciences. He is currently engaged in teaching and research in physical oceanography. His research encompasses coastal dynamics, sediment transport, numerical modeling, and data assimilation, as well as emergency information support for maritime incidents such as search-and-rescue and oil spill responses. His specific focus lies in model parameter optimization through data assimilation, model performance improvement, and the simulation and forecasting of oceanic dynamic processes.

Jungang Yang

Jungang Yang is a Research Fellow at the First Institute of Oceanography, Ministry of Natural Resources, Qingdao, China. His research interests include marine dynamic environment remote sensing, satellite altimeter data processing and applications (waves, tides, mesoscale eddies, circulation, and sea level changes, etc.), scatterometer data application technology, and marine underwater dynamic environment intelligent detection. He is dedicated to obtaining high-spatial-and-temporal-resolution sea surface and 3D ocean dynamic environment data and comprehending oceanic mesoscale and sub-mesoscale dynamic process with remote sensing technology. He has expertise in the data processing of altimeters and the reconstruction of 3D ocean temperature and salinity using artificial intelligence technology with sea surface remote sensing data.

remote sensing

MDPI

Article

Phenological Characteristics of the Yellow Sea Spring Bloom: A Comparative Evaluation of Multiple Diagnostic Methods

Kangjie Jin [1], Chen Dong [1], Xihan Liu [2], Yan Sun [1], Jibo Liu [1] and Lei Lin [1,3,*]

[1] College of Ocean Science and Engineering, Shandong University of Science and Technology, Qingdao 266590, China
[2] Hebei Technology Innovation Center for Geographic Information Application, Institute of Geographical Sciences, Hebei Academy of Sciences, Shijiazhuang 050011, China
[3] State Key Laboratory of Estuarine and Coastal Research, East China Normal University, Shanghai 200241, China
* Correspondence: llin@sdust.edu.cn; Tel.: +86-0532-85078327

Abstract

The phenological characteristics of the spring phytoplankton bloom in the mid- and high-latitude oceans, including its initiation, duration, and intensity, can be assessed using various diagnostic methods. However, there is currently a lack of systematic comparisons among these different methods. To elucidate the differences in spring bloom characteristics derived from different approaches and to identify suitable methods for shelf seas, this study comprehensively compares and evaluates the multiple methods for characterizing the spring bloom in the central Yellow Sea, based on satellite-derived chlorophyll-a (Chl-a) data from 2003 to 2020. The methods examined include concentration threshold (CT), cumulative concentration threshold (CCT), rate of change (RoC), and curve-fitting methods for determining bloom initiation; threshold and symmetric methods for estimating duration; and peak, mean, integral, and relative intensity index methods for assessing intensity. The results show that the bloom initiation determined by the CT method occurs earliest (average: Day of Year (DOY) 64), whereas the RoC method identifies a notably later initiation (average: DOY 100), approximately 40 days later. The CCT method yields an intermediate bloom initiation (average: DOY 70), with minimal interannual variability. Notably, curve-fitting methods often produce outliers (e.g., DOY 1) due to the fluctuations in Chl-a time series during winter. The threshold method yields a shorter bloom duration (average: 70 days), while the symmetric method results in a duration of more than 10 days longer. The four intensity assessment methods indicate that bloom intensity initially increased and subsequently decreased from 2003 to 2020, but the peak year varies depending on the method used. Overall, the CCT, symmetric, and relative index methods are more suitable for the Yellow Sea, as their computational results exhibit fewer outliers and relatively low standard deviations. The interannual variations in spring bloom characteristics assessed by different methods display distinct patterns and weak correlations, indicating that methodological choices can lead to divergent interpretations of spring bloom dynamics. Therefore, it is essential to carefully select methods based on research objectives and dataset characteristics.

Keywords: spring bloom; phytoplankton; satellite; yellow sea

1. Introduction

Phytoplankton, as primary producers in marine ecosystems, account for approximately half of global primary productivity. Variations in their biomass significantly influence global marine biogeochemical cycles [1–5]. In temperate and subpolar seas, seasonal changes in biomass are primarily driven by fluctuations in light availability and nutrient supply, resulting in seasonal phytoplankton blooms. Among these phenomena, the spring phytoplankton bloom (hereafter referred to as the spring bloom), characterized by a substantial increase in phytoplankton populations, was one of the earliest recognized marine phenomena [6] and continues to be a prominent feature of the annual phytoplankton biomass cycle [7–9]. The bloom has a significant impact on ecosystem energy flow and further influences the growth and reproduction of fish and other higher-trophic-level organisms [10–12]. Therefore, investigating and elucidating the phenological characteristics and variations in the spring bloom is essential for understanding marine ecosystem dynamics [8,13].

Recent advancements in satellite remote sensing have greatly expanded the spatial coverage and temporal range of marine data, significantly enhancing research on spring blooms [14–18]. Analyzing continuous satellite-derived sea surface chlorophyll-a (Chl-a) data is a key pathway for diagnosing spring bloom characteristics. However, different studies employ varying methods and criteria when quantifying seasonal bloom features [19,20]. For instance, Siegel et al. [21] defined the initiation of the North Atlantic bloom as the point at which Chl-a concentrations exceeded by 5% of the annual median. In contrast, Niu et al. [17] established a threshold of the annual median plus 15% to determine the initiation of the spring bloom in the South Yellow Sea. Ueyama et al. [22], in their study of seasonal blooms in the North Atlantic, utilized a sigmoidal curve fit to cumulative Chl-a variance time series to identify bloom initiation and termination. Meanwhile, Zhai et al. [23], investigating phytoplankton bloom phenology in the Northwest Atlantic, employed Gauss curve fitting to analyze Chl-a concentrations for assessing bloom characteristics and interannual variations. Brody et al. [24], using satellite Chl-a data, compared three methods—rate of change, concentration threshold, and cumulative concentration threshold—for determining the initiation of the spring bloom in the North Atlantic and reported significant discrepancies (up to 20 days) among the results. This highlights the potential for different diagnostic methods to yield varying conclusions in studies of seasonal blooms. Key characteristics of the spring bloom include initiation, duration, and intensity [13,17,25,26], each of which can be evaluated using multiple calculation methods. However, there is a notable lack of comprehensive summarizations and comparisons of these diagnostic methods. Thus, the differences and limitations among these methods have not been fully understood yet.

This study aims to elucidate the differences in the characteristics of spring bloom obtained by various methods and to identify a suitable method for a typical shelf sea—the Yellow Sea (Figure 1). First, we systematically reviewed the primary methods currently employed for diagnosing spring bloom characteristics. Subsequently, we applied these multiple methods to calculate bloom initiation, duration, and intensity in the central Yellow Sea using satellite Chl-a data in order to assess the differences and consist in spring bloom characteristics. Finally, we conducted a comparative analysis of the results obtained from different methods, highlighting their respective strengths and limitations. Our findings may serve as a valuable reference for future research on spring bloom phenology.

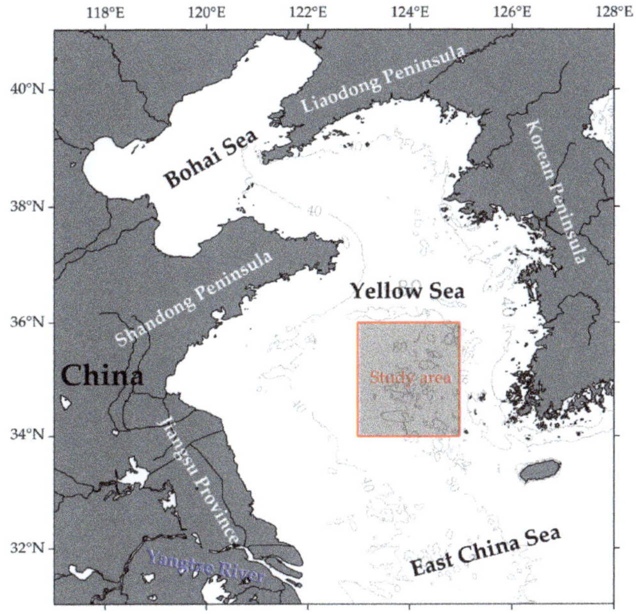

Figure 1. Topography of the Yellow Sea. The gray solid lines represent isobaths (in meters), and the shaded area indicates the study region (34–36°N, 123–125°E).

2. Study Area

The Yellow Sea, a typical temperate shelf sea located in the western Pacific Ocean, spans an area of approximately 380,000 km^2 with an average depth of 44 m (Figure 1). The spring bloom is one of the most significant phenomena in the annual cycle of phytoplankton in the Yellow Sea, accounting for approximately 40% of the Yellow Sea's annual primary productivity [27] and exerting a considerable impact on its fishery resources. Existing studies suggest [28–30] that the spring bloom in the Yellow Sea typically commences in March and concludes in early May, with a peak occurring in April. However, variations in physical conditions and nutrient availability lead to interannual variability in the characteristics of the bloom [29]. Several methods—including the concentration threshold method, cumulative concentration threshold method, Gauss curve fitting, and sigmoidal curve fitting—have been employed to determine bloom initiation in the Yellow Sea [15,25,29,31]. For example, Shi et al. [25] utilized the cumulative concentration threshold method to identify the initiation of the spring bloom in the central South Yellow Sea, while also employing the Peak Method and symmetric method to estimate bloom intensity and duration. Liu et al. [15] determined bloom initiation and termination by fitting a sigmoidal curve to Chl-a time series and assessed bloom intensity using a relative intensity index. Meanwhile, Kim et al. [31] applied Gauss curve fitting in their investigation of phytoplankton blooms in the South Yellow Sea to define bloom initiation. These studies illustrate that various methods have been utilized to characterize the spring bloom in the Yellow Sea; however, their consistency and discrepancies have yet to be thoroughly examined.

3. Materials and Methods

This section summarizes the primary computational methods for diagnosing bloom characteristics, with a focus on three key characteristics: bloom initiation, duration, and intensity.

3.1. Bloom Initiation

Common methods for determining bloom initiation using Chl-a time series mainly include the concentration threshold method, cumulative concentration threshold method, curve-fitting method, and rate of change method (Table 1).

3.1.1. Concentration Threshold Method (CT)

The concentration threshold method (hereafter referred to as the CT Method) identifies the initiation of blooms by establishing a specific threshold concentration of Chl-a. The bloom initiation is designated when Chl-a levels surpass this threshold [24]. However, the selection of concentration thresholds varies among studies. A prevalent method is to define the threshold as a percentage of the annual median Chl-a concentration [21,32–35]. For instance, Siegel et al. [21] established the threshold at 5% above the annual median Chl-a concentration (i.e., threshold = annual median $\times$ 105%) in their investigation of the North Atlantic spring bloom using satellite data (hereafter referred to as the CT-S Method). They tested thresholds ranging from 1% to 30% above the median and found minimal differences in the resulting outcomes. This method is regarded as a reliable indicator for bloom initiation, as it effectively captures the initial growth phase of blooms rather than relying solely on peak concentrations [21]. Henson et al. [32,33] enhanced this method by requiring that Chl-a concentrations remain above the threshold for three consecutive days to mitigate the effects of short-term extreme events (hereafter referred to as the CT-H Method). Brody et al. [24] utilized a similar method but with a modification: they began from the annual maximum Chl-a concentration and searched backward to identify the first instance where Chl-a exceeded the threshold, provided it was preceded by two consecutive days of sub-threshold concentrations (hereafter referred to as the CT-B Method). Some studies have employed fixed empirical Chl-a concentration values as thresholds. For example, Fleming et al. [36] defined the initiation of the spring bloom in the Baltic Sea as occurring when Chl-a concentrations first exceeded 5 $\mu g\ L^{-1}$. However, due to significant regional variations in baseline Chl-a levels, this fixed-threshold method lacks universality [32].

Table 1. An overview of methods for calculating bloom characteristics (more details in Section 3). Threshold = 105% of the median annual Chl-a concentration.

Bloom Characteristics	Method	Principle	Reference
Initiation	CT-S	When the concentration of Chl-a first exceeds the threshold.	[21]
	CT-H	When the concentration of Chl-a exceeds the threshold for three consecutive days, the first day of these three days is selected.	[32,33]
	CT-B	When Chl-a first exceeds the threshold following two consecutive days of values below this threshold (searching backward from the peak).	[24]
	CCT-X	When the daily integrated Chl-a concentration reaches 25% of the total integration from January to August.	[25,37,38]
	CCT-Y	Consistent with CCT-X, but the starting point for the integration is the date of minimum Chl-a concentration rather than DOY 1.	[24]
	RoC	The point at which the rate of change in Chl-a (dChl-a/dt) reaches its maximum.	[24,39]
	Gauss curve-fitting	Based on the time series of Chl-a fitted using a Gauss model, initiation was defined as t_Peak $-$ 2σ (the peak time and standard deviation of the Gauss fitting curve, respectively).	[40]
	Sigmoidal curve-fitting	The point at which the slope of the sigmoidal fitting curve first equals 20% of the maximum slope.	[22]

Table 1. *Cont.*

Bloom Characteristics	Method	Principle	Reference
Duration	Threshold	The bloom ends when the Chl-a concentration falls below the threshold for the first time after reaching its peak. The duration is the period from its initiation to its termination.	[17,19]
	Symmetric	Twice the interval between bloom initiation and peak.	[25]
Intensity	Peak	Peak concentration of Chl-a during the bloom.	[13,28,29]
	Integral	The integral value of Chl-a concentration during the bloom duration.	[41]
	Mean	The average Chl-a concentration during the bloom duration.	[41]
	Index	The ratio of the mean Chl-a concentration during the bloom period to the non-bloom period.	[22]

3.1.2. Cumulative Concentration Threshold Method (CCT)

The cumulative concentration threshold method (hereafter referred to as the CCT Method) involves integrating the Chl-a concentration time series, with bloom initiation defined as the point at which the integrated Chl-a value exceeds a specified threshold (CCT-X Method) [24,37,38]. This threshold is typically established as a percentage of the total cumulative Chl-a biomass. For example, Shi et al. [25] aggregated daily Chl-a concentration data from January to August and proposed that the initiation of spring blooms occurs when the cumulative value exceeds 30% of the total Chl-a biomass. Brody et al. [24] modified the preprocessing step of this method by utilizing the minimum value in the Chl-a time series as the starting point (before the peak occurs) for integration (CCT-Y Method), a process referred to as "series offset." The authors suggested that for subtropical regions, the optimal threshold for bloom initiation falls within the 10–15% range, while for subpolar regions, a threshold of 15–20% is more appropriate. In this study, we observed minimal sensitivity to threshold selection and consequently adopted a 25% threshold.

3.1.3. Curve-Fitting Methods

When employing curve-fitting methods to determine bloom initiation, it is essential to select an appropriate fitting function based on the characteristics of the Chl-a time series, including its waveform pattern. Commonly used fitting functions include Gauss and sigmoidal functions.

1. Gauss fitting method

Yamada and Ishizaka introduced a method for characterizing phytoplankton blooms based on a Gauss curve, which they applied to the analysis of spring blooms in the Japan Sea [40]. This method has since been utilized by numerous researchers to investigate bloom characteristics across various marine regions [31,42–44]. The fitting equation is given by:

$$B(t) = B_0 + Peak \times exp\,[-\,(t - t\_Peak)^2/(2\sigma^2)] \tag{1}$$

where B_0 represents the background Chl-a concentration (mg/m^3), *t_Peak* indicates the timing of peak Chl-a concentration (in days), and *Peak* denotes the maximum Chl-a concentration in the fitted curve. $B(t)$ is the fitted Chl-a concentration at time t, and σ is the standard deviation of the Gauss curve, which defines the width of the peak [25]. The bloom initiation is defined as the time when $t = t\_Peak - 2\sigma$.

2. Sigmoidal fitting method

Ueyama and Monger [22] developed a method based on cumulative variance and a sigmoidal curve to explore the relationship between bloom initiation/intensity and wind forcing in the North Atlantic (10°N–70°N, 90°W–10°E). The processing workflow involves the following steps: (1) applying a three-point median filter to Chl-a data for outlier removal; (2) performing gap-filling using linear interpolation; (3) temporally aligning the time series to center the most pronounced seasonal bloom (i.e., positioning the peak at the midpoint of the series); (4) applying three-point smoothing to the processed Chl-a data, followed by the calculation of 5-day moving window variances centered on each daily observation; and (5) cumulatively summing the resulting variance time series. The cumulative variance time series is then fitted with a sigmoidal function:

$$f(t) = C_1/(1.0 + exp\,(C_2 - C_3t)) + C_4 \tag{2}$$

where t represents time (in days) and C_1–C_4 are fitting coefficients. Bloom initiation is defined as the time when the curve's slope reaches one-twentieth of its maximum slope.

3.1.4. Rate of Change Method (RoC)

Alternative methods for determining bloom initiation utilize thresholds based on rates of chlorophyll variation. For example, Sharples et al. [39] identified the initiation of spring blooms as the timing of the maximum daily Chl-a variation rate in surface waters, suggesting that the period of most rapid biomass accumulation marks the initiation of blooms. Brody et al. [24] refined this method by first processing Chl-a time series using Harmonic Analysis of Time Series coupled with Fast Fourier Transform (HANTS-FFT) [45], subsequently identifying the time point of the maximum rate of change (referred to as the Rate of Change (RoC) Method). This preprocessing constrains the initiation to occur prior to Chl-a peaks, thereby minimizing the potential for anomalous detections [24].

More complex statistical techniques have also been employed. Friedland et al. [46] utilized the Sequential T-test Analysis of Regime Shifts (STARS) [45,47,48] to investigate the dynamics of spring blooms and zooplankton biomass responses in the Northeast U.S. continental shelf ecosystem. This method identifies abrupt changes in Chl-a time series to determine bloom initiation, termination, and duration. While some studies have utilized relationships between phytoplankton growth and respiration rates or other biological indicators for bloom detection [49], the present study focuses exclusively on Chl-a time series analyses, thus excluding these alternative methods.

3.2. *Bloom Duration*

Bloom duration is defined as the time interval from the initiation to the termination of a phytoplankton bloom. While bloom initiation can be determined using the aforementioned methods, accurately identifying bloom termination is crucial for calculating bloom duration. Two primary methods exist for determining termination time: the first employs the same concentration threshold method used for identifying bloom initiation (termed the threshold method) [17,19]; the second approximates termination time as the point symmetric to the initiation time relative to the chlorophyll peak [25], whereby bloom duration is calculated as twice the interval from bloom initiation to peak time (termed the symmetric method).

3.2.1. Threshold Method

Several studies have utilized threshold criteria analogous to those applied for bloom initiation to ascertain bloom termination, thereby determining the total duration of the bloom. A notable example is the method employed by Racault et al. [19], who established a consistent threshold at 5% above the annual median chlorophyll concentration to identify both bloom initiation and termination, subsequently calculating the overall bloom duration.

3.2.2. Symmetric Method

The symmetric method is particularly well-suited for the Gauss and sigmoidal fitting techniques described in Section 3.1.3, as both fitting curves exhibit symmetry around the central point of the time series. Consequently, the bloom initiation and termination times determined by these fitting functions are also symmetric about this central point. In the Gauss fitting method, bloom initiation is defined as the time corresponding to $t\_Peak - 2\sigma$, while termination corresponds to $t\_Peak + 2\sigma$, resulting in a total bloom duration of 4σ [25]. For the sigmoidal fitting method, bloom initiation and termination are identified as the time points when the curve's slope reaches one-twentieth of its maximum slope, with the interval between these points constituting the bloom duration [15].

3.3. Bloom Intensity

Bloom intensity is commonly employed to characterize the magnitude of phytoplankton blooms. Three primary computational methods exist: (1) utilizing the peak Chl-a concentration during the bloom period as the intensity metric; (2) deriving intensity from either the mean or integrated Chl-a concentration over the bloom duration; and (3) calculating a relative intensity index.

3.3.1. Peak Method

Platt et al. [13] defined bloom intensity as the maximum Chl-a concentration within the bloom period in their investigation of North Atlantic spring blooms. This method necessitates the identification of either the observed peak concentration or the fitted maximum value from previously described curve-fitting methods, with the magnitude of this peak serving as the intensity measure [28,29].

3.3.2. Integral/Mean Method

An alternative method quantifies intensity through either integrated Chl-a concentration (the summation of daily spatial average Chl-a concentrations over the bloom duration) or mean concentration. The integral method, sometimes referred to as "bloom magnitude" [41], inherently incorporates both bloom duration and Chl-a concentration levels, thereby providing a comprehensive assessment of bloom conditions while reducing sensitivity to extreme events. However, some studies contend that biomass concentration alone better reflects bloom intensity, leading to the preference for mean Chl-a concentration (integrated concentration divided by the duration in days) as the preferred metric [41].

3.3.3. Relative Intensity Index

Ueyama and Monger [22] developed a relative intensity index (hereafter referred to as the Index Method), defined as the ratio of mean Chl-a concentration during bloom versus non-bloom periods. This index effectively normalizes spatial and interannual variability in background Chl-a levels, providing a more standardized measure of intensity [15,22].

3.4. Data Sources

The study employed Chl-a data from the Yellow Sea, derived from a moderate resolution imaging spectroradiometer (MODIS) remote sensing reflectance dataset covering the period from 2003 to 2020. This dataset was developed by Wang et al. [50] using a Generalized Additive Model (GAM) algorithm, which has been rigorously validated against a comprehensive set of in situ measurements, demonstrating high accuracy. This high level of accuracy is further supported by its extensive application in research related to the Yellow Sea [51–53]. The original dataset featured a temporal resolution of 7 days, which we processed using linear interpolation to achieve daily resolution while maintaining a spatial resolution of approximately 4 km $\times$ 4 km.

The study utilized the climatological mean Chl-a to quantify the climatological characteristics of the spring bloom. Climatological mean values were calculated by averaging the Chl-a of each day from 2003 to 2020. Subsequently, the daily Chl-a concentration of the climatological mean was used to represent the climatological seasonal variation in Chl-a and to diagnose the climatological characteristics of the spring bloom in the Yellow Sea.

3.5. Correlation Analysis

To investigate the relationships between different characteristics calculated using various methods, a correlation analysis was conducted on the bloom initiation, intensity, and duration derived by these methods. The Pearson correlation coefficient (r) was employed to quantify the strength of the correlations between the variables; an absolute value of r closer to 1 indicates a stronger correlation between the two variables. The p-value from the t-test was used to assess the significance level of the correlations, with $p < 0.05$ indicating a statistically significant relationship between the variables.

4. Results

Based on the aforementioned diagnostic methods, we conducted comprehensive analyses of the bloom initiation, duration, and intensity in the central Yellow Sea (indicated by the shaded area in Figure 1) using satellite-derived Chl-a data. To minimize potential interference from autumn blooms (September-December) [29], our analysis exclusively utilized Chl-a concentration data from January to August for the central Yellow Sea region.

In our analysis, we found that the CT-H Method produced results nearly identical to those of the CT-S Method. This similarity is likely attributable to the tendency of Chl-a concentrations to remain above the threshold for several subsequent days in our dataset once they exceeded it. This observation may be linked to the relatively low temporal resolution (7 days) of the original Chl-a data. Consequently, the CT-H Method was not employed in subsequent calculations.

4.1. Bloom Initiation of the Yellow Sea

We employed four distinct methods—the CT Method, CCT Method, RoC Method, and curve-fitting methods (Gauss and sigmoidal)—to systematically assess the initiation of spring blooms in the central Yellow Sea. In particular, since the minimum values of the Chl-a time series occur on DOY 1, both the CCT-X and CCT-Y methods yield identical results. Therefore, we represent the results using the "CCT Method" in Figure 2.

Analyses of climatological mean Chl-a time series revealed significant methodological discrepancies, with differences in bloom initiation estimates exceeding three months (Figure 2). The CT-S Method indicated the earliest bloom initiation on Day of Year (DOY) 28, while its modified version (CT-B Method) delayed detection by 12 days (Figure 2a), yet still exhibited substantial divergence from the other methods. Both implementations of the CCT Method consistently identified DOY 70 as the initiation date (Figure 2 b). The RoC Method produced the latest initiation estimates (DOY 93), while the Gauss and sigmoidal fittings yielded intermediate values of DOY 81 and DOY 77, respectively.

Interannual variability analysis (Table 2) indicated that the CT-S Method produced anomalously early initiation estimates (mean: DOY 17). Given that peak Chl-a concentrations typically occur in April (averaging around DOY 111), we imposed an additional constraint in the CT-B Method by limiting the peak detection window to DOY 75–130 (late March to early May). This temporal constraint effectively filtered out spurious chlorophyll maxima occurring outside the typical spring bloom period, thereby enhancing the phenological relevance of the identified initiation. The CT-B Method yielded significantly later initiation estimates compared to the CT-S Method (mean: DOY 68 vs. 17), but with an 8-day larger standard deviation, indicating greater interannual variability in the results. This underscores the influence of threshold selection criteria in the CT Method on bloom initiation estimates. The CCT-X Method exhibited notable temporal stability (mean: DOY 70 ± 8), with minor adjustments (CCT-Y, mean: DOY 75 ± 8) observed when employing series offset initialization.

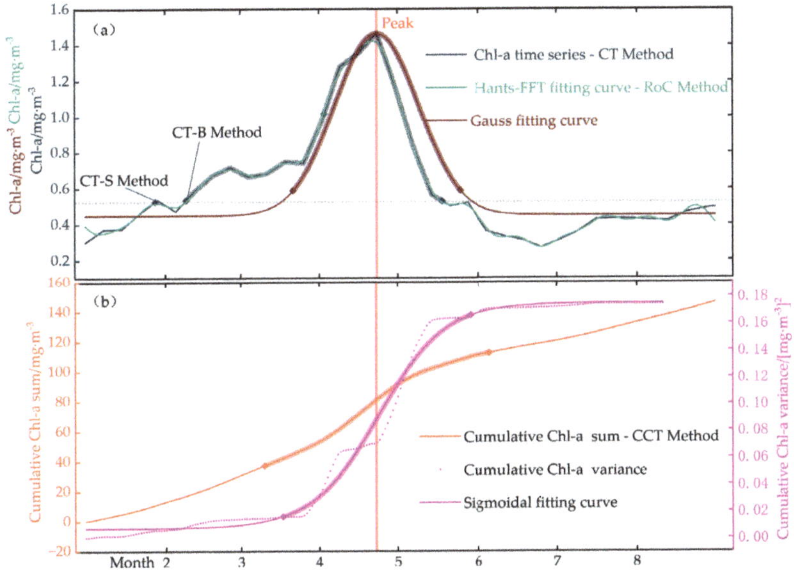

Figure 2. The calculated results of bloom initiation are presented, with diamonds indicating the initiation and termination of blooms; the shaded area between markers represents bloom duration. (**a**) The CT, RoC, and Gauss fitting method and (**b**) the CCT method and sigmoidal fitting method. The gray dashed line in (**a**) denotes the threshold value established by the CT Method.

Table 2. The calculated bloom initiation (in days) presents results from four methods: CT-S and CT-B represent the CT Method as implemented by Siegel et al. [21] and Brody et al. [24], respectively, while CCT-X and CCT-Y denote the CCT Method applied to the original time series and offset time series, correspondingly. σ represents the standard deviation.

Years	CT Method		CCT Method		RoC Method	Curve-Fitting Methods	
	CT-S	CT-B	CCT-X	CCT-Y		Gauss	Sigmoidal
2003	13	68	72	72	95	83	44
2004	1	76	58	83	1	−18	1
2005	26	69	74	77	100	93	81
2006	13	60	74	74	142	117	92
2007	16	34	68	68	91	87	83
2008	11	46	79	86	91	83	80
2009	29	72	73	73	108	97	64
2010	20	32	55	55	120	−25	37
2011	40	81	85	85	106	101	96
2012	23	94	75	75	112	104	37
2013	33	86	73	78	112	106	101
2014	1	79	64	80	1	−11	50
2015	13	47	64	64	108	104	102
2016	25	44	75	80	24	64	1
2017	1	86	59	67	97	83	1
2018	1	42	68	84	106	101	100
2019	11	53	66	66	96	52	38
2020	22	74	75	79	83	82	10
Mean ± σ	17 ± 11	64 ± 19	70 ± 8	75 ± 8	89 ± 39	72 ± 43	57 ± 36

The results from the RoC Method (mean: DOY 89, or DOY 100 when excluding outliers from 2004 and 2014, Table 2) consistently aligned with the peak timing of chlorophyll concentrations (mean: DOY 111), which is significantly later than the estimates from other methods. The curve-fitting methods demonstrated the highest interannual variability in bloom initiation estimates, with Gauss and sigmoidal fittings exhibiting substantially greater standard deviations (43 and 36 days, respectively) compared to the other methods. These estimates also deviated significantly from climatological mean values. The performance of both fitting algorithms was particularly compromised in years characterized by multiple short-lived bloom events (e.g., 2014, 2016, and 2017), where increased variability in Chl-a time series hindered successful curve convergence, rendering initiation identification ineffective. Additionally, the Gauss fitting method produced non-physical negative initiation values in certain years, an artifact resulting from compressed inter-bloom intervals that led to excessively broad fitted peaks extending beyond the temporal domain of the time series. Collectively, these limitations suggest that curve-fitting methods are more suitable for analyzing smoothed climatological data than for investigating interannual variability in spring bloom phenology.

In summary, comparative analysis indicates that while the CT-B Method and CCT Method yield comparable mean initiation estimates, the former exhibits greater interannual variability. The late initiation estimates from the RoC Method and the excessive variability from the curve-fitting methods limit their applicability for interannual studies.

We conducted a correlation analysis on the results presented in Table 2 (Figure 3), excluding years with outliers (e.g., 2004, 2014) to mitigate their influence on the analysis. The results revealed positive correlations among methods based on the same framework. Specifically, the variants CT-S and CT-B of the CT Method exhibited a significant correlation (r = 0.7, $p < 0.01$), as did the variants CCT-X and CCT-Y of the CCT Method (r = 0.94, $p < 0.01$). The high correlation observed among different implementations of the CCT Method indicates that time series offset adjustments have a minimal impact. Consequently, we excluded series offsets from subsequent analyses. Significant positive correlations were also found between the CT-B Method and the CCT-X Method, as well as between the results of the Gauss fitting and RoC Method. No significant correlations were detected among other methodological pairs. These findings underscore the importance of the chosen analytical method, which can significantly influence the derived interannual variability patterns in bloom initiation, potentially leading to varying ecological interpretations depending on the selected method.

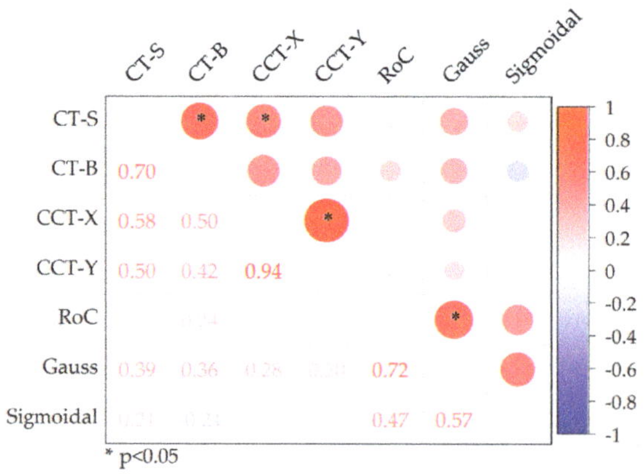

Figure 3. Correlation among different bloom initiation quantification methods. The data is sourced from Table 2. CT-S and CT-B represent the CT Method as implemented by Siegel et al. [21] and Brody et al. [24], respectively, while CCT-X and CCT-Y denote the CCT Method applied to the original time series and offset time series, correspondingly.

4.2. Bloom Duration of the Yellow Sea

This section utilized both the symmetric method and the threshold method to calculate the climatological mean duration of spring blooms in the central Yellow Sea from 2003 to 2020 (Figure 2). Analysis of climatologically averaged Chl-a data revealed method-dependent variations in bloom duration estimates: the symmetric method yielded 87 days, the threshold method produced 99 days, the Gauss fitting resulted in 80 days, and the sigmoidal fitting provided the shortest estimate of 73 days. The maximum inter-method discrepancy was one month, with the threshold method generating the longest durations and sigmoidal fitting yielding the shortest.

The analysis of interannual variability (Figure 4), which excluded curve-fitting methods due to their susceptibility to outliers (Section 4.1), demonstrated that durations derived from the threshold method exhibited the lowest variability (mean: 71 ± 21 days). When using initiation dates determined by the CT-B Method, the symmetric method produced substantially longer duration estimates (mean: 92 ± 38 days), representing a 17-day increase compared to direct threshold method calculations (α in Figure 4), along with greater interannual fluctuations. For initiations derived from the CCT-X Method, symmetric method durations averaged 80 ± 26 days, differing by more than 10 days from other estimates. A significant correlation was observed between the threshold and symmetric methods for identical initiation dates ($r = 0.87$, $p < 0.001$), although the threshold method exhibited a standard deviation that was 17 days lower. No significant correlations were found when comparing durations derived from different initiation methods.

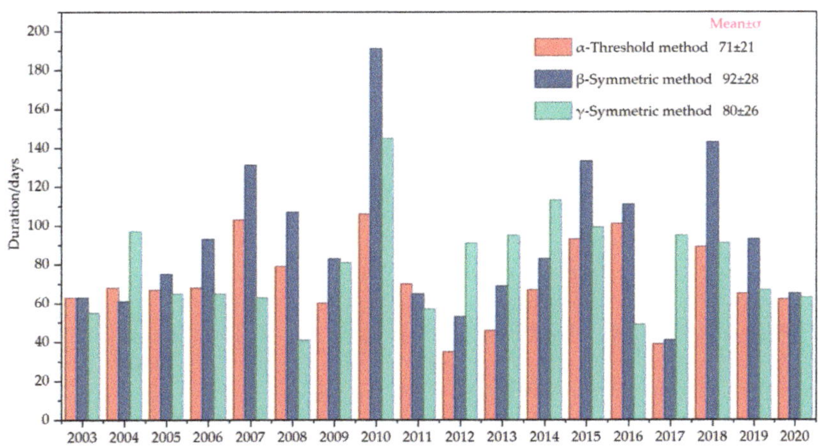

Figure 4. Estimates of bloom duration (in days) by the three methods. Bars labeled α and β were derived using the initiation from the CT-B method (Table 2) along with the threshold and symmetric methods, respectively, while bars labeled γ were derived using the initiation from the CCT-X method (Table 2) along with the symmetric methods. σ represents the standard deviation of the values across all years.

4.3. Bloom Intensity of the Yellow Sea

This section employed four methods to quantify bloom intensity: the Peak Method, Integral Method, Mean Method, and Index Method. The latter three methods required bloom duration inputs, which were calculated using the three distinct duration estimates presented in Figure 4, with results visualized in Figure 5.

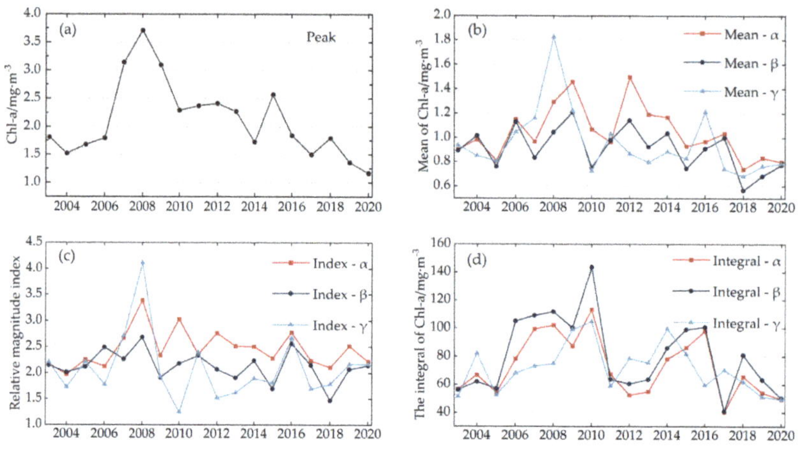

Figure 5. Bloom intensities of the spring bloom in the central Yellow Sea calculated by the four methods. Subfigures (**a–d**) correspond to the Peak, Mean, Index, and Integral Methods, respectively. Additionally, α, β, and γ denote the duration metrics specified in Figure 4 that are employed for the intensity calculations. Durations α and β were calculated using initiation data from the CT-B method, whereas duration γ was calculated using initiation data from the CCT-X method.

The interannual variability of bloom intensity exhibited notable differences across calculation methods. From 2003 to 2006, the Mean and Integral Methods demonstrated similar patterns of initial increase followed by a decrease and subsequent increase, whereas the Peak and Index Methods displayed an opposite trend of initial decrease followed by an increase. From 2007 to 2010, the Peak and Index Methods produced consistent temporal trends, while the Mean and Integral Methods yielded divergent patterns. After 2014, all methods indicated an overall declining trend in bloom intensity, with the Index Method showing the weakest decline. Since the Index Method incorporates both bloom and non-bloom period Chl-a concentrations, substantial differences in relative intensity indices may arise even when mean concentrations are similar (e.g., 2013 vs. 2014). While the Peak, Mean, and Index Methods all identified 2008 as the year of maximum intensity, the Integral Method indicated 2010 as the peak year. These results demonstrate that, for a given duration calculation method, different intensity estimation methods can yield substantially different results, whereas using the same method with varying duration inputs produces relatively consistent interannual patterns.

Correlation analysis revealed relationships among various methods (Figure 6). The Peak Method demonstrated significant correlations with the Mean Method, Integral Method, and Index Method when using durations derived from the threshold method (α in Figure 4) ($r > 0.5$, $p < 0.05$). Additionally, the Integral Method and Index Method exhibited a strong positive correlation ($r > 0.5$, $p < 0.05$). In contrast, for durations calculated using the symmetric method (β in Figure 4), only the Integral Method and Peak Method maintained a significant correlation ($r = 0.58$, $p < 0.05$). Notably, when employing durations derived from the symmetric method (γ in Figure 4), the Mean Method and Index Method displayed particularly strong correlation ($r = 0.85$, $p < 0.05$), but no correlation for other durations. Overall, for varying duration inputs, intensity values derived from the same calculation method usually showed high correlation; for example, the three results obtained from the Integral Method demonstrated significant positive correlations with each other (all $r > 0.5$, $p < 0.05$). However, the correlations between different methods were weak and non-significant, suggesting that different methods may capture different interannual variations in spring bloom intensity.

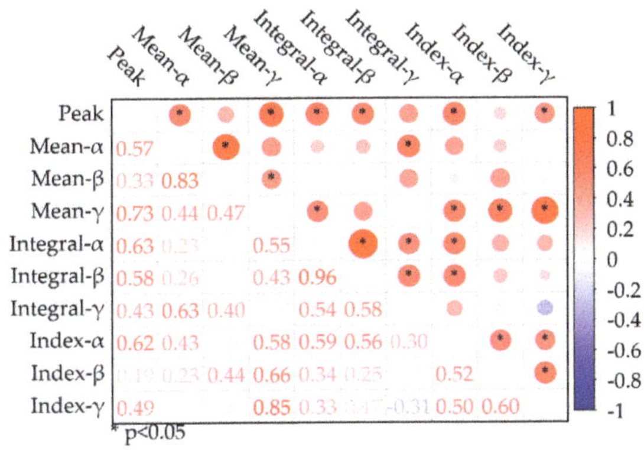

Figure 6. Correlations of bloom intensities among various methods, with α, β, and γ denoting the different duration metrics referenced in Figure 4 that were employed in the intensity calculations. Durations α and β were calculated using initiation data of the CT-B method, whereas duration γ was calculated using initiation data of the CCT-X method.

5. Discussion

5.1. Performances of Various Methods

In our study, we summarized the performances of different methods and compiled them into a table (Table 3). The initiation for the CT-S/H calculation was relatively early (average of DOY 17) and was significantly influenced by high Chl-a concentrations during winter. The cause of this issue is explained in detail in Section 5.2. In contrast, the initiation for the CT-B calculation was noticeably later (approximately 50 days later on average) but exhibited a larger standard deviation (19 days). The standard deviation for the initiation calculated using the CCT-X method was the smallest (8 days). The fitting methods (Gauss and Sigmoidal) were prone to outliers (e.g., values of DOY 1 or negative values) and are not recommended for use. Regarding duration, although the threshold method produced a smaller standard deviation (21 days), the symmetric method is more convenient for calculations. For bloom intensity, the different methods reflect varying characteristics of phytoplankton biomass (measured as Chl-a) during the bloom period, with further details presented in Table 3.

Table 3. Features of different calculation methods.

Bloom Characteristics	Method	Features
Initiation	CT-S	While it is straightforward to calculate, this method is also significantly affected by elevated Chl-a levels during winter, resulting in the presence of outliers.
	CT-H	Similar to the CT-S Method.
	CT-B	In comparison to CT-S/H, the impact of high concentrations of Chl-a during winter is relatively minor.
	CCT-X	Relatively stable with few outliers.
	CCT-Y	Similar to the CCT-X method, but with a more complex calculation.
	RoC	The calculated results were very late, approaching the time of peak.
	Gauss curve-fitting	Restricted by the distribution of the Chl-a time series, outliers (even negative values) occur.
	Sigmoidal curve-fitting	Similar to the Gauss curve-fitting method.
Duration	Threshold	Easily calculated, but influenced by the setup of threshold(s).
	Symmetric	It can be conveniently calculated based on the bloom initiation.

Table 3. *Cont.*

Bloom Characteristics	Method	Features
Intensity	Peak	Not affected by calculations of initiation and duration, but strongly influenced by extreme values.
	Integral/Mean	Reflects the overall levels of Chl-a concentration during bloom periods, but is influenced by the Accuracy of duration calculation.
	Relative Index	Accounts for Chl-a concentration during non-bloom periods, thereby providing an estimate of bloom intensity based on the complete seasonal variation.

5.2. Influence of Winter Chl-a Elevation on the CT Method

The CT Method exhibited anomalous bloom initiation (DOY 1) in specific years (e.g., 2004, 2014), as illustrated in Table 2. This artifact persisted across a range of threshold adjustments (5–30% above the median). An analysis of the Chl-a time series for these years (Figure 7) revealed wintertime peaks that significantly exceeded thresholds, leading to false initiation signals. Two primary mechanisms account for these elevated winter values: (1) ephemeral stabilization of the water column under favorable meteorological conditions may allow minor bloom events to occur, and (2) vigorous winter mixing can induce sediment resuspension, potentially inflating satellite-derived Chl-a estimates [39].

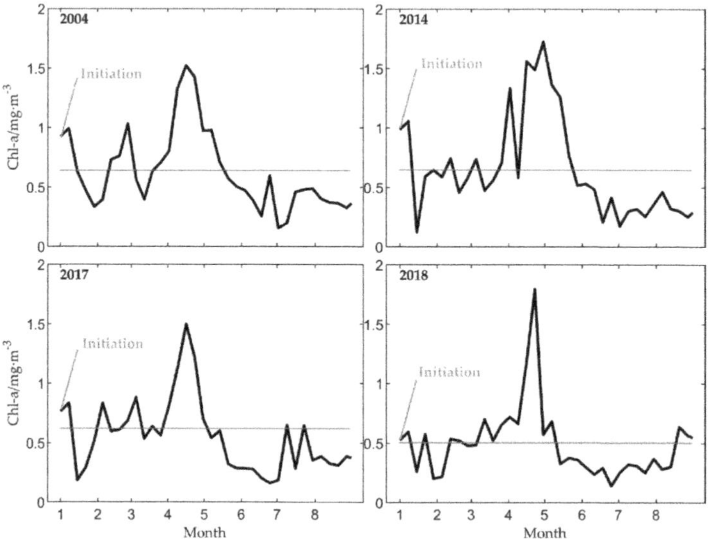

Figure 7. Time series of Chl-a in the central Yellow Sea from January to August in 2004, 2014, 2017, and 2018. The red solid line indicates the concentration threshold. The gray font indicates the initiation of spring blooms in winter diagnosed by the CT method for the corresponding years in Table 2.

The supplemental criterion of the CT-H Method, which required sustained threshold exceedance for three days, proved ineffective, as winter anomalies typically persisted beyond this duration, yielding results nearly identical to those of the CT-S Method. Conversely, the CT-B Method's backward-search algorithm—initiating from spring bloom peaks and retroactively identifying threshold crossings—effectively mitigated winter interference. Therefore, the modified CT Method proposed by Brody et al. [24] demonstrates superior reliability for detecting bloom initiation.

5.3. Interrelationships Between Spring Bloom Characteristics

The intrinsic definition of bloom duration as the interval between initiation and termination establishes a fundamental interdependence with initiation. Correlation analyses (Figure 8a) reveal significant negative correlations between initiation and duration across all methods (threshold method with CT-derived initiation [α]: r = −0.91; symmetric method with CT-derived initiation [β]: r = −0.90; symmetric method with CCT-derived initiation [γ]: r = −0.76; all $p < 0.01$), confirming that earlier initiation is consistently associated with extended bloom durations.

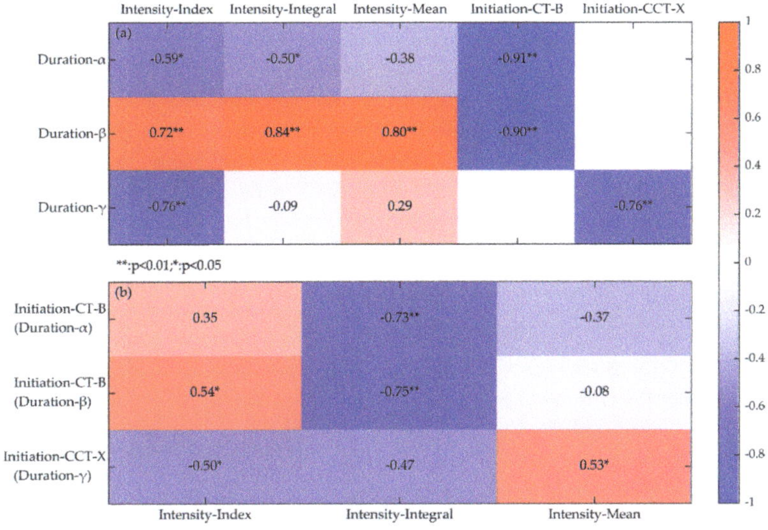

Figure 8. Correlations among bloom characteristics. (**a**) Relationships between bloom duration and intensity/initiation, with duration-α, β, and γ corresponding to the duration metrics in Figure 4, while initiation-CT-B and CCT-X indicate the initiation points from Table 2. (**b**) Relationships between initiation and bloom intensity. The y-axis represents the initiations corresponding to duration-α, β, and γ, while the x-axis indicates the bloom intensity calculated using different methods.

The relationship between bloom intensity and duration exhibits substantial methodological dependence. For threshold durations (α, Figure 8a), both the Integral Method and Index Method show strong duration dependence ($|r| > 0.5$), indicating a significant influence of temporal extent on intensity metrics (excluding the Peak Method). In contrast, the Mean Method demonstrates a weaker correlation. However, for symmetric durations (β and γ, Figure 8a), all intensity metrics (Mean, Integral, and Index Methods) exhibit significant correlations ($r > 0.7$, $p < 0.01$). This suggests that when employing the symmetric method, the calculated results for bloom intensity are more reliant on duration. Notably, for the relative intensity index method, the duration calculated by either method significantly affects the intensity calculations ($|r| > 0.5$).

The direct influence of initiation on bloom duration subsequently induces secondary correlations between initiation and intensity metrics across various methods (Figure 8b). Notably, initiation dates derived from the CT-B Method demonstrate the strongest association with intensity values obtained from the Integral Method ($|r| > 0.7$, $p < 0.01$). For instance, in years characterized by early initiation (e.g., 2004, 2014), the intensity measured by the Integral Method tends to be elevated, primarily due to the extended duration of the bloom resulting from earlier initiation, which in turn increases the integral value.

Figure 8 illustrates significant variability in inter-characteristic correlations depending on the analytical methodology employed for quantifying bloom characteristics. Specifically, bloom intensities derived from the Index Method exhibit a significant negative correlation with durations calculated using the CT-B Method (α in Figure 4), while showing a positive correlation with durations derived from symmetric methods (β/γ in Figure 4). Similarly, intensities calculated using the Mean Method correlate positively with initiation values from CCT-X (Table 2), yet show no significant relationship with initiation from CT-B. These methodological discrepancies highlight how the selection of diagnostic methods fundamentally influences the interpreted relationships between phenological metrics, with direct implications for ecological conclusions drawn from bloom characteristic analyses.

The analysis presented above demonstrates strong correlations between the mean, integral, and index-based intensity methods and the duration of blooms. Based on the correlations, we conducted linear regressions to examine the relationships among bloom duration, intensity, and initiation. The results of these regressions are summarized in Table 4. These linear relations could provide a reference for predicting bloom characteristics in the Yellow Sea.

Table 4. Linear regression relationships among the characteristics of blooms. Duration-α, β, and γ correspond to the duration metrics in Figure 4, while initiation CT-B and CCT-X indicate the initiation points from Table 2. Only significant regression relationships are included.

	Duration-α	**Duration-β**	**Duration-γ**
Initiation-CT-B	$y = -0.9852x + 133.727$ **	$y = -1.7705x + 204.6519$ **	
Initiation-CCT-X			$y = -2.5442x + 257.224$ **
Intensity-Integral	$y = 0.82024x + 14.0847$ **	$y = 0.60759x + 24.7294$ **	$y = 0.47858x + 33.6684$ **
Intensity-Mean		$y = -0.0023191x + 1.1239$ *	$y = -0.0062359x + 1.4479$ *
Intensity-Index			$y = -0.018523x + 3.5654$ **
	Initiation CT-B (α)	Initiation CT-B (β)	Initiation CCT-X (γ)
Intensity-Integral	$y = -0.81034x + 123.9155$ **	$y = -1.0667x + 148.4955$ **	$y = -1.0553x + 145.4369$ *
Intensity-Mean		$y = 0.0048841x + 0.5999$ *	$y = 0.017888x + -0.29735$ *
Intensity-Index			$y = 0.043322x + -0.93355$ *

*: $p < 0.05$; **: $p < 0.01$.

6. Conclusions

This study systematically evaluates three key spring bloom characteristics—initiation, duration, and intensity—through a comparative analysis of multiple diagnostic methods applied to the central Yellow Sea.

Initiation calculations exhibit substantial methodological dependence. The unmodified CT-S Method produced initiation dates that were, on average, 40 days earlier than those derived from the CT-B Method, albeit with greater variability (σ = 19 days). The CCT Method demonstrated minimal interannual fluctuation (σ = 8 days), although results were sensitive to the initialization of the time series. The RoC Method's peak-proximity bias (mean 10-day offset from bloom maximum) limits its applicability in this region. Curve-fitting methods (Gauss/sigmoidal) displayed excessive variability (σ > 35 days) due to sensitivity to waveform shapes and frequent outlier generation. Collectively, threshold-based methods (CT-B/CCT-X) offered the most optimal balance between precision and robustness for determining initiation.

Duration calculations revealed a strong correlation between threshold and symmetric methods (r = 0.87) when using identical initiation inputs. However, the threshold method demonstrated superior temporal stability, exhibiting a 17-day lower standard deviation (σ). The duration calculated using the threshold method was the shortest, averaging 70 days, while the symmetric method yielded durations that were, on average, more than 10 days longer than those derived from the threshold method.

In assessing bloom intensity, all analytical methods reveal a consistent interannual trend of rising and then falling bloom intensity, yet the detailed temporal characteristics vary among methods. The Peak Method may be influenced by anomalous chlorophyll peaks resulting from short-term extreme events and is sensitive to the temporal resolution of chlorophyll data. Despite its unique independence from duration, the Peak Method maintained significant correlations ($|r| > 0.5$) with all other intensity metrics. The Mean Method, Integral Method, and Index Method are all impacted by bloom duration.

For bloom initiation, the CCT-X method is preferred due to the absence of outliers in the results. Regarding bloom duration, the symmetric method is recommended because it can be conveniently calculated based on the initiation of the bloom. For assessing bloom intensity, the relative intensity index method is advisable, as it accounts for Chl-a concentration during non-bloom periods, thereby offering an estimate of bloom intensity based on the complete seasonal variation. Additionally, we advise researchers to select the appropriate calculation method for bloom intensity based on their specific research needs.

Bloom characteristics exhibit interdependencies. An earlier initiation significantly prolongs bloom duration, which directly affects the calculated intensity of the bloom (excluding the Peak Method). These relationships exhibit varying characteristics across different calculation methods, necessitating careful consideration of methodological choices based on research objectives and data characteristics. Different methods may yield entirely different conclusions in the quantitative analysis and research of the spring bloom. Future efforts should focus on developing more unified and comprehensive indicators to quantify the characteristics of spring blooms.

Author Contributions: Conceptualization, K.J. and L.L.; methodology, K.J., C.D. and X.L.; formal analysis, X.L., Y.S. and J.L.; writing—original draft preparation, K.J.; writing—review and editing, K.J. and L.L.; visualization, K.J., Y.S. and J.L.; funding acquisition, L.L. All authors have read and agreed to the published version of the manuscript.

Funding: The study was financially supported by the National Natural Science Foundation of China (42030402), Open Research Fund of State Key Laboratory of Estuarine and Coastal Research (SKLEC-KF202404), Pilot Project of Basic Research Operating Expenses System of Hebei Academy of Sciences (2025PF07), and the Open Research Fund of Shandong Key Laboratory of Marine Ecological Environment and Disaster Prevention and Mitigation (202305).

Data Availability Statement: The satellite data can be downloaded from the U.S. National Aeronautics and Space Administration (NASA) website (http://oceancolor.gsfc.nasa.gov) (accessed on 15 June 2022).

Acknowledgments: We thank Yueqi Wang from Yantai Institute of Coastal Zone Research, Chinese Academy of Sciences, for his support regarding the satellite data.

Conflicts of Interest: The authors declare no conflicts of interest.

References

1. Behrenfeld, M.J.; Randerson, J.T.; McClain, C.R.; Feldman, G.C.; Los, S.O.; Tucker, C.J.; Falkowski, P.G.; Field, C.B.; Frouin, R.; Esaias, W.E.; et al. Biospheric Primary Production During an ENSO Transition. *Science* **2001**, *291*, 2594–2597. [CrossRef]
2. Falkowski, P.G.; Barber, R.T.; Smetacek, V. Biogeochemical Controls and Feedbacks on Ocean Primary Production. *Science* **1998**, *281*, 200–206. [CrossRef] [PubMed]
3. Kim, Y.; Youn, S.-H.; Oh, H.-J.; Joo, H.; Jang, H.-K.; Kang, J.-J.; Lee, D.; Jo, N.; Kim, K.; Park, S.; et al. Seasonal Compositions of Size-Fractionated Surface Phytoplankton Communities in the Yellow Sea. *J. Mar. Sci. Eng.* **2022**, *10*, 1087. [CrossRef]
4. Hutchins, D.A.; Alessandro, T. Feedbacks between Phytoplankton and Nutrient Cycles in a Warming Ocean. *Nat. Geosci.* **2024**, *17*, 495–502. [CrossRef]

5. Uth, C.; Asmala, E.; Lewandowska, A.M. Phytoplankton Community Composition as a Driver of Annual Autochthonous Organic Carbon Dynamics in the Northern Coastal Baltic Sea. *Mar. Ecol. Prog. Ser.* **2024**, *745*, 13–24. [CrossRef]
6. Record, N.R.; Balch, W.M.; Stamieszkin, K. Century-scale changes in phytoplankton phenology in the Gulf of Maine. *PeerJ* **2019**, *7*, e6735. [CrossRef]
7. Legendre, L. The significance of microalgal blooms for fisheries and for the export of particulate organic carbon in oceans. *J. Plankton Res.* **1990**, *12*, 681–699. [CrossRef]
8. Salgado-Hernanz, P.M.; Racault, M.-F.; Font-Muñoz, J.S.; Basterretxea, G. Trends in phytoplankton phenology in the Mediterranean Sea based on ocean-colour remote sensing. *Remote Sens. Environ.* **2019**, *221*, 50–64. [CrossRef]
9. Sverdrup, H.U. On Conditions for the Vernal Blooming of Phytoplankton. *ICES J. Mar. Sci.* **1953**, *18*, 287–295. [CrossRef]
10. Hunt, G.L., Jr.; Stabeno, P.; Walters, G.; Sinclair, E.; Brodeur, R.D.; Napp, J.M.; Bond, N.A. Climate change and control of the southeastern Bering Sea pelagic ecosystem. *Deep. Sea Res. Part II Top. Stud. Oceanogr.* **2002**, *49*, 5821–5853. [CrossRef]
11. Atkinson, A.; Rossberg, A.G.; Gaedke, U.; Sprules, G.; Heneghan, R.F.; Batziakas, S.; Grigoratou, M.; Fileman, E.; Schmidt, K.; Frangoulis, C. Steeper size spectra with decreasing phytoplankton biomass indicate strong trophic amplification and future fish declines. *Nat. Commun.* **2024**, *15*, 381. [CrossRef] [PubMed]
12. Sari El Dine, Z.; Guinet, C.; Picard, B.; Thyssen, M.; Duforêt-Gaurier, L.; Hourany, R.E. Influence of the phytoplankton community structure on the southern elephant seals' foraging activity within the southern ocean. *Commun. Biol.* **2025**, *8*, 620. [CrossRef] [PubMed]
13. Platt, T.; Sathyendranath, S. Biological oceanography and fisheries management. *ICES J. Mar. Sci.* **2007**, *64*, 863–869. [CrossRef]
14. Deng, L.; Zhou, W.; Xu, J.; Gao, W.; Liao, J.; Zhao, J. Estimation of vertical size-fractionated phytoplankton primary production in the northern South China Sea. *Ecol. Indic.* **2022**, *135*, 108546. [CrossRef]
15. Liu, F.; Su, J.; Moll, A.; Krasemann, H.; Chen, X.; Pohlmann, T.; Wirtz, K. Assessment of the summer–autumn bloom in the Bohai Sea using satellite images to identify the roles of wind mixing and light conditions. *J. Mar. Syst.* **2014**, *129*, 303–317. [CrossRef]
16. Taylor, A.; Geider, R.; Gilbert, F. Seasonal and latitudinal dependencies of phytoplankton carbon-to-chlorophyll a ratios:results of a modelling study. *Mar. Ecol. Prog. Ser.* **1997**, *152*, 51–66. [CrossRef]
17. Niu, Y.; Liu, C.; Lu, X.; Zhu, L.; Sun, Q.; Wang, S. Phytoplankton blooms and its influencing environmental factors in the southern Yellow Sea. *Reg. Stud. Mar. Sci.* **2021**, *47*, 101916. [CrossRef]
18. Zhai, F.; Liu, Z.; Gu, Y.; He, S.; Hao, Q.; Li, P. Satellite-Observed Interannual Variations in Sea Surface Chlorophyll-a Concentration in the Yellow Sea Over the Past Two Decades. *J. Geophys. Res. Ocean.* **2023**, *128*, e2022JC019528. [CrossRef]
19. Racault, M.-F.; Le Quéré, C.; Buitenhuis, E.; Sathyendranath, S.; Platt, T. Phytoplankton phenology in the global ocean. *Ecol. Indic.* **2012**, *14*, 152–163. [CrossRef]
20. Malysheva, A.S.; Lobanova, P.V. Interannual Variability of Marine Phytoplankton Spring Bloom Phenology in the North European Basin. *Oceanology* **2024**, *64*, S107–S117. [CrossRef]
21. Siegel, D.A.; Doney, S.C.; Yoder, J.A. The North Atlantic Spring Phytoplankton Bloom and Sverdrup's Critical Depth Hypothesis. *Science* **2002**, *296*, 730–733. [CrossRef]
22. Ueyama, R.; Monger, B.C. Wind-induced modulation of seasonal phytoplankton blooms in the North Atlantic derived from satellite observations. *Limnol. Oceanogr.* **2005**, *50*, 1820–1829. [CrossRef]
23. Zhai, L.; Platt, T.; Tang, C.; Sathyendranath, S.; Hernández Walls, R. Phytoplankton phenology on the Scotian Shelf. *ICES J. Mar. Sci.* **2011**, *68*, 781–791. [CrossRef]
24. Brody, S.R.; Lozier, M.S.; Dunne, J.P. A comparison of methods to determine phytoplankton bloom initiation. *J. Geophys. Res. Ocean.* **2013**, *118*, 2345–2357. [CrossRef]
25. Shi, J.; Liu, Y.; Mao, X.; Guo, X.; Wei, H.; Gao, H. Interannual variation of spring phytoplankton bloom and response to turbulent energy generated by atmospheric forcing in the central Southern Yellow Sea of China: Satellite observations and numerical model study. *Cont. Shelf Res.* **2017**, *143*, 257–270. [CrossRef]
26. Friedland, K.D.; Hare, J.A.; Wood, G.B.; Col, L.A.; Buckley, L.J.; Mountain, D.G.; Kane, J.; Brodziak, J.; Lough, R.G.; Pilskaln, C.H. Does the fall phytoplankton bloom control recruitment of Georges Bank haddock, Melanogrammus aeglefinus, through parental condition? *Can. J. Fish. Aquat. Sci.* **2008**, *65*, 1076–1086. [CrossRef]
27. Fu, Q.; Yan, X.; Hong, Q.; Lin, L.; Zhang, Y. Variability of the Primary Productivity in the Yellow and Bohai Seas from 2003 to 2020 Based on the Estimate of Satellite Remote Sensing. *J. Mar. Sci. Eng.* **2023**, *11*, 684. [CrossRef]
28. Han, J. Numerical Study on the Physical Effect on Phytoplankton Bloom in The Yellow Sea. Ph.D. Thesis, Ocean University of China, Qingdao, China, 2008.
29. Lv, T.; Liu, D.; Zhou, P.; Lin, L.; Wang, Y.; Wang, Y. The coastal front modulates the timing and magnitude of spring phytoplankton bloom in the Yellow Sea. *Water Res.* **2022**, *220*, 118669. [CrossRef]
30. Liu, X.; Pu, X.; Qu, D.; Xu, Z. The role of wave-induced mixing in spring phytoplankton bloom in the south yellow sea. *Mar. Pollut. Bull.* **2025**, *211*, 117374. [CrossRef]

31. Kim, S.; Lee, D.; Kim, M.; Jang, H.-K.; Park, S.; Kim, Y.; Kim, J.; Park, J.-W.; Joo, H.; Lee, S.-H. Seasonal Patterns and Bloom Dynamics of Phytoplankton Based on Satellite-derived Chlorophyll-a in the Eastern Yellow Sea. *Mar. Environ. Res.* **2024**, *199*, 106605. [CrossRef]
32. Henson, S.; Sanders, R.; Holeton, C.; Allen, J.T. Timing of nutrient depletion, diatom dominance and a lower-boundary estimate of export production for Irminger Basin, North Atlantic. *Mar. Ecol. Prog. Ser.* **2006**, *313*, 73–84. [CrossRef]
33. Henson, S.A.; Dunne, J.P.; Sarmiento, J.L. Decadal variability in North Atlantic phytoplankton blooms. *J. Geophys. Res. Ocean.* **2009**, *114*, 2008JC005139. [CrossRef]
34. Thomalla, S.J.; Fauchereau, N.; Swart, S.; Monteiro, P.M.S. Regional scale characteristics of the seasonal cycle of chlorophyll in the Southern Ocean. *Biogeosciences* **2011**, *8*, 2849–2866. [CrossRef]
35. Cole, H.; Henson, S.; Martin, A.; Yool, A. Mind the gap: The impact of missing data on the calculation of phytoplankton phenology metrics. *J. Geophys. Res. Ocean.* **2012**, *117*, C08030. [CrossRef]
36. Fleming, V.; Kaitala, S. Phytoplankton Spring Bloom Intensity Index for the Baltic Sea Estimated for the years 1992 to 2004. *Hydrobiologia* **2006**, *554*, 57–65. [CrossRef]
37. Greve, W.; Prinage, S.; Zidowitz, H.; Nast, J.; Reiners, F. On the phenology of North Sea ichthyoplankton. *ICES J. Mar. Sci.* **2005**, *62*, 1216–1223. [CrossRef]
38. Mackas, D.L.; Greve, W.; Edwards, M.; Chiba, S.; Tadokoro, K.; Eloire, D.; Mazzocchi, M.G.; Batten, S.; Richardson, A.J.; Johnson, C.; et al. Changing zooplankton seasonality in a changing ocean: Comparing time series of zooplankton phenology. *Prog. Oceanogr.* **2012**, *97–100*, 31–62. [CrossRef]
39. Sharples, J.; Ross, O.; Scott, B.; Greenstreet, S.P.R.; Fraser, H. Inter-annual variability in the timing of stratification and the spring bloom in the North-western North Sea. *Cont. Shelf Res.* **2006**, *26*, 733–751. [CrossRef]
40. Yamada, K.; Ishizaka, J. Estimation of interdecadal change of spring bloom timing, in the case of the Japan Sea. *Geophys. Res. Lett.* **2006**, *33*, L02608. [CrossRef]
41. Friedland, K.D.; Mouw, C.B.; Asch, R.G.; Ferreira, A.; Henson, S.; Hyde, K.J.W.; Morse, R.E.; Thomas, A.C.; Brady, D.C. Phenology and time series trends of the dominant seasonal phytoplankton bloom across global scales. *Glob. Ecol. Biogeogr.* **2018**, *27*, 551–569. [CrossRef]
42. Jo, C.O.; Lee, J.; Park, K.; Kim, Y.H.; Kim, K.-R. Asian dust initiated early spring bloom in the northern East/Japan Sea. *Geophys. Res. Lett.* **2007**, *34*, L05602. [CrossRef]
43. Navarro, G.; Caballero, I.; Prieto, L.; Vázquez, A.; Flecha, S.; Huertas, I.E.; Ruiz, J. Seasonal-to-interannual variability of chlorophyll-a bloom timing associated with physical forcing in the Gulf of Cádiz. *Adv. Space Res.* **2012**, *50*, 1164–1172. [CrossRef]
44. Platt, T.; White, G.N.; Zhai, L.; Sathyendranath, S.; Roy, S. The phenology of phytoplankton blooms: Ecosystem indicators from remote sensing. *Ecol. Model.* **2009**, *220*, 3057–3069. [CrossRef]
45. Roerink, G.J.; Menenti, M.; Verhoef, W. Reconstructing cloudfree NDVI composites using Fourier analysis of time series. *Int. J. Remote Sens.* **2000**, *21*, 1911–1917. [CrossRef]
46. Friedland, K.D.; Leaf, R.T.; Kane, J.; Tommasi, D.; Asch, R.G.; Rebuck, N.; Ji, R.; Large, S.I.; Stock, C.; Saba, V.S. Spring bloom dynamics and zooplankton biomass response on the US Northeast Continental Shelf. *Cont. Shelf Res.* **2015**, *102*, 47–61. [CrossRef]
47. Rodionov, S.N. A sequential algorithm for testing climate regime shifts. *Geophys. Res. Lett.* **2004**, *31*, L09204. [CrossRef]
48. Rodionov, S.N. Use of prewhitening in climate regime shift detection. *Geophys. Res. Lett.* **2006**, *33*, L12707. [CrossRef]
49. Behrenfeld, M.J.; Boss, E.S. Student's tutorial on bloom hypotheses in the context of phytoplankton annual cycles. *Glob. Change Biol.* **2018**, *24*, 55–77. [CrossRef]
50. Wang, Y.; Liu, D.; Tang, D. Application of a generalized additive model (GAM) for estimating chlorophyll- a concentration from MODIS data in the Bohai and Yellow Seas, China. *Int. J. Remote Sens.* **2017**, *38*, 639–661. [CrossRef]
51. Lin, L.; Sun, Z.; Lv, T.; Wang, Y.; Tan, W.; Jin, K.; Xiang, D. In-Phase Pacific Decadal Oscillation and El Niño–Southern Oscillation Enhance the Interannual Variability of Spring Blooms in the Yellow Sea. *Ocean. Land Atmos. Res.* **2025**, *4*, 0083. [CrossRef]
52. Lin, L.; Liu, D.; Wang, Y.; Lv, T.; Zhao, Y.; Tan, W. Effect of wind on summer chlorophyll-a variability in the Yellow Sea. *Front. Mar. Sci.* **2023**, *9*, 1104258. [CrossRef]
53. Lin, L.; Wang, Y.; Liu, D. Vertical average irradiance shapes the spatial pattern of winter chlorophyll-a in the Yellow Sea. *Estuar. Coast. Shelf Sci.* **2019**, *224*, 11–19. [CrossRef]

remote sensing

MDPI

Article

Spatiotemporal Dynamics of Total Suspended Solids in the Yellow River Estuary Under New Water-Sediment Regulation: Insights from Sentinel-3 OLCI

Yafei Luo [1], Zhengyu Hou [2,*], Yanxia Liu [3], David Doxaran [4], Dongyang Fu [1], Liwen Yan [3] and Haijun Huang [3]

[1] College of Electronics and Information Engineering, Guangdong Ocean University, Zhanjiang 524088, China; luoyafei@gdou.edu.cn (Y.L.)
[2] School of Ocean Engineering and Technology, Sun Yat-sen University & Southern Marine Science and Engineering Guangdong Laboratory (Zhuhai), Zhuhai 519000, China
[3] Key Laboratory of Marine Geology and Environment, Institute of Oceanology, Chinese Academy of Sciences, Qingdao 266071, China
[4] Laboratoire d'Océanographie de Villefranche, UMR 7093—CNRS/SU, 06230 Villefranche-sur-Mer, France
* Correspondence: houzhy7@mail.sysu.edu.cn

Abstract

The Water and Sediment Regulation Scheme (WSRS), implemented since 2002, has been essential for controlling water flow and mitigating sediment siltation in the lower Yellow River. However, WSRS was suspended for the first time in 2016 and 2017 due to extremely low water flow. The rapid floodwater discharge over roughly 20 days conducted by WSRS strongly impacts total suspended solids (TSS) distribution in the Yellow River Estuary (YRE). This study employs high-frequency Sentinel-3 OLCI satellite imagery to investigate intraday TSS variations in the YRE under new water-sediment regulation conditions from 2016 to 2023. TSS concentrations were generally low during the 2016 and 2017 flood seasons, but increased markedly after WSRS resumed in 2018. Peak TSS values occurred in July or August, sometimes extending into September and October during autumn floods. A moderately strong positive correlation was observed between TSS concentrations at the river mouth and sediment load at Lijin Station during the flood seasons. The 2018 WSRS event generated an extensive river plume, with average TSS concentrations at the river mouth exceeding 400 $g \cdot m^{-3}$. From 2018 to 2023, TSS concentrations exhibited a declining trend during flood seasons, attributed to reduced sediment discharge and ongoing sediment accretion in the Yellow River Delta. Our findings highlight Sentinel-3 OLCI as a powerful tool to resolve WSRS-driven sediment dynamics, offering critical guidance for estuarine management.

Keywords: water and sediment regulation scheme; total suspended solids; Sentinel-3 OLCI imagery; Yellow River estuary

1. Introduction

With an annual sediment load of approximately ~1.08×10^9 tons transported into the Bohai Sea, the Yellow River exhibits one of the highest total suspended solids (TSS) concentrations globally [1]. Lower reaches of the river are particularly prone to sediment deposition due to an imbalance between water flow and sediment transport, which leads to riverbed elevation and increases the risk of flooding. To address this water-sediment imbalance, the Yellow River Water Conservancy Commission of the Ministry of Water

Resources has implemented the Water-Sediment Regulation Scheme (WSRS) for 13 consecutive years since 2002, then interrupted it in 2016 and 2017. The implementation of WSRS advanced high monthly water discharge and sediment load to as early as June, occurring at least two months prior to the typical high-flow season from August to October [2,3]. During the WSRS, both runoff and sediment loads are increased, facilitating the transport of substantial volumes of water and sediment to the sea. The WSRS is conducted in two phases: the first phase involves the release of water from the Xiaolangdi Reservoir, while the second phase is focused on flushing sediment from the reservoir [4]. The duration of WSRS operations averaged approximately 20 days per year, contributing to 27.6% and 48.9% of the annual water and sediment delivery to the sea, respectively [5]. Between 2002 and 2015, approximately 1.98×10^9 tons of sediments entered the Bohai Sea, with 1.31×10^9 tons being deposited along the coast of the Yellow River Estuary (YRE) [6]. By influencing suspended sediment transport, the WSRS directly affects both the ecological environment and the geomorphological evolution of the YRE [7]. In recent years, however, due to the combined effects of climate change and human activities, such as WSRS interventions, the hydrological conditions of the Yellow River have shifted, transitioning from a state of "less water and more sediment" to "less water and less sediment." Especially, in 2016 and 2017, the sediment load of the Yellow River experienced a significant decline, prompting the interruption of the WSRS for two years before its resumption in 2018. This disruption may have had new and profound effects on the ecological environment of both the delta and adjacent coastal areas.

The investigation of the spatiotemporal distribution of TSS in the YRE under the evolving conditions of water-sediment regulation is of considerable scientific significance. Such studies provide crucial theoretical support for understanding changes in marine sedimentation processes and the physical environment of the YRE. Traditionally, in-situ measurements have been the primary method for quantifying TSS. However, remote sensing techniques now provide an indispensable way to monitor large-scale TSS variations in estuarine and coastal waters.

Previous studies have frequently utilized Landsat imagery to investigate monthly and annual TSS variations in the YRE and to assess key influencing factors, including wind, currents, runoff, and sediment load [8–13]. Other higher resolution sensors, such as Sentinel 2-MSI [10] and Gaofen (GF) satellite images [14], were also used to map the spatiotemporal variations of the total suspended particulate matter in the Yellow River estuary. While these studies have provided robust insights into long-term TSS dynamics, Landsat's 16-day revisit cycle prevents monitoring interannual and daily variations. This constraint precludes comprehensive observation of shorter-term TSS fluctuations, such as those occurring during WSRS periods.

Alternatively, studies have shown that the ocean-color sensor, such as the Moderate Resolution Imaging Spectroradiometer (MODIS) and the Land Colour Instrument (OLCI), with a spatial resolution of 250 m/500 m/1000 m and 300 m, can be effectively applied for monitoring estuarine turbidity and TSS, and can also provide specific-term time series monitoring records [15]. For instance, Zhang et al. employed MODIS data with the red band to study river plume induced by heavy rainfall events in Lake Taihu [16]. Luo et al. utilized MODIS data to analyze the spatial distribution characteristics and influencing factors of different quantile trends of TSS in the Pearl River Estuary [17]. MODIS is prone to saturation in the highly turbid waters of YRE, probably due to high sensitivity or signal-to-noise ratio [8], posing persistent challenges for accurate TSS retrieval in such environments.

The daily revisit capability of the Sentinel-3 Ocean and Land Colour Instrument (OLCI) satellite sensors, launched on 16 February 2016 (Sentinel-3A) and 25 April 2018 (Sentinel-3B), coupled with its spatial resolution of 300 m, offers a promising solution to overcome these limitations. By providing near-daily observations of the YRE, Sentinel-3 OLCI (S3-OLCI) facilitates detailed analysis of TSS dynamics during WSRS events. This capability enables a more nuanced understanding of how TSS within the estuary responds to WSRS operations, addressing critical gaps in current research and advancing knowledge in this domain.

Studies have examined the long-term impacts of the WSRS on aspects such as the dispersion of freshwater and sediment in the YRE [5,18,19] and its effects on estuarine morphology, hydrodynamics, and biogeochemical processes [20–23]. Data on the daily variations in TSS distribution within the estuary remain limited [24,25]; even less is understood about the effects of WSRS under varying hydrological conditions on these daily TSS dynamics.

To enhance our understanding of the impacts of regulated water flow on ocean color in the YRE, particularly since the WSRS interruption in 2016, this study leverages eight years of S3-OLCI remote sensing satellite observations from 2016 to 2023. This research advances the understanding of WSRS impacts on TSS dynamics in the YRE, providing a framework for monitoring the current state and evolving trends of TSS not only during WSRS events but also over the post-event periods.

The objectives of the present study are to (i) propose a new algorithm to quantify TSS evolutions and distributions in the YRE using S3-OLCI images and validate the Dark Spectrum Fitting (DSF) atmospheric correction applied to S3-OLCI satellite data recorded over the YRE turbid estuarine waters; (ii) monitor TSS dynamics in the YRE before, during, and after the WSRS implementation, using S3-OLCI satellite data; (iii) provide a detailed understanding of WSRS impacts on TSS dynamics in the YRE by comparing TSS in the YRE with the river freshwater discharge and sediment load from the hydrologic station (Lijin Station on the Yellow River).

2. Materials and Methods

2.1. Study Area

The study area is the Yellow River Estuary (YRE), containing the Yellow River Mouth (YRM) and Laizhou Bay (LZB). It is located at the confluence of the Bohai Bay and Laizhou Bay (Figure 1). The isobaths in the near-shore area of the YRE are tortuous and irregularly distributed. The sediment transported from the Yellow River into the sea and accumulated in the northwest and northeast direction of the mouth of the estuary, and extends towards the offshore in the shape of a tongue [26].

The typical duration of WSRS operations prior to flooding usually spans approximately 20 days in June and July, though such events occasionally occur during flood periods in July and August. The start and end dates of the WSRS from 2016 to 2023 are detailed in Table 1. The data is recorded from the Yellow River Sediment Bulletin.

2.2. Data

2.2.1. In-Situ Data

The in-situ dataset was collected during a dedicated field campaign in the YRE from 21 to 22 and 25 to 27 June 2004 during the WSRS period. Water reflectance was measured, and surface water samples were collected to determine the concentration of suspended solids. Water sampling adhered strictly to national technical standards, specifically those outlined in GB/T13909-92 for marine geological and geophysical surveys. The TSS concentration was determined as the dried sediment mass divided by the known volume of the filtered water sample, resulting in values ranging from 1.7 to 1896.5 $g \cdot m^{-3}$. The remote sensing reflectance signal (Rrs, in sr^{-1}) was measured using an ASD FieldSpec Dual VNIR radiometer, covering the 350–1050 nm spectral range. In total, 44 pairs of high-quality co-located

TSS and *Rrs* values were obtained. This high-quality dataset was used to establish the TSS algorithm for retrieving TSS using S3-OLCI satellite data during the flood seasons. In addition, twelve in-situ *Rrs* measurements collected on 7–8 November 2024 and 17, 19 and 20 January 2025 were used to validate the *Rrs* values retrieved from S3-OLCI. Seven in-situ TSS measurements collected on 19–21 July 2019 and 9 July 2023 were used to validate the TSS model by comparing the S3-OLCI-retrieved TSS with the quasi-synchronous (±3 h) in-situ TSS. The geographical locations of the observation stations are shown in Figure 1, and related information is listed in Table 2.

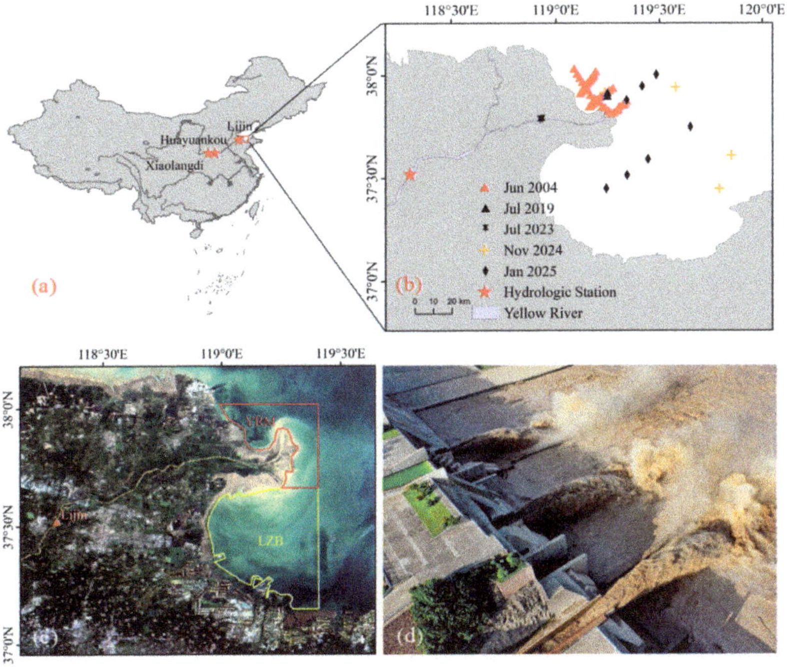

Figure 1. Geographic location and hydrometric stations of the study area (**a**). Sampling stations in the Yellow River Estuary (**b**). The study area (YRE), as shown with the S3-OLCI image (30 August 2023), is divided into the Yellow River Mouth (YRM) area and the Laizhou Bay (LZB) area, respectively in red polygon and in yellow polygon (**c**). The Yellow River Conservancy Commission has implemented the WSRS in the Yellow River since 2002 (**d**).

Table 1. The start and end times of the WSRS from 2016 to 2023.

Year	WSRS Before Flood		WSRS During Flood		Max. Q ($m^3 \cdot s^{-1}$) [1]/Day	Max. SC (10^3 $g \cdot m^{-3}$) [2]/Day	Water Discharge To The Sea		Sediment Load To The Sea	
	Starting	Ending	Starting	Ending			Q_w/(km^3)	(%) [3]	Q_s/(Mt)	(%) [4]
2016	-	-	-	-	1480/July 28	4.4/July 26	-	-	-	-
2017	-	-	-	-	813/December 8	3.14/April 28	-	-	-	-
2018	July 3	July 27	August 7	August 28	3670/July 11	48.1/July 20	8.9	26.7	164.7	55.5
2019	June 21	July 5	July 6	August 16	4060/August 12	26.3/July 21	11.3	36.2	163.3	60.3
2020	June 24	July 10	July 4	August 18	4580/July 3	24.1/July 30	12.1	33.6	168.1	53.5
2021	June 19	July 13	-	-	5240/October 8	37.9/July 10	6	13.6	48.9	20.1
2022	June 19	July 13	July 17	July 28	4030/July 2	39.2/July 24	7.3	28	71.4	57.1
2023	June 21	July 16	-	-	3780/July 3	37.9/July 12	5.87	25.9	53.5	55.2
Average							8.58	27.3	111.65	50.3

[1] Maximum water flow at Lijin station; [2] Maximum suspended sediment concentration at Lijin station; [3] Water discharge during the WSRS in percentage of annual water discharge to the sea; [4] Sediment load during the WSRS in percentage of annual sediment load to the sea; —No WSRS.

Table 2. Information on the in-situ measurements.

Date	Number of Stations	Measurements
21–22 and 25–27 June 2004	44	*Rrs*, TSS
19–21 July 2019	5	TSS
9 July 2023	2	TSS
7–8 November 2024	4	*Rrs*
17 and 19–20 January 2025	8	*Rrs*

2.2.2. Remote Sensing Dataset

The Sentinel-3A&B OLCI satellite data were used to derive TSS during the flood season from June to October since 2016. Both Sentinel-3A OLCI and Sentinel-3B OLCI images were obtained from the Copernicus Data Space Ecosystem server. OLCI revisit time with the two-satellite configuration is ~1 day at the latitude of 37°~38° for YRE. With 21 bands, a resolution of 300 m over all surfaces and a design optimized to minimize sun-glint effects, OLCI marks a new generation of measurements over the ocean and land. Cloud-free S3-OLCI images were considered during flood seasons (June to October) from 2016 to 2023.

Besides, Landsat-8 Operational Land Imager (OLI) satellite data with a spatial resolution of 30 m and a revisit time of 16 days were used. Landsat-8 OLI (L8-OLI) image recorded on 30 August 2023 over YRE at 10:53 UTC, 17 min before the OLCI image was collected.

2.2.3. Atmospheric Correction for OLCI

To retrieve the water-leaving reflectance, defined as $\rho = Rrs \times \pi$ (dimensionless), OLCI data were processed using the ACOLITE (v20221114.0) software, applying the Dark Spectrum Fitting (DSF) atmospheric correction method [27].

This DSF approach assumes negligible surface reflectance in at least one of the sensor bands, selected from the darkest pixels in the image. The DSF atmospheric correction algorithm was originally developed for the processing of metre-scale sensors for aquatic applications in inland and coastal waters [28,29], but it also performed well for decametre-scale sensors [30,31]. A new OLCI version of DSF was implemented in ACOLITE, and imagery was processed to water reflectances using a fixed aerosol optical thickness τ_a Vanhellemont and Ruddick (2021) [27].

The comparison between *Rrs* measured in situ and derived from S3-OLCI satellite data (3 × 3 pixels) corrected using the DSF method in the 560, 665 and 865 nm spectral bands and for the 665/560 nm band ratio was made. The in-situ *Rrs* were measured in November 2024 and January 2025 in the YRE. The station information can be found in Figure 1. From Figure 2, we can see that the comparison is overall good in red and green bands, even better when considering the band ratio between 665 and 560 nm, with a MAPE of 12% and close to a 1:1 line. The S3-OLCI-derived *Rrs* values in the NIR band are slightly lower than in situ measured *Rrs*.

ACOLITE has been shown to perform well in the highly turbid waters of the YRE, a finding previously validated through step-by-step comparisons with field measurements for Landsat data [8,10,32]. Cloud-free L8-OLI and S3-OLCI satellite images recorded on 30 August 2023, with a time difference of 17 min, were considered for the sensor-to-sensor *Rrs* validation. The two images were reprojected onto a common grid by resampling OLI data to a spatial resolution of 300 m, using mean averaging. The OLCI-derived water reflectance values at Green, Red, and NIR wavebands were then compared to the corresponding OLI water reflectance values.

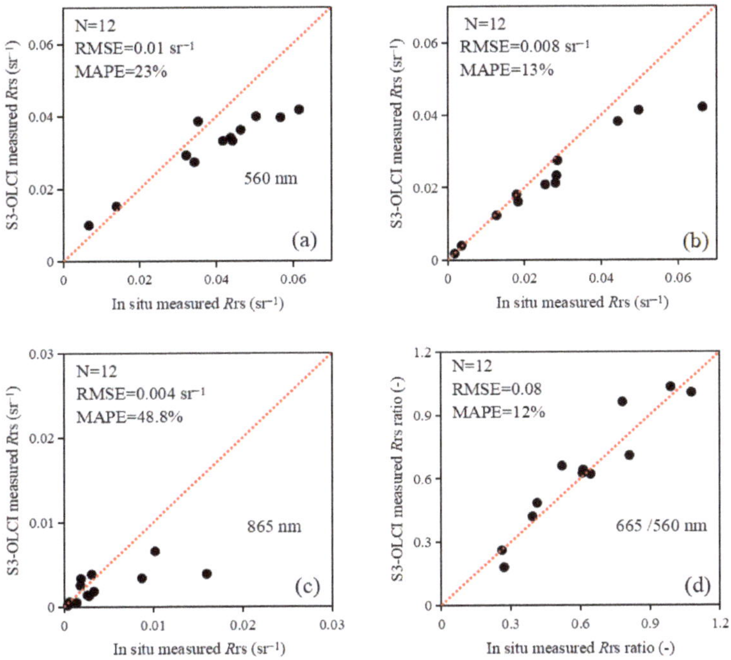

Figure 2. Scatterplots of S3-OLCI *Rrs* versus in situ *Rrs* measured in November 2024 and January 2025 at (**a**) 560 nm band, (**b**) 665 nm band, (**c**) 865 nm band, and (**d**) 665/560 band ratio.

2.2.4. Hydrologic Data

The Lijin Hydrologic Station (Figure 1a), located approximately 80 km upstream of the river mouth, is the last station along the Yellow River. Daily water discharge and Sediment Concentration (SC) data measured at 8:00 a.m. (local time) each day were sourced from the daily water condition reports published by the Yellow River Conservancy Commission of the Ministry of Water Resources. The SC data generally represent vertically-averaged or cross-sectionally averaged measurements. Monthly river discharge and sediment load data at Stations Lijin, Xiaolangdi and Huayuankou were retrieved from the Yellow River Sediment Bulletin (http://www.yrcc.gov.cn/ (accessed on 1 June 2024)).

2.3. Methods

2.3.1. TSS Inversion Algorithm

The TSS inversion algorithm relies on the range of the TSS observed and the form of the model equation [33]. TSS in the YRE can exceed 1000 $g{\cdot}m^{-3}$, as evidenced by in-situ measurements and previous studies, particularly during the WSRS. For such highly turbid waters, *Rrs* in visible bands, notably in blue and green bands, tends to saturate while TSS increases. *Rrs* in near-infrared (NIR) bands have basically low values when TSS is lower than 100 $g{\cdot}m^{-3}$, then increase proportionally to TSS up to concentrations of about 1500 $g{\cdot}m^{-3}$ [34,35].

To avoid the saturation of the water reflectance signal, a NIR spectral band is therefore required to estimate TSS concentrations over the wide range encountered in the YRE (0 to >1000 $g{\cdot}m^{-3}$) [35,36]. Also, according to the correlation analysis between TSS and *Rrs*, the correlation coefficients for the green, red and NIR bands are all higher than 0.5, especially for the NIR band, with a r of 0.94. Therefore, *Rrs*(865) is considered in our TSS modelling. Single-band algorithms are highly sensitive to particle type (size and composition) and environmental changes such as illumination conditions and glint effects, making them inadequate for comprehensively characterizing the spectral signatures of waters with varying suspended sediment concentrations. In contrast, band-ratio approaches can partially eliminate or mitigate the influence of particle size and illumination on the retrieval of suspended sediment concentrations [37]. Previous studies have shown that TSS retrieved using the Landsat images using the NIR band, combined with the ratio of red to green bands and best-fitted exponential relationships, exhibited the strongest agreement with in-situ TSS measurements in the YRE [10,36]. Consequently, we developed an empirical method based on the *Rrs*(665)/*Rrs*(560) ratio and the *Rrs*(865) band with a total of 44 coincident TSS and *Rrs* measurements, which were converted to OLCI equivalent *Rrs* values in each band by considering the corresponding spectral response functions (SRF).

$$log_{10}(TSS) = 1.9 \times X_1 + 9.96 \times X_2 - 0.006 \quad (1)$$

$$With\ X_1 = Rrs(665)/Rrs(560) \text{ and } X_2 = Rrs(865)$$

where, *Rrs*(560), *Rrs*(665) and *Rrs*(865) are the S3-LOCI equivalent *Rrs* values. Equation (1) yields a determination coefficient R^2 of 0.94, RMSE of 97 g·m^{-3}, and MAPE of 19% (Figure 3a). For low TSS concentrations (<27 g·m^{-3}), the relationship is also well-defined with a R^2 of 0.89, RMSE of 1.4 g·m^{-3}, and MAPE of 16.6% (Figure 3b). Figure 4 compares S3-OLCI estimated and in situ measured TSS concentrations drawn from seven observation stations over a 3 h time window between satellite overpasses and field measurements, during the WSRS periods of 19–21 July 2019 and 9 July 2023. The in-situ measured TSS concentrations range from 13 to 2019 g·m^{-3}. The results showed a good agreement with a highly significant linear relationship (R^2 of 0.96), with a MAPE of 29.6% and RMSE of 222 g·m^{-3} (Figure 4).

2.3.2. Accuracy Assessment

The Root Mean Square Error (RMSE), Mean Absolute Percentage Error (MAPE) and BIAS (Bias) were used to quantitatively compare the TSS or *Rrs* values:

$$\text{RMSE} = \sqrt{\frac{1}{N}\sum_{i=1}^{N}(x_i - y_i)^2} \quad (2)$$

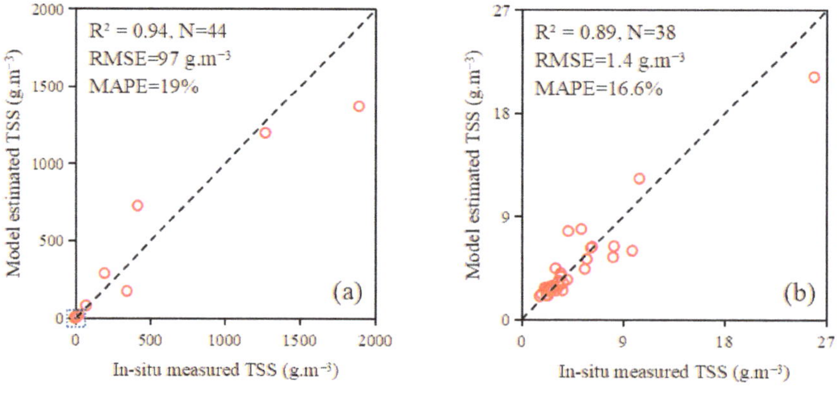

Figure 3. Comparison between estimated and in situ measurements of TSS values for the dataset generated during the WSRS period of June 2004 (**a**), with a zoom on TSS concentrations below 27 g·m^{-3} (blue dashed frame in Figure 3a) (**b**). The black dashed line indicates the 1:1 relationship.

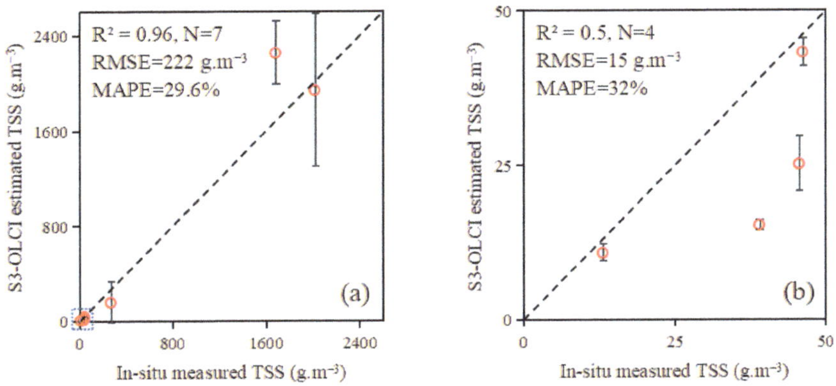

Figure 4. Comparison between S3-OLCI estimated TSS and in-situ measurements for the dataset generated during the 19–21 July 2019 and 9 July 2023 WSRS periods (**a**), with a zoom on TSS concentrations below 50 g·m^{-3} (blue dashed frame in Figure 4a) (**b**). The black dashed line indicates the 1:1 relationship. Error bars correspond to the associated standard deviations of 3 × 3 extracted satellite pixels.

$$\text{MAPE} = \sum_{i=1}^{N} \frac{|y_i - x_i|}{N \cdot x_i} \times 100\%, \tag{3}$$

$$\text{BIAS} = \frac{\sum_{i=1}^{N}(y_i - x_i)}{\sum_{i=1}^{N} x_i} \times 100\%, \tag{4}$$

where x_i and y_i are respectively the in-situ measured TSS or *Rrs* (including already-validated OLI products) and modelled TSS or *Rrs* retrieved from OLCI, N is the total number of samples.

2.3.3. Time-Series Analysis of TSS During the WSRS

To investigate the dynamic characteristics of TSS under new sediment transport conditions from 2016 to 2023, this study conducted a detailed analysis of TSS distribution during the Yellow River flood season (WSRS period) from June to October.

Time series of TSS data were obtained from regional mean values derived from the 300 m spatial resolution and daily revisit S3-OLCI imagery, considering separately the Yellow River Mouth (YRM), Laizhou Bay (LZB), and the Yellow River Estuary (YRE) as regions of interest (ROIs). In Figure 1, these ROIs are delineated by the red (YRM), yellow (LZB), and the combined red and yellow polygons (YRE), respectively. In total, 224 cloud-free images were considered in the daily TSS time series analysis. Visual inspection was conducted to ensure that the study area was not covered by clouds, and abnormal pixels were masked in the daily TSS time series analysis.

A systematic assessment of TSS was performed at daily and monthly time-scales for the YRM, LZB and YRE during the flood seasons. Mean TSS values during the annual floods were calculated using the images collected during the flood seasons (June to October) from 2016 to 2023. The variation in TSS during the flood season is primarily governed by river discharge and sediment load. To further explore the impact of WSRS on TSS dynamics, this study calculated Pearson correlation coefficients (r) and examined the relationships between daily and monthly sediment load, as well as the monthly discharge at the Lijin Station, with daily and monthly TSS in the YRM, Laizhou Bay, and over the entire study area.

3. Results

3.1. Trend of Daily TSS in the YRE During the Flood Seasons

From 2016 to 2017, the TSS at the Lijin Station consistently remained lower than 100 g·m^{-3}. Since the resumption of the WSRS in 2018, daily TSS has exhibited a pronounced variability within the flood season. WSRS events significantly increased TSS values. For instance, during the WSRS period from June to July 2018, the TSS in the YRM rose sharply, peaking at 446 g·m^{-3}. From July to October, the TSS displayed a fluctuating declining trend (Figure 5). Overall, TSS concentrations in the YRM were higher than those observed in the LZB and over the entire study region (YRE) (Figure 5).

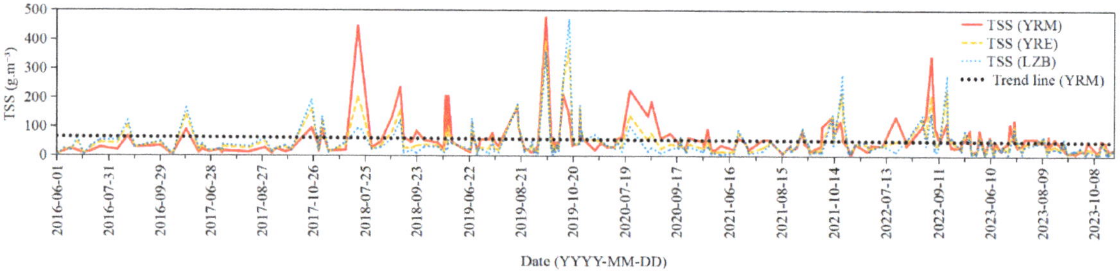

Figure 5. Time series of daily TSS in the river mouth area (YRM), the entire YRE and the Laizhou Bay area (LZB) during the flood seasons of 2016–2023. The black trend line indicates the linear fitting of TSS vs. Date.

3.2. Trends of Daily River Discharge and SC at the LIJIN STATION

Based on the daily records at station Lijin (2016–2023), 10 events of WSRS were identified (see also Table 1). These events were characterized by episodic increases in water discharge and SC, except in 2016 and 2017. The daily maximum river flow and maximum SC always occurred during the flood seasons, especially during the WSRS period and were observed to vary slightly from 2018 to 2023, with daily maximum river flow ranging from 3670 to 4580 m^3·s^{-1} and daily maximum SC ranging from 24.1 $\times$ 10^3 to 48.1 $\times$ 10^3 g·m^{-3} (Table 1, Figure 6). The WSRS contributed nearly 27.3% and 50.3% of the total annual water discharge and sediment load to the sea, respectively, within periods of 14 to 45 days (Table 1), which greatly modified the natural hydrological cycle.

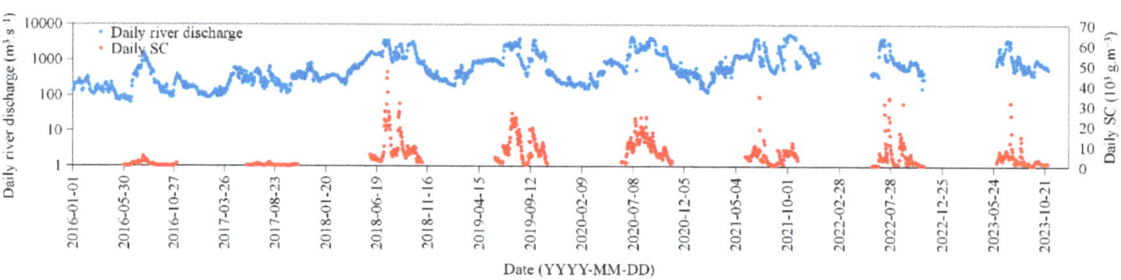

Figure 6. Trend of daily Sediment Concentration (SC) and river discharge at the Lijin Station from 2016 to 2023.

3.3. Spatiotemporal Variations of TSS Before, During, and After the Occurrence of a Typical WSRS Event in 2018

In 2018, the impact of a strong WSRS event on the TSS concentration in the coastal region of the Yellow River was significant. During the period from June 1 to July 2, prior to the strong WSRS event, OLCI imagery analysis revealed an average TSS concentration of less than 100 g·m^{-3} (Figure 7). However, during the event, particularly on July 16 and 20, the imagery showed a marked increase in TSS concentration in the river plume, reaching approximately 400 g·m^{-3}, which indicates a substantial influx of TSS from the Yellow River basin. By August 1, five days after the initial WSRS event, the average TSS concentration in the plume had decreased to approximately 30 g·m^{-3}. During the second WSRS event, on August 12 and 25, TSS concentrations in the plume increased to around 100 g·m^{-3}, significantly higher than in the surrounding coastal waters. Although the intensity and spatial extent of the river plume were considerably reduced in the OLCI images from September to October, compared to the observations on July 16 and 20, the plume was still observable (Figure 7).

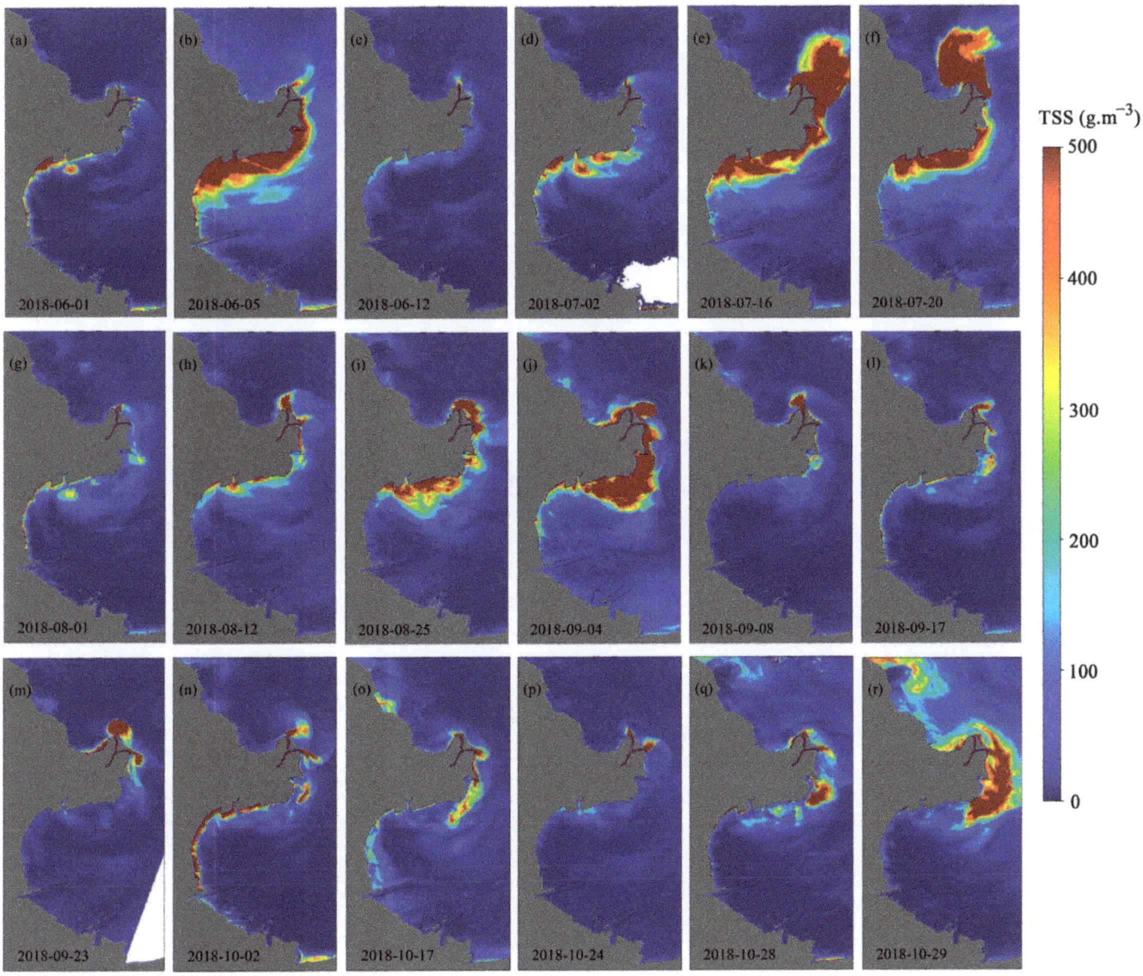

Figure 7. TSS concentrations derived from OLCI images before (**a–d**), during (**e–i**) and after (**j–r**) the typical WSRS event of 2018.

3.4. Interannual Variations of TSS During the Flood Seasons

Figure 8 illustrates the interannual variability of TSS from 2016 to 2023 during the flood season (June to October). Among the years considered, the peak average TSS concentration in the YRM was observed in 2018, reaching 117 g·m^{-3}. The analysis reveals that the average

TSS concentration in the YRM region remained relatively elevated, whereas that in Laizhou Bay was notably lower.

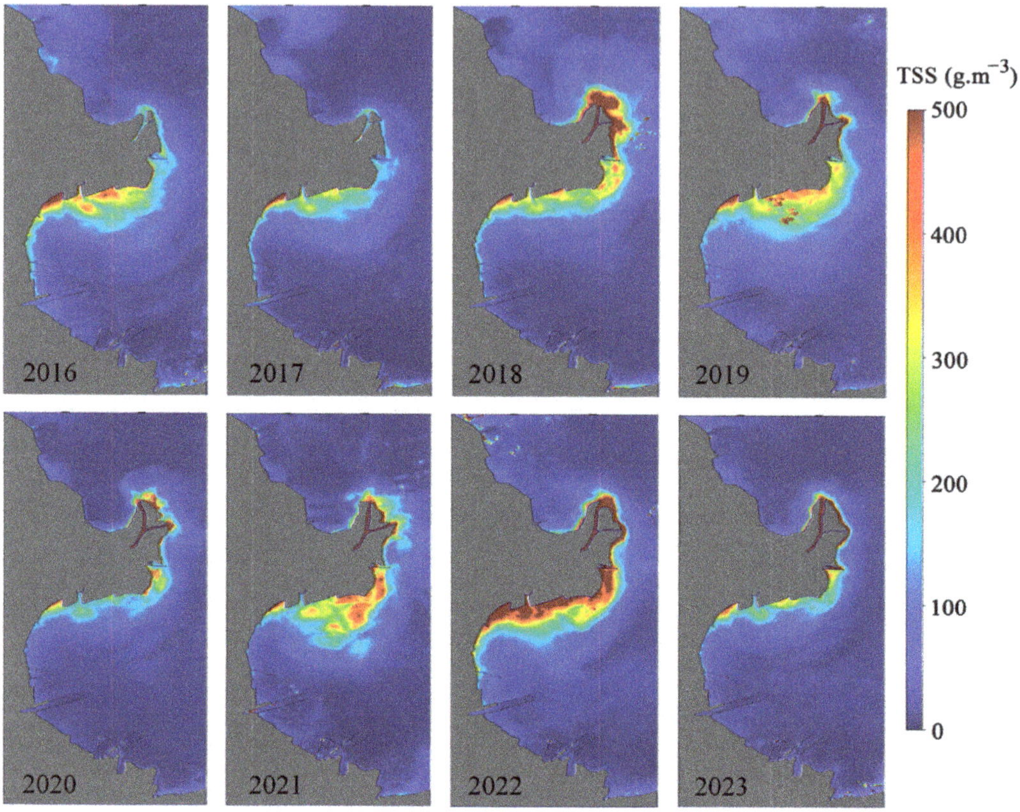

Figure 8. Average TSS maps in the YRE during the flood seasons from 2016 to 2023.

4. Discussion

4.1. Performance of Daily-Revisit OLCI Data in Monitoring TSS

The ACOLITE DSF atmospherically-corrected water reflectance values derived from S3-OLCI data exhibited a good consistency with the reference (OLI-derived) products (Figure 9a–c). Linear regressions with slopes close to 1 were obtained between OLCI and OLI water reflectance values for the green, red and NIR wavebands. The corresponding R^2 coefficients are 0.94, 0.97, and 0.90, with RMSE of 0.006, 0.013 and 0.006 and MAPE of 5.8%, 19.6% and 41.4%, respectively. OLCI-derived water reflectance values were, however, observed to be slightly lower than OLI-derived ones, with BIAS of −3.2%, −16% and −29%, respectively, in the green, red and NIR bands. The results showed that ACOLITE DSF atmospheric correction is hence appropriate for the retrieval of TSS in the highly turbid waters of the YRE. The disparity between the TSS derived from S3-OLCI (300 m) and that derived from Landsat-OLI (30 m) was not conspicuous, with MAPE of 11.47% (Figure 9d). The BIAS between the 30 and 300 m products is -1.8% (Figure 9d), indicating that 300 m imagery would only slightly underestimate TSS values in YRE. Therefore, the spatial resolution of OLCI may not be a disadvantage for monitoring TSS in YRE, as a limited difference between 300 m and 30 m resolution images was found. Compared to the 16-day revisit time of OLI, the daily revisit associated with OLCI data is a major advantage in order to observe and monitor the variation of TSS over short-time-scale events such as the WSRS in the YRE.

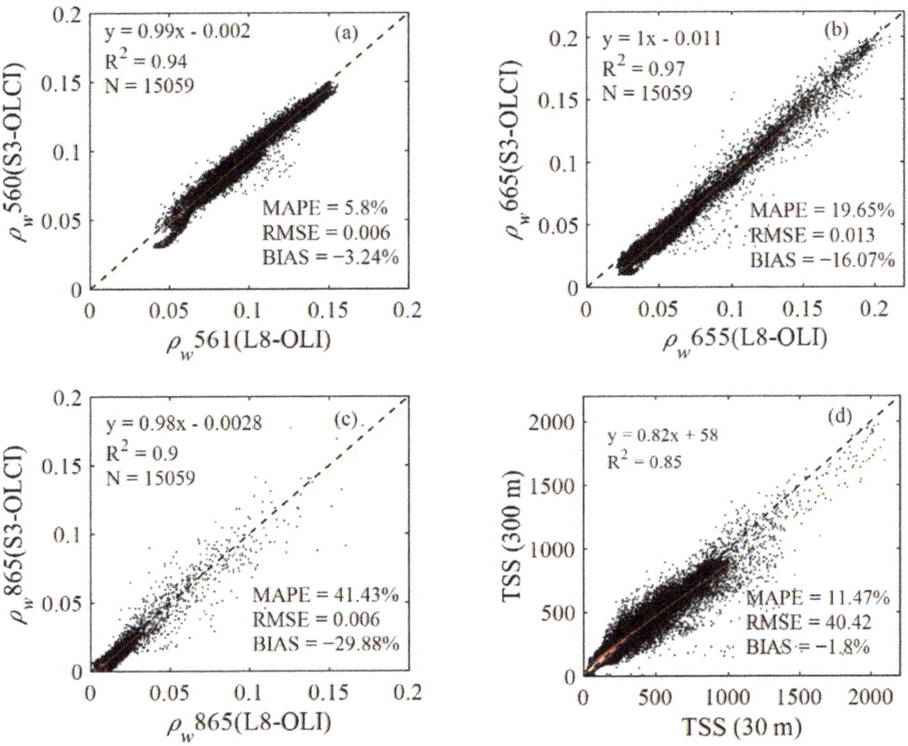

Figure 9. Comparisons between OLCI and OLI water reflectance values in the YRE on 30 August 2023 in the green (**a**), red (**b**) and NIR (**c**) wavebands. Comparison between the resulting TSS values derived from OLCI (300 m) and OLI (30 m) satellite data (**d**).

4.2. Daily TSS Fluctuation in the YRM and Daily SC at the Lijin Station

The variations of the diurnal TSS in the YRM and of the diurnal SC at the Lijin station from 2016 to 2023 are presented in Figure 10. Overall, the TSS concentration in the YRM during the flood seasons of 2016 and 2017 remained relatively stable, with average values mostly below 100 g·m^{-3} and no anomalously high peaks detected. This stability is primarily attributed to the cessation of the WSRS in 2016 and 2017, which substantially reduced the sediment load entering the marine environment.

From 2018 to 2023, significant changes in the daily SC at the Lijin station were observed from June to October. Prior to the implementation of the WSRS, the variations were not substantial, with daily SC values consistently below 10 × 10^3 g·m^{-3} and average TSS values in the YRM remaining below 100 g·m^{-3}. In contrast, during the WSRS period, daily SC exhibited a fluctuating increasing trend, with values reaching up to 20 × 10^3 g·m^{-3}. Concurrently, the average TSS concentration in the YRM showed an increasing trend, peaking in 2018 with an average value exceeding 400 g·m^{-3}. Following the implementation of the WSRS, daily SC at Lijin station decreased, and the average TSS concentration in the YRM gradually declined. The TSS values in 2019, 2021, and 2022 indicate occasional rebounds in TSS concentrations in September and October, surpassing levels observed during the WSRS period.

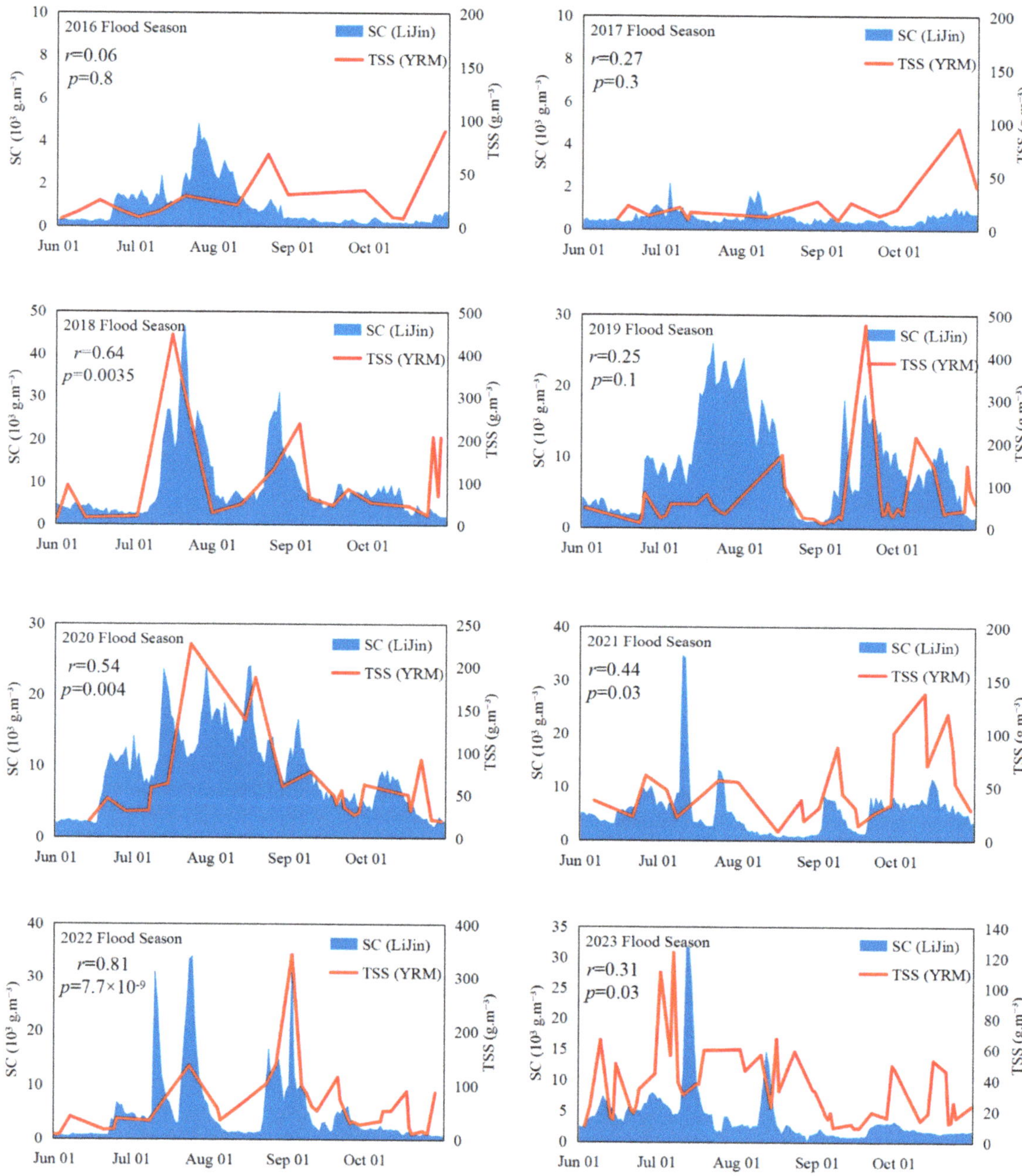

Figure 10. Daily TSS variations in the Yellow River Mouth (YRM) and daily SC variations at the Lijin Station during the flood seasons from 2016 to 2023.

As shown in Table 3, during the years 2016 and 2017, when the WSRS was not implemented, the correlation coefficient (r) between the daily SC at the Lijin station and TSS in the YRE was relatively low. However, following the reactivation of the WSRS from 2018 to 2023, a moderate to strong positive correlation emerged between the average TSS concentration in the YRM region and the daily SC at the Lijin station, with r ranging from

0.25 to 0.81. In contrast, the correlation between the average TSS concentration across the entire YRE and LZB region and the daily SC at the Lijin station was systematically lower.

Table 3. Correlation coefficient (*r*) between the diurnal variations of TSS concentration in the YRE and SC at the Lijin Station.

Year	In the YRM		In the YRE		In the LZB	
	r	*p*-Value	*r*	*p*-Value	*r*	*p*-Value
2016	0.06	>0.05	0.20	>0.05	0.24	>0.05
2017	0.27	>0.05	0.25	>0.05	0.24	>0.05
2018	0.64	<0.05	0.60	<0.05	0.42	>0.05
2019	0.25	>0.05	0.14	>0.05	0.08	>0.05
2020	0.54	<0.05	0.35	>0.05	0.04	>0.05
2021	0.44	<0.05	0.22	>0.05	0.16	>0.05
2022	0.81	<0.05	0.48	<0.05	0.21	>0.05
2023	0.31	<0.05	0.22	>0.05	0.14	>0.05

4.3. Correlation Between River Discharge, Sediment Load and TSS Concentration

During the 2016–2017 period, the suspension of the WSRS operations resulted in a lower monthly sediment load (consistently below 5.46 × 10^6 tons) and in a monthly runoff lower than 2.1 × 10^9 m^3. Correspondingly, the monthly average TSS in YRM remained generally below 70 $g{\cdot}m^{-3}$, with relatively stable variations during the flood seasons. In comparison to the river flow and sediment transport volumes of 2016 and 2017, the corresponding values from 2018 to 2023 showed a significant increase. During the 2018 WSRS period, the maximum monthly sediment load reached 1.4 × 10^8 tons. During the WSRS from 2018 to 2023, the highest river discharge and sediment transport rates typically occurred in July and August. Compared to the pre-WSRS period, where monthly sediment load was less than 32 × 10^6 tons, the monthly sediment load during the WSRS period significantly increased, surpassing 44.45 × 10^6 tons. Notably, between 2018 and 2020, the sediment load sometimes exceeded 100 × 10^6 tons.

Additionally, the concentration of TSS exhibited a marked upward trend during the WSRS. Specifically, as shown in Figure 11, the sediment load peaked in July 2018 during the WSRS, while the average TSS concentration in the YRM region in July reached 268 $g{\cdot}m^{-3}$.

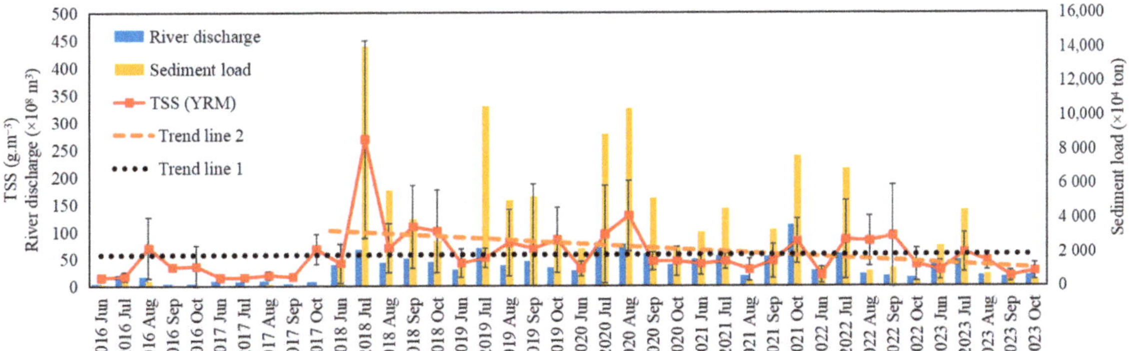

Figure 11. Monthly variations of river discharge and sediment load at the Lijin Station, together with monthly variations of TSS concentration in the YRM from June to October, from 2016 to 2023. Error bars correspond to the associated standard deviations. Trend lines 1 and 2 represent the linear fitting TSS variations vs. Date in the YRM from 2016–2023 and 2018–2023, respectively.

A similar pattern was observed in 2020, where, as depicted in Figures 8 and 12, the TSS concentration in the estuarine region peaked from late July to early August. During this period, both the daily SC and the monthly average sediment load at the Lijin station also reached their highest levels.

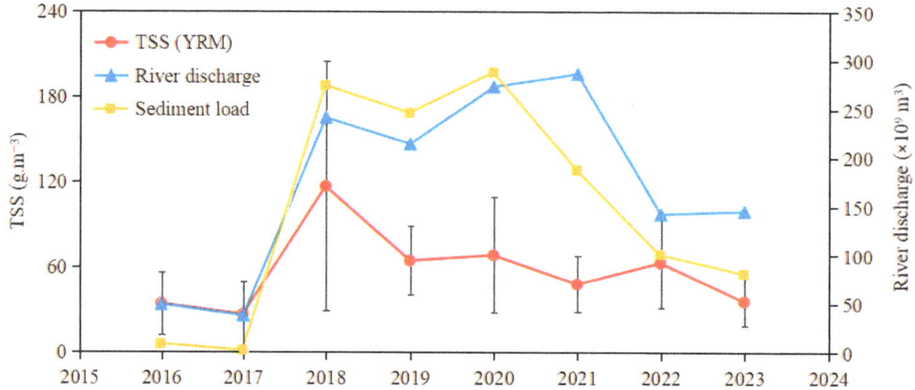

Figure 12. River discharge and sediment load at the Lijin Station and average TSS concentration in the YRM during flood seasons. Error bars correspond to the associated standard deviations.

In 2023, TSS concentrations remained consistently high during the WSRS period compared to before and after the event. In summary, the trends in daily SC and monthly average sediment load at the Lijin station in 2018, 2020 and 2023 were closely correlated with changes in TSS concentration, all showing an increase in TSS during the WSRS, followed by a decline post-scheme.

From June to October in the YRM region, a significant correlation was observed between the river discharge (and sediment load) and TSS (r = 0.52 and 0.73) (Table 4). During the WSRS period from June to August, TSS exhibited a more pronounced correlation with the sediment load (r = 0.79) and river discharge (r = 0.60). During June–October, river discharge and sediment load exhibited significant yet weak correlations with TSS in the YRE (r = 0.33 and 0.39, respectively). Conversely, in the LZB, sediment load showed no correlation with TSS (r = 0.06).

Table 4. Correlation coefficient (r) between the monthly-averaged TSS, river discharge and sediment load from 2016 to 2023.

Station	Period	YRM	YRE	LZB	YRM	YRE	LZB
		TSS vs. Sediment Load			TSS vs. River Discharge		
Lijin	June–October	0.73 $p < 0.05$	0.39 $p < 0.05$	0.06 $p > 0.05$	0.52 $p < 0.05$	0.33 $p < 0.05$	0.11 $p > 0.05$
	June–August	0.79 $p < 0.05$	0.45 $p < 0.05$	−0.009 $p > 0.05$	0.60 $p < 0.05$	0.27 $p > 0.05$	−0.12 $p > 0.05$
Xiaolangdi	June–October	0.55 $p < 0.05$	0.23 $p > 0.05$	−0.03 $p > 0.05$	0.42 $p < 0.05$	0.20 $p > 0.05$	0.014 $p > 0.05$
	June–August	0.63 $p < 0.05$	0.36 $p > 0.05$	0.009 $p > 0.05$	0.37 $p < 0.05$	0.09 $p > 0.05$	−0.18 $p > 0.05$
Huayuankou	June–October	0.72 $p < 0.05$	0.33 $p < 0.05$	−0.007 $p > 0.05$	0.45 $p < 0.05$	0.24 $p > 0.05$	0.04 $p > 0.05$
	June–August	0.80 $p < 0.05$	0.45 $p < 0.05$	0.004 $p > 0.05$	0.45 $p < 0.05$	0.12 $p > 0.05$	−0.20 $p > 0.05$

This suggests that the high TSS values at the river mouth are directly related to increased river discharge and sediment load due to WSRS events and autumn floods.

Less sediment has thus been supplied to the LZB located far away from the river mouth (Figure 7) [11].

Consequently, during the flood season, WSRS events involving large river discharge and sediment load play a dominant role in the TSS variations in the YRM. Additionally, rainfall, flooding, and wind also impact the TSS concentration and distribution in the YRE.

4.4. Correlation Between Rainfall, Autumn Flood, Wind Speed and TSS

In some instances, in the YRM, the TSS following WSRS events may exceed that during the WSRS period itself. For instance, in 2019, TSS concentrations in July and August were lower than those in September and October (Figure 10). This can be attributed to the sustained rainfall between September 17 and 24, in conjunction with higher wind speeds in September and October (not shown). Wind waves are one of the main factors that control local sediment resuspension in shallow waters [38].

In 2021, TSS concentrations in September and October were similarly elevated due to severe autumn flooding from late August to October. During the autumn period, the main stem of the Yellow River experienced three major flood events. As a result, the sediment load in October 2021 was the highest (Figure 11), with the corresponding TSS concentrations during the autumn flood period even surpassing those observed during the WSRS events (Figures 10 and 11).

The YRE is affected by the East Asian monsoon climate. Previous studies indicated that wave height (WH) and sea surface height (SSH) play a major role in sediment resuspension and dispersion [11,39,40]. Over the study period, the mean WH was about 0.8 m in the cold period, which was nearly twice the value in the warm season. In the summer, the south wind dominates with a moderate speed (3–5 m/s) [10,41]. As a consequence, the wave is gentler and the height of the wave is generally <1 m in the summer. As a result, the waves can cause the limited resuspension of bottom sediment in the warm flood seasons.

The tides in the YRE are irregular semi-diurnal, with an average tidal range of 0.6–0.8 m. Intraday TSS variation is influenced by tidal currents [24]. The shift in the tidal directions results in a shift in TSS plume direction and transportation of suspended sediment near groins [25]. The WSRS-induced increase in freshwater and sediment discharge to the river mouth enhanced the offshore extension of the turbid plume [24].

Current speeds show different seasonal patterns. In the summer, current speeds decrease rapidly to the southern coast of the estuary, suggesting that the sediment carried by the Yellow River is transported primarily to this area. Correspondingly, the contour lines with TSS > 100 $g \cdot m^{-3}$ display a narrow shape along the coast (Figure 7) [26].

4.5. Correlation Between WSRS and Sediment Transport

From 2018 to 2023, the monthly average TSS in the YRM manifested a downward trend (Figures 10 and 12). Despite the total sediment load during the flood seasons in 2019 and 2020 being comparable to that in 2018, the TSS concentration in the YRM in 2019–2020 was not as high as it was in 2018. Some of the sediments were retained within the estuary, leading to siltation; a part of the suspended sediments was also transported beyond the estuary, as shown in Figure 7, but this occurred only during the WSRS period. A part of suspended sediments from the Yellow River ended up deposited at the river mouth before entering the ocean (Figure 13).

This sedimentation situation was further exacerbated in 2023. Although the total sediment load during the flood seasons has decreased since 2021, the river mouth has undergone a seaward advancement due to the sediment accumulation during the flood seasons. TSS concentrations in the YRM during the flood seasons of 2023 were systematically lower than those during the flood seasons of 2018 to 2022. As a result, TSS in the

YRE showed a lower correlation with the daily SC at Lijin station. The Water-Sediment Regulation seems to be closely related to shoreline evolution. From 2000 to 2023, the shoreline has continuously advanced, indicating an overall accretion trend [42]. The sediment accumulation around the current river mouth would be a local geomorphic phenomenon (Figure 13). The WSRS of the Yellow River from 2018 to 2023 promoted the spread of TSS into the YRE, so that extremely high TSS concentrations were observed in the YRE, especially in the river mouth area. Due to the elevation of the Yellow River's lower reaches and the reduction in water and sediment supplies, the progradation rate of the active sub-delta at the river mouth will gradually slow down. As a result, the shoreline will tend to stabilize, and the changes in suspended sediments in the estuary will also become more stable.

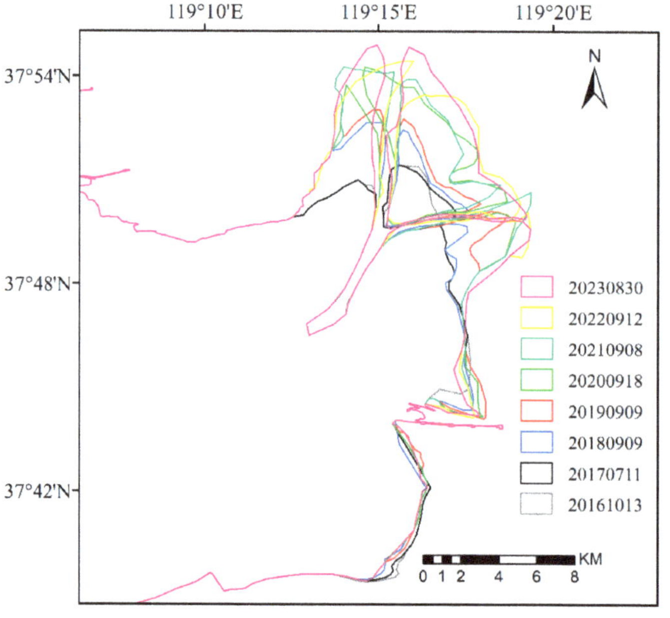

Figure 13. The coastlines of the YRM from 2016 to 2023 were obtained with Sentinel2-MSI (Multi-Spectral Instrument) and Landsat multi-spectral images during the flood seasons, obtained using Normalized Difference Water Index (NDWI) combined with visual interpretation.

In the future, climate change and impacts of human activities are expected to result in the elevation buildup of the Yellow River's tail channel and the reduction of incoming water and sediment load, and the progradation rate of the subaqueous delta at the active river mouth will progressively decrease. The shoreline will gradually stabilize, and suspended sediment dynamics in the estuary should also tend to reach equilibrium [42]. The water–sediment regime of the Yellow River has shifted from the former state of "low water and high sediment" to the current condition of "low water and low sediment." It indicates that WSRS will have less effect on the delta evolution with less river sediment supply. Therefore, the coastal zone of the Yellow River Estuary will face further erosion. This process is expected to favor the stabilization of the existing estuarine channel [6].

4.6. Future Improvements of the Present Study

In the present study, the use of in-situ data collected in 2004 to establish a model to retrieve TSS from satellite data recorded from 2016 to 2023 may raise questions about temporal representativeness. However, the key point is that the field dataset was generated in 2004 during the WSRS period, so it is expected to be suitable for our study. To verify whether the model is reliable, an independent validation was made using S3-OLCI estimated TSS values and TSS in situ measurements for the dataset measured during the WSRS period on 19–21 July 2019 and 9 July 2023. The results show a good agreement with a highly significant linear relationship (R^2 of 0.96), with a MAPE of 29.6% and RMSE of 222 $g \cdot m^{-3}$ (Figure 4). Therefore, in this case, the 2004 dataset is a valuable choice for the TSS retrieval. Future efforts should prioritize real-time in situ observations during WSRS periods, enabling more robust TSS validation for a comprehensive monitoring of sediment dynamics and characteristics in the YRE.

OLCI satellite data have a 300-m resolution, which may limit TSS retrievals in channels and small estuaries. However, the MAPE between the 30 and 300 m TSS products in the YRE is 11.5% (Figure 9d). The improvement from 30 m OLI to 300 m OLCI is not that obvious in the highly turbid waters [43]. Discrepancies in spatial scale between the 300 m resolution remote-sensing data and in situ suspended sediment concentration measurements from hydrologic stations may also introduce inconsistencies in the correlation analysis between daily TSS fluctuations in the YRE and daily SC at the Lijin Station.

One limitation in our study is the temporal discontinuity caused by selecting cloud-free images. Although this approach ensures data quality, it may limit the detection of short-duration extreme events and be biased in terms of seasonal coverage. Future work could integrate harmonized datasets (e.g., GOCI) to obtain not only daily but also hourly variations of TSS. Tides and currents also affect the short-term dispersal pattern of freshwater and sediment plume [24,25]. Further developments should include the influence of coastal currents, wave energy, and bathymetry, determined by numerical modeling, to generate a complete assessment of the littoral areas exposed to the Yellow River turbid plume.

5. Conclusions

In order to investigate the variations of total suspended solids in the YRE under the new conditions imposed by the WSRS, S3-OLCI images recorded from 2016 to 2023 during flood periods in the YRE were processed and analyzed. The response of TSS in the YRE to the WSRS was discussed, combining satellite observations and in-situ data from the Lijin Station.

1. Based on the correlation analysis between OLCI-derived *Rrs* values and in situ measured TSS, an exponential model was established combining the *Rrs*(665)/*Rrs*(560) ratio and the *Rrs*(865) band (R^2= 0.94). OLCI images processed using the ACOLITE DSF atmospheric correction showed good agreement with both in situ *Rrs* and the previously validated OLI images. The results showed that daily OLCI images can be used to monitor the dynamics of terrigenous and anthropogenic TSS in the YRE over short time-scale events such as the WSRS, notably over the recent period of new regulations of the WSRS.

2. Satellite-derived TSS concentrations were generally low with no significant variations as the WSRS was suspended in 2016 and 2017. After the resumption of the WSRS, the daily TSS increased and varied significantly during the flood seasons from 2018 to 2023. The TSS was generally low before the WSRS in June and gradually rose over time due to the large inputs of water and sediment in July and August, then returned to normal values once the WSRS ended. The TSS in the river mouth was observed to be significantly correlated with the sediment load in the river, suggesting that the sediment and river discharges during the WSRS period from June to August were the determining factors controlling the river plume.

3. The flood events, including those caused by the WSRS, gave rise to a significant river plume and outputs of a substantial amount of TSS in adjacent coastal waters. A significant

part of the suspended sediments most probably deposited in the Yellow River delta, leading to continuous shoreline progradation with accretion from 2016 to 2023. The maximum TSS increase was observed during the flood season in 2018, then demonstrated a downward trend from 2018 to 2023.

This study demonstrates the capability of S3-OLCI data to capture short-term WSRS-driven TSS dynamics in the YRE, providing a robust tool for evaluating the scheme's environmental and ecological impacts. Future efforts should prioritize real-time in situ observations during WSRS periods, enabling more robust validation and comprehensive monitoring of sediment characteristics across different water column strata in the YRE.

Author Contributions: Conceptualization, Y.L. (Yafei Luo) and Z.H.; methodology, Y.L. (Yafei Luo) and D.D.; validation, Y.L. (Yafei Luo), Y.L. (Yanxia Liu) and L.Y.; formal analysis, Y.L. (Yafei Luo), Y.L. (Yanxia Liu) and L.Y.; investigation, Y.L. (Yanxia Liu) and L.Y.; resources, Y.L. (Yanxia Liu), D.F. and H.H.; data curation, Y.L. (Yanxia Liu) and L.Y.; writing—original draft preparation, Y.L. (Yafei Luo) and D.D.; writing—review and editing, Y.L. (Yafei Luo) and D.D.; supervision, Z.H., D.F. and H.H.; project administration, Y.L. (Yafei Luo) and Z.H.; funding acquisition, Y.L. (Yafei Luo) and Z.H. All authors have read and agreed to the published version of the manuscript.

Funding: This research was funded by the National Natural Science Foundation of China (Grant No. 42206187 and 12374428), the Open Fund of the Key Laboratory of Marine Geology and Environment, Chinese Academy of Sciences (Grant No. MGE2020KG12), the program for scientific research start-up funds of Guangdong Ocean University (Grant No. R20010) and the Southern Marine Science and Engineering Guangdong Laboratory (Zhuhai) (Grant No. 311024018 and SML2024SP006).

Acknowledgments: We appreciate Junwu Tang, Wei Zhao, Qingjun Song, and Xiaomei Wang in the National Satellite Ocean Application Service and Song Jin in Yantai Institute of Coastal Zone Research, Chinese Academy of Sciences, for collecting in situ data. Acknowledgement for the data support from "National Earth System ScienceData Center, National Science and Technology Infrastructure of China (https://www.geodata.cn (accessed on 29 April 2025))".

Conflicts of Interest: The authors declare no conflicts of interest.

References

1. Milliman, J.; Meade, R. World-wide delivery of river sediments to the oceans. *J. Geol.* **1983**, *91*, 1–21. [CrossRef]
2. Liu, S.M.; Li, L.W.; Zhang, G.L.; Liu, Z.; Yu, Z.; Ren, J. Impacts of human activities on nutrient transports in the Huanghe (Yellow River) estuary. *J. Hydrol.* **2012**, *430*, 103–110. [CrossRef]
3. Liu, S.M. Response of nutrient transports to water–sediment regulation events in the Huanghe basin and its impact on the biogeochemistry of the Bohai. *J. Mar. Syst.* **2015**, *141*, 59–70. [CrossRef]
4. Wang, H.; Wu, X.; Bi, N.; Li, S.; Yuan, P.; Wang, A.; Syvitski, J.; Saito, Y.; Yang, Z.; Liu, S.; et al. Impacts of the dam-orientated water-sediment regulation scheme on the lower reaches and delta of the Yellow River (Huanghe): A review. *Glob. Planet Change* **2017**, *157*, 93–113. [CrossRef]
5. Yu, Y.; Shi, X.; Wang, H.; Yue, C.; Chen, S.; Liu, Y.; Hu, L.; Qiao, S. Effects of dams on water and sediment delivery to the sea by the Huanghe (Yellow River): The special role of Water-Sediment Modulation. *Anthropocene* **2013**, *3*, 72–82. [CrossRef]
6. Chen, S.; Gu, S.; Ji, H.; Xu, C. Processes of the Yellow River Mouth on new water and sediment condition. *J. Sediment. Res.* **2019**, *44*, 61–67. (In Chinese)
7. Fan, H.; Huang, H.; Zeng, T.; Wang, K. River mouth bar formation, riverbed aggradation and channel migration in the modern Huanghe (Yellow) River delta, China. *Geomorphology* **2006**, *74*, 124–136. [CrossRef]
8. Qiu, Z.F.; Xiao, C.; Perrie, W.; Sun, D.; Wang, S.; Shen, H.; Yang, D.; He, Y. Using Landsat 8 data to estimate suspended particulate matter in the Yellow River estuary. *J. Geophys. Res. Oceans.* **2017**, *122*, 276–290. [CrossRef]
9. Li, J.; Hao, Y.; Zhang, Z.; Li, Z.; Yu, R.; Sun, Y. Analyzing the distribution and variation of Suspended Particulate Matter (SPM) in the Yellow River Estuary (YRE) using Landsat 8 OLI. *Reg. Stud. Mar. Sci.* **2021**, *48*, 102064. [CrossRef]
10. Li, P.; Ke, Y.; Bai, J.; Zhang, S.; Chen, M.; Zhou, D. Spatiotemporal dynamics of suspended particulate matter in the Yellow River Estuary, China during the past two decades based on time-series Landsat and Sentinel-2 data. *Mar. Pollut. Bull.* **2019**, *149*, 110518. [CrossRef]

11. Jin, S.; Fagherazzi, S.; Fichot, C.G.; Wu, X.; Liu, Y.; Zheng, X.; Zou, T.; Xing, Q. Drivers of suspended sediment dynamics along the shorelines of the Yellow River Delta detected from satellite data. *Earth Surf. Process. Landf.* **2023**, *48*, 3091–3102. [CrossRef]
12. Jin, S.; Zou, T.; Xing, Q.; Zheng, X.; Fagherazzi, S. Particle Size Distribution Slope Changes along the Yellow River Delta Observed from Sentinel 3A/B OLCI Images. *Remote Sens.* **2024**, *16*, 938. [CrossRef]
13. Zhang, M.; Dong, Q.; Cui, T.; Xue, C.; Zhang, S. Suspended sediment monitoring and assessment for Yellow River estuary from Landsat TM and ETM+ imagery. *Remote Sens. Environ.* **2014**, *146*, 136–147. [CrossRef]
14. Yu, Z.; Zhang, J.; Chen, Z.; Hu, Y.; Shum, C.K.; Ma, C.; Song, Q.; Yuan, X.; Wang, B.; Zhou, B. Spatiotemporal evolutions of the suspended particulate matter in the Yellow River Estuary, Bohai Sea and characterized by Gaofen imagery. *Remote Sens.* **2023**, *15*, 4769. [CrossRef]
15. Constantin, S.; Doxaran, D.; Constantinescu, Ş. Estimation of water turbidity and analysis of its spatio-temporal variability in the Danube River plume (Black Sea) using MODIS satellite data. *Cont. Shelf Res.* **2016**, *112*, 14–30. [CrossRef]
16. Zhang, Y.; Shi, K.; Zhou, Y.; Liu, X.; Qin, B. Monitoring the river plume induced by heavy rainfall events in large, shallow, Lake Taihu using MODIS 250 m imagery. *Remote Sens. Environ.* **2016**, *173*, 109–121. [CrossRef]
17. Luo, X.; Wang, J.; Chen, G.; Zhu, S.; Huo, Z.; Liu, X.; Deng, W. Research on Distribution of Suspended Sediment Concentration in Long Time Series Based on Mid-Resolution Imaging Spectrometer Data and Quantile Trend Analysis. *Reg. Stud. Mar. Sci.* **2020**, *39*, 101399. [CrossRef]
18. Wang, K. Impact and evaluation of water and sediment regulation in the Yellow River on the estuary and its delta. *J. Sediment. Res.* **2005**, *6*, 31–35. (In Chinese)
19. Wang, Q.; Li, Y. Phosphorus adsorption and desorption behavior on sediments of different origins. *J. Soils Sediments* **2010**, *10*, 1159–1173. [CrossRef]
20. Bi, N.; Yang, Z.; Wang, H.; Xu, C.; Guo, Z. Impact of artificial water and sediment discharge regulation in the Huanghe (Yellow River) on the transport of particulate heavy metals to the sea. *Catena* **2014**, *121*, e232–e240. [CrossRef]
21. Xu, B.; Yang, D.; Burnett, W.C.; Ran, X.; Yu, Z.; Gao, M.; Diao, S.; Jiang, X. Artificial water sediment regulation scheme influences morphology, hydrodynamics and nutrient behavior in the Yellow River estuary. *J. Hydrol.* **2016**, *539*, 102–112. [CrossRef]
22. Wang, Y.; Liu, D.; Lee, K.; Dong, Z.; Di, B.; Wang, Y.; Zhang, J. Impact of Water-Sediment Regulation Scheme on seasonal and spatial variations of biogeochemical factors in the Yellow River estuary. *Estuar. Coast. Shelf Sci.* **2017**, *198*, 92–105. [CrossRef]
23. Wu, X.; Bi, N.; Xu, J.; Nittrouer, J.; Yang, Z.; Saito, Y.; Wang, H. Stepwise morphological evolution of the active Yellow River (Huanghe) delta lobe (1976–2013): Dominant roles of riverine discharge and sediment grain size. *Geomorphology* **2017**, *292*, 115–127. [CrossRef]
24. Guo, K.; Zou, T.; Jiang, D.; Tang, C.; Zhang, H. Variability of Yellow River turbid plume detected with satellite remote sensing during water-sediment regulation. *Cont. Shelf Res.* **2017**, *135*, 74–85. [CrossRef]
25. Li, P.; Ke, Y.; Wang, D.; Ji, H.; Chen, S.; Chen, M.; Lyu, M.; Zhou, D. Human impact on suspended particulate matter in the Yellow River Estuary, China: Evidence from remote sensing data fusion using an improved spatiotemporal fusion method. *Sci. Total Environ.* **2021**, *750*, 141612. [CrossRef]
26. Liu, Y.; Huang, H.; Yang, X. The transportation and deposition of suspended sediment and its dynamic mechanism analysis based on Landsat images in the Laizhou Bay. *Acta Oceanolog. Sin.* **2013**, *35*, 43–53. (In Chinese)
27. Vanhellemont, Q.; Ruddick, K. Atmospheric correction of Sentinel-3/OLCI data for mapping of suspended particulate matter and chlorophyll-a concentration in Belgian turbid coastal waters. *Remote Sens. Environ.* **2021**, *256*, 112284. [CrossRef]
28. Vanhellemont, Q.; Ruddick, K. Atmospheric correction of metre-scale optical satellite data for inland and coastal water applications. *Remote Sens. Environ.* **2018**, *216*, 586–597. [CrossRef]
29. Vanhellemont, Q. Daily metre-scale mapping of water turbidity using CubeSat imagery. *Opt. Express* **2019**, *27*, A1372–A1399. [CrossRef] [PubMed]
30. Vanhellemont, Q. Adaptation of the dark spectrum fitting atmospheric correction for aquatic applications of the Landsat and Sentinel-2 archives. *Remote Sens. Environ.* **2019**, *225*, 175–192. [CrossRef]
31. Vanhellemont, Q. Sensitivity analysis of the dark Spectrum fitting atmospheric correction for metre- and decametre-scale satellite imagery using autonomous hyperspectral radiometry. *Opt. Express* **2020**, *27*, A1372–A1399. [CrossRef] [PubMed]
32. A, R.; Qing, S.; Bao, Y. The inspection and application of atmospheric correction algorithm in Landsat-8 OLI data. *Mar. Sci.* **2018**, *42*, 107–115. (In Chinese)
33. Zhan, C.; Yu, J.; Wang, Q.; Li, Y.; Zhou, D.; Xing, Q.; Chu, X. Remote sensing retrieval of surface suspended sediment concentration in the Yellow River Estuary. *Chin. Geogr. Sci.* **2017**, *27*, 934–947. [CrossRef]
34. Shen, F.; Salama, M.H.D.S.; Zhou, Y.X.; Li, J.F.; Su, Z.; Kuang, D.B. Remote-sensing reflectance characteristics of highly turbid estuarine waters—A comparative experiment of the Yangtze River and the Yellow River. *Int. J. Remote Sens.* **2010**, *31*, 639–2654. [CrossRef]
35. Luo, Y.; Doxaran, D.; Ruddick, K.; Shen, F.; Gentili, B.; Yan, L.; Huang, H. Saturation of water reflectance in extremely turbid media based on field measurements, satellite data and bio-optical modelling. *Opt. Express.* **2018**, *26*, 10435–10451. [CrossRef]

36. Zhou, Y.; Hao, Y.; Liu, D.; Cui, T.; Yu, R.; Zhang, Z. Estimation of suspended particulate matter concentration based on Landsat 8 data in the Yellow River Estuary. *J. Mar. Sci.* **2018**, *36*, 35–45. (In Chinese)
37. Doxaran, D.; Froidefond, J.M.; Lavender, S.; Castaing, P. Spectral signature of highly turbid waters: Application with SPOT data to quantify suspended particulate matter concentrations. *Remote Sens. Environ.* **2002**, *81*, 149–161. [CrossRef]
38. Zhao, G.; Jiang, W.; Wang, T.; Chen, S.; Bian, C. Decadal variation and regulation mechanisms of the suspended sediment concentration in the Bohai Sea, China. *J. Geophys. Res. Ocean.* **2022**, *127*, e2021JC017699. [CrossRef]
39. Wen, M.; Shan, H.; Zhang, S.; Liu, X.; Jia, Y. Contribution of waves and currents to sediment resuspension in the Yellow River Delta. *Mar. Georesour. Geotechnol.* **2018**, *37*, 96–102. [CrossRef]
40. Wang, H.; Yang, Z.; Li, G.; Jiang, W. Wave climate modeling on the abandoned Huanghe (Yellow River) Delta lobe and related deltaic erosion. *J. Coast. Res.* **2006**, *22*, 906–918. [CrossRef]
41. Yang, Z.; Ji, Y.; Bi, N.; Lei, K.; Wang, H. Sediment transport off the Huanghe (Yellow River) delta and in the adjacent Bohai Sea in winter and seasonal comparison. *Estuar. Coast. Shelf Sci.* **2011**, *93*, 173–181. [CrossRef]
42. Zuo, F.; Cao, Y.; Fan, Z.; Yin, P.; Li, R.; Qian, Z.; Wang, H.; Zhan, C.; Wang, Q. Research on the stability of river mouth shoreline in the Yellow River Delta from 1996 to 2022. *Mar. Sci. Bull.* **2024**, *43*, 426–436. (In Chinese)
43. Luo, Y.; Doxaran, D.; Vanhellemont, Q. Retrieval and validation of water turbidity at metre-scale using Pléiades satellite data: A case study in the Gironde estuary. *Remote Sens.* **2020**, *12*, 946. [CrossRef]

remote sensing

MDPI

Article

Multiscale Evaluation and Error Characterization of HY-2B Fused Sea Surface Temperature Data

Xiaomin Chang [1,2,*], Lei Ji [1,2], Guangyu Zuo [3,4], Yuchen Wang [1,3], Siyu Ma [1,2] and Yinke Dou [1,3]

1 Key Laboratory of Cleaner Intelligent Control on Coal & Electricity, Ministry of Education and College of Electrical and Power Engineering, Taiyuan University of Technology, Taiyuan 030024, China; 2023520909@link.tyut.edu (L.J.); wangyuchen0050@link.tyut.edu.cn (Y.W.); 2024520893@link.tyut.edu (S.M.); douyinke@tyut.edu.cn (Y.D.)

2 College of Water Conservancy Science and Engineering, Taiyuan University of Technology, Taiyuan 030024, China

3 College of Electrical and Power Engineering, Taiyuan University of Technology, Taiyuan 030024, China; zuoguangyu@tyut.edu.cn

4 Key Laboratory of Polar Ecosystem and Climate Change, Shanghai Key Laboratory of Polar Life and Environment Sciences, School of Oceanography, Ministry of Education, Shanghai Jiao Tong University, Shanghai 200030, China

* Correspondence: changxiaomin@tyut.edu.cn

Abstract

The Haiyang-2B (HY-2B) satellite, launched on 25 October 2018, carries both active and passive microwave sensors, including a scanning microwave Radiometer (SMR), to deliver high-precision, all-weather global observations. Sea surface temperature (SST) is among its key products. We evaluated the HY-2B SMR Level-4A (L4A) SST (25 km resolution) over the North Pacific (0–60°N, 120°E–100°W) for the period 1 October 2023 to 31 March 2025 using the extended triple collocation (ETC) and dual-pairing methods. These comparisons were made against the Remote Sensing System (RSS) microwave and infrared (MWIR) fused SST product and the National Oceanic and Atmospheric Administration (NOAA) in situ SST Quality Monitor (iQuam) observations. Relative to iQuam, HY-2B SST has a mean bias of −0.002 °C and a root mean square error (RMSE) of 0.279 °C. Compared to the MWIR product, the mean bias is 0.009 °C with an RMSE of 0.270 °C, indicating high accuracy. ETC yields an equivalent standard deviation (ESD) of 0.163 °C for HY-2B, compared to 0.157 °C for iQuam and 0.196 °C for MWIR. Platform-specific ESDs are lowest for drifters (0.124 °C) and tropical moored buoys (0.088 °C) and highest for ship and coastal moored buoys (both 0.238 °C). Both the HY-2B and MWIR products exhibit increasing ESD and RMSE toward higher latitudes, primarily driven by stronger winds, higher columnar water vapor, and elevated cloud liquid water. Overall, HY-2B SST performs reliably under most conditions, but incurs larger errors under extreme environments. This analysis provides a robust basis for its application and future refinement.

Keywords: sea surface temperature; HY-2B; iQuam; validation; extended triple collocation

1. Introduction

Sea surface temperature (SST) is a critical parameter for monitoring, understanding, and forecasting heat, momentum, and gas fluxes across various spatial and temporal scales [1–5]. These fluxes govern the intricate coupling between the atmosphere and the ocean. SST not only influences air–sea moisture and heat exchange but also serves as a fundamental indicator of ocean circulation patterns, frontal zones, water mass distributions, and other dynamic processes [6–8]. SST data are primarily obtained from in situ measurements (ships and buoys) and satellite remote sensing. In situ measurements offer high accuracy but have limited spatial coverage and resolution, and such datasets often exhibit quality and completeness issues due to occasional outliers [9]. In contrast, satellite-derived SST products provide high spatiotemporal resolution and near-synoptic global coverage, making them increasingly important in oceanographic research.

Spaceborne scanning microwave radiometers provide continuous, all-weather observations and have been used to generate a global SST record for over 40 years [10,11]. However, differences in instrument design and calibration among microwave radiometers can introduce biases in brightness temperature measurements, affecting the accuracy of SST retrievals [12,13]. In addition, sensor performance can degrade over time, leading to time-dependent errors in long-term SST observations [14]. Therefore, obtaining more comprehensive and accurate SST products requires fusing observations from multiple satellite sensors, which improves spatial coverage and consistency.

The Haiyang-2B (HY-2B) satellite, launched in October 2018, carries the Scanning Microwave Radiometer (SMR), which continuously measures SST, wind speed, atmospheric water vapor, cloud liquid water, and rainfall intensity [15,16]. The addition of SMR significantly expands the spatiotemporal coverage and timeliness of SST observations. Previous studies have validated the accuracy of HY-2B Level-2B (L2B) SST products. Zhou et al. [17] used NOAA iQuam in situ data to assess HY-2B SST, reporting a bias of −0.02 °C and an RMSE of 0.80 °C. Similarly, Zhang et al. [18] reported a bias of 0.09 °C and an RMSE of 0.72 °C for these products.

Since its launch, HY-2B has been operational for over six years, far exceeding its design life. However, no validation study has yet focused on the HY-2B L4A SST fusion product. Compared to traditional single-sensor products, fused SST products offer substantially improved spatial coverage, enabling near-global observations. In generating fused SST products, different institutions may combine varied data sources (e.g., infrared SST, microwave SST, and in situ data) and processing methods to meet specific objectives. These sources differ in format, spatial/temporal resolution, cloud detection algorithms, and quality control procedures. Additionally, differences in fusion algorithms, initial field generation, land–sea boundary handling, and ice-masking techniques can lead to discrepancies in the final product performance [19].

In this study, we apply extended triple collocation (ETC) and direct pairwise comparison methods to evaluate the operational HY-2B L4A fused SST product for the period from 1 October 2023 to 31 March 2025. Section 2 describes the data and methods. Sections 3 and 4 present the results and discussion, respectively. Section 5 concludes the study. The flowchart of the step-by-step research is shown in Figure 1.

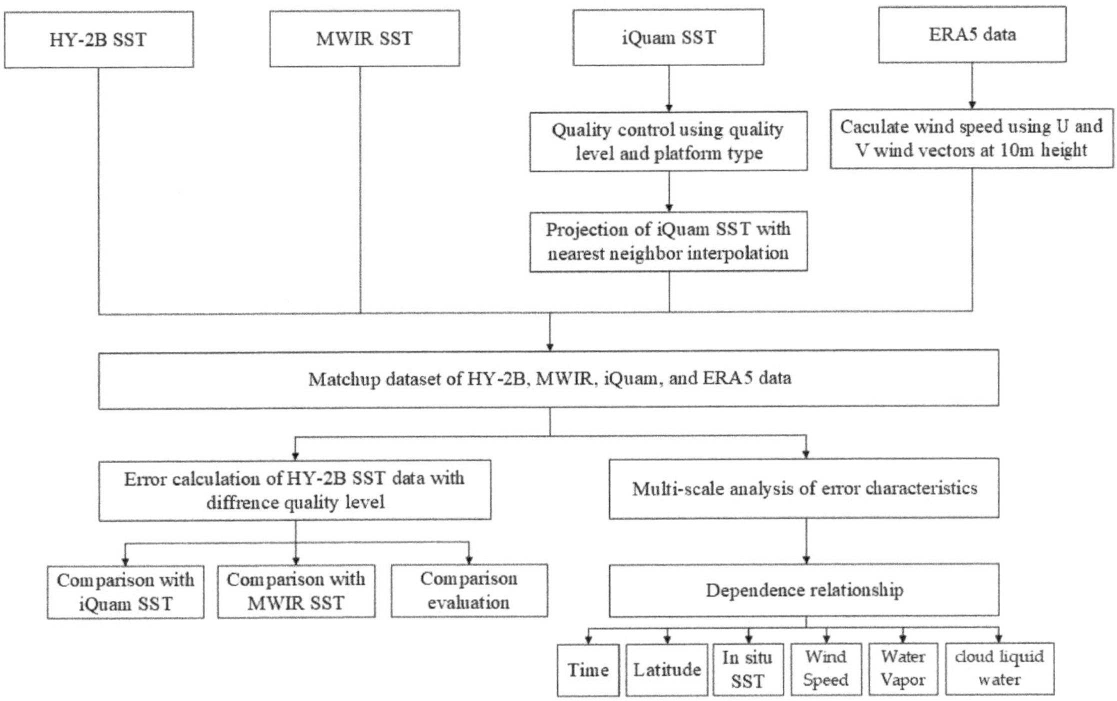

Figure 1. Flowchart of the research methods.

2. Materials and Methods

This study employs multiple remote sensing SST datasets along with comparative analyses and multi-scale validation to assess the observational performance of the HY-2B SMR over the North Pacific. The datasets include the HY-2B L4A fused SST product, the RSS MWIR fused SST product, NOAA iQuam in situ observations for accuracy assessment, and ERA5 reanalysis data for environmental context analysis.

2.1. Study Area

The study area covers the North Pacific (0–60°N, 120°E–100°W) (Figure 2). This region spans from the tropics to mid and high latitudes and includes dynamic features such as the Equatorial Warm Pool, the Kuroshio and California Currents, and the North Pacific Subtropical High. Sea surface temperature in the North Pacific exhibits pronounced spatio-temporal variability and strong seasonality, making it an ideal region for evaluating satellite SST accuracy and stability across diverse climate regimes [20,21]. Moreover, a dense network of buoys in this area provides ample collocated observations for validation.

2.2. Datasets

2.2.1. HY-2B SST

The HY-2B L4A fused sea surface temperature (SST) product used in this study is an operational daily (24 h) fusion field produced by the National Satellite Ocean Application Service (NSOAS) and distributed via the NSOAS Ocean Dynamics portal. The exact product employed here is "MUL_OPER_SST_L4A_FU_01D_20250101_dps_250_10_sst" (product level L4A). Files are provided in NetCDF-4 format with a nominal spatial resolution of ~25 km; the file name convention indicates a daily aggregation ("01D"), which we used for all collocations and analyses. The L4A fusion combines multiple satellite retrievals (microwave and infrared) through NSOAS's operational fusion procedures (spatiotemporal interpolation, quality control, bias correction and weighting) to produce continuous, gap-filled SST fields suitable for large-scale validation and applications. We regridded/processed the daily HY-2B L4A fields as described in Section 2.3. to match the temporal and spatial collocation scheme used for MWIR and iQuam. The HY-2B SST dataset spans October 2023–March 2025.

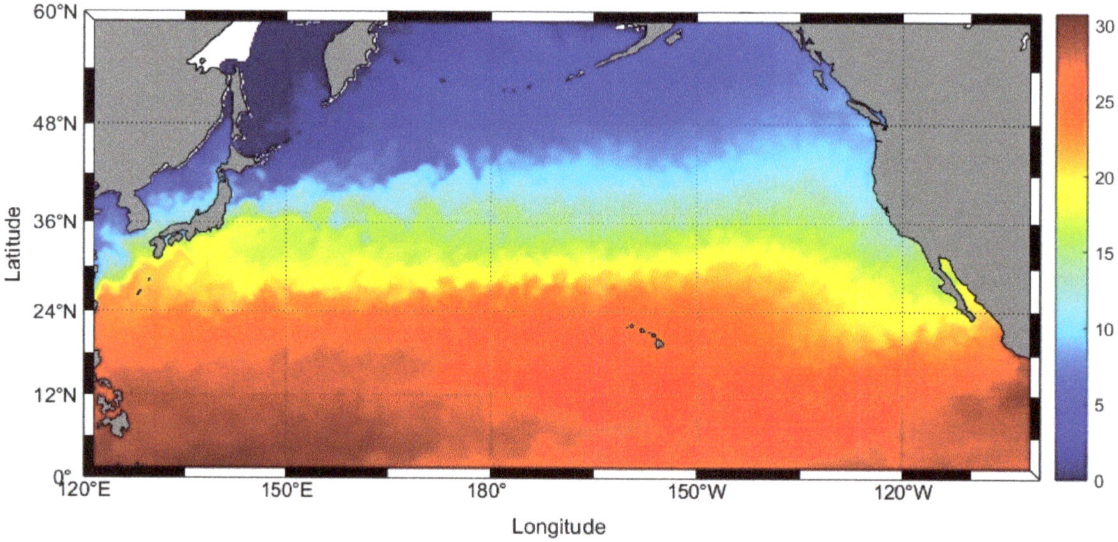

Figure 2. HY-2B L4A SST distribution on 1 March 2025 over the North Pacific.

2.2.2. MWIR SST

The MWIR product (version 5.1) is a daily fused SST dataset provided by Remote Sensing Systems (RSSs). It employs an optimized interpolation scheme to map daily SST estimates onto a 0.09° global grid, with records covering 2002 to the present. MWIR integrates data from multiple sensors: microwave radiometers, including the GPM Microwave Imager (GMI), TRMM Microwave Imager (TMI), Advanced Microwave Scanning Radiometer for EOS (AMSR-E), Advanced Microwave Scanning Radiometer 2 (AMSR2), and WindSat, and infrared (IR) radiometers, including Moderate Resolution Imaging Spectroradiometer (MODIS), on the Aqua and Terra satellites and Visible Infrared Imaging Radiometer Suite (VIIRS) on Suomi NPP and NOAA-20. IR sensors provide high spatial resolution but require clear-sky conditions, whereas microwave sensors can retrieve SST through clouds. In that the MWIR dataset does not make use of HY-2B, it serves as an independent reference. A diurnal cycle correction is applied using modeled base temperature adjustments to mitigate daily thermal expansion/contraction effects. Although MWIR does not directly ingest in situ data, its constituent microwave sensors are buoy-calibrated. The MWIR dataset serves as an independent reference for cross-comparing the HY-2B fusion SST product. MWIR data are available from the RSS data repository.

2.2.3. In Situ Observation Data

High-precision in situ SST observations are essential for satellite SST validation. The NOAA in situ SST Quality Monitor (iQuam) dataset, developed by NOAA's Satellite Applications and Research Center, is widely used for this purpose. The current iQuam version (2.10) provides quality control (QC) of in situ SST measurements, online monitoring of QC-processed data, and reformatted SST output with quality flags [22]. iQuam incorporates SST data from commercial vessels, drifting buoys, tropical moored buoys (T-M), coastal moored buoys (C-M), and Argo floats, and is extensively used for satellite validation [23–25].

The iQuam QC procedures include pre-screening, plausibility checks, internal consistency checks, and cross-consistency checks. Each observation is assigned a quality level from 1 (poorest) to 5 (highest); this study uses only level 5 (highest quality) observations from ships, drifting buoys, and moored buoys. All iQuam data are provided in NetCDF format. Xu et al. used triple collocation to estimate platform-specific observation errors in iQuam [26]: their results (standard deviations of 0.75 K for ships, 0.21–0.22 K for drifting buoys/Argo, 0.17 K for tropical moorings, and 0.40 K for coastal moorings) indicate relatively low random errors across all platforms. This underscores the reliability of high-quality iQuam data for validation.

2.2.4. ERA5 Reanalysis Data

ERA5 is the fifth-generation global climate reanalysis produced by the European Centre for Medium-Range Weather Forecasts (ECMWF) under the Copernicus Climate Change Service (C3S), superseding ERA-Interim [27,28]. ERA5 provides hourly global fields on a ~31 km horizontal grid with 137 vertical levels, encompassing over 30 atmospheric, land, and oceanic variables. In this study, we employ daily-mean ERA5 single-layer data for auxiliary environmental parameters: total column water vapor, total column cloud liquid water, and the 10 m eastward (U) and northward (V) wind components. Notably, ERA5 assimilates a wide range of observations (over 200 data types, including satellite and conventional upper-air measurements) but excludes HY-2B sensor data. It also directly assimilates in situ ship and buoy observations (e.g., 10 m wind, 2 m humidity, sea surface pressure) [28]. These ERA5 variables are used to analyze HY-2B SST error characteristics under varying environmental conditions.

2.3. *Methods*

2.3.1. Data Matching and Quality Control

To ensure comparability between the HY-2B, MWIR, and iQuam datasets, we employ a unified spatiotemporal collocation and quality control framework. First, the MWIR SST dataset was re-gridded to a 0.25° latitude–longitude grid using two-dimensional spatial interpolation. For spatial collocation, each satellite grid point must lie within 25 km of an in situ observation. Hourly in situ observations are averaged to daily means to match the temporal resolution of the satellite products. Within each 0.25° grid cell and day, all available observations from ships, drifting buoys, tropical moored (T-M) buoys, and coastal moored (C-M) buoys are collocated.

All initial matchups undergo strict quality control: any records with missing values or SST outside the 0–34 °C range are discarded, and only iQuam observations with quality level 5 (highest quality) are retained. Figure 3 illustrates the spatial distribution of the resulting collocated points. Finally, we tally the number of samples by platform type to form platform-specific datasets for subsequent analysis. These collocation and QC procedures ensure consistency and reliability across all SST data sources.

2.3.2. Comparison Methods

Comparing satellite-retrieved SST with in situ SST from ships and buoys is a standard approach for evaluating retrieval accuracy [29–32]. Such comparisons assume that in situ observations represent the true sea surface temperature; however, in situ measurements contain both random and systematic errors [33,34]. Moreover, satellite sensors observe different layers of the surface: infrared radiometers sense the near-surface "skin" temperature (top few micrometers), while microwave radiometers retrieve the sub-skin SST (roughly 1 mm depth) [35–37]. Cold-skin and diurnal warming effects can create significant temperature differences between the skin, sub-skin, and bulk layers [38,39]. These factors introduce inherent biases and uncertainties in direct comparisons.

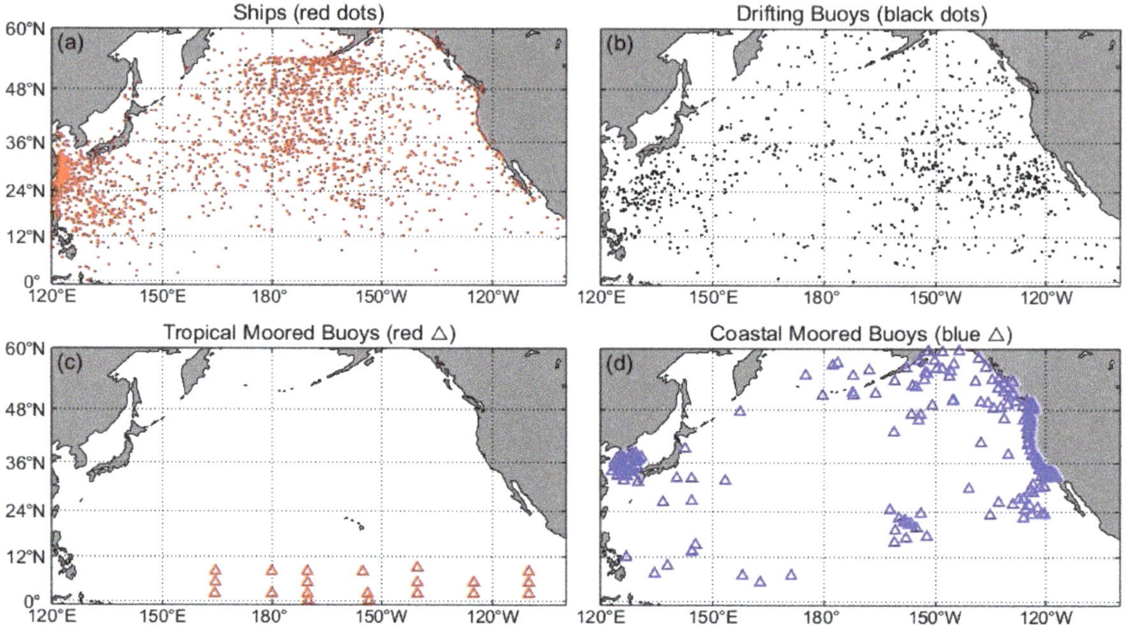

Figure 3. Spatial distribution of all matched collocations (October 2023–March 2025). (**a**) Ships. (**b**) Drifting buoys. (**c**) Tropical moored buoys (**d**) Coastal moored buoys.

To address these limitations, triple collocation (TC) methods have been developed to estimate the error variances of three mutually independent datasets without assuming any single dataset as truth [40,41]. TC has been widely applied to validate soil moisture, precipitation, sea surface salinity, sea surface temperature, and sea surface wind and waves [42–48]. Building on the TC method, the extended triple collocation (ETC) method was proposed. ETC adopts the same basic assumptions as TC but derives additional performance metrics, including the correlation coefficient between each product and the unknown target, as well as the scaled unbiased signal-to-noise ratio (SNR_{sub}). Compared to standard deviation, ETC provides a complementary perspective on product performance [49].

2.3.3. Statistical Indicators and Analysis

To quantify SST accuracy, we compute the mean bias, root mean square error (RMSE), and Pearson correlation coefficient (R) between each satellite product and the reference (in situ) SST. Bias is defined as the average difference between the satellite SST and the reference SST. RMSE represents the standard deviation of these differences and is sensitive to outliers, serving as a robust measure of overall precision. The Pearson correlation coefficient R indicates the strength of the linear relationship between the satellite and reference SST values. The mathematical formulas for these metrics are given by

$$\text{Bias} = \frac{1}{N}\sum_{i=1}^{N}(S_i - O_i) \tag{1}$$

$$\text{RMSE} = \sqrt{\frac{1}{N}\sum_{i=1}^{N}(S_i - O_i)^2} \tag{2}$$

$$R = \frac{\sum_{i=1}^{N} (S_i - \overline{S})(O_i - \overline{O})}{\sqrt{\sum_{i=1}^{N} (S_i - \overline{S})^2 \times \sum_{i=1}^{N} (O_i - \overline{O})^2}} \tag{3}$$

In the formula, N denotes the total number of matching data points between remote sensing observations of SST and iQuam in situ SST, while S and O denote the satellite data set and reference data set being verified, respectively.

2.3.4. Extended Triple Collocation (ETC)

ETC is a technique for estimating the noise error variance (errVar) and correlation coefficient (rho) of three measurement systems (e.g., satellite, in situ, and model-based products) relative to the unknown true values of the measured variables (e.g., sea surface temperature, soil moisture, wind speed) [49]. ETC applies to both absolute and anomaly values without requiring rescaling to a reference system.

Traditional approaches (e.g., linear regression) assume in situ data as truth, yet these observations contain both random and systematic errors. Therefore, using extended triple collocation (ETC) allows for independent estimation of the error variance for each measurement system. ETC enables independent estimation of each system's error variance, avoiding reliance on a single reference dataset. Unlike traditional methods, ETC does not assume any dataset as truth, thus providing independent error variance estimates for all systems [50]. ETC uses the same assumptions as TC but provides an additional validation parameter, namely the coefficient of determination (ρ^2_{t,X_i}) relative to the unknown "true" value of the measured variable.

Assume that three independent measurement values are linearly related to the unknown true value T. The affine error model between the measurement values and T can be expressed as follows:

$$X_i = X_i' + \varepsilon_i = \alpha_i + \beta_i t + \varepsilon_i \tag{4}$$

where $X_i (i \in \{1,2,3\})$ is a set of three spatially and temporally co-located datasets (X_1: system-based products; X_2: model-based products; X_3: in situ products), each with additive random error ε_i; t is the unknown true value (or true state) and α_i and β_i are the multiplicative bias and proportionality factor of dataset X_i relative to the true value.

In this study, we adopted the covariance combination principle among the three products without rescaling [40,49]. The covariance between different products is

$$\mathrm{Cov}(X_i, X_j) = \mathrm{E}(X_i X_j) - \mathrm{E}(X_i)\mathrm{E}(X_j) = \beta_i \beta_j \sigma_t^2 + \beta_i \mathrm{Cov}(t, \varepsilon_j) + \beta_j \mathrm{Cov}(t, \varepsilon_i) + \mathrm{Cov}(\varepsilon_i, \varepsilon_j) \tag{5}$$

$$\sigma_t^2 = \mathrm{Var(t)} \tag{6}$$

The basic assumptions of ETC are that (i) the expected value of errors from independent data sources is zero ($\mathrm{E}(\varepsilon_i) = 0$), (ii) they are mutually independent ($\mathrm{Cov}(\varepsilon_i, \varepsilon_j) = 0$, for $\mathrm{i} \neq \mathrm{j}$), and (iii) the errors of the three products are independent of each other and unrelated to the true value t ($\mathrm{Cov}(t, \varepsilon_i) = 0$).

The covariance of these three products can be simplified as follows:

$$\mathrm{Q}_{ij} \equiv \mathrm{Cov}(X_i, X_j) = \begin{cases} \beta_i \, \beta_j \, \sigma_t^2, & \text{for } i \neq j \\ \beta_i^2 \, \sigma_t^2 + \sigma_{\varepsilon_i}^2, & \text{for } i = j \end{cases} \tag{7}$$

The error variance $\sigma_{\varepsilon_i}{}^2 = \text{Var}(\varepsilon_i)$. The error standard deviation (ESD) of the three products can be determined from the square root of the error variance

$$\sigma_{\varepsilon_1} = \begin{bmatrix} \sqrt{Q_{11} - \frac{Q_{12}\,Q_{13}}{Q_{23}}} \\ \sqrt{Q_{22} - \frac{Q_{12}\,Q_{23}}{Q_{13}}} \\ \sqrt{Q_{33} - \frac{Q_{13}\,Q_{23}}{Q_{12}}} \end{bmatrix} \tag{8}$$

The correlation coefficient between t and X expressed in terms of covariance values is

$$\rho_{t,X} = \pm \begin{bmatrix} \sqrt{\frac{Q_{12}\,Q_{13}}{Q_{11}\,Q_{23}}} \\ \text{sign}(Q_{13}\,Q_{23})\sqrt{\frac{Q_{12}\,Q_{23}}{Q_{22}\,Q_{13}}} \\ \text{sign}(Q_{12}\,Q_{23})\sqrt{\frac{Q_{13}\,Q_{23}}{Q_{33}\,Q_{12}}} \end{bmatrix} \tag{9}$$

It is worth noting that the $\rho_{t,X}$ provided by ETC has signed ambiguity; however, in practical applications, $\rho_{t,X}$ is always positive. To describe the combined effects of product sensitivity (β_i), true signal variability (σ_t), and measurement error variability (σ_ε), the formula for calculating the correlation coefficient squared (the scaled unbiased signal-to-noise ratio SNR_{sub}) is as follows:

$$\text{SNR}_{\text{sub}} = \pm \begin{bmatrix} \sqrt{\frac{Q_{12}\,Q_{13}}{Q_{11}\,Q_{23}}} \\ \text{sign}(Q_{13}\,Q_{23})\sqrt{\frac{Q_{12}\,Q_{23}}{Q_{22}\,Q_{13}}} \\ \text{sign}(Q_{12}\,Q_{23})\sqrt{\frac{Q_{13}\,Q_{23}}{Q_{33}\,Q_{12}}} \end{bmatrix}^2 \tag{10}$$

where SNR_{sub} represents the proportional unbiased signal-to-noise ratio. It contains information about product sensitivity and can be used to evaluate whether the noise level of the system is suitable for detecting changes in target variables. We apply ETC to three datasets—HY-2B, MWIR, and iQuam—across four platforms (Ship, Drifter, T-M, and C-M) to derive their random errors (ESD) and SNR_{sub} values. To elucidate error sources, we examine the interplay between sensor characteristics and environmental conditions: infrared (IR) sensors offer high spatial resolution but are hindered by cloud cover and atmospheric water vapor, whereas microwave sensors provide all-weather SST retrievals at the cost of coarser resolution and greater susceptibility to surface roughness and wind waves. The MWIR product integrates data from two thermal-infrared sensors (MODIS, AVHRR) and three passive microwave sensors (AMSR-E, AMSR2, WindSat) to optimize the trade-off between observational accuracy and spatial coverage [51].

3. Results

3.1. Comparison with In Situ SST

Figure 4 shows scatter plots and bias histograms comparing the HY-2B and MWIR SST products with the iQuam in situ observations over the period of October 2023–March 2025. In each scatter plot, the data cluster tightly around the 1:1 line, with regression slopes very close to unity and coefficients of determination (R^2) exceeding 0.99. The bias histograms for both products are approximately normal and centered near zero, with most differences within ±0.5 °C. These results demonstrate strong consistency between the satellite SST products and the in situ measurements.

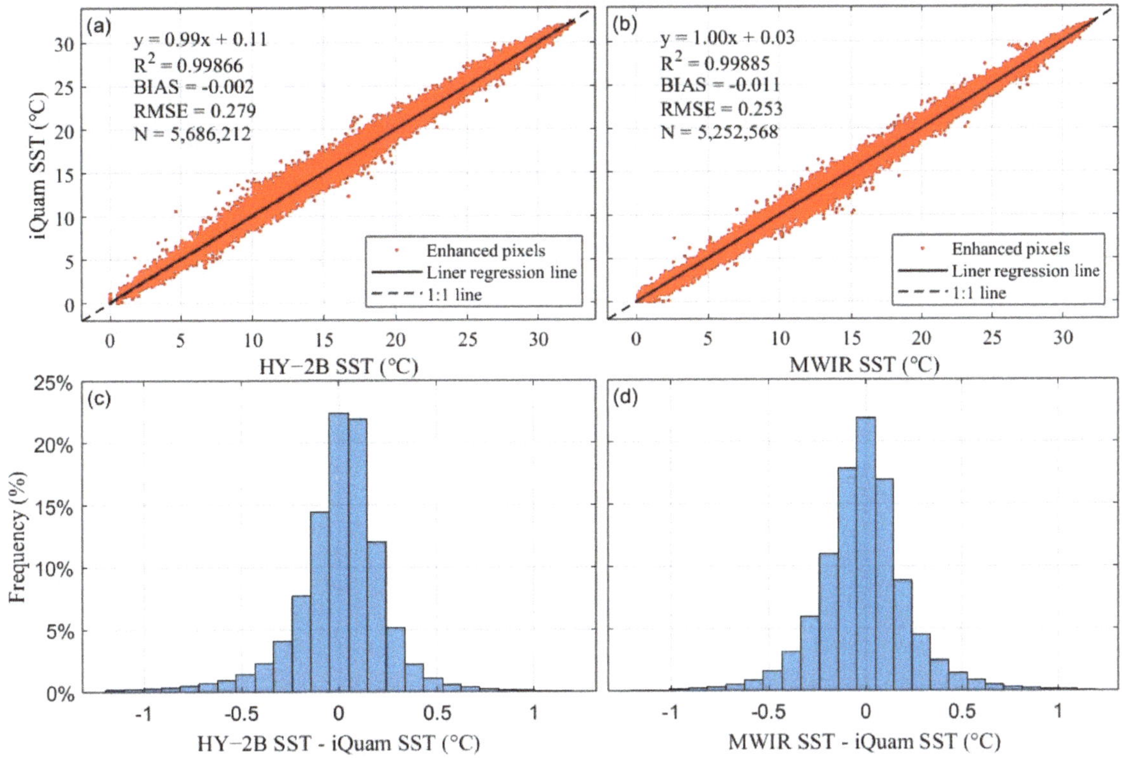

Figure 4. Comparison results between HY-2B and MWIR with iQuam SST. (**a**) Scatter plot of HY-2B and iQuam. (**b**) Scatter plot of MWIR and iQuam. (**c**) HY-2B bias histogram. (**d**) MWIR bias histogram.

Table 1 summarizes the comparison statistics (bias, RMSE, R^2) for HY-2B and MWIR versus iQuam, broken down by platform (All, Ship, Drifter, T-M, C-M). For HY-2B versus iQuam (all platforms), the mean bias is –0.002 °C, RMSE is 0.279 °C, and R^2 is 0.9987. The lowest RMSE values are found for drifters (0.177 °C) and tropical moored buoys (0.145 °C), with corresponding R^2 values of 0.9994 and 0.9903. In contrast, ship and coastal moored observations yield higher RMSEs (0.427 °C and 0.376 °C, respectively). For MWIR versus iQuam, the overall bias is –0.011 °C and RMSE is 0.253 °C. MWIR also performs best on drifters (RMSE 0.205 °C) and tropical moored buoys (0.175 °C), while ships and coastal moorings have RMSEs of 0.337 °C and 0.350 °C, respectively. Both satellite products achieve similar overall accuracy (R^2 > 0.99), performing particularly well on drifter and tropical buoy platforms.

These results suggest that drifter and tropical moored buoys, which sample relatively uniform open-ocean waters, yield the highest agreement with satellite SST (lowest representativeness error). By contrast, ships and coastal moorings often operate in regions with strong SST gradients or local disturbances, leading to larger point errors. In those heterogeneous environments, representativeness errors increase. Notably, on drifter and T m platforms, HY-2B exhibits the smallest random errors, whereas its errors are larger for ship-based data. Overall, the HY-2B and MWIR products demonstrate comparable accuracy (R^2 > 0.99), but platform-specific differences highlight distinct observational characteristics.

Table 1. Bias, RMSE, and R^2 of HY-2B and RSS MWIR SST against iQuam in situ observations by platform.

Match Type	Indicator	iQuam Platforms				
		All	**Ship**	**Drifter**	**T-M**	**C-M**
HY-2B VS iQuam	Bias (°C)	−0.002	−0.090	0.042	0.007	−0.063
	RMSE (°C)	0.279	0.427	0.177	0.145	0.376
	R-squared	0.9987	0.9974	0.9994	0.9903	0.9969
	Number	5,686,212	535,382	3,359,875	126,826	1,663,980
MWIR VS iQuam	Bias (°C)	−0.011	−0.017	0.002	−0.057	−0.042
	RMSE°C	0.253	0.337	0.205	0.175	0.350
	R-squared	0.9989	0.9983	0.9991	0.9871	0.9970
	Number	5,252,568	200,550	3,635,475	135,713	1,280,670

Figure 5 displays the spatial distribution of HY-2B SST error metrics (bias, RMSE, and Pearson's R) on a 2° × 2° grid (HY-2B vs. iQuam). In mid- to high-latitude open-ocean regions, biases cluster near zero and RMSE remains below 0.25 °C, with R exceeding 0.99, indicating excellent agreement where sea conditions are homogeneous and sampling is dense. In contrast, coastal zones and data-sparse equatorial areas exhibit pronounced negative biases (down to –0.1 °C), elevated RMSE (up to 0.4 °C), and slightly reduced R (around 0.95), reflecting increased representativeness error and local disturbances. Notably, these spatial patterns show no clear seasonal drift: overall, bias and RMSE stay low and uniform in well-sampled open waters, whereas accuracy degrades in heterogeneous or sparsely observed regions.

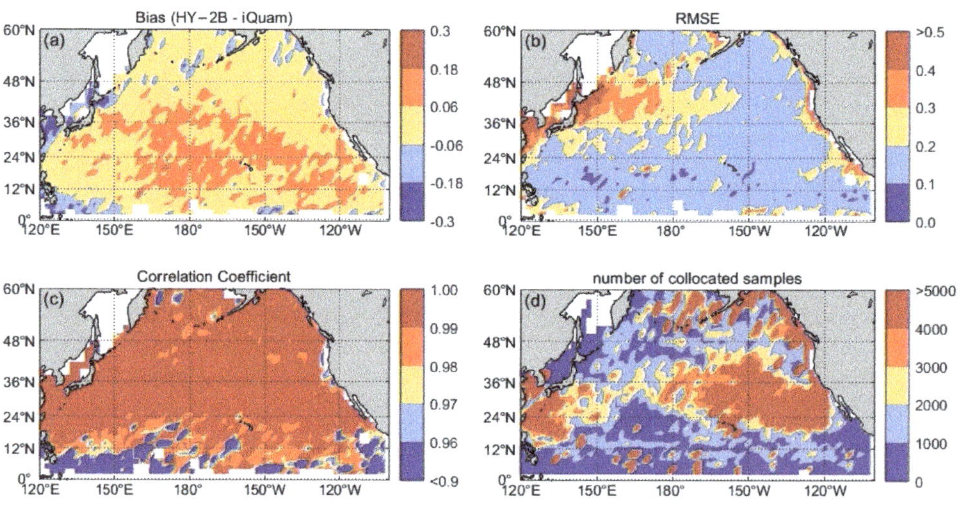

Figure 5. Spatial distribution of HY-2B SST versus iQuam SST on a 2° × 2° grid. (**a**) Mean bias, (**b**) RMSE, (**c**) Pearson's correlation coefficient (R), and (**d**) number of collocated samples.

3.2. Comparison of Fusion SST Products

Figure 6 compares the HY-2B and RSS MWIR SST fusion products directly. The scatter density plot shows the HY-2B SST on the x-axis versus the MWIR SST on the y-axis. The data points cluster tightly around the 1:1 reference line. A linear fit yields y = 0.99x + 0.15 °C with R^2 = 0.999, indicating extremely strong agreement. The overall mean bias (HY-2B minus MWIR) is only 0.009 °C, with an RMSE of 0.287 °C. The bias distribution is approximately normal and centered near zero, with the majority of errors within ±0.5 °C. In summary, HY-2B and MWIR fusion products show negligible systematic and random differences under large-sample conditions, confirming their exceptional consistency and reliability.

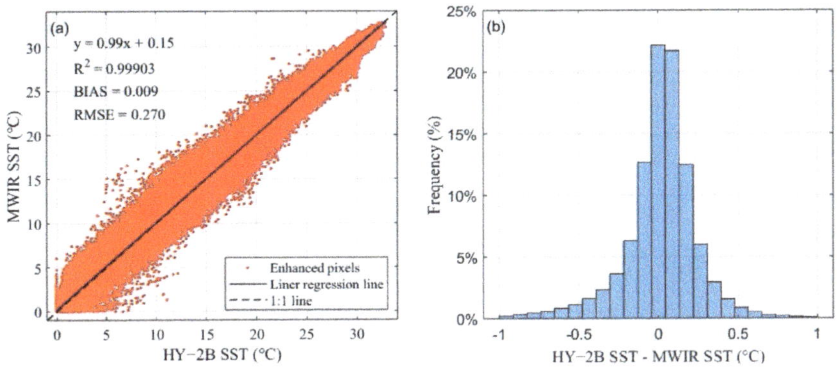

Figure 6. HY-2B and MWIR SST comparison. (**a**) Scatter density plot. (**b**) Bias histogram.

Figures 7 and 8 present monthly maps of the HY-2B minus MWIR bias from October 2023 to March 2025. In general, these maps reveal no systematic seasonal drift. Equatorial regions remain near zero bias year-round, with no obvious zonal trends. Small alternating positive/negative bias patterns appear in mid- and high-latitude seas. For example, the Sea of Okhotsk exhibits a positive bias from January to July and a negative bias from August to December. During boreal summer (June–August), bias hotspots emerge in the northwestern Pacific, the Sea of Okhotsk, and the Sea of Japan—likely associated with the Kuroshio Extension and other regional dynamics [52]. In boreal winter (December–February), these regions show negative biases (HY-2B cooler than MWIR). Overall, biases are concentrated in high-latitude and coastal areas, while open-ocean biases remain minimal and uniform. These monthly maps confirm the absence of any significant seasonal drift: errors appear randomly distributed, and the relatively short time span of this study precludes definitive identification of longer-term seasonal or regional biases.

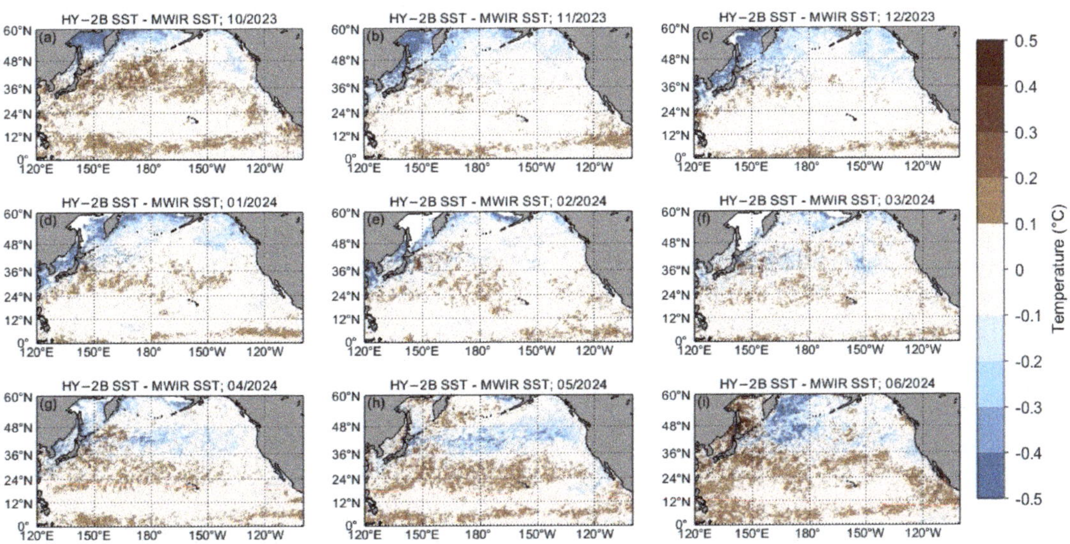

Figure 7. Monthly mean bias map of HY-2B minus RSS MWIR SST from October 2023 to June 2024.

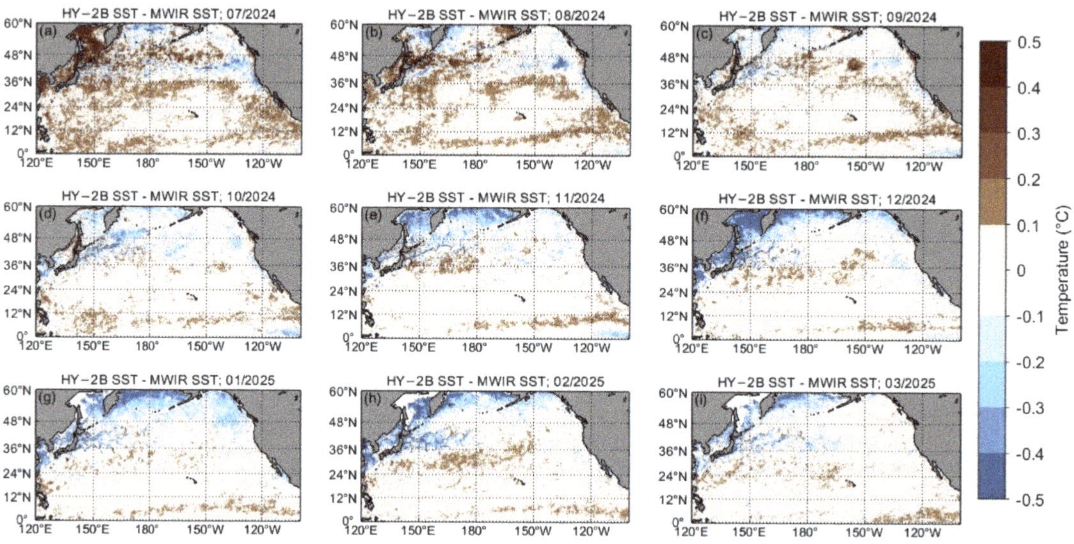

Figure 8. Monthly mean bias map of HY-2B minus RSS MWIR SST from July 2024 to March 2025.

3.3. ETC Analysis Results

Extended triple collocation (ETC) was used to evaluate the performance of the three independent SST datasets (HY-2B, MWIR, and iQuam) on each platform. ETC separates the random error variance of each system from representativeness error. The resulting error standard deviation (ESD) quantifies the unbiased random error, and the sub-sampled signal-to-noise ratio (SNR_{sub}) indicates the relative strength of the true signal. Table 2 lists the ETC analysis results for HY-2B, MWIR, and iQuam data across various platforms. From this, we observe a significant difference between ETC-based ESD and conventional RMSE, as ESD excludes the influence of representativeness errors [53,54]. At the aggregate level (all platforms), iQuam has the smallest ESD (0.157 °C), followed by HY-2B (0.163 °C), and MWIR has the largest (0.196 °C). The corresponding SNR_{sub} values are very close to 1 (0.9996 for iQuam and HY-2B, 0.9993 for MWIR), indicating that all three systems have strong detection capability for true SST variations.

Table 2. ETC analysis results of HY-2B, MWIR, and iQuam on different platforms.

Data	ETC	ALL	Ship	Drifting	T-M	C-M
HY-2B	ESD (°C)	0.163	0.238	0.120	0.088	0.238
	SNR_{sub}	0.9996	0.9991	0.9997	0.9962	0.9986
MWIR	ESD (°C)	0.196	0.252	0.164	0.115	0.267
	SNR_{sub}	0.9993	0.9990	0.9994	0.9933	0.9983
iQuam	ESD (°C)	0.157	0.220	0.123	0.111	0.218
	SNR_{sub}	0.9996	0.9992	0.9997	0.9941	0.9988

Platform-specific results reveal that the lowest errors occur on drifter and tropical moored buoy platforms. For drifters and tropical moorings (T-M), the ESDs are as follows: iQuam = 0.123 °C and 0.111 °C (T-M); HY-2B = 0.120 °C and 0.088 °C; MWIR = 0.164 °C and 0.115 °C, respectively. In contrast, ships and coastal moorings show higher ESD values for all datasets. This pattern likely arises because drifters and T-M buoys are generally located in the open ocean, where conditions are more stable and representativeness errors are low. Notably, on the drifter and T-M platforms, HY-2B's ESD is slightly lower than iQuam's, indicating that HY-2B SST achieves exceptionally high precision under those conditions. Overall, iQuam exhibits the highest accuracy (smallest ESD) as expected, while HY-2B provides better accuracy than MWIR across all platforms.

4. Discussion

Our results indicate that the HY-2B L4A fusion SST product performs exceptionally well. The scatter plots (Figures 4 and 6) show that both HY-2B and MWIR products are highly consistent with each other and with in situ observations, with most errors randomly distributed within ±0.5 °C. In situ validation reveals that HY-2B has essentially zero mean bias over the study period and a low RMSE (~0.2–0.3 °C). This represents a substantial improvement over earlier HY-2B Level-2B products: for example, previous work reported a bias of 0.1 °C and an RMSE of 0.87 °C for the HY-2B L2 SST [17]. Thus, the L4 fusion processing appears to significantly reduce systematic errors and improve accuracy. Overall, the HY-2B L4A product exhibits very small systematic errors and high correlation with in situ SST.

Previous studies have shown that environmental parameters such as sea surface wind speed, atmospheric water vapor, and cloud liquid water can introduce errors in SST retrieval. As sea surface wind speed increases, sea surface roughness and foam increase, altering microwave emission characteristics and thereby increasing SST retrieval uncertainty. Additionally, water vapor and cloud liquid water in the atmosphere both absorb and scatter microwave signals, increasing the complexity of atmospheric correction. These factors can induce inter-channel crosstalk effects, significantly impacting SST retrieval accuracy [55].

We analyzed the variation patterns of HY-2B SST error metrics (ESD, SNR_{sub}, bias, and RMSE) with respect to environmental variables (time, latitude, sea surface temperature, wind speed, columnar water vapor, and columnar cloud liquid water) to comprehensively evaluate its performance.

4.1. Time-Dependent Characteristics of SST Errors

Figure 9 shows the evolution of error metrics from October 2023 to March 2025 (10-day averages). All products remain generally stable over time. In Figure 9a, HY-2B and iQuam maintain low ESD (<0.2 °C) and high SNR_{sub} (>0.999) throughout the period, with their curves nearly overlapping. MWIR has a slightly higher ESD and marginally lower SNR_{sub}, but its temporal trend closely follows HY-2B. Figure 9b shows that SNR_{sub} stays above 0.998 for all products, indicating a consistently high degree of agreement.

Figure 9c,d indicate that the HY-2B bias relative to iQuam fluctuates around 0 °C with no obvious trend, and RMSE fluctuates around 0.2 °C. The bias and RMSE time series for MWIR are similarly flat. Notably, both satellite products exhibit somewhat higher ESD and RMSE during boreal summer (June–August); MWIR in particular shows a noticeable bias increase in that period, possibly due to stronger solar heating affecting the sensors. Over the entire period, however, the consistency among HY-2B, MWIR, and iQuam remains high. This suggests that both satellite products have stable time-series quality, with no significant drift or degradation over the six-month span.

4.2. SST Error Characteristics at Different Latitudes

Figure 10 shows error metrics as a function of latitude (0–58°N). In Figure 10a, the ESD for all three products increases with latitude from 0° up to about 35°N. Beyond 35°N, the MWIR ESD remains roughly between 0.2 and 0.3 °C, while iQuam and HY-2B ESD gently decline with further latitude. Figure 10b shows that SNR_{sub} stays above 0.99 at all latitudes, with only small dips around 8°N and 44°N. Above 50°N, SNR_{sub} fluctuates more significantly for all products, reflecting increased random error. These high-latitude fluctuations are likely due to stronger winds and more variable conditions, which degrade SST accuracy.

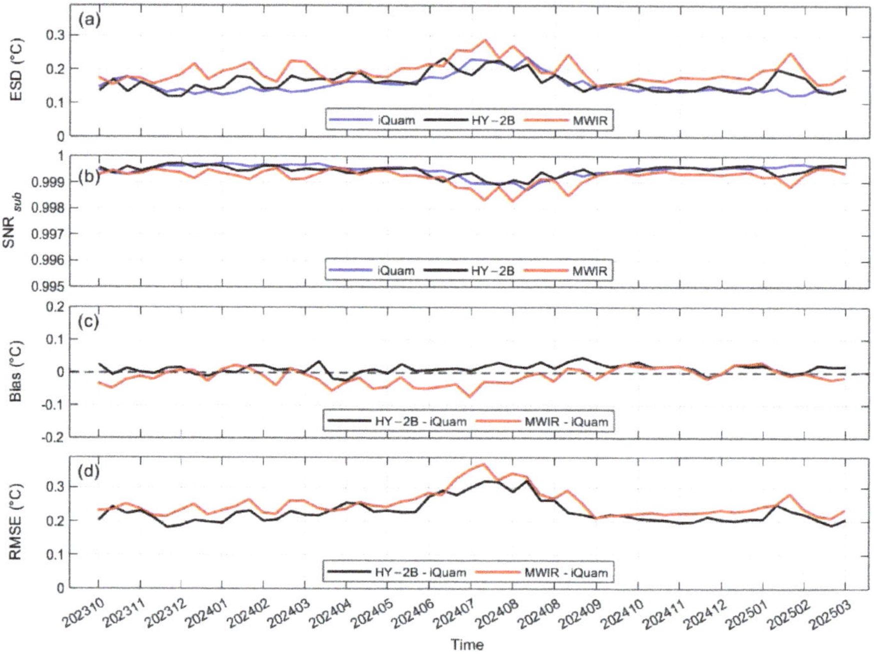

Figure 9. Time variation of error characteristics (10-day averages). (**a**) ESD. (**b**) SNR_{sub}. (**c**) Bias (relative to iQuam). (**d**) RMSE.

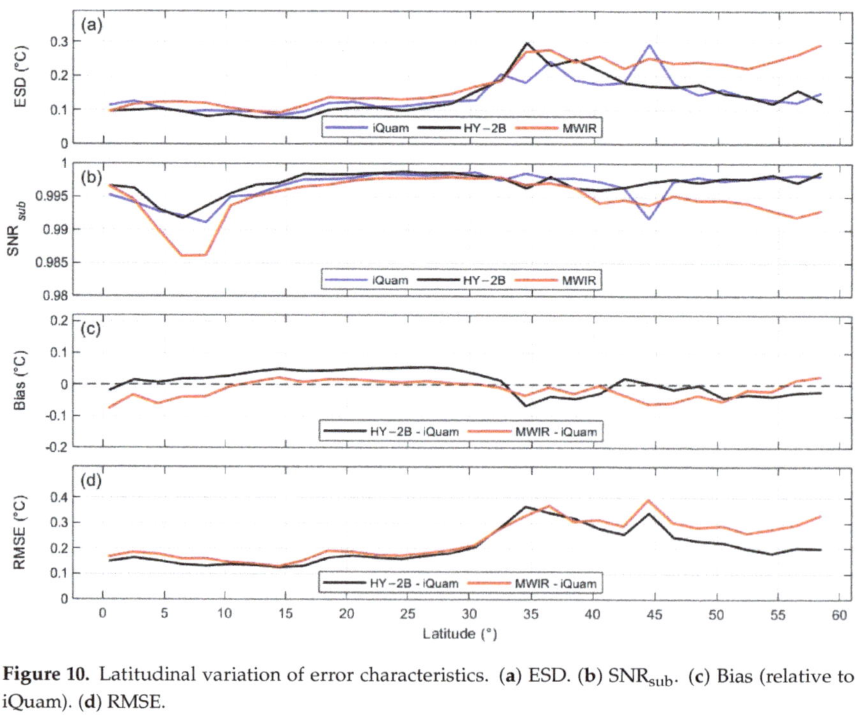

Figure 10. Latitudinal variation of error characteristics. (**a**) ESD. (**b**) SNR_{sub}. (**c**) Bias (relative to iQuam). (**d**) RMSE.

Figure 10c indicates that HY-2B bias (relative to iQuam) closely mirrors MWIR bias across latitudes: slight positive biases occur in low latitudes and slight negative biases in mid-to-high latitudes. Over 0–55°N, the difference in bias between HY-2B and MWIR remains within ±0.1 °C. Figure 10d shows that both satellite products have increasing RMSE (and ESD) with latitude up to 35°N, then a gradual decrease. The similarity in latitudinal error trends suggests that both products respond similarly to environmental changes with latitude. In summary, SST errors generally grow with latitude into mid-latitudes due to harsher conditions (higher winds, etc.) but then stabilize, and both HY-2B and MWIR exhibit comparable error behavior as a function of latitude.

4.3. Changes in SST Error Characteristics at Different Temperatures

Figure 11 shows the changes in error characteristics within the 2 °C in situ sea surface temperature interval, ranging from 0 to 30 °C. Since there are few data matching points above 30 °C, these data will be excluded in this paper. The ESD changes of HY-2B and iQuam are similar. Within the 0 to 14 °C range, the ESD of both systems increases with rising temperature. In contrast, MWIR starts at a higher ESD value (0.3 °C). Subsequently, the ESD of all three products decreases with increasing sea surface temperature, from about 0.25 °C to 0.1 °C. Where HY-2B and iQuam have similar ESD values, both lower than MWIR, it indicates that the random error of HY-2B is smaller than that of MWIR in the low temperature range. The SNR_{sub} of iQuam and HY-2B changes similarly with SST. Within the SST range of 0 to 14 °C, SNR_{sub} decreases with increasing SST, while MWIR's SNR_{sub} slightly increases with rising SST. However, as sea surface temperature increases, the SNR_{sub} of all three products gradually increases to above 0.99. As shown in Figure 11, under extreme high-temperature conditions, the errors of all three products increase.

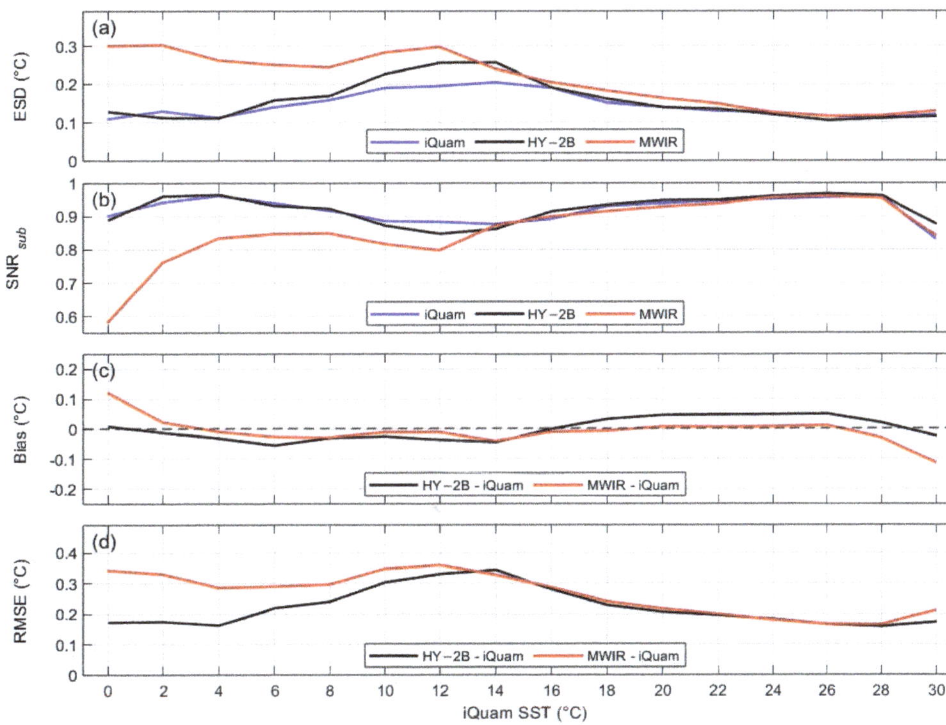

Figure 11. Error characteristics as a function of iQuam SST. (**a**) ESD. (**b**) SNR_{sub}. (**c**) Bias (relative to iQuam). (**d**) RMSE.

As shown in Figure 11c,d, within the range excluding extreme low and high temperatures, the bias of HY-2B and MWIR relative to iQuam SST remains around 0 degrees. Compared to HY-2B, MWIR has a higher RMSE relative to iQuam at low temperatures (0.35 °C), but as SST increases, the two become increasingly similar. The overall performance is similar to the ESD variation, with both showing a decreasing trend with increasing SST after 14 °C.

4.4. Variations in SST Error Under Wind Speed Influence

Sea surface wind speed is an important environmental parameter affecting the accuracy of sea surface temperature measurements, largely due to the "cold skin effect" [36]. High wind speeds increase water mixing, thereby reducing temperature differences between different depths. Additionally, high sea surface wind speeds can cause white caps, thereby increasing the uncertainty of sea temperature measurements.

Figure 12 shows error metrics as a function of surface wind speed (0–19 m/s). As wind increases, MWIR's ESD begins to rise noticeably above 10 m/s, and its SNR_{sub} correspondingly decreases (Figure 12a,b). By contrast, HY-2B and iQuam show no obvious trend in ESD or SNR_{sub} with wind speed; both remain relatively stable with only small fluctuations. Both products' biases stay near 0 °C across wind speeds (Figure 12c). The RMSE of HY-2B remains low and stable up to 10 m/s, with only slight increases in variability beyond that (Figure 12d). MWIR's RMSE, however, clearly increases with wind speeds above 10 m/s. These observations are consistent with the cool-skin interpretation [56,57]: increasing wind speeds enhance mixing and reduce the skin–subskin temperature difference, thereby diminishing the cool-skin contribution to satellite minus in situ differences. Since HY-2B SMR retrieves a sub-skin temperature (~1 mm), the direct cool-skin contribution to HY-2B minus typical in situ near-surface differences is small in our aggregated comparisons; nevertheless, wind-driven roughness and foam remain important contributors to the increased uncertainty at high winds.

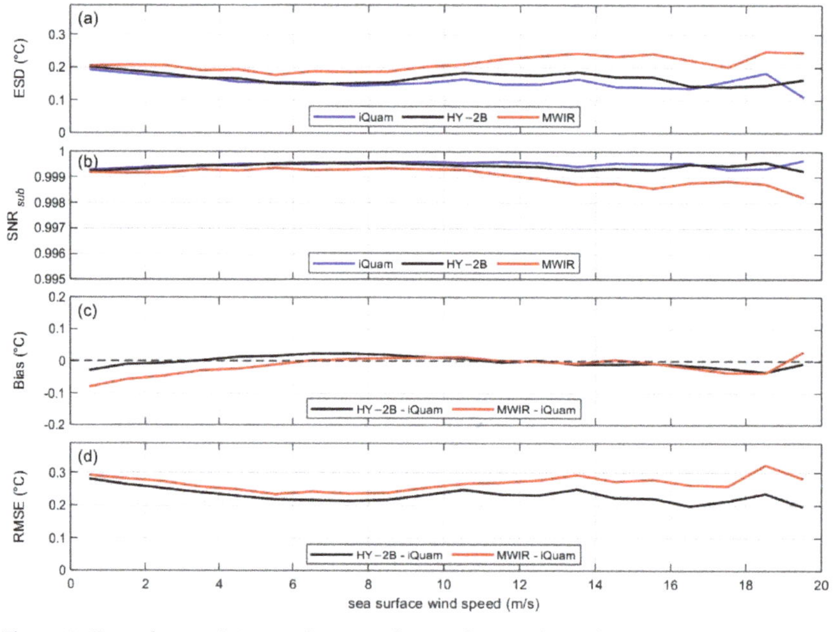

Figure 12. Error characteristics as a function of sea surface wind speed. (**a**) ESD. (**b**) SNR_{sub}. (**c**) Bias (relative to iQuam). (**d**) RMSE.

4.5. Variations in SST Error Under Columnar Water Vapor Conditions

The attenuation and radiation of water vapor in the atmosphere can affect the signals received by microwave radiometers, thereby increasing the uncertainty in sea surface temperature retrieval [58]. This study investigates the relationship between the SST error characteristics of HY-2B and columnar water vapor (Figure 13). Since columnar water vapor does not directly affect in situ data, the analysis in this section does not include iQuam SST.

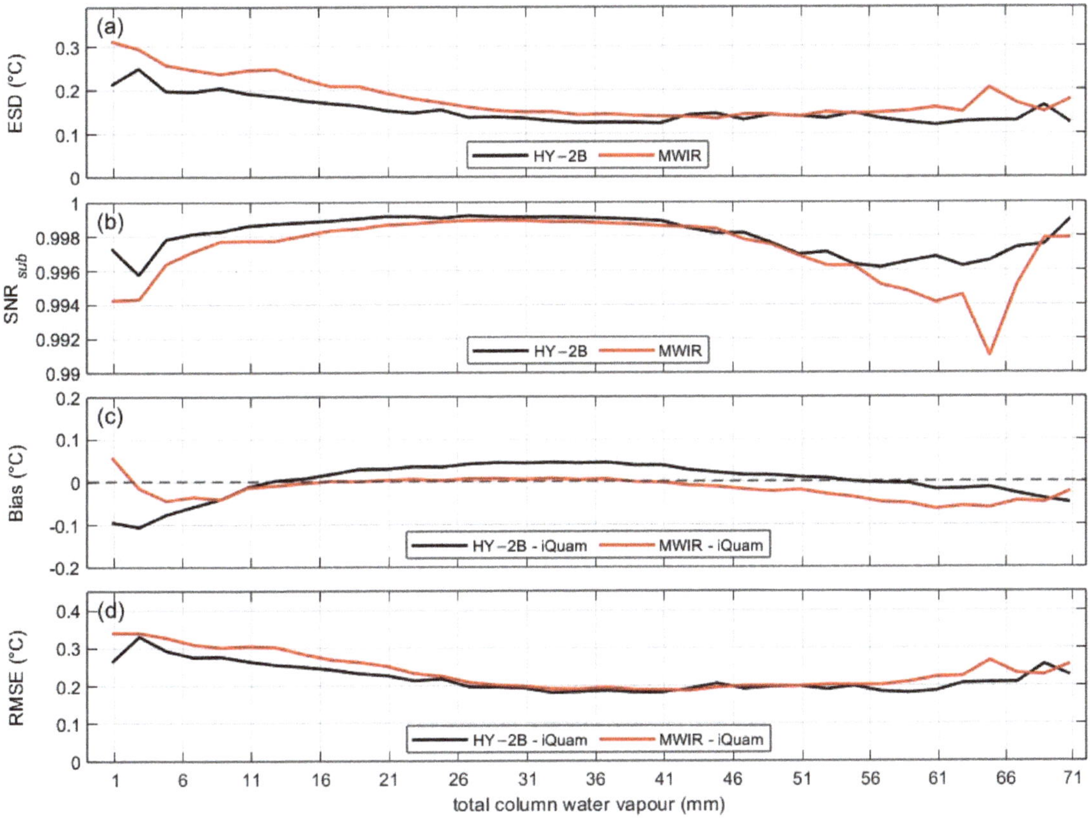

Figure 13. Error characteristics as a function of columnar water vapor. (**a**) ESD. (**b**) SNR_{sub}. (**c**) Bias (relative to iQuam). (**d**) RMSE.

Figure 13 analyzes HY-2B and MWIR errors versus total column water vapor (1–75 mm). Between 1 and 35 mm of vapor, both HY-2B and MWIR ESD decrease, reaching about 0.14 °C (Figure 13a), after which the ESD remains roughly constant; minor oscillations occur at very high vapor (>60 mm). The SNR_{sub} for both products (Figure 13b) drops when vapor is below 10 mm or above 40 mm (reaching minima) and then increases again at very high vapor (>65 mm). In terms of bias, HY-2B tends to have a slight positive bias in the mid range (11–55 mm), whereas MWIR shows a slight negative bias at very low and very high vapor, being near zero otherwise (Figure 13c). The RMSE trends (Figure 13d) largely mirror those of ESD: RMSE declines with increasing vapor up to 35 mm and then fluctuates. These patterns suggest that moderate amounts of water vapor improve retrieval (reducing random error), but extremes of low or high vapor content degrade accuracy through increased attenuation.

4.6. Impact of Cloud Liquid Water Content on SST Error Characteristics

Similar to water vapor, the presence of liquid water in clouds also increases the uncertainty of microwave SST retrieval. Figure 14 shows the effect of columnar cloud liquid water (0–1.0 mm) on HY-2B and MWIR SST errors. As liquid water content rises from 0 to ~0.5 mm, the ESD of both HY-2B and MWIR increases substantially (Figure 14a), with MWIR's ESD exceeding that of HY-2B. Beyond 0.5 mm, the errors tend to saturate or even decrease slightly. Both products maintain generally negative biases as liquid water increases (Figure 14c), and the RMSE (Figure 14d) follows a similar upward trend. This behavior indicates that cloud liquid water exacerbates SST retrieval errors (due to increased microwave attenuation and scattering), although beyond a certain point the incremental effect levels off.

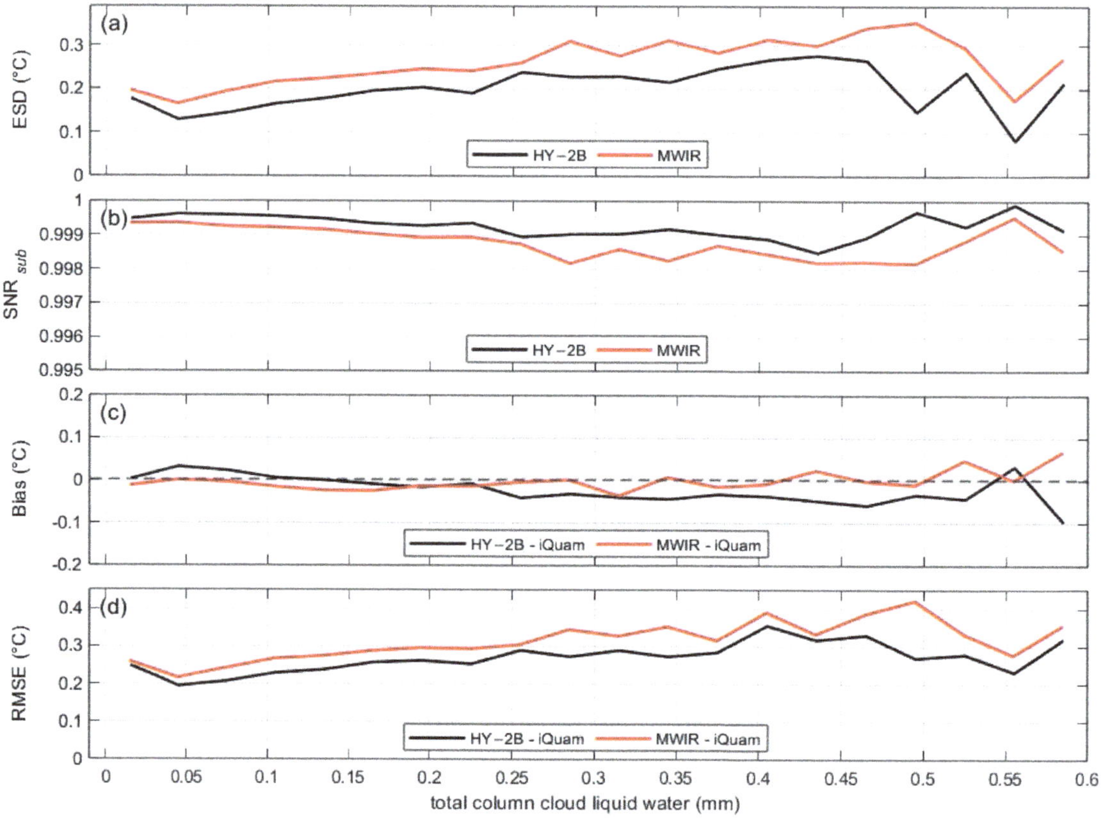

Figure 14. Error characteristics as a function of liquid water content in columnar clouds. (**a**) ESD. (**b**) SNR$_{sub}$. (**c**) Bias (relative to iQuam). (**d**) RMSE.

Our analysis indicates that HY-2B SST errors are most pronounced under extreme conditions. For example, at high latitudes (around 60°N), we observe HY-2B biases up to 0.13 °C and RMSE 0.36 °C. The harsher environments there (strong winds, sea ice) contribute to higher retrieval errors. In contrast, tropical and subtropical waters have more uniform conditions (weaker winds, less vapor variability), leading to lower and more stable errors. Overall, HY-2B SST performs reliably under most conditions, but errors increase under conditions of high winds, high latitudes, and heavy atmospheric moisture.

5. Conclusions

This study systematically assessed the observational performance of the HY-2B L4A fusion SST product by comparing it with RSS MWIR satellite SST and NOAA iQuam in situ data from October 2023 to March 2025. All datasets were collocated in time and space for a fair comparison. The HY-2B and iQuam comparison yielded a near-zero mean bias (−0.002 °C), an RMSE of 0.279 °C, and an R^2 = 0.9987, indicating negligible systematic error and very high accuracy. Platform-specific analysis showed that HY-2B has the lowest RMSE (0.145–0.177 °C) on drifters and tropical moored buoys, outperforming MWIR on those platforms. Ships and coastal moorings exhibited higher RMSE (0.376–0.427 °C) for HY-2B, as expected for these heterogeneous environments. These differences arise because drifters and tropical moorings are located in open ocean with uniform SST, resulting in low representativeness error, whereas ships and coastal moorings often operate in areas with large temperature gradients or local disturbances.

In a direct comparison between the two satellite fusion products, HY-2B and MWIR agree very closely: the average bias is 0.009 °C and the RMSE is 0.270 °C, with most errors within ±0.5 °C. This demonstrates the consistency and reliability of both products under broad conditions. Extended triple collocation analysis confirms these findings. Drifter and tropical moored platforms have the lowest ESD values (e.g., HY-2B ESD of 0.088–0.120 °C), while ships and coastal moorings have higher ESD. Overall, HY-2B's aggregate ESD (0.163 °C) lies between iQuam (0.157 °C) and MWIR (0.196 °C). The corresponding signal-to-noise ratios are nearly identical for all three products, indicating robust detection of true SST signals. These results suggest that HY-2B has slightly better random error control than MWIR, which contains more noise.

Finally, our analysis revealed that HY-2B SST errors increase under extreme environmental conditions: high winds, high latitudes, and high water vapor or cloud content. For instance, above 50°N, random errors and SNR_{sub} fluctuations increase due to colder sea and rougher conditions. At wind speeds above 10 m/s, HY-2B RMSE begins to rise (the "cold skin effect"), whereas MWIR errors increase more steadily. Very high or very low total water vapor causes a modest bias increase (0.05 °C) and higher RMSE, reflecting increased attenuation. Similarly, columnar cloud water up to 0.5 mm raises HY-2B and MWIR ESD and RMSE. These findings suggest that atmospheric attenuation and cloud radiation notably affect microwave SST retrieval. Future work could incorporate environmental corrections or advanced multi-source fusion to improve SST accuracy under such challenging conditions.

Overall, the HY-2B L4A fusion SST product demonstrates high accuracy and consistency across most conditions, providing a reliable resource for oceanographic applications.

Author Contributions: Conceptualization, L.J. and G.Z.; methodology, L.J.; software, L.J.; validation, L.J., G.Z. and X.C.; formal analysis, Y.W.; investigation, Y.D.; resources, Y.W.; data curation, S.M.; writing—original draft preparation, L.J.; writing—review and editing, L.J.; visualization, S.M.; supervision, X.C.; project administration, X.C.; funding acquisition, Y.D. All authors have read and agreed to the published version of the manuscript.

Funding: This research was funded by the National Key Research and Development Program of China, grant number 2024YFC2813602; the National Natural Science Foundation of China, grant numbers 42306260 and U23A20649; and the Shanghai Frontiers Science Center of Polar Science (SCOPS), grant number SOO2025-04.

Data Availability Statement: All the data and products are publicly available through the websites of the respective organizations. The HY-2B L4A data are available at https://osdds.nsoas.org.cn/OceanDynamics (accessed 1 April 2025). The MWIR data are available at https://data.remss.com/SST/daily/mw_ir/v05.1/ (accessed 31 April 2025). The iQuam data are available at https:

//www.star.nesdis.noaa.gov/socd/sst/iquam/ (accessed 31 April 2025). The ERA5 data are available at https://doi.org/10.24381/cds.4991cf48 (accessed 1 April 2025).

Acknowledgments: The authors would like to thank the National Satellite Ocean Application Service (NSOAS), Remote Sensing System (RSS), and National Oceanic, Atmospheric Administration (NOAA) and European Centre for Medium-Range Weather Forecasts (ECMWF) for providing the data used in this study.

Conflicts of Interest: The authors declare no conflicts of interest.

References

1. Pisano, A.; Marullo, S.; Artale, V.; Falcini, F.; Yang, C.; Leonelli, F.E.; Santoleri, R.; Buongiorno Nardelli, B. New evidence of Mediterranean climate change and variability from sea surface temperature observations. *Remote Sens.* **2020**, *12*, 132. [CrossRef]
2. Ateweberhan, M.; McClanahan, T.R. Relationship between historical sea-surface temperature variability and climate change–induced coral mortality in the western Indian Ocean. *Mar. Pollut. Bull.* **2010**, *60*, 964–970. [CrossRef]
3. Enfield, D.B.; Mestas-Nuñez, A.M. Multiscale variabilities in global sea surface temperatures and their relationships with tropospheric climate patterns. *J. Clim.* **1999**, *12*, 2719–2733. [CrossRef]
4. Ruela, R.; Sousa, M.C.; de Castro, M.; Dias, J.M. Global and regional evolution of sea surface temperature under climate change. *Glob. Planet. Change* **2020**, *190*, 103190. [CrossRef]
5. O'Carroll, A.G.; Armstrong, E.M.; Beggs, H.M.; Bouali, M.; Casey, K.S.; Corlett, G.K.; Dash, P.; Donlon, C.J.; Gentemann, C.L.; Høyer, J.L.; et al. Observational needs of sea surface temperature. *Front. Mar. Sci.* **2019**, *6*, 420. [CrossRef]
6. Hausmann, U.; Czaja, A. The observed signature of mesoscale eddies in sea surface temperature and the associated heat transport. *Deep-Sea Res. Part I Oceanogr. Res. Pap.* **2012**, *70*, 60–72. [CrossRef]
7. Wang, Y.; Yu, Y.; Zhang, Y.; Zhang, H.-R.; Chai, F. Distribution and variability of sea surface temperature fronts in the South China Sea. *Estuar. Coast. Shelf Sci.* **2020**, *240*, 106793. [CrossRef]
8. Tandeo, P.; Chapron, B.; Ba, S.; Autret, E.; Fablet, R. Segmentation of mesoscale ocean surface dynamics using satellite SST and SSH observations. *IEEE Trans. Geosci. Remote Sens.* **2014**, *52*, 4227–4235. [CrossRef]
9. Zhang, H.; Ignatov, A. A completeness and complementarity analysis of the data sources in the NOAA in situ sea surface temperature quality monitor (iQuam) system. *Remote Sens.* **2021**, *13*, 3741. [CrossRef]
10. Minnett, P.J.; Alvera-Azcárate, A.; Chin, T.M.; Corlett, G.K.; Gentemann, C.L.; Karagali, I.; Li, X.; Marsouin, A.; Marullo, S.; Maturi, E.; et al. Half a century of satellite remote sensing of sea-surface temperature. *Remote Sens. Environ.* **2019**, *233*, 111366. [CrossRef]
11. Wentz, F.J.; Gentemann, C.; Smith, D.; Chelton, D. Satellite measurements of sea surface temperature through clouds. *Science* **2000**, *288*, 847–850. [CrossRef]
12. Biswas, S.K.; Gopalan, K.; Jones, W.L.; Bilanow, S. Correction of time-varying radiometric errors in TRMM Microwave Imager calibrated brightness temperature products. *IEEE Geosci. Remote Sens. Lett.* **2010**, *7*, 851–855. [CrossRef]
13. Wentz, F.J. A 17-yr climate record of environmental parameters derived from the Tropical Rainfall Measuring Mission (TRMM) Microwave Imager. *J. Clim.* **2015**, *28*, 6882–6902. [CrossRef]
14. Stammer, D.; Wentz, F.; Gentemann, C. Validation of microwave sea surface temperature measurements for climate purposes. *J. Clim.* **2003**, *16*, 73–87. [CrossRef]
15. Qian, G. China launches HY-2B satellite atop a LM-4B. *Aerosp. China* **2018**, *19*, 57.
16. Liu, S.B.; Cui, X.D.; Li, Y.N.; Jin, X.; Zhou, W.; Dang, H.X.; Li, H. Retrieval of sea surface temperature from the scanning microwave radiometer aboard HY-2B. *Int. J. Remote Sens.* **2021**, *42*, 4621–4643. [CrossRef]
17. Zhou, W.; Lin, M.; Yin, X.; Ma, X.; Huang, L.; Wang, S.; Ma, C.; Zhang, Y. Preliminary estimate of sea surface temperature from the scanning microwave radiometer onboard HY-2B satellite. In Proceedings of the IGARSS 2019—2019 IEEE International Geoscience and Remote Sensing Symposium, Yokohama, Japan, 28 July–2 August 2019; pp. 8173–8176. [CrossRef]
18. Zhang, L.; Yu, H.; Wang, Z.; Yin, X.; Yang, L.; Du, H.; Li, B.; Wang, Y.; Zhou, W. Evaluation of the initial sea surface temperature from the HY-2B scanning microwave radiometer. *IEEE Geosci. Remote Sens. Lett.* **2021**, *18*, 137–141. [CrossRef]
19. Dash, P.; Ignatov, A.; Martin, M.; Donlon, C.; Brasnett, B.; Reynolds, R.W.; Banzon, V.; Beggs, H.; Cayula, J.; Chao, Y.; et al. Group for High Resolution Sea Surface Temperature (GHRSST) analysis fields inter-comparisons—Part 2: Near real time web-based level 4 SST Quality Monitor (L4-SQUAM). *Deep Sea Res. Part II Top. Stud. Oceanogr.* **2012**, *77–80*, 31–43. [CrossRef]
20. Casey, K.S.; Adamec, D. Sea surface temperature and sea surface height variability in the North Pacific Ocean from 1993 to 1999. *J. Geophys. Res.* **2002**, *107*, 14-1–14-12. [CrossRef]
21. Werb, B.E.; Rudnick, D.L. Remarkable changes in the dominant modes of North Pacific sea surface temperature. *Geophys. Res. Lett.* **2023**, *50*, e2022GL101078. [CrossRef]
22. Xu, F.; Ignatov, A. In situ SST Quality Monitor (iQuam). *J. Atmos. Ocean. Technol.* **2014**, *31*, 164–180. [CrossRef]

23. Wang, H.; Lin, M.; Ma, C.; Yin, X.; Guan, L. Evaluation of sea surface temperature from HY-1C data. In Proceedings of the IGARSS 2020—2020 IEEE International Geoscience and Remote Sensing Symposium, Waikoloa, HI, USA, 26 September–2 October 2020; pp. 5897–5900. [CrossRef]
24. Tu, Q.; Hao, Z. Validation of sea surface temperature derived from Himawari-8 by JAXA. *IEEE J. Sel. Top. Appl. Earth Obs. Remote Sens.* **2020**, *13*, 448–459. [CrossRef]
25. Chen, Y.; Qu, L.; Guan, L. Evaluation of NOAA/AVHRR sea surface temperature at full HRPT resolution in the Northwest Pacific Ocean. *J. Ocean Univ. China* **2021**, *20*, 1431–1439. [CrossRef]
26. Xu, F.; Ignatov, A. Error characterization in iQuam SSTs using triple collocations with satellite measurements. *Geophys. Res. Lett.* **2016**, *43*, 10826–10834. [CrossRef]
27. Bell, B.; Hersbach, H.; Simmons, A.; Berrisford, P.; Dahlgren, P.; Horányi, A.; Muñoz-Sabater, J.; Nicolas, J.; Radu, R.; Schepers, D.; et al. The ERA5 global reanalysis: Preliminary extension to 1950. *Q. J. R. Meteorol. Soc.* **2021**, *147*, 4186–4227. [CrossRef]
28. Hersbach, H.; Bell, B.; Berrisford, P.; Hirahara, S.; Horányi, A.; Muñoz-Sabater, J.; Nicolas, J.; Peubey, C.; Radu, R.; Schepers, D.; et al. The ERA5 global reanalysis. *Q. J. R. Meteorol. Soc.* **2020**, *146*, 1999–2049. [CrossRef]
29. Gentemann, C.L.; Wentz, F.J.; Mears, C.A.; Smith, D.K. In situ validation of Tropical Rainfall Measuring Mission microwave sea surface temperatures. *J. Geophys. Res. Oceans* **2004**, *109*, C04021. [CrossRef]
30. Gentemann, C.L.; Hilburn, K.A. In situ validation of sea surface temperatures from the GCOM-W1 AMSR2 RSS calibrated brightness temperatures. *J. Geophys. Res. Oceans* **2015**, *120*, 3567–3585. [CrossRef]
31. Zhao, Y.; Zhu, J.; Lin, M.; Chen, C.; Huang, X.; Wang, H.; Zhang, Y.; Peng, H. Assessment of the initial sea surface temperature product of the scanning microwave radiometer aboard on HY-2 satellite. *Acta Oceanol. Sin.* **2014**, *33*, 109–113. [CrossRef]
32. Woo, H.-J.; Park, K.-A. Inter-comparisons of daily sea surface temperatures and in-situ temperatures in the coastal regions. *Remote Sens.* **2020**, *12*, 1592. [CrossRef]
33. Kennedy, J.J. A review of uncertainty in in situ measurements and data sets of sea surface temperature. *Rev. Geophys.* **2014**, *52*, 1–32. [CrossRef]
34. Kennedy, J.J.; Rayner, N.A.; Smith, R.O.; Parker, D.E.; Saunby, M. Reassessing biases and other uncertainties in sea surface temperature observations measured in situ since 1850: 1. Measurement and sampling uncertainties. *J. Geophys. Res. Atmos.* **2011**, *116*, D14103. [CrossRef]
35. Dong, S.; Gille, S.T.; Sprintall, J.; Gentemann, C. Validation of the Advanced Microwave Scanning Radiometer for the Earth Observing System (AMSR-E) sea surface temperature in the Southern Ocean. *J. Geophys. Res. Oceans* **2006**, *111*, C04002. [CrossRef]
36. Donlon, C.J.; Minnett, P.J.; Gentemann, C.; Nightingale, T.J.; Barton, I.J.; Ward, B.; Murray, M.J. Toward improved validation of satellite sea surface skin temperature measurements for climate research. *J. Clim.* **2002**, *15*, 353–369. [CrossRef]
37. Minnett, P.J. Radiometric measurements of the sea-surface skin temperature: The competing roles of the diurnal thermocline and the cool skin. *Int. J. Remote Sens.* **2003**, *24*, 5033–5047. [CrossRef]
38. Fairall, C.W.; Bradley, E.F.; Godfrey, J.S.; Wick, G.A.; Edson, J.B.; Young, G.S. Cool-skin and warm-layer effects on sea surface temperature. *J. Geophys. Res. Oceans* **1996**, *101*, 1295–1308. [CrossRef]
39. Liu, Z.; Yang, M.; Qu, L.; Guan, L. Regional study on the oceanic cool skin and diurnal warming effects: Observing and modeling. *Remote Sens.* **2023**, *15*, 3814. [CrossRef]
40. Stoffelen, A. Toward the true near-surface wind speed: Error modeling and calibration using triple collocation. *J. Geophys. Res. Oceans* **1998**, *103*, 7755–7766. [CrossRef]
41. Gruber, A.; Su, C.-H.; Zwieback, S.; Crow, W.; Dorigo, W.; Wagner, W. Recent advances in (soil moisture) triple collocation analysis. *Int. J. Appl. Earth Obs. Geoinf.* **2016**, *45*, 200–211. [CrossRef]
42. Draper, C.; Reichle, R.; de Jeu, R.; Naeimi, V.; Parinussa, R.; Wagner, W. Estimating root mean square errors in remotely sensed soil moisture over continental scale domains. *Remote Sens. Environ.* **2013**, *137*, 288–298. [CrossRef]
43. Peng, J.; Misra, S.; Piepmeier, J.R.; Dinnat, E.P.; Yueh, S.H.; Meissner, T.; Vine, D.M.L.; Shelton, K.E.; Freedman, A.P.; Dunbar, R.S.; et al. Soil Moisture Active/Passive (SMAP) L-Band Microwave Radiometer post-launch calibration upgrade. *IEEE J. Sel. Top. Appl. Earth Obs. Remote Sens.* **2019**, *12*, 1647–1657. [CrossRef]
44. Hong, Z.; Moreno, H.A.; Li, Z.; Li, S.; Greene, J.S.; Hong, Y.; Alvarez, L.V. Triple collocation of ground-, satellite- and land surface model-based surface soil moisture products in Oklahoma—Part I: Individual product assessment. *Remote Sens.* **2022**, *14*, 5641. [CrossRef]
45. Li, C.; Tang, G.; Hong, Y. Cross-evaluation of ground-based, multi-satellite and reanalysis precipitation products: Applicability of the triple collocation method across mainland China. *J. Hydrol.* **2018**, *562*, 71–83. [CrossRef]
46. Wild, A.; Chua, Z.-W.; Kuleshov, Y. Triple collocation analysis of satellite precipitation estimates over Australia. *Remote Sens.* **2022**, *14*, 2724. [CrossRef]
47. Hoareau, N.; Portabella, M.; Lin, W.; Ballabrera-Poy, J.; Turiel, A. Error characterization of sea surface salinity products using triple collocation analysis. *IEEE Trans. Geosci. Remote Sens.* **2018**, *56*, 5160–5168. [CrossRef]

48. Caires, S.; Sterl, A. Validation of ocean wind and wave data using triple collocation. *J. Geophys. Res. Oceans* **2003**, *108*, 3098. [CrossRef]
49. McColl, K.A.; Vogelzang, J.; Konings, A.G.; Entekhabi, D.; Piles, M.; Stoffelen, A. Extended triple collocation: Estimating errors and correlation coefficients with respect to an unknown target. *Geophys. Res. Lett.* **2014**, *41*, 6229–6236. [CrossRef]
50. Saha, K.; Dash, P.; Zhao, X.; Zhang, H.-m. Error estimation of Pathfinder version 5.3 Level-3C SST using extended triple collocation analysis. *Remote Sens.* **2020**, *12*, 590. [CrossRef]
51. Cao, M.; Mao, K.; Yan, Y.; Shi, J.; Wang, H.; Xu, T.; Fang, S.; Yuan, Z. A new global gridded sea surface temperature data product based on multisource data. *Earth Syst. Sci. Data* **2021**, *13*, 2111–2134. [CrossRef]
52. Hosoda, K.; Kawamura, H. Seasonal variation of space/time statistics of short-term sea surface temperature variability in the Kuroshio region. *J. Oceanogr.* **2005**, *61*, 709–720. [CrossRef]
53. Wu, X.; Xiao, Q.; Wen, J.; You, D. Direct comparison and triple collocation: Which is more reliable in the validation of coarse-scale satellite surface albedo products. *J. Geophys. Res. Atmos.* **2019**, *124*, 5198–5213. [CrossRef]
54. Wu, X.; Lu, G.; Wu, Z.; He, H.; Scanlon, T.; Dorigo, W. Triple collocation-based assessment of satellite soil moisture products with in situ measurements in China: Understanding the error sources. *Remote Sens.* **2020**, *12*, 2275. [CrossRef]
55. Liu, P.; Zhao, Y.; Zhou, W.; Wang, S. Evaluation of HY-2B SMR sea surface temperature products from 2019 to 2024. *Remote Sens.* **2025**, *17*, 300. [CrossRef]
56. Merchant, C.J.; Le Borgne, P. Retrieval of sea surface temperature from space based on modeling of infrared radiative transfer: Capabilities and limitations. *J. Atmos. Ocean. Technol.* **2004**, *21*, 1734–1746. [CrossRef]
57. Price, J.F.; Weller, R.A.; Pinkel, R. Diurnal cycling: Observations and models of the upper ocean response to diurnal heating, cooling, and wind mixing. *J. Geophys. Res. Oceans* **1986**, *91*, 8411–8427. [CrossRef]
58. Anding, D.; Kauth, R. Estimation of sea surface temperature from space. *Remote Sens. Environ.* **1970**, *1*, 217–220. [CrossRef]

remote sensing

MDPI

Article

Air–Sea Interaction During Ocean Frontal Passage: A Case Study from the Northern South China Sea

Ruichen Zhu [1,*], Jingjie Yu [2], Xingzhi Zhang [1], Haiyuan Yang [2,3] and Xin Ma [2,3]

1 Laoshan Laboratory, Qingdao 266237, China; xzzhang@qnlm.ac

2 Frontier Science Center for Deep Ocean Multi-Spheres and Earth System (FDOMES) and Physical Oceanography Laboratory, Ocean University of China, Qingdao 266100, China; yujingjie@ouc.edu.cn (J.Y.); yanghaiyuan@ouc.edu.cn (H.Y.); maxin@ouc.edu.cn (X.M.)

3 Laboratory for Ocean Dynamics and Climate, Qingdao Marine Science and Technology Center, Qingdao 266237, China

* Correspondence: rczhu@qnlm.ac

Abstract

The northern South China Sea has abundant frontal systems near coastal and island regions, which play crucial roles in regional ocean dynamics and ecosystem. While previous studies have established preliminary understanding of their spatial distribution, seasonal variability, and dynamic characteristics, the atmospheric response to these frontal systems remains poorly understood. This study integrates observations from a moored buoy deployed on the continental shelf of the South China Sea with satellite remote sensing data to analyze oceanic and atmospheric variations during frontal passage. The results reveal that the ocean front can not only induce pronounced oceanic changes characterized by significant cooling, saltiness, and surface current acceleration, but also exert substantial influence on the overlying atmosphere, with consistent decreasing trends in air temperature, humidity, and atmospheric pressure, all of which rapidly recovered following frontal retreat. Notably, when the front directly traversed the buoy location, diurnal temperature cycles were markedly suppressed, while turbulent heat flux and downfront wind-stress curl reached peak magnitudes. These findings demonstrate that ocean fronts and associated sea surface temperature gradients can trigger intense air–sea exchange processes at the ocean–atmosphere interface.

Keywords: air–sea interaction; oceanic front; South China Sea

1. Introduction

The oceanic front is defined as narrow transitional zones marking the boundaries between distinct water masses with contrasting physical, chemical, and biological properties [1–3]. These frontal systems have spatial scales ranging from $O(1)$ to $O(1000)$ km and serve as regions of intense dynamical activity where energy transfers strongly from large-scale to small-scale motions through processes including frontal instabilities, internal wave generation, and intense turbulent mixing [4–6]. The enhanced vertical circulation associated with frontal dynamics facilitates the upward transport of heat and material from ocean interior into surface layer, creating conditions that significantly enhance local biological productivity [2,7,8].

Beyond their oceanic impacts, frontal systems can also exert considerable influence on atmospheric boundary-layer processes through multiple mechanisms [9–13]. Based on satellite observations and model simulations, ocean mesoscale fronts and eddies have been suggested to drive substantial variabilities in local surface wind [14–17], precipitation [13,18], and cloud formation [14,19]. The impacts of these fronts are not confined to the local area and can propagate to remote regions by forcing planetary waves [20]. The mechanisms for this air–sea interaction typically involve changes in the atmospheric vertical mixing or surface pressure above a strong sea surface temperature (SST) gradient in the frontal region [9,10,16,21]. Submesoscale fronts are also associated with a strong SST gradient and could potentially drive significant changes in the atmosphere, especially in the marine atmospheric boundary layer (MABL) [22,23].

The South China Sea (SCS) is the largest marginal sea in the Western Pacific Ocean, encompassing a complex oceanographic environment [24–30]. This semi-enclosed basin serves as a critical component of the global ocean circulation system, where multiple oceanic and atmospheric forcing mechanisms converge to create intricate patterns of water mass distribution and frontal dynamics [31–35]. The northern region of the South China Sea exhibits typical frontal systems that demonstrate significant spatiotemporal variability under the influence of multiple physical processes. These include monsoon forcing, tidal mixing, river plume dynamics, coastal upwelling, and topographic interactions [36–39]. The interplay of these diverse mechanisms results in the formation of distinct water masses with contrasting thermohaline properties along coastal regions and in the vicinity of islands. The winter frontal formation is predominantly driven by the interaction between two distinct water masses: the cooler Guangdong coastal current induced by the northeast monsoon and the warmer waters transported shoreward through Ekman transport mechanisms [40–42]. Conversely, during the summer monsoon season, frontal activity diminishes significantly. The southwest monsoon generates coastal upwelling along the Vietnamese coast and the eastern side of Hainan Island, bringing cooler, nutrient-rich subsurface waters to the surface and forming oceanic fronts [36,42].

Despite considerable advances in understanding the spatial distribution, seasonal variability, and dynamical characteristics of these coastal frontal systems, a significant knowledge gap remains regarding their potential impact on atmospheric processes in marginal sea environments. Previous satellite observations and reanalysis data have revealed statistically significant positive SST–wind correlations in the northern South China Sea coastal region, with coupling coefficients that exceed those observed at mid-latitude oceans but remain smaller than equatorial regions [43]. Oceanic fronts can induce baroclinic adjustment of perturbation pressure within the marine atmospheric boundary layer, with pressure gradient adjustments representing the dominant term in horizontal momentum budgets over frontal zones [44]. However, the lack of high-frequency observational data has hindered our ability to further investigate the instantaneous variations in atmospheric responses to these fontal processes. In this study, we use observational data collected from a moored buoy system deployed along the eastern coast of Hainan Island to investigate the influence of oceanic frontal systems on air–sea interaction. The in situ measurements from hydrographic instrumentation were combined with satellite remote sensing data to successfully identify and characterize a frontal passage event. By examining atmospheric changes occurring before and after the frontal passage, it is found that the oceanic front can influence local atmospheric conditions and contribute to air–sea interaction processes in complex coastal environments.

The rest of this paper is organized as follows. In Section 2, the buoy and satellite observations are introduced, including detailed descriptions of instrumentation, data processing methodologies, and quality control procedures. In Section 3, the frontal passage process and its significant atmospheric impacts are described. The conclusion and discussion are presented in Section 4.

2. Materials and Methods

2.1. Buoy Observation

On 10 April 2025, a buoy system was deployed at a water depth of ~105 m, approximately 50 km east of Hainan Province, China (Figure 1). The buoy configuration was designed to capture both atmospheric and oceanic parameters across multiple depths to facilitate detailed analysis of air–sea interaction processes. The atmospheric measurement suite was positioned 10 m above the sea surface and comprised sensors for temperature, humidity, atmospheric pressure, and wind speed. The sample interval of all sensors was set to one minute. To minimize measurement uncertainties and random noise effects, all meteorological data were subsequently processed to hourly averages. Below the sea surface, five Conductivity–Temperature–Depth (CTD) sensors and five single-point current meters were positioned at depths of 5 m, 20 m, 30 m, 50 m, and 75 m with a sampling interval of one hour. Unfortunately, several technical issues affected data availability during the deployment period. The CTD at 30 m depth experienced an unidentified malfunction prior to 25 April, resulting in the absence of valid measurements from this level. Additionally, the current meter deployed at 75 m depth failed shortly after deployment. After applying a 24 h sliding window analysis for each measured variable, all observations exceeding 3 standard deviations from the local mean within each window were flagged as anomalous values and subsequently removed from the dataset. Finally, the remaining 65% of dataset extended from 10 April to 10 May are used in this study.

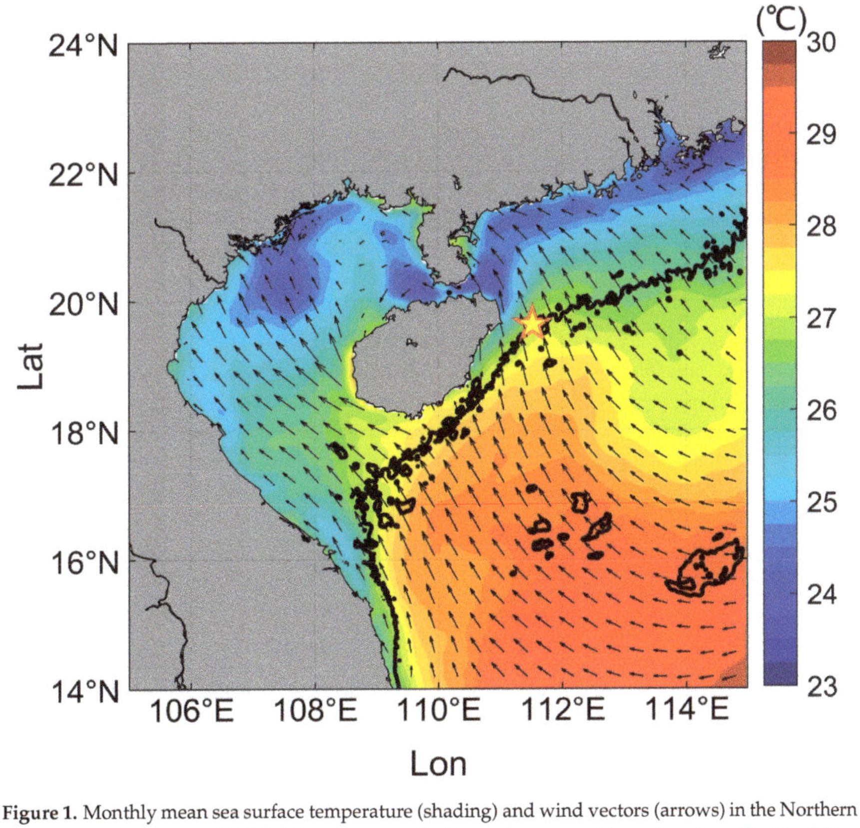

Figure 1. Monthly mean sea surface temperature (shading) and wind vectors (arrows) in the Northern South China Sea for April 2025. The yellow star indicates the moored buoy location. The black line denotes 100 m isobath.

2.2. Satellite Observation

To clarify the regional oceanic conditions surrounding the deployment site, SST fields from the Operational Sea Surface Temperature and Sea Ice Analysis (OSTIA) product were incorporated here. The OSTIA dataset provides a global, daily, gap-free foundation SST analysis at a high horizontal resolution of 0.05° on a regular grid, providing sufficient detail to resolve small-scare oceanic features such as fronts and eddies in the vicinity of coastal area. This level-4 (L4) product, covering the period from 1 October 1981, integrates multiple satellite observations alongside in situ measurements. The processing chain involves extracting and quality-controlling observations, applying satellite bias adjustments, assimilating the corrected inputs along with reprocessed sea ice concentration data, and performing an objective analysis. Each daily analysis incorporates data from a three-day window centered on the analysis date, with reduced weighting for surrounding days, resulting in global maps of analyzed SST and its uncertainty field. It should be noted that the three-day temporal averaging may dampen small-scale and short-lived oceanic events, such as frontal passages. This temporal smoothing acts as a low-pass filter, potentially reducing the amplitude and sharpness of rapid SST changes compared to point measurements from buoys or other high-frequency observations. Additionally, hourly wind-field data from scatterometer observations and the European Centre for Medium-Range Weather Forecasts (ECMWF) reanalysis were utilized to characterize the wind variability in the vicinity of the buoy deployment site. The following discussion is within the domain of 17.5–21.5°N and 109.5–113.5°E.

2.3. Numerical Model Simulation

To further investigate the atmospheric response to ocean fronts, we conducted high-resolution numerical simulations using the Weather Research and Forecasting Model (WRF) version 3.4, developed by the National Center for Atmospheric Research. The model domain was configured over 18.5–21.5°N and 115–118°E, with a horizontal resolution of 1 km and 30 vertical layers (Figure 2). Initial and boundary conditions were derived from ERA5 reanalysis data, temporally and spatially averaged over the corresponding study region during April 2025. For the bottom boundary conditions, we prescribed an idealized north–south oriented sea surface temperature (SST) front, with temperatures of 21 °C and 23 °C on the western and eastern sides, respectively. The transition zone was set to 4 km width, corresponding to a thermal gradient intensity of 5×10^{-4} °C m^{-1}. Initial wind conditions were specified as southerly flow. The model was integrated with hourly output frequency, reaching steady states after approximately three days. Our analysis focuses on the averaged results from the subsequent five-day period to ensure statistical robustness.

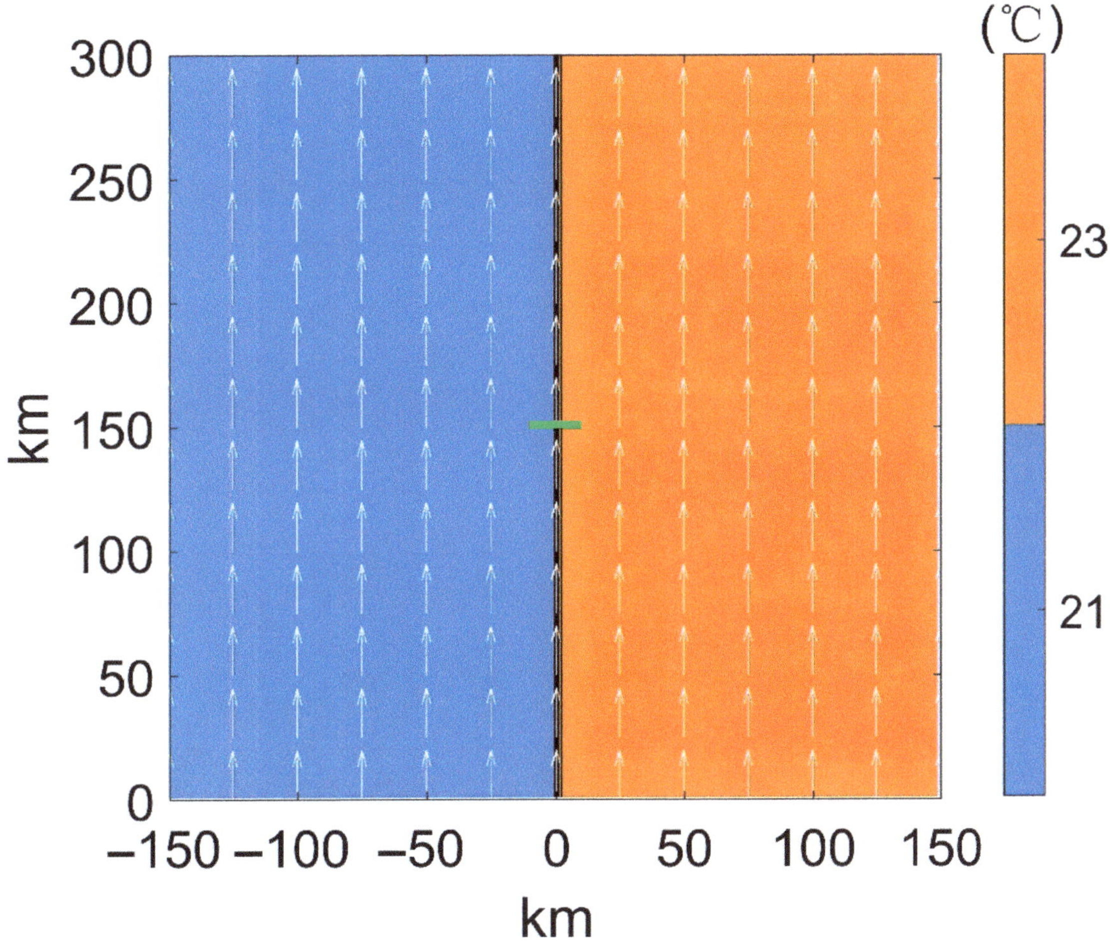

Figure 2. The horizontal distribution of Sea surface temperature for numerical model experiment. White arrows represent initial wind vectors.

3. Results

3.1. Moored Buoy and Satellite Observations of an Oceanic Front

During the one-month observation period from 10 April to 10 May, the CTD measurements from the buoy reveal distinct vertical stratification and temporal variability of upper 75 m of water mass (Figure 3). Prior to 12 April, the potential temperature values remained relatively stable around 22–25 °C across all depths, while salinity showed minimal variation around 34.2–34.4 psu, and potential density anomaly (σ_θ) maintained consistent values near 23.8–24.0 kg m^{-3}, indicating a well-mixed condition in the whole water column. Subsequently, the water temperature at 15 m progressively increased from approximately 24 °C in mid-April to nearly 29 °C by 10 May, representing a remarkable 5 °C warming over less than one month (Figure 3a). Concurrently, a significant freshening process occurred, with salinity decreasing from 34.3 psu to approximately 33.0 psu (Figure 3b). In contrast, the temperature and salinity at 50–75 m remained consistently around 22–23 °C and 34.5 psu, respectively, until the onset of fluctuations in late April, which suggests the development of a strong thermocline that effectively isolated the surface mixed layer from deeper waters.

The combined effects of surface warming and freshening resulted in a pronounced decrease in σ_θ at shallow depths, declining from approximately 23.8 kg m^{-3} to values approaching 21.0 kg m^{-3} by early May (Figure 3c). The resulting density stratification represents a classic example of ocean stratification, transitioning from the well-mixed winter conditions to strongly stratified summer conditions.

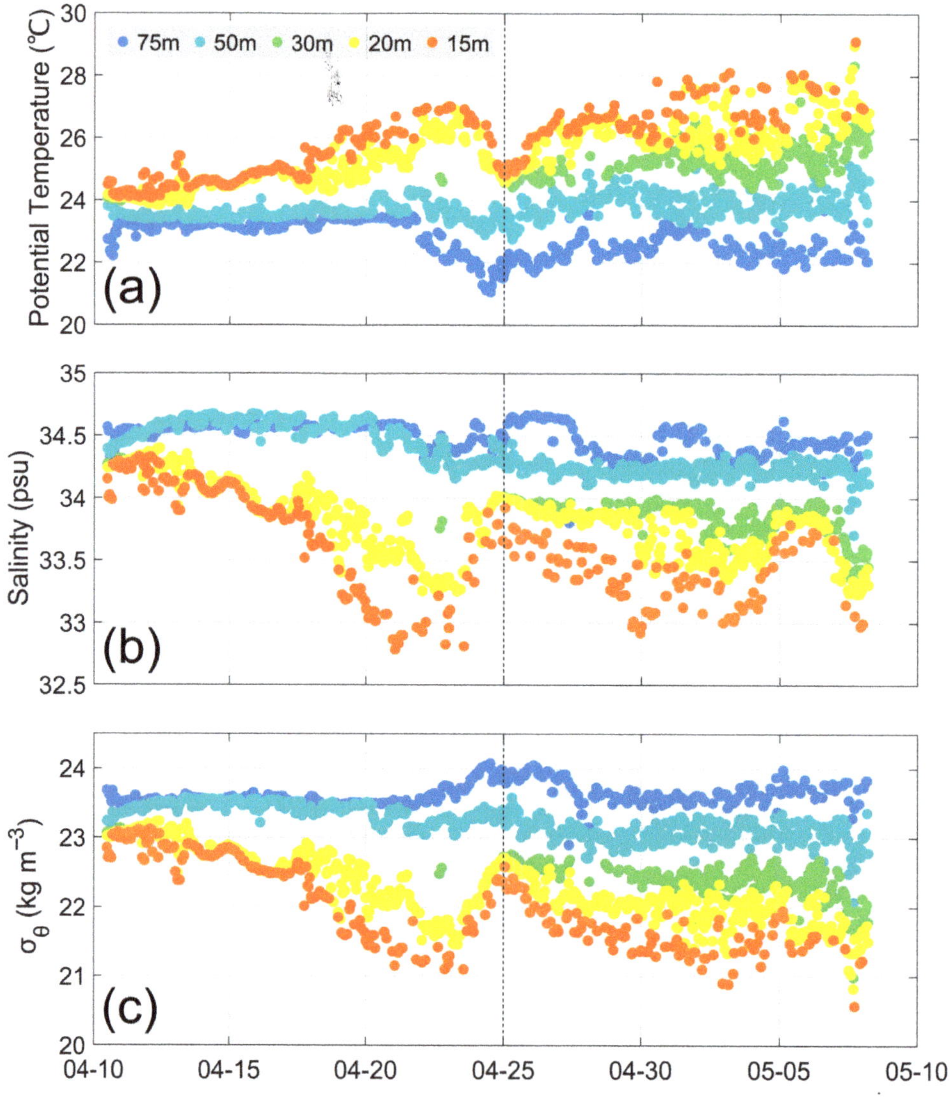

Figure 3. (**a**) Potential temperature, (**b**) salinity and (**c**) σ_θ measured by five CTD at multiple depth layers. Noted that the CTD deployed at 30 m depth achieved stable operation and initiated continuous data logging from 25 April. The black dashed line represents the timing when frontal passage causes near-surface temperature to decrease and salinity and density to increase.

To eliminate temperature and salinity disturbances caused by vertical motions of CTDs on the buoy, we interpolate all variables onto two reference depths (13 m and 68 m) (Figure 4). These two depths are obtained by averaging the time-averaged pressure values from the upper two CTDs and lower two CTDs, respectively. This approach minimizes the difference between the measured pressure and the chosen constant reference depths [45]. Beyond the overall warming trend and stratification processes, the temperature time series clearly show a pronounced cooling event occurring around 25 April, characterized by temperature decreases of approximately 2 °C at both depths within less than two days. Accompanying the temperature anomaly, significant changes in salinity and density patterns, with their values at 13 m increased by 1 psu and 1 kg m^{-3}, respectively. At the 68 m, salinity and density also increased, though the magnitude was considerably smaller, to only 20–30% of the 13 m values. After 2 days, all variables at both depth levels began to return toward their pre-event levels, implying that this abrupt and short-lived cooling event may be not caused by a simple surface cooling process such as atmospheric heat flux modification, but rather involved the advection or upwelling of cooler water masses from laterally or below into the observation area.

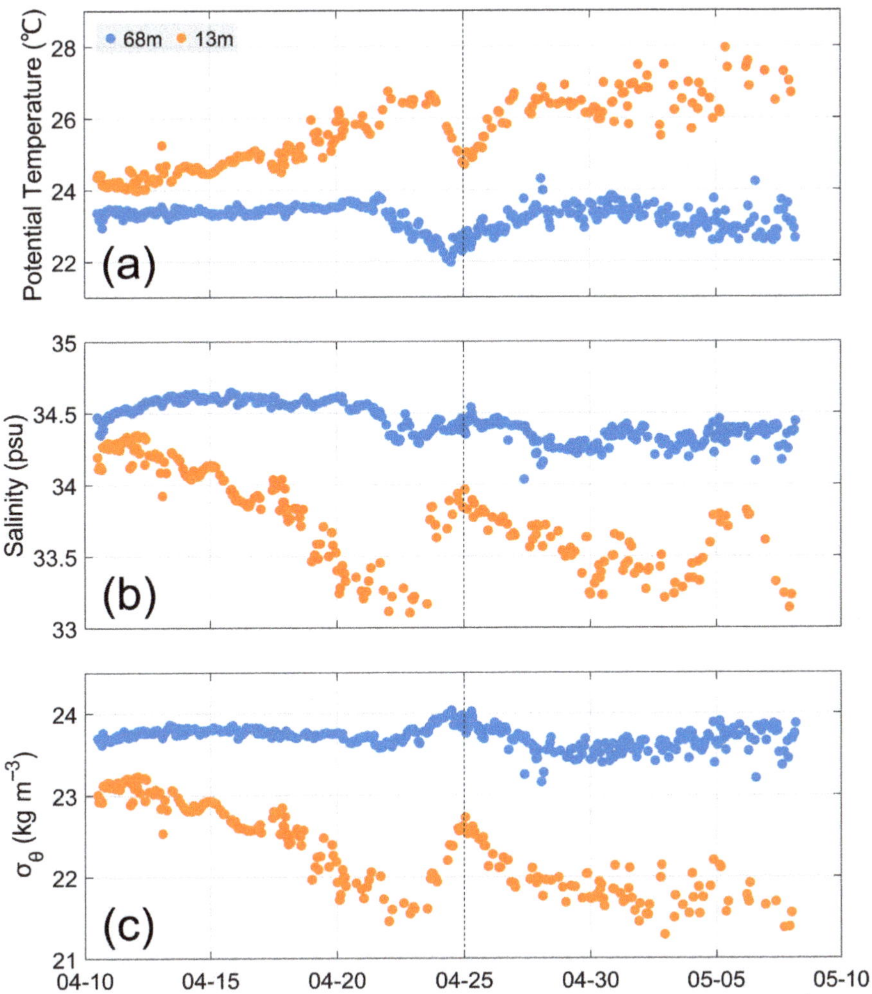

Figure 4. Measurements of the same values in Figure 3, but for the interpolated values at 13 m and 68 m, respectively. The black dashed line represents the timing when frontal passage causes near-surface temperature to decrease and salinity and density to increase.

The current velocity measurements from the upper 50 m also show a significant flow change that preceded the cooling event by approximately two days (Figure 5). Throughout most of the observation period, the current magnitudes remained relatively consistent across all depth levels, exhibiting a gradual strengthening trend with predominantly southward flow directions. However, on 21 April, the 15 m current speed increases from approximately 0.5 m/s to nearly 1.0 m/s within 25 h, representing an acceleration rate that far exceeded the background trend. Noted that the flow rotates from southward to eastward direction during the acceleration process, which provides crucial insight into the potential source region of the anomalous water mass, suggesting that the cooler, more saline waters originated from the continental shelf or nearshore coastal areas to the buoy location. Similarly to the rapid return of water properties, the current velocities drop back to the previous levels by 25 April, confirming that the entire sequence of events represents a coherent, short-lived perturbation to the local ocean state. Only energetic mesoscale or smaller-scale features such as oceanic fronts or eddies possess the capacity to generate such dramatic and rapid changes in both the velocity field and water mass properties.

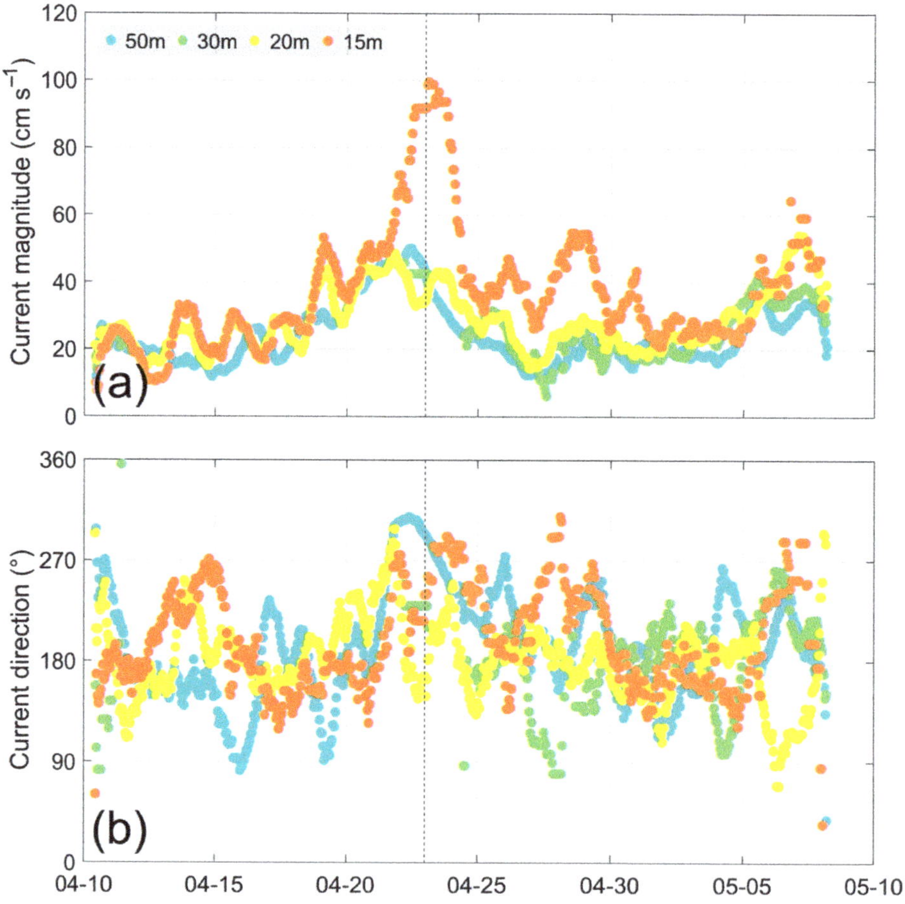

Figure 5. (**a**) Current magnitude and (**b**) direction measured by four current meters at multiple depth layers. The black dashed line represents the timing of frontal passage when current velocity increases, which occurs two days earlier than the changes in hydrographic properties.

An examination of satellite-derived SST data provides a broader spatial perspective on the regional oceanographic conditions during the observation period (Figure 6). Consistent with the CTD measurement, the SST field reveals a coherent warming trend throughout the surrounding region over the one-month period. However, the spatial distribution of SST exhibits a distinct north–south gradient, with temperatures systematically decreasing from south to north due to the presence of coastal currents that transport cooler, saltier waters along the continental margin. When these cooler, saltier coastal waters encounter the warmer, fresher offshore waters characteristic of the South China Sea, the resulting density contrasts create favorable conditions for the development and intensification of frontal systems [42,46]. On 20 April, a well-defined oceanic front oriented in a northeast-southwest direction was formed to approximately 30 km northwest of the buoy deployment site (Figure 6a,d). This front manifests as a sharp gradient in SST, with temperature differences across the frontal zone reaching ~2 °C over distances of only 15–20 km. Subsequently, the intensification of northeasterly winds drives the coastal current system to transport increasingly larger volumes of cold water toward the Hainan Island, leading to a progressive strengthening of the frontal system and its gradual offshore migration (Figure 6b,e). By 25 April, the frontal system reaches its maximum intensity and extends to the buoy location, spanning approximately 200 km in length with an average cross-frontal temperature gradient reaching magnitudes of 2×10^{-4} °C m^{-1} (Figure 6e). Following the peak intensity on 25 April, a shift from northeasterly to southeasterly winds leads to a retreat of the frontal system back toward the coast, allowing the warmer offshore waters to reoccupy the buoy location and explaining the subsequent recovery of temperature and salinity values(Figure 6c,f). The satellite observations thus provide compelling evidence that the rapid cooling event, current acceleration, and water mass property changes recorded at the buoy resulted from the passage of this well-defined oceanic front, rather than from local surface cooling or other small-scale processes.

3.2. Atmospheric Response to the Frontal Passage

Although the moored buoy system cannot directly measure the spatial structure of the oceanic front, the satellite-derived evidence of frontal migration toward the buoy location enables us to treat this event as a cross-frontal observation, offering valuable insights into whether oceanic fronts can exert measurable influence on the overlying atmospheric boundary layer. Figure 7 displays the hourly evolution of atmospheric variables from 22 to 25 April during the frontal approach and passage toward the buoy location. The air temperature time series reveal a progressive cooling trend that closely mirrors the underlying SST evolution (Figure 7a), demonstrating strong coupling between oceanic and atmospheric thermal structures. Initially maintaining relatively stable values around 27 °C on 22–23 April, the air temperature begins a systematic decline as the oceanic front approaches the observation site. Most remarkably, the typical diurnal temperature cycle, which remained clearly visible during 22–23 April with amplitude variations of approximately 0.5 °C, becomes virtually eliminated on 24 April, indicating that the oceanic forcing associated with the frontal passage overwhelms the normal solar heating cycle and creates anomalous atmospheric conditions. The air temperature reaches its minimum value of approximately 26.2 °C at 06:00 UTC on 25 April, coinciding with the timing when satellite observations show the oceanic front achieving maximum intensity and closest proximity to the buoy location, thus establishing a clear temporal correlation between oceanic and atmospheric thermal anomalies. Following this minimum, the atmospheric temperature begins a gradual recovery that parallels the retreat of the oceanic front and the subsequent reoccupation of warm waters.

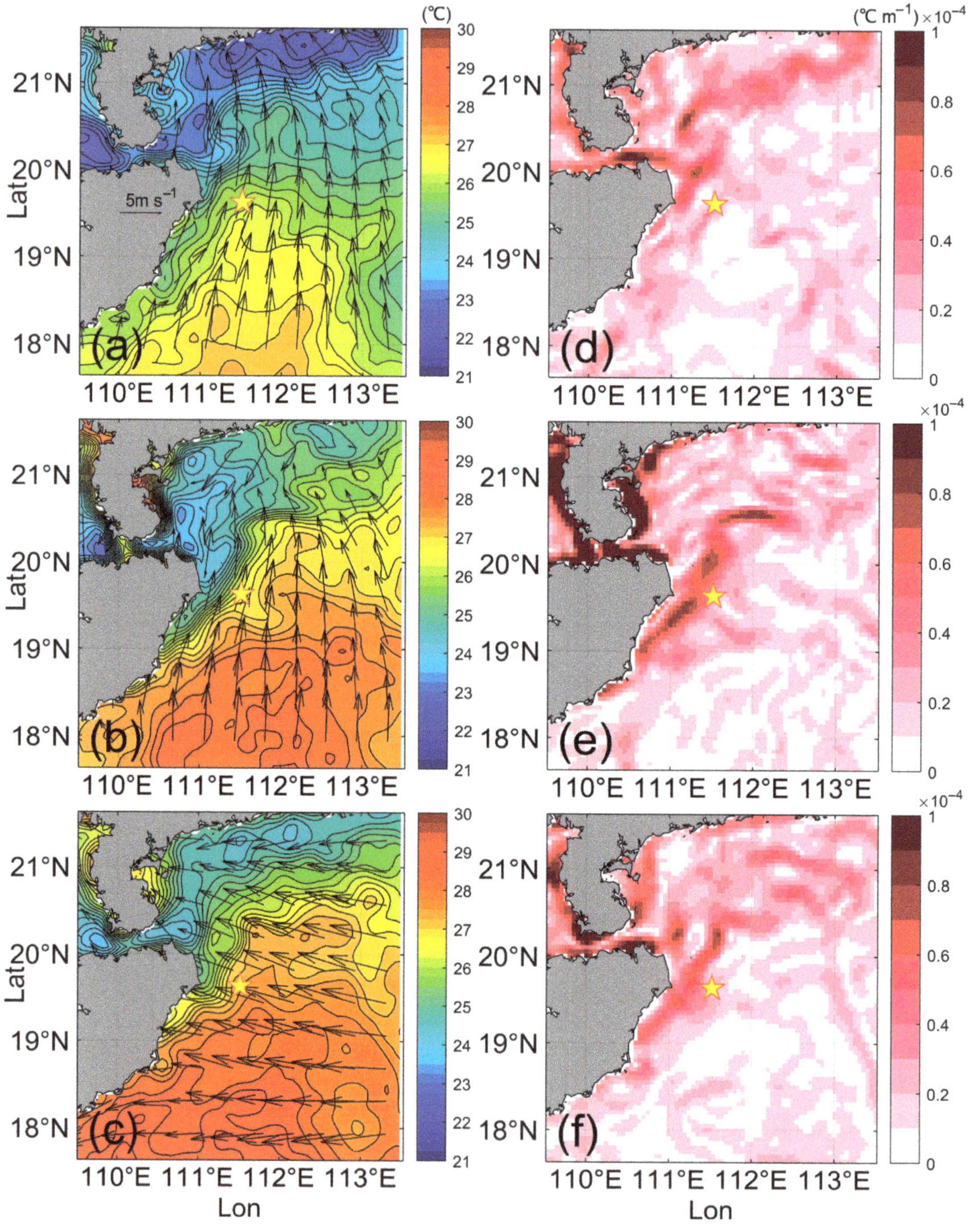

Figure 6. (**a–c**) Satellite-derived SST (shading) and surface wind field (vectors) on 20, 25, and 28 April 2025. (**d–f**) are same as (**a–c**), but for the horizontal SST gradient. The yellow star indicates the moored buoy location. The black line denotes 100 m isobath.

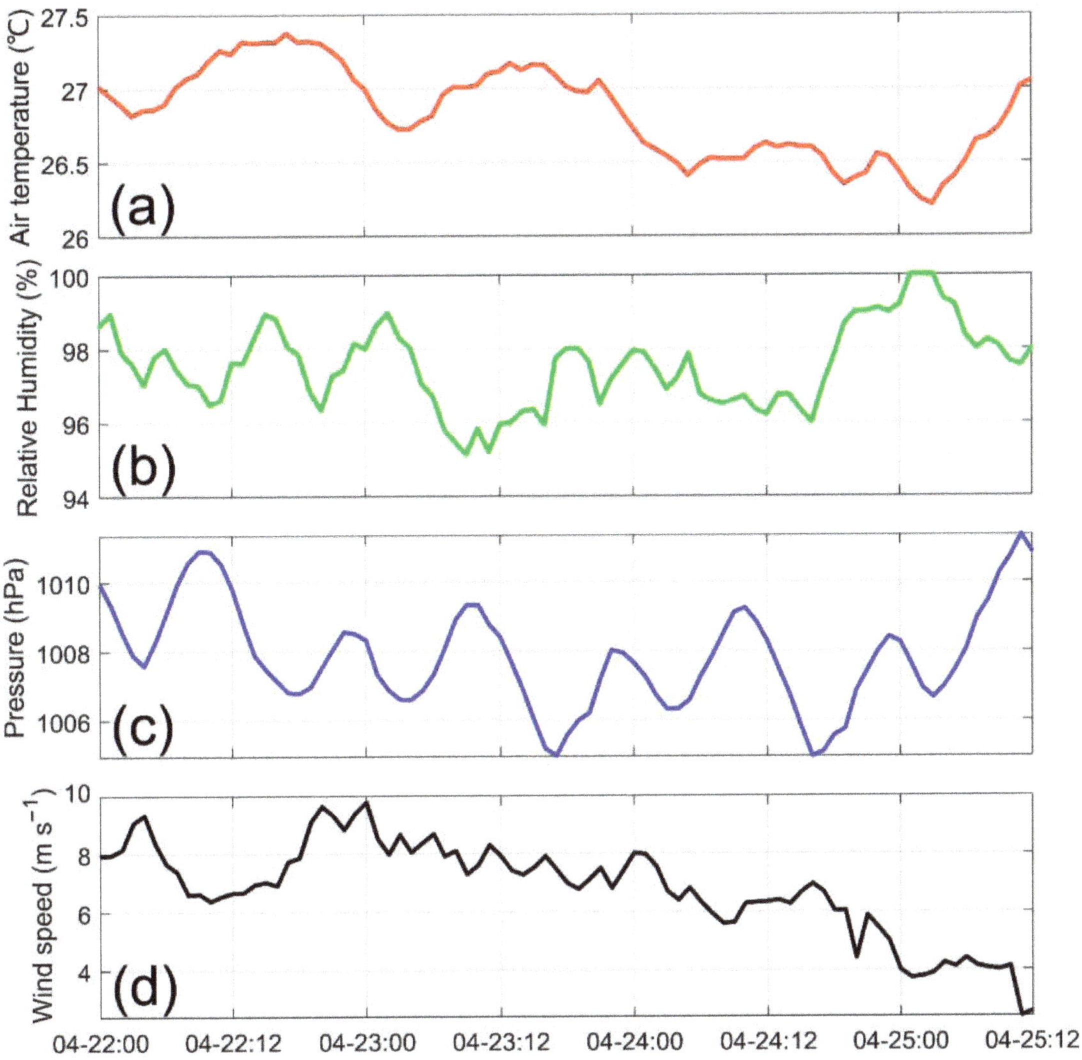

Figure 7. (**a**) Air temperature, (**b**) relative humidity, (**c**) pressure and (**d**) wind speed measurements at 10 m height above the sea surface.

The relative humidity and atmospheric pressure fields (Figure 7b,c) exhibit similarly coherent responses to the frontal passage, but their minimum values occur approximately 15 h earlier than the temperature minimum, at around 15:00 UTC on 24 April. This temporal offset suggests that humidity and pressure respond more rapidly to the initial approach of the cooler, denser air mass associated with the oceanic front. The wind speed had a sustained weakening trend throughout the entire observation period, with a reduction in more than 50% (Figure 7d). Although the wind speed does not recover to its initial values following the frontal passage, a pronounced wind speed perturbation occurs at 20:00 UTC on 24 April. This wind speed anomaly likely represents the instantaneous atmospheric response generated by the sharp horizontal temperature gradients associated with the oceanic front.

The modifications to the air–sea exchange processes across the front are further quantified by the surface heat flux based on COARE 3.0 algorithm [47]. The temporal evolution of sensible heat flux (SHF), latent heat flux (LHF), and turbulent heat flux (THF) from 22 to 25 April demonstrates that the heat exchange is predominantly governed by sensible heat transfer (Figure 8). Before 22–23 April, the turbulent heat fluxes display characteristic diurnal cycling patterns, with THF reaching maximum positive values of approximately 12–13 W m^{-2} during nighttime hours around 05:00 UTC when the air–sea temperature differential favors upward heat transfer from the relatively warm ocean surface to the cooler atmospheric boundary layer. The diurnal minima typically occur during daytime hours around 14:00 UTC when solar heating elevates air temperature, resulting in sign reversal of flux values. However, the approach and passage of the oceanic front on 24 April fundamentally disrupts these established heat exchange patterns. The THF maintains persistently negative values, indicating a sustained period of oceanic heat absorption from the atmosphere. The magnitude of this thermal forcing becomes particularly evident when examining the timing and intensity of the daily flux maximum, which undergoes both temporal displacement and substantial amplification during the frontal passage. The trough heat flux on 24 April is delayed by approximately 6 h, not reaching its minimum until around 20:00 UTC on 24 April. This temporal shift coincides with the period of maximum frontal intensity and closest proximity to the observation site. More remarkably, the magnitude of this delayed trough reaches approximately −40 W m^{-2}, representing nearly a two-fold amplification compared to the typical values of 10–20 W m^{-2}. This doubling of heat flux intensity underscores the extraordinary efficiency of oceanic fronts in enhancing air–sea exchange processes through their ability to maintain steep temperature gradients and sustained thermal disequilibrium between oceanic and atmospheric boundary layers.

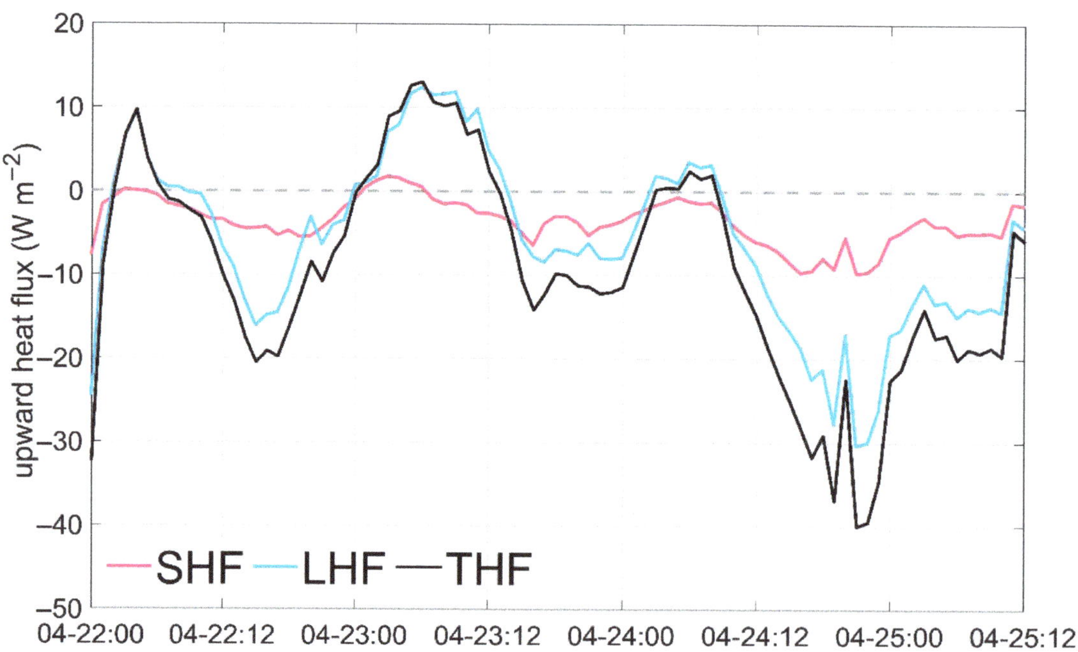

Figure 8. Turbulent, sensible and latent heat flux estimated using COARE 3.0 algorithm.

An additional complexity in the heat flux evolution appears at 19:00 UTC on 24 April, precisely coinciding with the sudden wind speed reduction documented in the atmospheric measurements. The wind speed perturbation at this time likely represents the atmospheric boundary-layer response to the passage of the most intense portion of the oceanic front. Given that the oceanic front maintained a relatively consistent northeast-southwest orientation throughout its approach and passage, with satellite analyses revealing an average frontal orientation of 30° east of north during 22–25 April, we projected the measured wind vectors onto this front-relative coordinate system to isolate the downfront and cross-front wind components, thereby enabling a more precise characterization of the atmospheric response to frontal forcing (Figure 9a,b). This coordinate transformation shows that the wind speed reduction manifests predominantly in the downfront component rather than the cross-front component. To quantify the dynamical implications of these wind-field modifications and assess their relationship to oceanic frontal forcing, we computed the temporal derivatives of wind-stress components and interpreted these quantities as approximations for the cross-front components of wind-stress curl and divergence, respectively (Figure 9c,d). The calculated curl component (Figure 9c) reveals a dramatic intensification between 18:00 and 22:00 on 24 April, with magnitude increases of approximately three times of magnitude compared to pre-frontal conditions. Although the values oscillate between positive and negative extremes reflects the limitation of calculating only one component of the wind-stress curl, it is somewhat indicative of a positive correlation between the cross-front SST gradient intensity and curl magnitude, consist with the previous satellite observation studies [17,43,48].

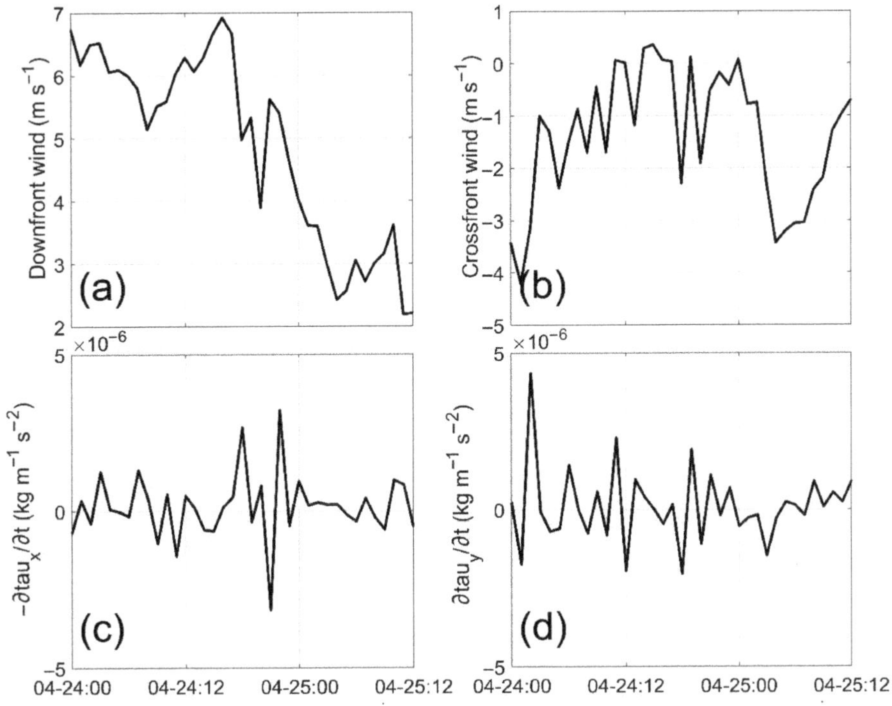

Figure 9. (**a**) Downfront wind speed, (**b**) cross-front wind speed, and temporal change rate of (**c**) downfront wind stress and (**d**) cross-front wind stress.

The simulation results provide compelling evidence for the atmospheric response to ocean fronts. Figure 10 shows the cross-frontal SST and atmospheric variations. The air temperature and THF exhibit a sharp transition coinciding with the SST front at x = 0 km, demonstrating the strong coupling between oceanic and atmospheric thermal structures. Although the magnitude of LHF is only half of the observed values, the THF increase is still primarily contributed by LHF rather than SHF. The LHF shows a dramatic peak near the frontal position, reaching approximately -15 W m^{-2}, while the SHF contribution remains relatively modest throughout the cross-frontal section. Furthermore, the wind-stress curl intensity indeed shows significant amplification at the frontal location, consistent with our observational findings. The downfront wind-stress gradient in the cross-front direction ($-\partial\tau_x/\partial y$) dominates the wind-stress curl magnitude. This enhanced wind-stress curl generation is attributed to the differential surface roughness and stability conditions across the thermal front. Therefore, these results substantiate that the observed atmospheric anomalies can be attributed to the presence and dynamics of oceanic fronts.

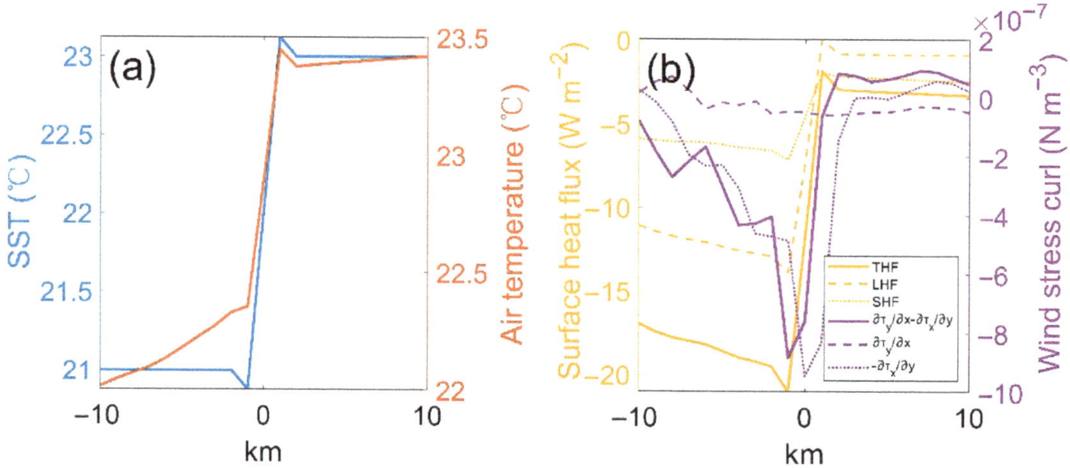

Figure 10. The distribution of (**a**) sea surface temperature and air temperature; (**b**) surface heat flux and wind-stress curl along the cross-frontal transect (green line in Figure 2).

4. Conclusions and Discussion

This study presents a comprehensive analysis of oceanic frontal impacts on air–sea interface processes in the northern South China Sea using in situ buoy observations combined with satellite-derived data. Through a one-month deployment period, we successfully captured and characterized the passage of a well-defined oceanic front and its subsequent effects on both water properties and atmospheric boundary-layer conditions. The main conclusions are summarized as follows:

(a) The buoy recorded a distinct cooling and saltiness event characterized by rapid temperature decreases of approximately 2 °C and salinity increases of 1 psu at 13 m depth. The current velocity measurements revealed a significant flow enhancement occurring two days prior to the cooling event, with speeds accelerating from 0.5 m/s to nearly 1.0 m/s, accompanied by a directional shift from southward to eastward flow. This temporal sequence indicates that oceanic dynamical processes, rather than atmosphere forcing, dominated the change in water mass properties.

(b) Satellite-derived SST fields confirmed that the observed anomalies resulted from the passage of an oceanic front. The frontal system, driven by northeasterly winds, transported cooler, saltier coastal waters offshore, reaching maximum intensity on 25 April with cross-frontal temperature gradients of 2×10^{-4} °C·m^{-1} spanning approximately 200 km. The subsequent wind direction changes to southeasterly caused frontal retreat and the restoration of pre-frontal conditions.

(c) The oceanic front exerted measurable influence on the overlying atmospheric boundary layer, manifesting through significant decreases in air temperature, humidity, and atmospheric pressure that closely tracked the underlying SST evolution. Most notably, the passage of the frontal system completely suppressed the normal diurnal cycle of air temperature on 24 April, indicating that oceanic forcing overwhelmed solar heating effects. The enhancement of turbulent heat flux and downfront wind stress curl collectively demonstrates the capacity of oceanic fronts to significantly modify air–sea exchange processes.

While our single-point moored buoy observations detected the temporal evolution of frontal passage, the inherently limited spatial coverage constrains our ability to fully characterize the three-dimensional structure and cross-frontal gradients of the oceanic front. The observed atmospheric responses also represent point measurements that may not accurately reflect the spatial heterogeneity of air–sea exchange processes across the frontal zone. Future investigations would benefit from spatially distributed observational arrays or cross-frontal transects to better resolve the horizontal scale dependence of frontal impacts on atmospheric boundary-layer dynamics. In addition, the observed frontal event occurred during the spring-to-summer transition period, when the South China Sea experiences significant changes in monsoon patterns and thermal stratification [40,42,43]. The relationship between frontal intensity and seasonal atmospheric forcing remains unclear from our limited temporal dataset.

Author Contributions: Conceptualization, H.Y. and R.Z.; software, R.Z.; validation, J.Y., X.Z. and H.Y.; data curation, X.M.; writing—original draft preparation, R.Z.; writing—review and editing, J.Y. and X.Z.; visualization, R.Z.; supervision, X.M.; project administration, H.Y.; funding acquisition, X.M. All authors have read and agreed to the published version of the manuscript.

Funding: This research was funded by National Key Research and Development Program of China (2022YFC3104801 and 2022YFC3104205), Laoshan Laboratory Science and Technology Innovation Project (LSKJ202202503) and National Natural Science Foundation of China (42306020).

Data Availability Statement: The OSITA sea surface temperature and surface wind-field data are taken from the CMEMS website (Copernicus Marine Environment Monitoring Service, https://doi.org/10.48670/moi-00165 and https://doi.org/10.48670/moi-00305, respectively) (accessed on 11 May 2025). Ocean-bottom depth data are obtained from ETOPO 2022 (https://www.ncei.noaa.gov/products/etopo-global-relief-model) (accessed on 1 October 2023). The buoy observation data are available from the corresponding author upon reasonable request.

Acknowledgments: All the figures were created using Matlab software version 24.2.0 (2024b) (https://www.mathworks.com; accessed on 18 January 2025).

Conflicts of Interest: The authors declare no conflicts of interest.

References

1. Bowman, M.J. Introduction and Historical Perspective. In *Proceedings of the Oceanic Fronts in Coastal Processes*; Bowman, M.J., Esaias, W.E., Eds.; Springer: Berlin/Heidelberg, Germany, 1978; pp. 2–5.
2. Woodson, C.B.; Litvin, S.Y. Ocean Fronts Drive Marine Fishery Production and Biogeochemical Cycling. *Proc. Natl. Acad. Sci. USA* **2015**, *112*, 1710–1715. [CrossRef]
3. Greer, A.T.; Cowen, R.K.; Guigand, C.M.; Hare, J.A. Fine-Scale Planktonic Habitat Partitioning at a Shelf-Slope Front Revealed by a High-Resolution Imaging System. *J. Mar. Syst.* **2015**, *142*, 111–125. [CrossRef]
4. Chu, P.C.; Wang, G. Seasonal Variability of Thermohaline Front in the Central South China Sea. *J. Oceanogr.* **2003**, *59*, 65–78. [CrossRef]

5. Thomas, L.N.; Taylor, J.R.; Ferrari, R.; Joyce, T.M. Symmetric Instability in the Gulf Stream. *Deep Sea Res. Part II* **2013**, *91*, 96–110. [CrossRef]
6. Qiu, C.; He, B.; Wang, D.; Zou, Z.; Tang, H. Mechanisms of a Shelf Submesoscale Front in the Northern South China Sea. *Deep Sea Res. Part I Oceanogr. Res. Pap.* **2023**, *202*, 104197. [CrossRef]
7. Belkin, I.M.; Cornillon, P.C.; Sherman, K. Fronts in Large Marine Ecosystems. *Prog. Oceanogr.* **2009**, *81*, 223–236. [CrossRef]
8. Zhu, R.; Yang, H.; Li, M.; Chen, Z.; Ma, X.; Cai, J.; Wu, L. Observations Reveal Vertical Transport Induced by Submesoscale Front. *Sci. Rep.* **2024**, *14*, 4407. [CrossRef]
9. Hayes, S.P.; McPhaden, M.J.; Wallace, J.M. The Influence of Sea-Surface Temperature on Surface Wind in the Eastern Equatorial Pacific: Weekly to Monthly Variability. *J. Clim.* **1989**, *2*, 1500–1506. [CrossRef]
10. Wallace, J.M.; Mitchell, T.P.; Deser, C. The Influence of Sea-Surface Temperature on Surface Wind in the Eastern Equatorial Pacific: Seasonal and Interannual Variability. *J. Clim.* **1989**, *2*, 1492–1499. [CrossRef]
11. Kelly, K.A.; Dickinson, S.; McPhaden, M.J.; Johnson, G.C. Ocean Currents Evident in Satellite Wind Data. *Geophys. Res. Lett.* **2001**, *28*, 2469–2472. [CrossRef]
12. Chelton, D.B.; Schlax, M.G.; Samelson, R.M. Summertime Coupling between Sea Surface Temperature and Wind Stress in the California Current System. *J. Phys. Oceanogr.* **2007**, *37*, 495–517. [CrossRef]
13. Small, R.J.; deSzoeke, S.P.; Xie, S.P.; O'Neill, L.; Seo, H.; Song, Q.; Cornillon, P.; Spall, M.; Minobe, S. Air–Sea Interaction over Ocean Fronts and Eddies. *Dyn. Atmos. Ocean.* **2008**, *45*, 274–319. [CrossRef]
14. Lindzen, R.S.; Nigam, S. On the Role of Sea Surface Temperature Gradients in Forcing Low-Level Winds and Convergence in the Tropics. *J. Atmos. Sci.* **1987**, *44*, 2418–2436. [CrossRef]
15. Xie, S.-P. Satellite Observations of Cool Ocean–Atmosphere Interaction. *Bull. Am. Meteorol. Soc.* **2004**, *85*, 195–208. [CrossRef]
16. Tanimoto, Y.; Kanenari, T.; Tokinaga, H.; Xie, S.-P. Sea Level Pressure Minimum along the Kuroshio and Its Extension. *J. Clim.* **2011**, *24*, 4419–4434. [CrossRef]
17. O'Neill, L.W.; Chelton, D.B.; Esbensen, S.K. The Effects of SST-Induced Surface Wind Speed and Direction Gradients on Midlatitude Surface Vorticity and Divergence. *J. Clim.* **2010**, *23*, 255–281. [CrossRef]
18. Minobe, S.; Kuwano-Yoshida, A.; Komori, N.; Xie, S.-P.; Small, R.J. Influence of the Gulf Stream on the Troposphere. *Nature* **2008**, *452*, 206–209. [CrossRef]
19. Young, G.S.; Sikora, T.D. Mesoscale Stratocumulus Bands Caused by Gulf Stream Meanders. *Mon. Weather Rev.* **2003**, *131*, 2177–2191. [CrossRef]
20. Held, I.M.; Ting, M.; Wang, H. Northern Winter Stationary Waves: Theory and Modeling. *J. Clim.* **2002**, *15*, 2125–2144. [CrossRef]
21. Smahrt, L.; Vickers, D.; Moore, E. Flow Adjustments Across Sea-Surface Temperature Changes. *Bound. -Layer Meteorol.* **2004**, *111*, 553–564. [CrossRef]
22. Yang, H.; Chen, Z.; Sun, S.; Li, M.; Cai, W.; Wu, L.; Cai, J.; Sun, B.; Ma, K.; Ma, X.; et al. Observations Reveal Intense Air-Sea Exchanges Over Submesoscale Ocean Front. *Geophys. Res. Lett.* **2024**, *51*, e2023GL106840. [CrossRef]
23. Zhu, R.; Li, M.; Yang, H.; Ma, X.; Chen, Z. Oceanic Eddy with Submesoscale Edge Drives Intense Air-Sea Exchanges and Beyond. *Sci. Rep.* **2024**, *14*, 25183. [CrossRef]
24. Qu, T.; Du, Y.; Gan, J.; Wang, D. Mean Seasonal Cycle of Isothermal Depth in the South China Sea. *J. Geophys. Res.* **2007**, *112*, C02020. [CrossRef]
25. Liu, Q.; Kaneko, A.; Jilan, S. Recent Progress in Studies of the South China Sea Circulation. *J. Oceanogr.* **2008**, *64*, 753–762. [CrossRef]
26. Tian, J.; Yang, Q.; Zhao, W. Enhanced Diapycnal Mixing in the South China Sea. *J. Phys. Oceanogr.* **2009**, *39*, 3191–3203. [CrossRef]
27. Ramp, S.R.; Yang, Y.J.; Bahr, F.L. Characterizing the Nonlinear Internal Wave Climate in the Northeastern South China Sea. *Nonlinear Process. Geophys.* **2010**, *17*, 481–498. [CrossRef]
28. Chen, G.; Hou, Y.; Chu, X. Mesoscale Eddies in the South China Sea: Mean Properties, Spatiotemporal Variability, and Impact on Thermohaline Structure. *J. Geophys. Res.* **2011**, *116*, C06018. [CrossRef]
29. Lin, X.; Dong, C.; Chen, D.; Liu, Y.; Yang, J.; Zou, B.; Guan, Y. Three-Dimensional Properties of Mesoscale Eddies in the South China Sea Based on Eddy-Resolving Model Output. *Deep Sea Res. Part I Oceanogr. Res. Pap.* **2015**, *99*, 46–64. [CrossRef]
30. Zhang, Z.; Tian, J.; Qiu, B.; Zhao, W.; Chang, P.; Wu, D.; Wan, X. Observed 3D Structure, Generation, and Dissipation of Oceanic Mesoscale Eddies in the South China Sea. *Sci. Rep.* **2016**, *6*, 24349. [CrossRef] [PubMed]
31. Fang, G.H. A Survey of Studies on the South China Sea Upper Ocean Circulation. *Acta Oceanogr. Taiwanica* **1998**, *37*, 1–16.
32. Wang, G.; Li, J.; Wang, C.; Yan, Y. Interactions among the Winter Monsoon, Ocean Eddy and Ocean Thermal Front in the South China Sea. *J. Geophys. Res.* **2012**, *117*, C08002. [CrossRef]
33. Liu, K.-K.; Chao, S.-Y.; Shaw, P.-T.; Gong, G.-C.; Chen, C.-C.; Tang, T.Y. Monsoon-Forced Chlorophyll Distribution and Primary Production in the South China Sea: Observations and a Numerical Study. *Deep Sea Res. Part I Oceanogr. Res. Pap.* **2002**, *49*, 1387–1412. [CrossRef]

34. Dong, J.; Zhong, Y. Submesoscale Fronts Observed by Satellites over the Northern South China Sea Shelf. *Dyn. Atmos. Ocean.* **2020**, *91*, 101161. [CrossRef]
35. Wang, D. Air-Sea Interaction in the South China Sea. In *Ocean Circulation and Air-Sea Interaction in the South China Sea*; Wang, D., Ed.; Springer Nature Singapore: Singapore, 2022; pp. 307–394. ISBN 978-981-19-6262-2.
36. Wang, D.; Zhuang, W.; Xie, S.-P.; Hu, J.; Shu, Y.; Wu, R. Coastal Upwelling in Summer 2000 in the Northeastern South China Sea. *J. Geophys. Res.* **2012**, *117*, C04009. [CrossRef]
37. Hu, J.Y.; Kawamura, H.; Tang, D.L. Tidal Front around the Hainan Island, Northwest of the South China Sea. *J. Geophys. Res.* **2003**, *108*, 3342. [CrossRef]
38. Wang, D.; Liu, Y.; Qi, Y.; Shi, P. Seasonal Variability of Thermal Fronts in the Northern South China Sea from Satellite Data. *Geophys. Res. Lett.* **2001**, *28*, 3963–3966. [CrossRef]
39. Jing, Z.; Qi, Y.; Du, Y.; Zhang, S.; Xie, L. Summer Upwelling and Thermal Fronts in the Northwestern South China Sea: Observational Analysis of Two Mesoscale Mapping Surveys. *J. Geophys. Res. Ocean.* **2015**, *120*, 1993–2006. [CrossRef]
40. Jing, Z.; Qi, Y.; Fox-Kemper, B.; Du, Y.; Lian, S. Seasonal Thermal Fronts on the Northern South China Sea Shelf: Satellite Measurements and Three Repeated Field Surveys. *J. Geophys. Res. Ocean.* **2016**, *121*, 1914–1930. [CrossRef]
41. Ren, S.; Zhu, X.; Drevillon, M.; Wang, H.; Zhang, Y.; Zu, Z.; Li, A. Detection of SST Fronts from a High-Resolution Model and Its Preliminary Results in the South China Sea. *J. Atmos. Ocean. Technol.* **2021**, *38*, 387–403. [CrossRef]
42. Chen, J.; Hu, Z. Seasonal Variability in Spatial Patterns of Sea Surface Cold- and Warm Fronts over the Continental Shelf of the Northern South China Sea. *Front. Mar. Sci.* **2023**, *9*, 1100772. [CrossRef]
43. Shi, R.; Guo, X.; Wang, D.; Zeng, L.; Chen, J. Seasonal Variability in Coastal Fronts and Its Influence on Sea Surface Wind in the Northern South China Sea. *Deep Sea Res. Part II: Top. Stud. Oceanogr.* **2015**, *119*, 30–39. [CrossRef]
44. Shi, R.; Chen, J.; Guo, X.; Zeng, L.; Li, J.; Xie, Q.; Wang, X.; Wang, D. Ship Observations and Numerical Simulation of the Marine Atmospheric Boundary Layer over the Spring Oceanic Front in the Northwestern South China Sea. *J. Geophys. Res. Atmos.* **2017**, *122*, 3733–3753. [CrossRef]
45. Hogg, N.G. On the Correction of Temperature and Velocity Time Series for Mooring Motion. *J. Atmos. Ocean. Technol.* **1986**, *3*, 204–214. [CrossRef]
46. Wang, Y.; Yu, Y.; Zhang, Y.; Zhang, H.-R.; Chai, F. Distribution and Variability of Sea Surface Temperature Fronts in the South China Sea. *Estuar. Coast. Shelf Sci.* **2020**, *240*, 106793. [CrossRef]
47. Edson, J.B.; Jampana, V.; Weller, R.A.; Bigorre, S.P.; Plueddemann, A.J.; Fairall, C.W.; Miller, S.D.; Mahrt, L.; Vickers, D.; Hersbach, H. On the Exchange of Momentum over the Open Ocean. *J. Phys. Oceanogr.* **2013**, *43*, 1589–1610. [CrossRef]
48. Chelton, D.B.; Schlax, M.G.; Freilich, M.H.; Milliff, R.F. Satellite Measurements Reveal Persistent Small-Scale Features in Ocean Winds. *Science* **2004**, *303*, 978–983. [CrossRef] [PubMed]

remote sensing

MDPI

Article

Coastal Eddy Detection in the Balearic Sea: SWOT Capabilities

Laura Fortunato [1,*], Laura Gómez-Navarro [2], Vincent Combes [2,3], Yuri Cotroneo [1], Giuseppe Aulicino [1] and Ananda Pascual [2]

1 Dipartimento di Scienze e Tecnologie, Università degli Studi di Napoli Parthenope, 80143 Naples, Italy; yuri.cotroneo@uniparthenope.it (Y.C.); giuseppe.aulicino@uniparthenope.it (G.A.)
2 Institut Mediterrani d'Estudis Avançats (IMEDEA), Spanish National Research Council–University of the Balearic Islands (CSIC–UIB), 07190 Esporles, Spain; laura.gomez@uib.es (L.G.-N.); vcombes@imedea.uib-csic.es (V.C.); ananda.pascual@imedea.uib-csic.es (A.P.)
3 Departamento de Física, Universitat de les Illes Balears, 07120 Palma de Mallorca, Spain
* Correspondence: laura.fortunato001@studenti.uniparthenope.it

Abstract

Mesoscale coastal eddies are key components of ocean circulation, mediating the transport of heat, nutrients, and marine debris. The Surface Water and Ocean Topography (SWOT) mission provides high-resolution sea surface height data, offering a novel opportunity to improve the observation and characterization of these features, especially in coastal regions where conventional altimetry is limited. In this study, we investigate a mesoscale anticyclonic coastal eddy observed southwest of Mallorca Island, in the Balearic Sea, to assess the impact of SWOT-enhanced altimetry in resolving its structure and dynamics. Initial eddy identification is performed using satellite ocean color imagery, followed by a qualitative and quantitative comparison of multiple altimetric datasets, ranging from conventional nadir altimetry to wide-swath products derived from SWOT. We analyze multiple altimetric variables—Sea Level Anomaly, Absolute Dynamic Topography, Velocity Magnitude, Eddy Kinetic Energy, and Relative Vorticity—highlighting substantial differences in spatial detail and intensity. Our results show that SWOT-enhanced observations significantly improve the spatial characterization and dynamical depiction of the eddy. Furthermore, Lagrangian transport simulations reveal how altimetric resolution influences modeled transport pathways and retention patterns. These findings underline the critical role of SWOT in advancing the monitoring of coastal mesoscale processes and improving our ability to model oceanic transport mechanisms.

Keywords: mesoscale eddy; ocean dynamics; satellite altimetry; Surface Water and Ocean Topography (SWOT); Lagrangian simulations

1. Introduction

Mesoscale and submesoscale oceanic structures—such as eddies, fronts, and filaments—play a crucial role in transporting heat, salt, and biogeochemical properties between coastal regions and the open ocean [1–11]. These dynamic features account for a significant portion of the ocean's kinetic energy, and they contribute to the redistribution of tracers, such as nutrients, heat, and carbon, across different spatial and temporal scales [12–17]. Beyond their role in physical circulation, eddies play a significant part in ecological and environmental processes by facilitating the transport and retention of pollutants [18], plankton [19–21], and marine larvae [22–24]. Moreover, eddies influence the behavior and migration patterns of marine animals, including turtles [25], sharks [26], and birds [27–29].

Eddies are commonly detected using satellite altimetry data, which provides valuable observations of Sea Surface Height (SSH) at a global scale. Altimetric data has proven highly effective in tracking large-scale oceanic eddies and other mesoscale features [1,30]. In addition to SSH, satellite-derived sea surface temperature and ocean color imagery offer useful visual cues for identifying eddies. However, these surface signatures often provide only qualitative insights into the underlying dynamics [31,32]. While these observations help identify eddy structures, further quantitative analyses are needed to understand the dynamics of the associated currents and subsurface features.

Despite the advantages of altimetry, conventional nadir altimetric products are often limited by spatial resolution, particularly in coastal and regional areas where mesoscale features are more pronounced [33]. The lower resolution of these datasets can result in the underrepresentation of small-scale structures such as coastal eddies, which are critical for understanding local ocean dynamics [34]. To mitigate this issue, the integration of data from multiple altimetry missions has proven to be an effective strategy for enhancing the resolution and accuracy of sea level measurements. Studies in the Mediterranean Sea, for instance, have shown that merging multi-mission altimetric data significantly improves the detection and characterization of mesoscale features [35,36].

The SWOT (Surface Water and Ocean Topography) mission represents a major technological advancement, offering unprecedented spatial resolution capable of resolving mesoscale and submesoscale ocean structures, including coastal zones [37,38]. Previous efforts, including the FaSt-SWOT project, have validated SWOT data in the western Mediterranean Sea, demonstrating its ability to capture fine-scale coastal dynamics with high accuracy through combined observational and numerical approaches [39–43].

In light of these advances, the present study focuses on the Balearic Sea, a region characterized by a variety of mesoscale oceanic structures, including anticyclonic eddies previously identified in the literature [44–49]. This study specifically examines a coastal eddy southwest of Mallorca Island, utilizing the high-resolution capabilities of the SWOT mission. By uniquely combining high-resolution SWOT altimetry with cloud-free ocean color imagery over a focused 5-day period, we investigate the dynamics of this well-defined coastal eddy. Integrating these complementary datasets allows us to provide novel insights into fine-scale variability and demonstrate SWOT's added value compared to traditional altimetric products in this region. The 5-day period from 23 June 2023 to 27 June 2023 was carefully selected to ensure uninterrupted satellite coverage, essential for a consistent and meaningful comparison between altimetric and optical data. The eddy under investigation is clearly visible in the ocean color imagery throughout this period, justifying its selection for detailed analysis.

While the core analysis focuses on these five days, the eddy was detectable in altimetric velocity fields as early as 11 June and remained coherent until around 1 July, with signs of weakening thereafter. The selected period corresponds to its mature phase, during which the structure and trapping behavior were most clearly expressed. The aim is to demonstrate the added value of SWOT observations over traditional altimetric products in resolving small-scale ocean features [50,51]. The aim is to demonstrate the added value of SWOT observations over traditional altimetric products in resolving small-scale ocean features [50,51].

The paper is organized as follows: First, we present the characteristics of the study area. Second, we describe the data and methods used, including satellite ocean color images, altimetric data, and Lagrangian simulations. Third, we present the results of our analysis, focusing on the identification and dynamics of the coastal eddy southwest of Mallorca. Finally, we discuss the implications of these results and the advantages and limitations of the different altimetric products in resolving coastal eddies and their associated transport dynamics.

2. Study Area

The Balearic Sea (Figure 1a) is a sub-basin of the western Mediterranean, located between the Iberian Peninsula, the Gulf of Lion, and the Balearic Islands (Ibiza, Mallorca, and Menorca). This region was selected for the present SWOT-based study because it hosts a range of mesoscale and submesoscale oceanic structures, including a well-defined coastal eddy southwest of Mallorca that exhibits clear surface signatures in both altimetric and ocean color data. The dynamic complexity of this coastal area, shaped by interactions between ocean currents and bathymetry, makes it a challenging environment for traditional altimetric products and thus ideal to highlight SWOT's enhanced spatial resolution and ability to resolve fine-scale coastal features.

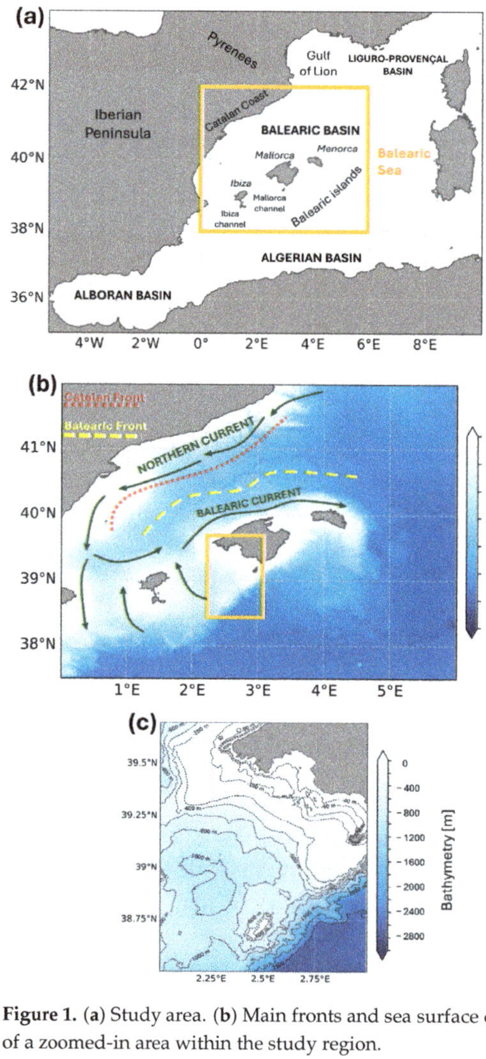

Figure 1. (**a**) Study area. (**b**) Main fronts and sea surface currents in the Balearic Sea. (**c**) Bathymetry of a zoomed-in area within the study region.

To the north, it is bounded by the Liguro–Provençal Basin, an area influenced by strong atmospheric forcings, while to the south, it is bounded by the Algerian Basin, which is dominated by intense mesoscale eddies and their interactions with the unstable Algerian current [46–48,52]. Consequently, the Balearic Basin can be considered a transitional region between the northern (Gulf of Lion) and southern (Algerian Sea) sections of the Western Basin's cyclonic [52–54].

The general surface circulation of the Balearic Sea is mainly controlled by two permanent density fronts: the Catalan front, over the Iberian Peninsula slope, and the Balearic front, located over the insular slope (Figure 1b). The Catalan Front is a shelf/slope front that separates old Atlantic Water (AW) in the central Balearic subbasin from the less dense water transported by the Northern Current (NC), which, while also consisting of old AW, is influenced by the influx of continental freshwater from the Gulf of Lion and the Catalan shelves. The NC is a density-driven coastal current that flows southwestward from the Ligurian Sea into the Balearic Sea, where it either exits through the Ibiza Channel or retroflects cyclonically over the insular slope, forming the Balearic Current (BC). This current flows southward along the continental slope and is further fed by warm, fresh, recently modified AW from the Algerian Basin, entering through the Mallorca and Ibiza channels. The Balearic Front is a slope front formed by the recently modified AW entering the basin through the southern channels [55], separating older AW in the central basin from the less dense waters transported by the Balearic Current.

Beyond the general circulation at the basin scale, the Balearic subbasin is characterized by distinct frontal dynamics near the slope regions, particularly between the BC and the NC. These dynamics include the formation of mesoscale eddies [56–58], filaments, and changes in shelf-slope flows [55,59]. Such processes are known to influence not only local dynamics, leading to significant vertical motions [60], but also broader circulation patterns, as demonstrated by Pascual et al. [48] in their study of the blocking effect caused by a large anti-cyclonic eddy. Smaller-scale coastal eddies, such as the one analyzed in this study, often originate from instabilities in these frontal zones or interactions with complex bathymetry along the insular and continental slopes. These features are short-lived and spatially confined, yet they can play a disproportionate role in shaping local transport and mixing. Their detection and analysis require high-resolution observations, as their dynamics are often missed or smoothed out in conventional altimetric products.

3. Materials and Methods

3.1. Satellite Ocean Color

The approach of this study began with a qualitative analysis using ocean color data, particularly chlorophyll-a (Chla) concentration, to observe the presence of eddies. This analysis was based on a multi-modal satellite dataset from the Copernicus Marine Environment Monitoring Service (CMEMS, marine.copernicus.eu), specifically the *Mediterranean Sea, Bio-Geo-Chemical, L3, daily Satellite Observations* product (OCEANCOLOUR_MED_BGC_L3_MY_009_143), which provides biogeochemical data at a Level 3 (L3) processing level [60]. The dataset, characterized by daily images at a 1 km spatial resolution, integrates observations from several widely used satellites, including Sea-Viewing Wide Field-of-View Sensor (SeaWiFS); Moderate-Resolution Imaging Spectroradiometer (MODIS); Medium-Resolution Imaging Spectrometer (MERIS); Visible Infrared Imaging Radiometer Suite–Suomi National Polar-orbiting Partnership (VIIRS-SNPP); Joint Polar Satellite System-1 (JPSS1); and Ocean and Land Colour Instrument—Sentinel-3A and Sentinel-3B (OLCI-S3A and S3B). The product is available through DOI: https://doi.org/10.48670/moi-00299.

3.2. Satellite Altimetry

Satellite altimetry data were used to derive the mean surface geostrophic circulation and analyze mesoscale features using a combination of conventional and SWOT-enhanced datasets. The study utilized multiple Level 3 and Level 4 products, including along-track and gridded datasets, encompassing near-real-time and experimental data incorporating SWOT's Ka-band Radar Interferometer (KaRIn) wide-swath observations.

- The conventional product used is the CMEMS product *European Seas Gridded L4 Sea Surface Heights and Derived Variables NRT* (SEALEVEL_EUR_PHY_L4_NRT_008_060, DOI: https://doi.org/10.48670/moi-00142), processed by the DUACS (Data Unification and Altimeter Combination System) multimission altimeter data processing system. This product provides daily gridded Level-4 (L4) Sea Level Anomaly (SLA) maps for European Seas at $1/8° \times 1/8°$ spatial resolution, computed relative to a 20-year reference period (1993–2012). The processing methodology is based on Optimal Interpolation, integrating measurements from 10 altimeter missions (TOPEX/Poseidon, Jason series, ERS, ENVISAT, GFO, CryoSat-2, Saral/AltiKa, Haiyang-2A) as detailed in Pujol et al. [61].
- The *Experimental multimission gridded L4 sea level heights and velocities with SWOT* is a gridded product derived from the along-track (or Level-3) SEA LEVEL products (DOI: doi.org/10.48670/moi-00147) provided by CMEMS for the satellites SARAL/AltiKa, CryoSat-2, HaiYang-2B, Jason-3, Copernicus Sentinel-3A & 3B, Sentinel-6A, SWOT nadir, and SWOT Level-3 KaRIn sea level products (DOI: https://doi.org/10.24400/527896/A01-2023.018). The product is processed by SSALTO/DUACS and distributed by AVISO (https://www.aviso.altimetry.fr, accessed on 20 November 2024), supported by CNES (version 1.0.0) (DOI: https://doi.org/10.24400/527896/A01-2024.007). This dataset has daily temporal resolution and a spatial resolution of $1/10° \times 1/10°$. It uses the MIOST (Multiscale Interpolation Ocean Science Topography) approach [62,63], which models various modes of ocean surface topography variability to improve the representation of mesoscale ocean variability [64].
- Additionally, the study used *European Seas Along Track L3 Sea Surface Heights Reprocessed 1993–Ongoing Tailored For Data Assimilation* (SEALEVEL_EUR_PHY_L3_MY_008_061, DOI: https://doi.org/10.48670/moi-00139), processed by the DUACS multimission altimeter data processing system. This product provides along-track, Level-3 (L3) sea surface height (SSH) observations reprocessed for consistency and optimized for data assimilation applications. It processes data from all altimeter missions available (e.g., TOPEX/Poseidon, Jason-1/2/3, Sentinel-3A/B, HaiYang-2A/B, etc.) and covers the European Seas from 1993 to the present. This dataset includes SWOT-nadir Calibration/Validation (CalVal) data used to validate SWOT KaRIn measurements and support mesoscale structure interpretation.
- Finally, the *SWOT Expert Level 3 Low-Rate Sea Surface Height* (L3_LR_SSH_Expert) product (DOI: https://doi.org/10.24400/527896/A01-2023.018) is a gridded ocean topography dataset derived from Level-2 KaRIn and nadir altimetry observations collected by the SWOT satellite. This expert-level product includes a variety of key variables for oceanographic and geodetic research, such as Sea Surface Height Anomaly (SSHA), Mean Dynamic Topography (MDT), geostrophic currents (both absolute and anomalies), backscatter coefficient (sigma0), and the Mean Sea Surface (MSS). Additionally, it integrates quality flags, altimetric corrections, and external model outputs as separate layers, ensuring data accuracy and flexibility. The data are mapped onto a regular grid with a spatial resolution of approximately 0.05° (~5 km), covering the KaRIn swath. This product is derived from the L2 SWOT KaRIn low-rate ocean data products (NASA/JPL and CNES) and is produced and made freely available by the

AVISO and DUACS teams as part of the DESMOS Science Team project. It is designed to offer high-resolution global ocean surface topography measurements, which are particularly useful for studying mesoscale ocean variability.

For clarity and consistency throughout this study, the four satellite altimetry datasets will be referred to as follows: *DUACS CMEMS* for the near-real-time gridded L4 product (SEALEVEL_EUR_PHY_L4_NRT_008_060); *MIOST AVISO+SWOT* for the experimental gridded L4 product incorporating SWOT KaRIn data; *Along-Track SWOT L3* for the reprocessed along-track L3 product; and *SWOT L3* for the SWOT Level 3 Ocean Products. For a detailed summary of the datasets used in this study, see Table 1.

Table 1. Summary of satellite altimetry datasets used in this study. See text for further details.

Abbreviation	Full Product Name	Processing Level	Temporal Resolution	Spatial Resolution	Source/Processor
DUACS CMEMS	European Seas Gridded L4 Sea Surface Heights and Derived Variables (NRT)	L4	Daily	$1/8° \times 1/8°$	CMEMS/DUACS
MIOST AVISO+SWOT	Experimental multimission gridded L4 sea level heights and velocities with SWOT	L4	Daily	$1/10° \times 1/10°$	CMEMS, AVISO, SSALTO/DUACS, CNES
Along-Track SWOT L3	European Seas Along Track L3 Sea Surface Heights Reprocessed—Tailored for Data Assimilation	L3	Along-track (varies by altimeter)	Along-track (varies by altimeter)	CMEMS, DUACS, AVISO, CNES
SWOT L3	SWOT Level3 experimental products	L3	Daily	2 km $\times$ 2 km, covering the KaRIn swath	AVISO, DUACS, NASA/JPL, CNES

3.3. Altimetric Data Coverage and Uncertainty

To better illustrate the spatial coverage and quality of the altimetric data used in this study, Figure 2 provides an overview of the observations available in the Balearic Sea during the study period (23–26 June 2023). The first panel (Figure 2a) shows the distribution of altimetric tracks from the four satellites active in the region at that time—SWOT, HaiYang-2B, Sentinel-3B, and Sentinel-6A. The color of each point represents the Sea Level Anomaly (SLA) value recorded along the satellite ground tracks. This visualization highlights the dense spatial sampling achieved during the observation window, particularly due to the contribution of SWOT's wide-swath coverage. Notably, the presence of SWOT enhances the resolution near coastal areas, where traditional altimetry often struggles with sparse sampling and land contamination.

The observations from these satellites directly contribute to the gridded SLA field from the DUACS CMEMS product shown in the second panel (Figure 2b), which represents the midpoint of the analysis period (25 June 2023). This multi-mission product combines data from all available satellite tracks to generate a continuous representation of sea level anomalies. While the overall circulation patterns are coherent, the product offers a relatively smooth representation of mesoscale features. Fine-scale gradients and eddy structures, particularly in the vicinity of the coast, appear less distinct compared to the along-track data, suggesting some limitations in the gridded product's ability to resolve submesoscale variability. The third panel (Figure 2c) displays the corresponding gridded error field for the DUACS CMEMS product. Here, increased uncertainty is evident along coastal zones and dynamically active regions, confirming the known limitations of conventional altimetry in areas with complex topography and proximity to land. These spatial patterns of observational uncertainty further reinforce the need for higher-resolution observations, such as those provided by SWOT, in order to improve the detection and analysis of coastal eddy dynamics.

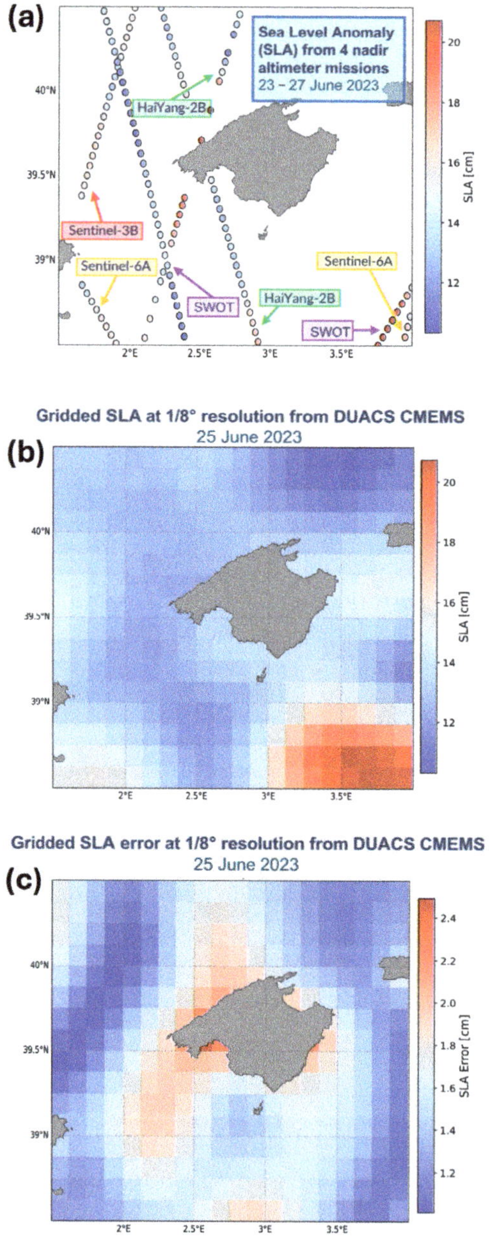

Figure 2. (**a**) Sea Level Anomaly (SLA) from 4 nadir altimeter missions from 23 to 27 June. (**b**) Gridded SLA at 1/8° resolution from DUACS CMEMS product. (**c**) Gridded SLA error at 1/8° resolution from DUACS CMEMS product.

3.4. Lagrangian Simulator for Particle Tracking

Particle tracking was performed using OceanParcels (Probably A Really Computationally Efficient Lagrangian Simulator), a Python toolbox specifically designed for tracking particles using output from ocean general circulation models [65]. This tool can simulate the tracking of both passive particles, such as water and plastic, and active particles, like plankton and fish.

OceanParcels v3.1.1 [66] was employed to simulate surface particle trajectories in the northwestern Mediterranean Sea, between 2.0°E and 3.0°E and between 38.5°N and 39.7°N. Particles were released once on 23 June 2023 at 00:00 UTC and advected forward in time until 27 June 2023 at 00:00 UTC, for a total of 5 days.

Particles were initialized randomly across the sea surface, with 10 particles per sea-grid cell, resulting in a total of 10,180 particles. Sea-cells were defined using a custom sea-mask based on Absolute Dynamic Topography (ADT) to exclude land-influenced areas. The velocity field used in the simulation was derived from ADT altimetry data, and the simulation was restricted to the surface layer (2D simulation).

The simulation proceeded by using the advection-only method without incorporating any diffusion terms. Particles were advected with a timestep (Δt) of 1 min, meaning each particle's position was updated every 60 s based on the velocity fields. The particles were advected according to the Eulerian velocity field, and their displacement was calculated using the fourth-order Runge–Kutta (RK4) method [67], which ensures accurate interpolation of the velocity fields and smooth particle tracking. The advection equation is defined as

$$x(t + \Delta t) = x(t) + \int_{t}^{t+\Delta t} v(x(\tau), \tau)\, d\tau \tag{1}$$

where $x(t)$ represents the position of the particle at time t and v is the velocity field (u, v, w) interpolated from the ocean data. This equation is solved using the RK4 advection scheme as implemented in OceanParcels [66]. The simulation setup does not include explicit diffusion terms; thus, particle motion is solely governed by advection from the geostrophic velocity field derived from altimetry. However, subgrid-scale dispersion is implicitly represented through the particle release strategy by releasing 10 particles per grid cell. Particles reaching land were halted and remained stationary. It is important to note that OceanParcels was used primarily to illustrate the trapping behavior of particles within the mesoscale eddy. The simulations were limited to the surface layer (2D) and involved passive particles without vertical mixing, diffusion, or active behaviors. These simplifications may underestimate actual dispersion, particularly in coastal zones affected by turbulence and vertical processes. Moreover, the particle trajectories strongly depend on the quality of the underlying geostrophic velocity fields derived from ADT. For mesoscale eddies with Ro > 0.1, cyclostrophic corrections can reach 0.5 m/s and significantly improve velocity estimates [68], indicating a potential source of error in our trajectory simulation.

Non-geostrophic components such as wind-driven Ekman flows, tidal currents, and unbalanced motions (e.g., cyclostrophic flows or submesoscale dynamics) were not included due to data limitations. Such processes may influence near-surface transport, particularly in coastal and submesoscale regimes [69,70], and their exclusion represents a potential source of uncertainty in particle trajectory estimates [71].

While these processes may influence transport in the study region, especially near the coast or in shallow areas, their omission does not compromise the main objective of the Lagrangian analysis, which was to illustrate the eddy's retention capability.

4. Results

4.1. Comparison Between Altimetric Products

Several eddies were observed in the Balearic Sea using Chla ocean color imagery from 23 to 27 June 2023. Further analyses were performed using two satellite altimetry products (DUACS CMEMS and MIOST AVISO+SWOT) to investigate a mesoscale anticyclonic coastal eddy located southwest of Mallorca Island. In this section, we compare the performance of both products in capturing fine-scale oceanic features by examining key dynamic parameters: Velocity Magnitude, Eddy Kinetic Energy (EKE), relative vorticity, and Absolute Dynamic Topography (ADT) gradient. This comparison assesses the capability of each product to represent mesoscale and submesoscale dynamics, with a particular focus on the observed coastal eddy. Although the full analysis spans five days (23–27 June), we present results for two representative dates (23 and 26 June) to highlight the eddy's temporal evolution. In addition to differences in magnitude and resolution, differences in the direction of geostrophic velocity vectors are also evident between the two products, particularly at the eddy boundaries and in coastal regions, further supporting the added value of higher-resolution datasets. Chla imagery for the entire study period (Figure S1) is provided in the Supplementary Material.

4.1.1. Absolute Dynamic Topography (ADT)

The comparison between MIOST AVISO+SWOT and DUACS CMEMS data reveals significant differences in the geostrophic Velocity Magnitude pattern. Further comparison with chlorophyll-a concentrations from ocean color imagery (Figure 3a,b) shows spatial coherence between the presence of the eddy and surface chlorophyll distribution, supporting the hypothesis that MIOST AVISO+SWOT is more effective in detecting coastal eddies, where ocean dynamics interact with coastal topography. Specifically, MIOST AVISO+SWOT shows a higher presence of coastal eddies, likely due to its superior spatial resolution, which enables the detection of smaller-scale oceanic structures. This is particularly evident in a well-defined eddy near the southwestern coastal area of Mallorca, visible in the MIOST AVISO+SWOT product but almost absent in the DUACS CMEMS data (Figure 3c,d). In the area around the coastal eddy, the average geostrophic Velocity Magnitude from the MIOST AVISO+SWOT product is 10 cm/s, which is higher than the 6 cm/s recorded by DUACS CMEMS. This difference represents an increase of 4 cm/s, or approximately 72.4%.

This suggests that DUACS CMEMS, with its lower resolution, may not fully capture small-scale oceanic features or may have limited capability to detect fine-scale dynamic details. In terms of Velocity Magnitude, the MIOST AVISO+SWOT data exhibit more pronounced velocity gradients, indicating a higher sensitivity in detecting current variations. The DUACS CMEMS product, on the other hand, shows a less detailed velocity field, with weaker gradients and less-defined structures. While the main eddies are visible in DUACS CMEMS, their intensity and size appear less distinct. In contrast, MIOST AVISO+SWOT provides a more detailed representation, with well-defined contours of ADT anomalies associated with vortical structures and sharper velocity gradients. High-velocity areas (>0.3 m/s) are clearer and better localized in the MIOST AVISO+SWOT maps.

Temporal analysis of the coastal eddy identified in the MIOST AVISO+SWOT data shows spatial stability over several days, suggesting that it is a dynamically relevant structure. These results suggest that MIOST AVISO+SWOT excels in capturing small-scale oceanic variability, particularly in coastal regions. Complete daily maps for the entire study period (Figure S2) are provided in the Supplementary Material.

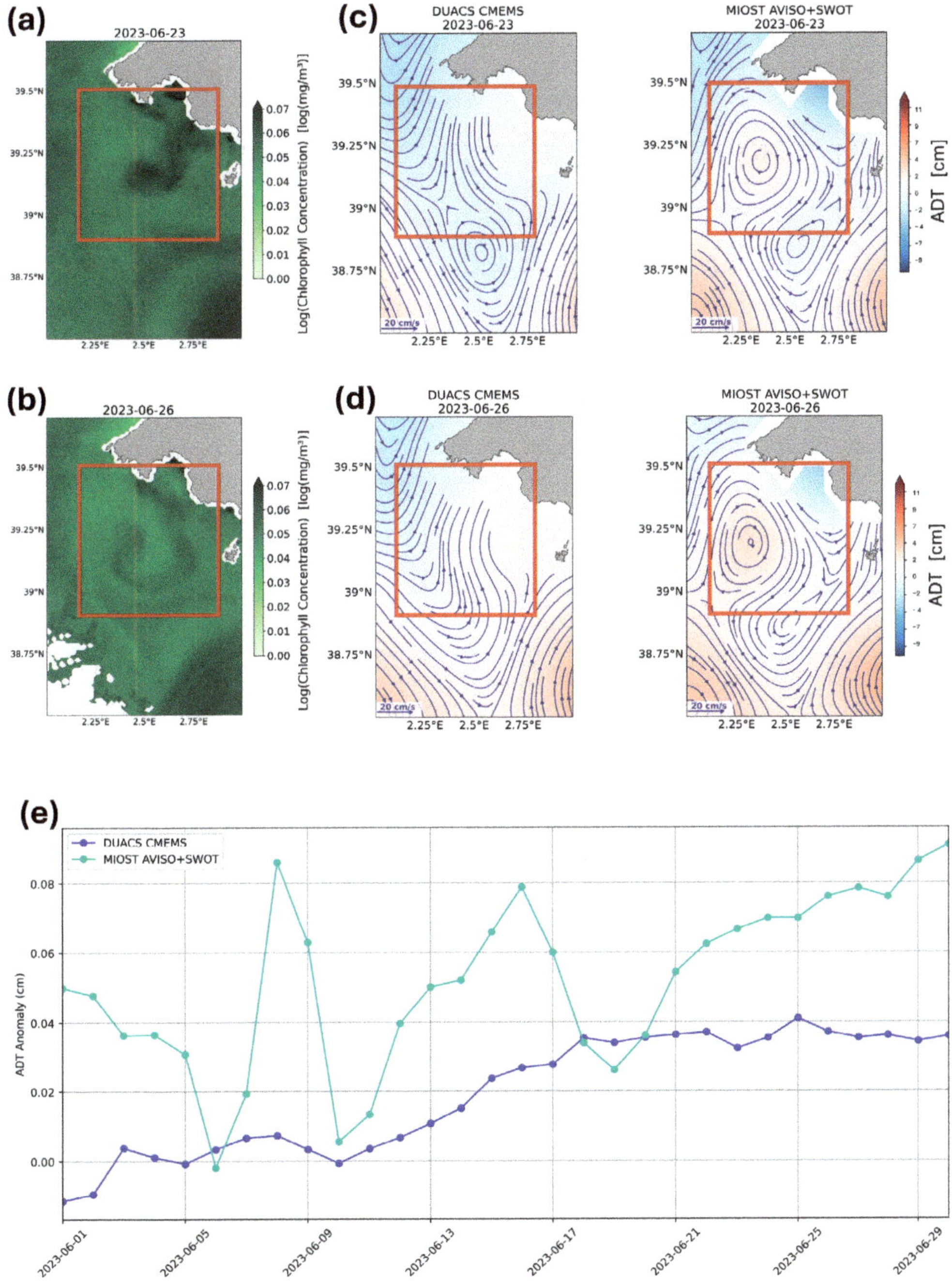

Figure 3. (**a**,**b**) Daily chlorophyll-a concentrations from ocean color imagery on 23 and 26 June 2023 in the Balearic Sea. (**c**,**d**) ADT fields from DUACS CMEMS and MIOST AVISO+SWOT altimetric products for the corresponding dates. (**e**) Time series of the mean ADT within the eddy region (red rectangle) comparing the two altimetric products. The MIOST AVISO+SWOT data show a marked increase in ADT values from 23 June to 27 June, consistent with the presence of an anticyclonic eddy, in contrast to the DUACS CMEMS product, which shows lower and more stable ADT values.

These spatial observations are further supported by the temporal evolution of ADT values within the eddy region, as shown in the time series plot in Figure 3e. Additional analysis of the mean ADT values within the red-boxed eddy region (shown in Figure 3a,b) highlights further discrepancies between the two altimetric products (Figure 3e).

Quantitatively, over the full month of June, the ADT values from the MIOST AVISO+SWOT product within the eddy region ranged from 0.03 to 0.13 cm, with a mean of 0.09 cm and a standard deviation of 0.03 cm. The maximum daily increase recorded was approximately 0.07 cm/day, indicating short-term dynamic variability consistent with the development and evolution of mesoscale features. In contrast, DUACS CMEMS ADT values remained lower and more stable, fluctuating between −0.01 and 0.04 cm, with a mean of 0.02 cm and a standard deviation of 0.02 cm. These numerical descriptors strengthen the visual evidence of enhanced detection capabilities in the MIOST AVISO+SWOT product for coastal and mesoscale dynamics. Between 23 June and 27 June, when the signature of an anticyclonic eddy was clearest, the MIOST AVISO+SWOT product showed the most pronounced ADT increase, reinforcing the spatial evidence of a well-defined eddy structure (Figure 3c,d). On the other hand, the DUACS CMEMS data maintained low and relatively flat ADT throughout this period, reflecting a limited capability to resolve finer eddy signals, likely due to its lower spatial resolution.

4.1.2. Eddy Kinetic Energy

The EKE maps (Figure 4) provide quantitative information on the dynamics of the currents in the study domain. The maps representing the distribution of EKE reveal significant differences in both intensity and spatial details. The comparative analysis shows that MIOST AVISO+SWOT offers higher spatial resolution compared to DUACS CMEMS, capturing with greater detail the energy structures associated with eddies, particularly along the southern coast, around ~2°E–39.5°N, and southwest of Mallorca, where the coastal eddy studied in this article is clearly visible. The coastal eddy visible southwest of Mallorca represents an interesting element for comparing the performance of the DUACS CMEMS and MIOST AVISO+SWOT products. This is especially relevant in the study area, where land proximity modulates eddy dynamics. As previously mentioned, coastal interactions play a crucial role in the observed differences between the two datasets, especially in regions where the proximity to the coastline affects the eddy's structure and energy distribution. The land–sea interface can introduce additional complexities that are better captured by MIOST AVISO+SWOT, which has higher spatial resolution and is more sensitive to such local variations, while DUACS CMEMS, with its coarser resolution, tends to smooth out these effects. From the maps, it is evident that the eddy is detected differently by the two datasets. In the MIOST AVISO+SWOT product, the eddy is depicted with a well-defined spatial structure, characterized by sharp gradients and higher EKE values. Thanks to its higher resolution, MIOST AVISO+SWOT captures sub-mesoscale details, including spatial variations in energy and eddy asymmetry. In contrast, DUACS CMEMS presents a more diffuse representation of the eddy, with lower EKE values and less defined contours. This difference is likely due to a stronger spatial filtering effect in the DUACS CMEMS product, which tends to remove smaller spatial scales and, consequently, reduces sensitivity to sub-mesoscale dynamics and local variations near the coast. While DUACS CMEMS can capture the general location and stability of the eddy, it provides a less detailed view of the energy dynamics, which may be more suitable for large-scale studies. On the other hand, MIOST AVISO+SWOT is better suited for studies requiring the precise monitoring of local energy dynamics, such as those focusing on coastal interactions.

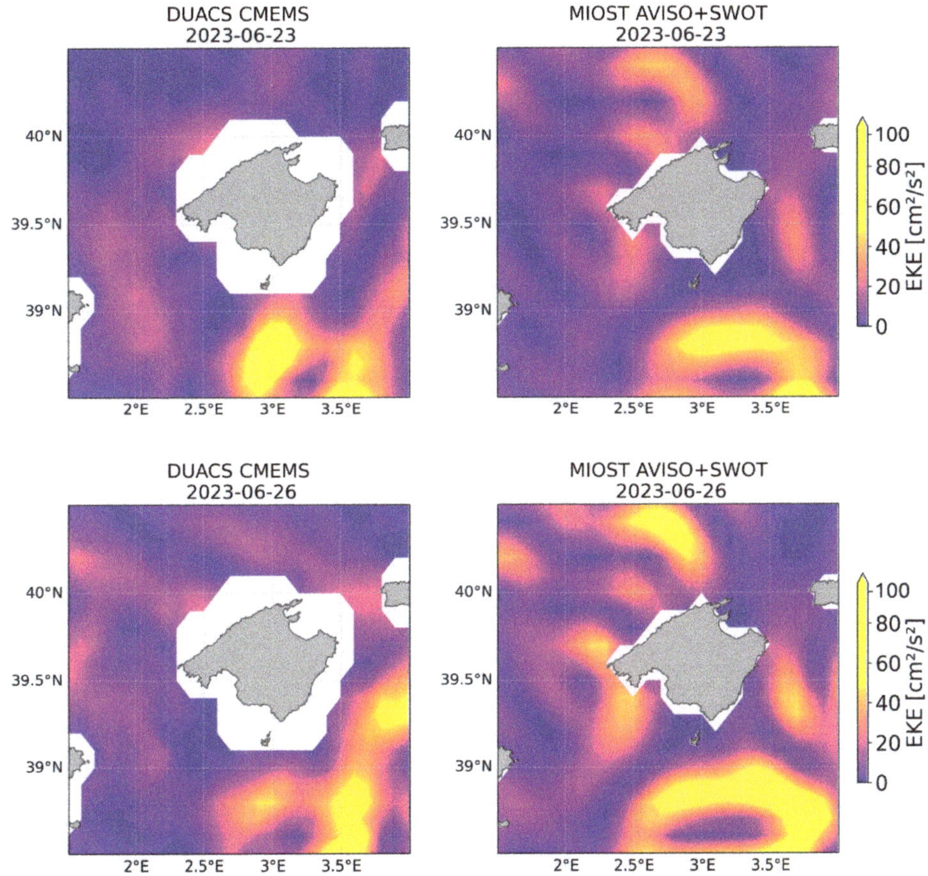

Figure 4. Eddy Kinetic Energy (EKE) maps from 23 June 2023 to 27 June 2023 from DUACS CMEMS and MIOST AVISO+SWOT altimetric products. The white area near the coast corresponds to regions where data are unavailable due to the altimetric product's lower spatial resolution in coastal zones.

In the area southwest of Mallorca Island, around the coastal eddy, the average Eddy Kinetic Energy (EKE) from the MIOST AVISO+SWOT product is 0.87 cm^2/s^2, which is higher than the 0.75 cm^2/s^2 recorded by DUACS CMEMS. This difference corresponds to an increase of 0.12 cm^2/s^2, or approximately 16%. The enhanced spatial resolution of MIOST AVISO+SWOT allows for a more detailed representation of local energy structures, revealing stronger eddy dynamics. This finding further reinforces the conclusion that MIOST AVISO+SWOT is more sensitive to sub-mesoscale features compared to DUACS CMEMS. The improved detection of local variations in energy intensity in MIOST AVISO+SWOT is particularly valuable for studies focused on coastal interactions, where fine-scale dynamics are critical. Conversely, DUACS CMEMS provides a more generalized depiction of the eddy, with lower EKE values and less detailed spatial features. Complete daily maps for the entire study period (Figure S3) are available in the Supplementary Material.

4.1.3. Relative Vorticity

Relative vorticity is a key parameter for diagnosing the rotational properties of ocean flows and for identifying mesoscale and submesoscale structures such as eddies and filaments [1]. The comparison of relative vorticity fields from the two altimetry products

(Figure 5) highlights substantial differences in the representation of ocean dynamics. The relative vorticity field from DUACS CMEMS exhibits a more homogeneous and less structured distribution, with smoother transitions between regions of positive and negative vorticity. In contrast, the MIOST AVISO+SWOT product captures greater spatial variability, with more defined vorticity structures and enhanced contrasts, and exhibits more pronounced extreme values of vorticity compared to DUACS CMEMS, indicating a greater sensitivity to local variations in the geostrophic velocity field. DUACS CMEMS, while effective at providing a more uniform vorticity field, is less detailed, with smoother transitions that limit its ability to resolve small-scale vorticity structures. This suggests that MIOST AVISO+SWOT is better suited for studies requiring high-resolution data to analyze fine-scale oceanic dynamics. Another notable difference is observed in coastal regions and around Mallorca Island. MIOST AVISO SWOT highlights more distinct and detailed vorticity structures, suggesting a superior capability in detecting mesoscale and sub-mesoscale phenomena, potentially related to current–topography interactions. Conversely, DUACS CMEMS presents more gradual transitions, offering a less detailed representation of these dynamic features. Complete daily maps for the entire study period (Figure S4) are provided in the Supplementary Material.

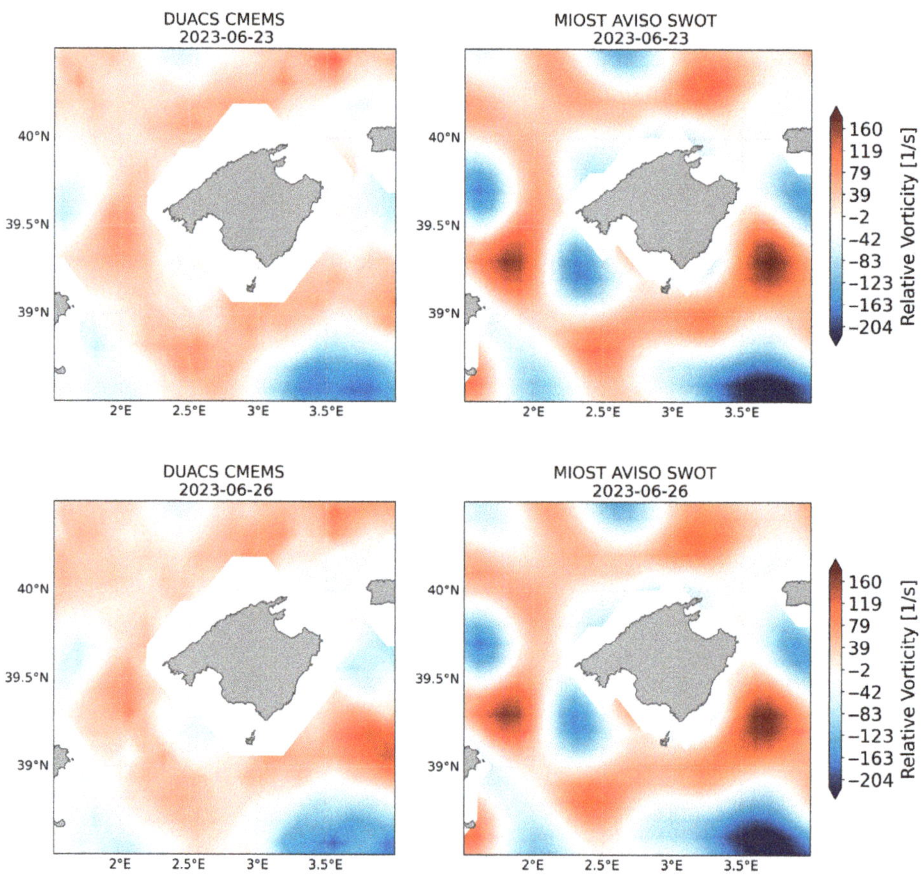

Figure 5. Relative Vorticity maps on 23 and 26 June 2023 from the DUACS CMEMS and MIOST AVISO+SWOT altimetric products. The white area near the coast corresponds to regions where data are unavailable due to the altimetric product's lower spatial resolution in coastal zones.

4.2. Comparison Between Gridded and Along-Track SWOT Data

Given the interesting performance of the MIOST AVISO+SWOT gridded (L4) product in resolving coastal structures, an additional comparison was conducted with the SWOT nadir CalVal along-track product (L3). This analysis aims to investigate the consistency between gridded and along-track datasets and to explore the potential of native-resolution SWOT observations in capturing fine-scale coastal dynamics.

The maps in Figure 6 illustrate two moments of the daily evolution of the ADT, derived from the MIOST AVISO+SWOT product, along with orthogonal geostrophic velocity fields obtained from the Along-Track L3 SWOT data over the period from 23 to 27 June 2023. The continuous background shading represents the gridded ADT contours, while the along-track data from the SWOT nadir CalVal product show both the along-track ADT anomalies and their associated orthogonal geostrophic velocities.

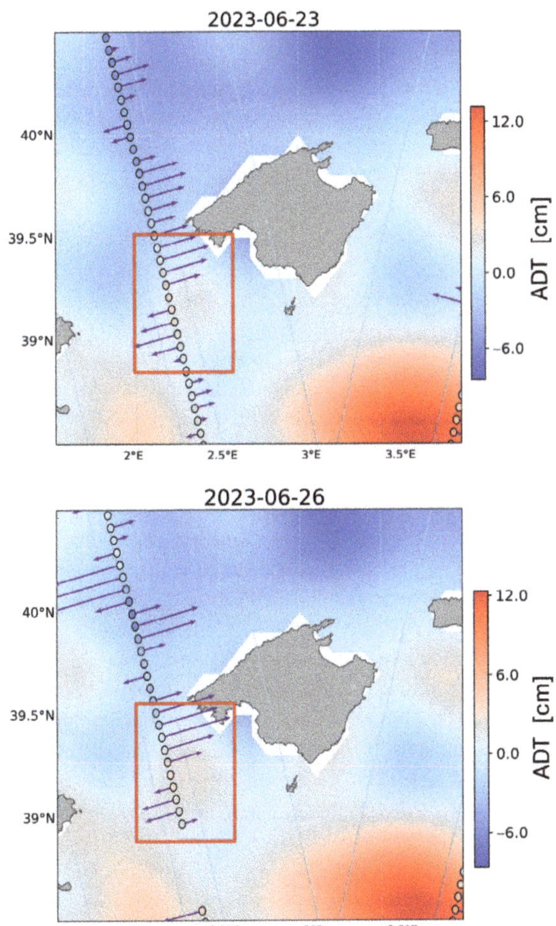

Figure 6. Evolution of the ADT, derived from the MIOST AVISO+SWOT product, along with orthogonal geostrophic velocity (purple arrows) and ADT fields obtained from the Along-Track L3 SWOT data for 23 and 26 June 2023. A mesoscale anticyclonic eddy is consistently identified southwest of Mallorca Island, with a well-defined structure (red rectangle).

The data reveal the presence of a coastal anticyclonic eddy southwest of Mallorca, which was consistently identified throughout the observed period. In particular, the 27 June dataset exhibits the most pronounced and coherent eddy structure, with a positive ADT centered around 2.5°E and 39°N and a velocity pattern indicative of clockwise rotation, consistent with an anticyclonic eddy in the Northern Hemisphere. Earlier days show progressively weaker but coherent signals, indicating a developing or translating eddy. This visual comparison between gridded and along-track datasets allows for an evaluation of the consistency between the two data representations and highlights the capability of SWOT's native-resolution observations to resolve sub-mesoscale features in coastal environments. The combined analysis underlines not only the high spatial detail provided by the along-track dataset but also its potential contribution to a more accurate characterization of coastal eddy dynamics and associated transport processes. Complete daily maps for the entire study period (Figure S5) are provided in the Supplementary Material.

4.3. Comparison Between SWOT L3 Product and Ocean Color Imagery

Figure 7 presents the Velocity Magnitude and Sea Surface Height Anomaly (SSHA) derived from the SWOT L3 product, used to assess the presence of the anticyclonic coastal eddy southwest of Mallorca. The analysis compares altimetric data with ocean color imagery. The analysis focuses on 23 and 26 June, with additional analyses for other days available in the Supplementary Material (see Figures S6–S8).

In the SSHA maps (Figure 7b), a well-defined positive anomaly appears near 2.5°E and 39°N, a typical signature of anticyclonic eddies. This anomaly is spatially consistent with the Velocity Magnitude patterns (Figure 7a), which exhibit a circular flow structure consistent with eddy dynamics.

Interestingly, the ocean color map (Figure 3a,b) reveals a localized patch of elevated chlorophyll-a concentration within the eddy core. While anticyclonic eddies are typically associated with downwelling and reduced surface productivity, enhanced chlorophyll concentrations can occasionally be observed in their center, particularly in coastal or shelf regions. This may result from eddy trapping of nutrient-rich coastal water, interactions with bathymetric features, or transient upwelling events [1,4].

In support of this, several physical–biogeochemical mechanisms have been proposed to explain such anomalies in productivity. One prominent process is eddy trapping, where the eddy isolates and retains coastal or shelf waters rich in nutrients and phytoplankton, allowing biological activity to persist or intensify over time [13,72]. Additionally, when eddies interact with complex bathymetry, they may induce localized upwelling—even in anticyclonic systems—by perturbing isopycnal surfaces and generating vertical motions [11, 73]. Another possible factor is the presence of submesoscale frontal instabilities along the eddy periphery, which enhances vertical nutrient fluxes toward the interior, potentially sustaining elevated chlorophyll concentrations [12]. These mechanisms are particularly relevant in semi-enclosed coastal basins like the Balearic Sea, where lateral stirring, shelf–edge interactions, and water mass contrasts are intensified.

The spatial concurrence between chlorophyll enhancement, positive SSHA, and rotational velocity patterns strongly supports the presence of a mesoscale anticyclonic eddy. Despite being an experimental product, SWOT L3 demonstrates a notable capacity to resolve fine-scale coastal features, consistent with independent satellite ocean color observations. This alignment further supports the utility of SWOT L3 data in studying mesoscale and sub-mesoscale coastal processes.

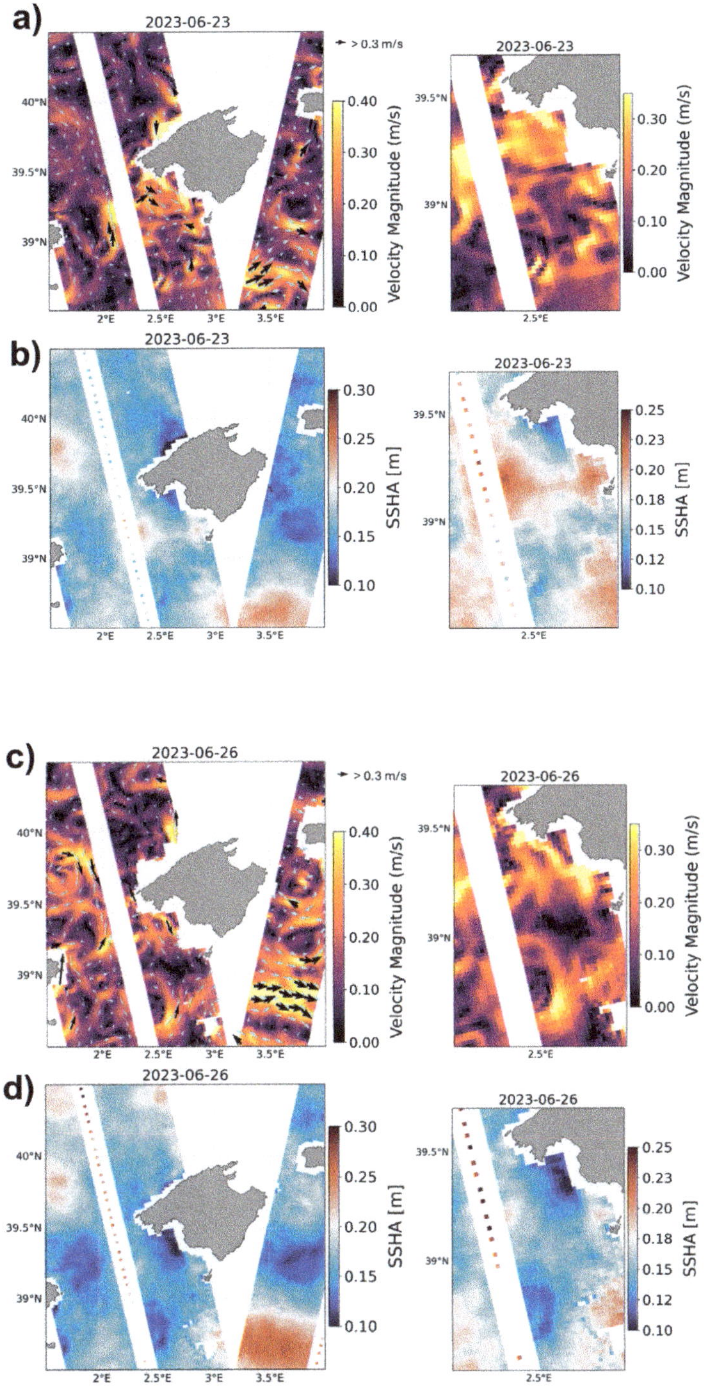

Figure 7. (**a**) Surface Velocity Magnitude [m/s] and (**b**) Sea Surface Height Anomaly (SSHA) [m] derived from the SWOT L3 product on 23 June 2023. (**c**) Surface Velocity Magnitude [m/s] and (**d**) Sea Surface Height Anomaly (SSHA) [m] derived from the SWOT L3 product on 26 June 2023.

4.4. Lagrangian Simulations and Eddy Dynamics

To assess the influence of altimetric resolution on coastal transport dynamics, Lagrangian simulations were performed using OceanParcels with two different geostrophic velocity fields: one derived from the ADT of the conventional DUACS CMEMS dataset and the other from the MIOST AVISO+SWOT product.

The simulation using DUACS CMEMS data (Figure 8) resulted in a less organized particle motion pattern. Although a general anticyclonic circulation was discernible, the eddy appeared more diffuse, with less-defined streamlines and weaker rotational trapping. The spatial distribution of beached particles (Figure 8c) was more scattered, with a broader and less coherent probability (Figure 8b) footprint along the Mallorca coast. The reduced resolution of the conventional altimetric product limited its ability to resolve the fine-scale eddy structure and its associated transport effects.

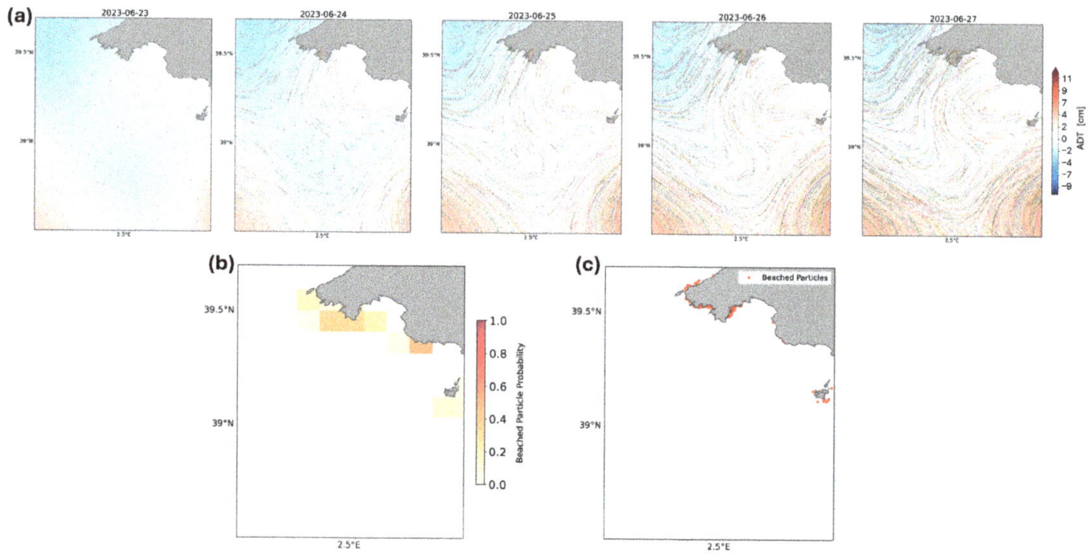

Figure 8. (**a**) Particle simulation trajectory with ADT pattern from the DUACS CMEMS product; (**b**) probability of particle beaching; (**c**) position of beached particles.

In contrast, in the MIOST AVISO+SWOT configuration (Figure 9), the particle trajectories clearly reflected the presence of a well-defined anticyclonic eddy southwest of Mallorca Island. Between 23 and 27 June 2023, the high-resolution ADT fields captured a compact, circular structure with tight streamlines and strong rotational flow. This coherent eddy structure effectively trapped particles within its core, indicating strong retention capabilities typical of mesoscale features. The beached particle probability map (Figure 9b) revealed concentrated stranding zones along the northern Mallorca coast, and particle landings were spatially clustered (Figure 9c), highlighting the improved predictability and spatial accuracy afforded by SWOT-derived fields.

The comparison between the two simulations underscores the critical influence of altimetric resolution on Lagrangian transport modeling in coastal settings, as evidenced by the distinct differences in transport behavior revealed by each dataset.

The analysis of particle travel distances from the Lagrangian simulations further highlights these differences quantitatively. The SWOT-enhanced product yields an average particle travel distance of 29.57 km, approximately 7.8% greater than the 27.44 km found in the DUACS simulation. Similarly, the maximum distance traveled is higher with SWOT data, reflecting the enhanced spatial resolution and ability to capture finer-scale ocean dynamics. These factors contribute to a more realistic representation of particle dispersion and transport pathways.

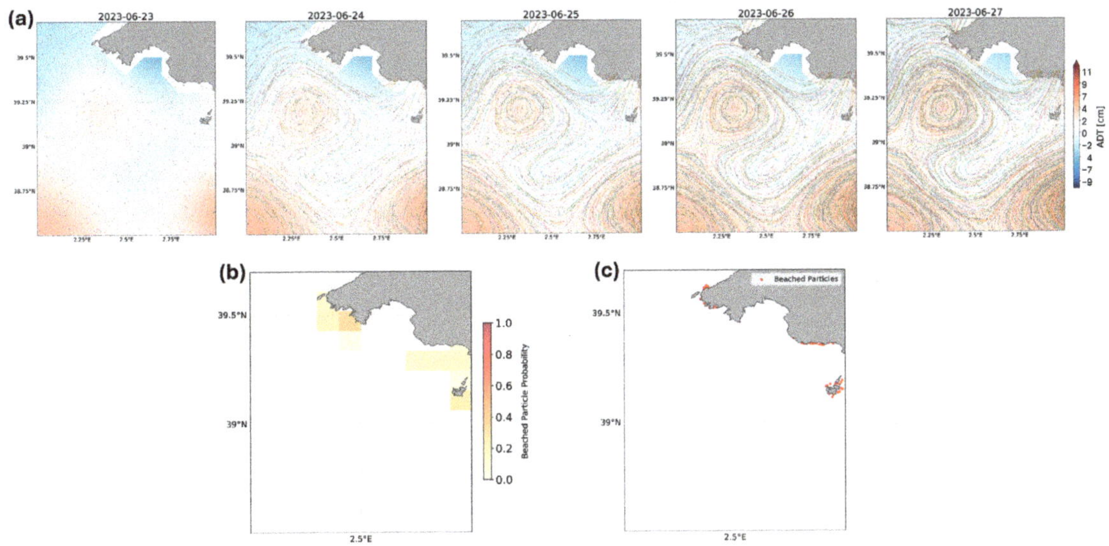

Figure 9. (**a**) Particle simulation trajectory with ADT pattern from the MIOST AVISO+SWOT product; (**b**) probability of particle beaching; (**c**) position of beached particles.

The inclusion of SWOT data significantly enhances the representation of mesoscale and sub-mesoscale features, which in turn affects the trapping efficiency and transport pathways of passive tracers. In coastal zones, where traditional altimetry suffers from lower accuracy due to land contamination [74,75] and sparse data coverage, the SWOT-enhanced product provides a more realistic dynamical background for particle tracking applications.

These results clearly demonstrate that the improved altimetric resolution provided by SWOT data leads to significantly different—and more realistic—Lagrangian transport patterns compared to conventional altimetric products. This improvement has critical implications for operational applications: the enhanced representation of mesoscale and sub-mesoscale eddy dynamics translates directly into more accurate predictions of particle pathways, trapping efficiency, and coastal deposition patterns. Consequently, Lagrangian transport modeling benefits greatly from SWOT-derived data, improving operational applications such as pollutant dispersion forecasting, biological connectivity studies, and search-and-rescue missions. In this sense, the differences in altimetric resolution are not merely academic but have a tangible impact on the quality and reliability of coastal transport predictions.

5. Discussion

A deeper understanding of eddy formation, evolution, and dissipation is essential for evaluating the ocean's role in the Earth's climate system. The analysis focuses on a 5-day period (23–27 June 2023) selected for optimal satellite data availability, ensuring continuous cloud-free coverage essential for integrating altimetric and ocean color imagery. While this timeframe limits the observation to the eddy's short-term behavior, it enables a robust and consistent comparison between datasets.

This study highlights the critical role of high-resolution satellite altimetry in observing and understanding coastal mesoscale and sub-mesoscale ocean dynamics. By leveraging ocean color imagery and comparing two altimetric products—conventional DUACS CMEMS and SWOT-enhanced MIOST AVISO+SWOT—we identified and characterized a mesoscale anticyclonic eddy southwest of Mallorca Island. The SWOT-enhanced dataset consistently outperformed DUACS CMEMS in resolving small-scale features across multiple dynamic variables, including Velocity Magnitude, eddy kinetic energy, relative vorticity, and Absolute Dynamic Topography gradients.

The enhanced spatial resolution of MIOST AVISO+SWOT allowed for a more detailed and accurate depiction of eddy structures, especially near the coast, where conventional altimetry often suffers from reduced reliability. These findings were supported by ocean color imagery and further reinforced through the comparison with SWOT L3 data, which revealed a strong spatial coherence between physical and biogeochemical signals—such as chlorophyll-a concentrations—further supporting the presence and persistence of the identified eddy.

Lagrangian particle simulations highlighted the strong dependence of transport modeling outcomes on the accuracy of the underlying velocity fields derived from altimetry. The MIOST AVISO+SWOT-based simulations reproduced more coherent and physically consistent particle trajectories, with well-defined trapping within the eddy core and localized beaching patterns along the Mallorca coast. In contrast, simulations using DUACS CMEMS resulted in more dispersed and less predictable transport behavior, underscoring the limitations of lower-resolution products for modeling coastal dispersion processes.

The Lagrangian simulations conducted with OceanParcels served primarily to demonstrate the trapping of particles within the eddy structure. While effective in illustrating retention, these simulations were restricted to surface-layer passive advection without specific diffusion or vertical processes, which may lead to an underestimation of dispersion in complex coastal environments. Nonetheless, the modeled particle trajectories coherently reflect the altimetric fields and capture the main dynamical features of the eddy. Future studies should include vertical mixing, diffusion, and other dynamical processes, such as wave effects, to better represent full coastal transport mechanisms.

It is also important to acknowledge that particle trajectories are highly sensitive to the quality of the velocity fields, as these directly drive the Lagrangian simulations. In this study, the focus was on geostrophic currents and on assessing the improvements brought by SWOT and KaRIn data. Non-geostrophic components—such as wind-driven Ekman flows, tidal currents, and unbalanced motions (e.g., cyclostrophic flows or submesoscale dynamics)—were not included due to data limitations. While these processes may contribute to coastal transport in the study region, especially near the coast, their omission does not compromise the main objective of our Lagrangian analysis, which is to demonstrate the eddy's capacity for particle retention during its mature stage. Future research should aim to incorporate these additional forcings to provide a more comprehensive representation of coastal transport and enhance predictive capabilities.

However, it is important to recognize the practical challenges associated with operationalizing SWOT data in near-real-time applications. These include limited data availability due to satellite revisit intervals and inherent data latency, the need for rigorous filtering and quality control to reduce noise and measurement errors, and seasonal variations that may influence observation frequency and reliability [36,43,76]. Overcoming these limitations is essential to fully leverage SWOT's capabilities for timely coastal monitoring and management.

Furthermore, although this study focuses on the Balearic Sea, the advantages demonstrated here for high-resolution altimetry in resolving mesoscale and sub-mesoscale coastal processes have broader implications. These findings can inform coastal monitoring and management strategies globally, especially in regions with complex coastal dynamics where conventional altimetry faces limitations [51,76–78].

Future efforts integrating SWOT data with complementary observations and models are expected to enhance global coastal oceanographic applications [79–81].

Our findings confirm that SWOT-enhanced altimetry markedly improves the detection and characterization of coastal dynamics, offering clear benefits for operational applications such as pollutant dispersion modelling and marine spatial planning.

In summary, this study demonstrates that high-resolution altimetry, particularly from SWOT-enhanced products, significantly improves the detection and understanding of coastal mesoscale dynamics. While limitations remain due to modeling assumptions and data constraints, these results lay a robust foundation for advancing coastal oceanographic research and operational monitoring in complex coastal regions.

Future research should aim to extend this analysis to broader spatial and temporal domains, integrate additional observational datasets, and develop hybrid methodologies that combine high-resolution altimetry with ocean color and in situ measurements to advance coastal oceanography.

6. Conclusions

SWOT-enhanced altimetry markedly improves the detection and analysis of coastal dynamics. By enabling more accurate velocity fields and supporting the identification of physical–biogeochemical coupling, this approach proves to be a valuable asset for coastal oceanography. Our results underscore the importance of integrating next-generation satellite altimetry into both research and operational ocean monitoring frameworks. Specifically, SWOT data have great potential in several applications. One is oil spill modeling, where precise current information is crucial for predicting pollutant dispersion. Another is the monitoring of coastal storm surges and flooding, which benefits from better characterization of nearshore circulation and sea level changes. SWOT data can also support the management and conservation of marine protected areas by detecting environmental changes that affect biodiversity. Finally, they can improve the prediction of nutrient or contaminant dispersion, helping safeguard water quality and ecosystem health. These examples highlight how SWOT-based altimetry can enhance decision-making and environmental management in coastal regions.

Supplementary Materials: The following supporting information can be downloaded at https://www.mdpi.com/article/10.3390/rs17152552/s1. Figure S1: Daily chlorophyll-a concentration derived from ocean color images from 23 June 2023 to 27 June 2023 in the Balearic Sea. The red rectangle highlights the mesoscale anticyclonic coastal eddy located southwest of Mallorca Island; Figure S2: Daily chlorophyll-a concentration derived from ocean color imagery from 23 June 2023 to 27 June 2023 in the Balearic Sea (left panels). ADT maps from the two altimetric products (right panels); Figure S3: Eddy Kinetic Energy (EKE) maps from 23 June 2023 to 27 June 2023; Figure S4: Relative Vorticity maps from 23 June 2023 to 27 June 2023; Figure S5: Daily evolution of the ADT, derived from the MIOST AVISO+SWOT product, along with orthogonal geostrophic velocity and ADT fields obtained from the Along-Track L3 SWOT data from June 23 to 27, 2023. A mesoscale anticyclonic eddy is consistently identified southwest of Mallorca Island, with a well-defined structure on 27 June (red rectangle); Figure S6: (a) Surface velocity magnitude [m/s] and (b) Sea Surface Height Anomaly (SSHA) [m] derived from the SWOT L3 product on 24 June 2023. (c) Chlorophyll-a concentration [log(mg/m^3)] derived from ocean color imagery on 24 June 2023; Figure S7: (a) Surface velocity magnitude [m/s] and (b) Sea Surface Height Anomaly (SSHA) [m] derived from the SWOT L3 product on 25 June 2023. (c) Chlorophyll-a concentration [log(mg/m^3)] derived from ocean color imagery on 25 June 2023; Figure S8: (a) Surface velocity magnitude [m/s] and (b) Sea Surface

Height Anomaly (SSHA) [m] derived from the SWOT L3 product on 27 June 2023. (c) Chlorophyll-a concentration [log(mg/m^3)] derived from ocean color imagery on 27 June 2023.

Author Contributions: L.F., V.C., A.P. and L.G.-N. conceptualized the study and designed the methodology. L.F. performed data curation and formal analysis. L.F. wrote the original draft of the manuscript. L.F., V.C., A.P., L.G.-N., Y.C. and G.A. contributed to the review and editing. A.P. supervised the project and acquired funding. All authors have read and agreed to the published version of the manuscript.

Funding: This research was supported by Laura Fortunato's PhD grant in the "Environmental Phenomena and Risks" doctoral program at the University Parthenope of Naples, funded by the PNRR, Mission 4, Component 1, Investment 3.4 "Enhancement of educational services: from nurseries to universities" and Investment 4.1 "University teaching and advanced skills", as well as "Extension of the number of PhD programs and innovative PhDs for public administration and cultural heritage." This PhD grant was funded under Ministerial Decree No. 351/2022, CUP I61I22000080007. This study has received funding from the Spanish Ministry of Science, Innovation, and Universities, the Spanish Research Agency, the European Regional Development Fund (MCIN/AEI/10.13039/501100011033/FUE) under Grant PID2021-122417NB-I00 (FaSt-SWOT project). No APC funding was received.

Data Availability Statement: All datasets used in this study are publicly available from open-access satellite data repositories. Ocean color data were obtained from the Copernicus Marine Environment Monitoring Service (CMEMS) under the product ID OCEANCOLOUR_MED_BGC_L3_MY_009_143, accessible at https://doi.org/10.48670/moi-00299. Conventional altimetry data were sourced from CMEMS, including the Level-4 gridded product SEALEVEL_EUR_PHY_L4_NRT_008_060 (https://doi.org/10.48670/moi-00142) and the along-track Level-3 product SEALEVEL_EUR_PHY_L3_MY_008_061 (https://doi.org/10.48670/moi-00139). SWOT-enhanced altimetry data, including the experimental L4 MIOST product and the Expert Level 3 Low-Rate Sea Surface Height dataset, were obtained from AVISO+ and are available via https://doi.org/10.24400/527896/A01-2024.007 and https://doi.org/10.24400/527896/A01-2023.018, respectively. All data are freely accessible and were used in compliance with the corresponding data usage licenses. The Lagrangian analyses were performed using the open-source OceanParcels framework (https://oceanparcels.org, accessed on 20 November 2024), version 3.1.1, which is available at https://github.com/OceanParcels/parcels (accessed on 20 November 2024) under the MIT license. Any custom scripts developed for preprocessing and analysis can be made available upon reasonable request to the corresponding author.

Acknowledgments: The authors acknowledge the internship opportunity of the first author at IMEDEA's María de Maeztu Centre of Excellence (CEX2021-001198), where part of this study was conducted. The authors would like to thank the project "Un approccio multidisciplinare alla contaminazione da idrocarburi nei mitili allevati nel Golfo di Pozzuoli," supported by the University of Naples Parthenope Local Research Project 2023, in which some of the authors are involved. Gratitude is also extended to the European Union—NextGenerationEU, National Recovery and Resilience Plan (PNRR), Mission 4, Component 2, Investment 1.4, "Strengthening research infrastructures and creation of national R&D sample centers on selected Key Enabling Technologies," Code CN00000023—Sustainable Mobility Center (Centro Nazionale per la Mobilità Sostenibile—CNMS), Spoke 3 "Waterways" and Spoke 7 "CCAM, Connected Networks and Smart Infrastructure," in which some of the authors also participate. The authors also acknowledge the Antarctic Circumpolar Current Eddies Survey and Simulations (ACCESS) project (PNRA 19_00032), part of the Italian National Antarctic Research Program (PNRA), with the authors' involvement. V. Combes also acknowledges support from the Spanish Ramón y Cajal Program (RYC2020-029306-I) through Grant AEI/UIB—10.13039/501100011033. Additional support comes from the Copernicus Marine Service (Sea-Level Thematic Center 24251L02-COP-TAC SL-2100).

Conflicts of Interest: The authors declare that the research was conducted in the absence of any commercial or financial relationships that could be construed as a potential conflict of interest.

References

1. Chelton, D.B.; Schlax, M.G.; Samelson, R.M. Global observations of nonlinear mesoscale eddies. *Prog. Oceanogr.* **2011**, *91*, 167–216. [CrossRef]
2. Cotroneo, Y.; Budillon, G.; Fusco, G.; Spezie, G. Cold core eddies and fronts of the Antarctic Circumpolar Current south of New Zealand from in situ and satellite data. *J. Geophys. Res. Ocean.* **2013**, *118*, 2653–2666. [CrossRef]
3. Dong, C.; McWilliams, J.C.; Liu, Y.; Chen, D. Global heat and salt transports by eddy movement. *Nat. Commun.* **2014**, *5*, 3294. [CrossRef] [PubMed]
4. Gaube, P.; McGillicuddy, D.J., Jr.; Chelton, D.B.; Behrenfeld, M.J.; Strutton, P.G. Regional variations in the influence of mesoscale eddies on near-surface chlorophyll. *J. Geophys. Res. Ocean.* **2014**, *119*, 8195–8220. [CrossRef]
5. Gaube, P.; McGillicuddy, D.J. The influence of Gulf Stream eddies and meanders on near-surface chlorophyll. *Deep Sea Res. Part I Oceanogr. Res. Pap.* **2017**, *122*, 1–16. [CrossRef]
6. Kahru, M.; Mitchell, B.G.; Gille, S.T.; Hewes, C.D.; Holm-Hansen, O. Eddies enhance biological production in the Weddell-Scotia confluence of the Southern Ocean. *Geophys. Res. Lett.* **2007**, *34*, L14601. [CrossRef]
7. Mahadevan, A.; Archer, D. Modeling the impact of fronts and mesoscale circulation on the nutrient supply and biogeochemistry of the upper ocean. *J. Geophys. Res.* **2000**, *105*, 1209–1225. [CrossRef]
8. Nagai, T.; Gruber, N.; Frenzel, H.; Lachkar, Z.; McWilliams, J.C.; Plattner, G.-K. Dominant role of eddies and filaments in the offshore transport of carbon and nutrients in the California Current System. *J. Geophys. Res. Ocean.* **2015**, *120*, 5318–5341. [CrossRef]
9. Siegel, D.A.; Peterson, P.; McGillicuddy, D.J., Jr.; Maritorena, S.; Nelson, N.B. Bio-optical footprints created by mesoscale eddies in the Sargasso Sea. *Geophys. Res. Lett.* **2011**, *38*, L13605. [CrossRef]
10. Wang, Y.; Zhang, H.-R.; Chai, F.; Yuan, Y. Impact of mesoscale eddies on chlorophyll variability off the coast of Chile. *PLoS ONE* **2018**, *13*, e0203598. [CrossRef] [PubMed]
11. Zhang, Z.; Wang, W.; Qiu, B. Oceanic mass transport by mesoscale eddies. *Science* **2014**, *345*, 322–324. [CrossRef] [PubMed]
12. Lévy, M. The modulation of biological production by oceanic mesoscale turbulence. In *Transport in Geophysical Flow: Ten Years After*; Weiss, J.B., Provenzale, A., Eds.; Springer: Berlin/Heidelberg, Germany, 2008; Volume 744, pp. 219–261. [CrossRef]
13. McGillicuddy, D.J., Jr.; Robinson, A.R.; Siegel, D.A.; Jannasch, H.W.; Johnson, R.; Dickey, T.D.; McNeil, J.; Michaels, A.F.; Knap, A.H. Influence of mesoscale eddies on new production in the Sargasso Sea. *Nature* **1998**, *394*, 263–265. [CrossRef]
14. Volkov, D.L.; Lee, T.; Fu, L.-L. Eddy-induced meridional heat transport in the ocean. *Geophys. Res. Lett.* **2008**, *35*, L20601. [CrossRef]
15. Bouffard, J.; Renault, L.; Ruiz, S.; Pascual, A.; Dufau, C.; Tintoré, J. Subsurface small-scale eddy dynamics from multi-sensor observations and modeling. *Prog. Oceanogr.* **2012**, *106*, 62–79. [CrossRef]
16. Fu, L.L.; Chelton, D.B.; Le Traon, Y.; Morrow, R. Eddy dynamics from satellite altimetry. *Oceanography* **2010**, *23*, 14–25. Available online: http://www.jstor.org/stable/24860859 (accessed on 20 November 2024). [CrossRef]
17. Fu, L.L.; Alsdorf, D.; Rodriguez, E.; Morrow, R.; Mognard, N.; Lambin, J.; Lafon, T. The SWOT (Surface Water and Ocean Topography) mission: Spaceborne radar interferometry for oceanographic and hydrological applications. In Proceedings of the OceanObs'09: Sustained Ocean Observations and Information for Society, Venice, Italy, 21–25 September 2009; Volume 9, pp. 21–25.
18. Brach, L.; Deixonne, P.; Bernard, M.-F.; Durand, E.; Desjean, M.-C.; Perez, E.; Lambert, C.; Sire, O.; Baudrimont, M.; Gasperi, J.; et al. Mesoscale anticyclonic eddies increase accumulation of microplastic in the North Atlantic subtropical gyre. *Mar. Pollut. Bull.* **2018**, *126*, 191–196. [CrossRef] [PubMed]
19. Condie, S.; Condie, R. Retention of plankton within ocean eddies. *Glob. Ecol. Biogeogr.* **2016**, *25*, 1274–1286. [CrossRef]
20. Krom, M.D.; Brenner, S.; Kress, N.; Neori, A.; Gordon, L.I. Nutrient dynamics and new production in a warm-core eddy from the eastern Mediterranean Sea. *Deep Sea Res. Part A Oceanogr. Res. Pap.* **1992**, *39*, 467–480. [CrossRef]
21. Salihoğlu, İ.; Saydam, C.; Başturk, O.; Yilmaz, K.; Gocmen, D.; Hatipoğlu, E.; Yilmaz, A. Transport and distribution of nutrients and chlorophyll-a by mesoscale eddies in the northeastern Mediterranean. *Mar. Chem.* **1990**, *29*, 375–390. [CrossRef]
22. Lobel, P.S.; Robinson, A.R. Transport and entrapment of fish larvae by ocean mesoscale eddies and currents in Hawaiian waters. *Deep Sea Res. Part A Oceanogr. Res. Pap.* **1986**, *33*, 483–500. [CrossRef]
23. Sabates, A.; Salat, J.; Palomera, I.; Emelianov, M.; Fernandez de Puelles, M.L.; Olivar, M.P. Advection of anchovy (*Engraulis encrasicolus*) larvae along the Catalan continental slope (NW Mediterranean). *Fish. Oceanogr.* **2007**, *16*, 130–141. [CrossRef]
24. Sabates, A.; Salat, J.; Raya, V.; Emelianov, M. Role of mesoscale eddies in shaping the spatial distribution of the coexisting *Engraulis encrasicolus* and *Sardinella aurita* larvae in the northwestern Mediterranean. *J. Mar. Syst.* **2013**, *111*, 108–119. [CrossRef]
25. Gaube, P.; Barceló, C.; McGillicuddy, D.J., Jr.; Domingo, A.; Miller, P.; Giffoni, B.; Marcovaldi, N.; Swimmer, Y. The use of mesoscale eddies by juvenile loggerhead sea turtles (*Caretta caretta*) in the southwestern Atlantic. *PLoS ONE* **2017**, *12*, e0172839. [CrossRef] [PubMed]

26. Gaube, P.; Braun, C.D.; Lawson, G.L.; McGillicuddy, D.J., Jr.; Penna, A.D.; Skomal, G.B.; Fischer, C.; Thorrold, S.R. Mesoscale eddies influence the movements of mature female white sharks in the Gulf Stream and Sargasso Sea. *Sci. Rep.* **2018**, *8*, 7363. [CrossRef] [PubMed]
27. Roda, S.B.; Rotger, A.; Santangeli, A.; Tavecchia, G.; Sola, A.; Catchot, S.; Sanz-Aguilar, A. Coping with ocean dynamics: Foraging strategy and implications for conservation of a small petrel. *Biol. Conserv.* **2025**, *302*, 110913. [CrossRef]
28. Sánchez-Román, A.; Gómez-Navarro, L.; Fablet, R.; Oro, D.; Mason, E.; Arcos, J.M.; Ruiz, S.; Pascual, A. Rafting behaviour of seabirds as a proxy to describe surface ocean currents in the Balearic Sea. *Sci. Rep.* **2019**, *9*, 17775. [CrossRef] [PubMed]
29. Gomez-Navarro, L.; Sánchez-Román, A.; Pascual, A.; Fablet, R.; Hernandez-Carrasco, I.; Mason, E.; Oro, D. Seabird drift as a proxy to estimate surface currents in the western Mediterranean? In *EGU General Assembly Conference Abstracts*; European Geosciences Union: Vienna, Austria, 2017; p. 9548. Available online: https://www.researchgate.net/publication/318211642_Seabird_drift_as_a_proxy_to_estimate_surface_currents_in_the_western_Mediterranean (accessed on 20 November 2024).
30. Morrow, R.; Le Traon, P.-Y. Recent advances in observing mesoscale ocean dynamics with satellite altimetry. *J. Adv. Space Res.* **2011**, *47*, 1067–1082. [CrossRef]
31. Lehahn, Y.; d'Ovidio, F.; Levy, M.; Heifetz, E. Stirring of the Northeast Atlantic spring bloom: A Lagrangian analysis based on multisatellite data. *J. Geophys. Res.* **2007**, *112*, C08005. [CrossRef]
32. Ducet, N.; Le Traon, P.Y.; Reverdin, G. Global high-resolution mapping of ocean circulation from TOPEX/Poseidon and ERS-1 and -2. *J. Geophys. Res. Ocean.* **2000**, *105*, 19477–19498. [CrossRef]
33. Dussurget, R.; Birol, F.; Morrow, R.A.; De Mey, P. Fine resolution altimetry data for a regional application in the Bay of Biscay. *Mar. Geod.* **2011**, *34*, 447–476. [CrossRef]
34. Pascual, A.; Pujol, M.I.; Larnicol, G.; Le Traon, P.Y.; Rio, M.H. Mesoscale mapping capabilities of multisatellite altimeter missions: First results with real data in the Mediterranean Sea. *J. Mar. Syst.* **2007**, *65*, 190–211. [CrossRef]
35. Pascual, A.; Faugère, Y.; Larnicol, G.; Le Traon, P.Y. Improved description of the ocean mesoscale variability by combining four satellite altimeters. *Geophys. Res. Lett.* **2006**, *33*. [CrossRef]
36. Morrow, R.; Fu, L.-L.; Ardhuin, F.; Benkiran, M.; Chapron, B.; Cosme, E.; D'Ovidio, F.; Farrar, J.T.; Gille, S.T.; Lapeyre, G.; et al. Global observations of fine-scale ocean surface topography with the surface water and ocean topography (SWOT) mission. *Front. Mar. Sci.* **2019**, *6*, 232. [CrossRef]
37. Fu, L.L.; Pavelsky, T.; Cretaux, J.F.; Morrow, R.; Farrar, J.T.; Vaze, P.; Dibarboure, G. The surface water and ocean topography mission: A breakthrough in radar remote sensing of the ocean and land surface water. *Geophys. Res. Lett.* **2024**, *51*, e2023GL107652. [CrossRef]
38. Verger Miralles, E.; Mourre, B.; Barceló Llull, B.; Casas Pérez, B.; Combes, V.; Cutolo, E.; Pascual Ascaso, A. Study of fine-scale dynamics in the Balearic Sea through high-resolution observations and SWOT satellite data. *Instrum. Viewp.* **2024**, *23*, 21–22. Available online: https://hdl.handle.net/2117/411406 (accessed on 20 November 2024).
39. Mourre, B.; Verger-Miralles, E.; Gómez-Navarro, L.; Barceló-Llull, B.; Casas, B.; Combes, V.; Cutolo, E.; Diaz-Barroso, L.; Garcia-Jove, M.; Lizarrán, I.; et al. FaSt-SWOT field campaigns combining high-resolution observations and modelling for SWOT validation in the Western Mediterranean Sea. In *EGU General Assembly Conference Abstracts*; European Geosciences Union: Vienna, Austria, 2024; p. 11336. [CrossRef]
40. Pascual, A.; Mourre, B.; Balaguer, P.; Barceló-Llull, B.; Calafat, N.; Casas, B.; Combes, V.; Cutolo, E.; Díaz-Barroso, L.; Fernández, J.G.; et al. *Fine-Scale Ocean Currents from Integrated Multi-Platform Experiments and Numerical Simulations: Contribution to the New SWOT Satellite Mission (FaSt-SWOT, PID2021-122417NB-I00)*; Cruise Plan; IMEDEA (CSIC-UIB): Esporles, Spain, 2023.
41. Gómez-Navarro, L.; Cosme, E.; Le Sommer, J.; Papadakis, N.; Pascual, A. Development of an image de-noising method in preparation for the surface water and ocean topography satellite mission. *Remote Sens.* **2020**, *12*, 734. [CrossRef]
42. Gómez-Navarro, L.; Fablet, R.; Mason, E.; Pascual, A.; Mourre, B.; Cosme, E.; Le Sommer, J. SWOT spatial scales in the western Mediterranean Sea derived from pseudo-observations and an Ad Hoc filtering. *Remote Sens.* **2018**, *10*, 599. [CrossRef]
43. Aguiar, E.; Mourre, B.; Alvera-Azcárate, A.; Pascual, A.; Mason, E.; Tintoré, J. Strong long-lived anticyclonic mesoscale eddies in the Balearic Sea: Formation, intensification, and thermal impact. *J. Geophys. Res. Ocean.* **2022**, *127*, e2021JC017589. [CrossRef]
44. Pessini, F.; Cotroneo, Y.; Olita, A.; Sorgente, R.; Ribotti, A.; Jendersie, S.; Perilli, A. Life history of an anticyclonic eddy in the Algerian basin from altimetry data, tracking algorithm and in situ observations. *J. Mar. Syst.* **2020**, *207*, 103346. [CrossRef]
45. Cotroneo, Y.; Aulicino, G.; Ruiz, S.; Sánchez Román, A.; Torner Tomàs, M.; Pascual, A.; Budillon, G. Glider data collected during the Algerian basin circulation unmanned survey. *Earth Syst. Sci. Data* **2019**, *11*, 147–161. [CrossRef]
46. Cotroneo, Y.; Aulicino, G.; Ruiz, S.; Pascual, A.; Budillon, G.; Fusco, G.; Tintoré, J. Glider and satellite high resolution monitoring of a mesoscale eddy in the Algerian basin: Effects on the mixed layer depth and biochemistry. *J. Mar. Syst.* **2016**, *162*, 73–88. [CrossRef]
47. Aulicino, G.; Cotroneo, Y.; Ruiz, S.; Román, A.S.; Pascual, A.; Fusco, G.; Tamburini, A.; Budillon, G. Glider, satellite altimetry and numerical simulations for the monitoring of the Algerian Basin along a SARAL/AltiKa track. *J. Mar. Syst.* **2018**, *179*, 55–71. [CrossRef]

48. Pascual, A.; Buongiorno Nardelli, B.; Larnicol, G.; Emelianov, M.; Gomis, D. A case of an intense anticyclonic eddy in the Balearic Sea (western Mediterranean). *J. Geophys. Res.* **2002**, *107*, 3183. [CrossRef]
49. Bouffard, J.; Pascual, A.; Ruiz, S.; Faugère, Y.; Tintoré, J. Coastal and mesoscale dynamics characterization in the Balearic Sea. *J. Geophys. Res.* **2010**, *115*, C10029. [CrossRef]
50. Malanotte-Rizzoli, P.; Artale, V.; Borzelli-Eusebi, G.L.; Brenner, S.; Crise, A.; Gacic, M.; Kress, N.; Marullo, S.; Ribera d'alcalà, M.; Sofianos, S.; et al. Physical forcing and physical/biochemical variability of the Mediterranean Sea: A review of unresolved issues and directions for future research. *Ocean Sci.* **2014**, *10*, 281–322. [CrossRef]
51. Millot, C. Circulation in the Western Mediterranean Sea. *J. Mar. Syst.* **1999**, *20*, 423–442. [CrossRef]
52. Millot, C.; Taupier-Letage, I. Circulation in the Mediterranean Sea. In *The Mediterranean Sea. Handbook of Environmental Chemistry*; Saliot, A., Ed.; Springer: Berlin/Heidelberg, Germany, 2005; Volume 5K, pp. 29–66. [CrossRef]
53. López García, M.J.; Millot, C.; Font, J.; García-Ladona, E. Surface circulation variability in the Balearic Basin. *J. Geophys. Res.* **1994**, *99*, 3285–3296. [CrossRef]
54. La Violette, P.E.; Tintoré, J.; Font, J. The surface circulation of the Balearic Sea. *J. Geophys. Res.* **1990**, *95*, 1559–1568. [CrossRef]
55. Wang, F.; Vieira, M.; Salat, J.; Tintoré, J.; La Violette, P.E. A shelf/slope filament off the Northeast Spanish Coast. *J. Mar. Res.* **1988**, *46*, 321–332. Available online: https://elischolar.library.yale.edu/journal_of_marine_research/1891 (accessed on 20 November 2024). [CrossRef]
56. Pinot, J.M.; López-Jurado, J.L.; Riera, M. The CANALES experiment (1996–1998). Interannual, seasonal, and mesoscale variability of the circulation in the Balearic Channels. *Prog. Oceanogr.* **2002**, *55*, 335–370. [CrossRef]
57. Tintoré, J.; Wang, D.P.; La Violette, P.E. Eddies and thermohaline intrusions of the shelf/slope front off Northeast Spain. *J. Geophys. Res.* **1990**, *95*, 1627–1633. [CrossRef]
58. Pascual, A.; Gomis, D.; Haney, R.L.; Ruiz, S. A quasigeostrophic analysis of a meander in the Palamós canyon: Vertical velocity, geopotential tendency and a relocation technique. *J. Phys. Oceanogr.* **2004**, *34*, 2274–2287. [CrossRef]
59. Rubio, A.; Barnier, B.; Jordà, G.; Espino, M.; Marsaleix, P. Origin and dynamics of mesoscale eddies in the Catalan Sea (NW Mediterranean): Insight from a numerical model study. *J. Geophys. Res.* **2009**, *114*, C06009. [CrossRef]
60. Volpe, G.; Colella, S.; Brando, V.E.; Forneris, V.; Padula, F.L.; Cicco, A.D.; Santoleri, R. Mediterranean ocean colour Level 3 operational multi-sensor processing. *Ocean Sci.* **2019**, *15*, 127–146. [CrossRef]
61. Pujol, M.-I.; Faugère, Y.; Taburet, G.; Dupuy, S.; Pelloquin, C.; Ablain, M.; Picot, N. DUACS DT2014: The new multi-mission altimeter data set reprocessed over 20 years. *Ocean Sci.* **2016**, *12*, 1067–1090. [CrossRef]
62. Ubelmann, C.; Dibarboure, G.; Gaultier, L.; Ponte, A.; Ardhuin, F.; Ballarotta, M.; Faugère, Y. Reconstructing ocean surface current combining altimetry and future spaceborne Doppler data. *J. Geophys. Res. Ocean.* **2021**, *126*, e2020JC016560. [CrossRef]
63. Ubelmann, C.; Carrere, L.; Durand, C.; Dibarboure, G.; Faugère, Y.; Ballarotta, M.; Briol, F.; Lyard, F. Simultaneous estimation of ocean mesoscale and coherent internal tide sea surface height signatures from the global altimetry record. *Ocean Sci.* **2022**, *18*, 469–481. [CrossRef]
64. Ballarotta, M.; Ubelmann, C.; Bellemin-Laponnaz, V.; Le Guillou, F.; Meda, G.; Anadon, C.; Dibarboure, G. Assimilation of wide-swath altimetry in level-4 multi-mission satellite maps. *Ocean Sci.* **2025**, *21*, 63–80. [CrossRef]
65. OceanParcels Project. What Is Oceanparcels? Available online: http://oceanparcels.org/ (accessed on 16 October 2019).
66. Delandmeter, P.; van Sebille, E. The Parcels v2.0 Lagrangian framework: New field interpolation schemes. *Geosci. Model Dev.* **2019**, *12*, 3571–3584. [CrossRef]
67. Butcher, J.C. *Numerical Methods for Ordinary Differential Equations*, 3rd ed.; John Wiley & Sons: Hoboken, NJ, USA, 2016. [CrossRef]
68. Ioannou, A.; Zervakis, V.; Tzortziou, M. Cyclostrophic corrections of AVISO/DUACS surface velocities and its application to mesoscale eddies in the Mediterranean Sea. *J. Geophys. Res. Ocean.* **2019**, *124*, 8995–9012. [CrossRef]
69. Lévy, M.; Klein, P.; Treguier, A.-M. Impact of sub-mesoscale physics on production and subduction of phytoplankton in an oligotrophic regime. *J. Mar. Res.* **2001**, *59*, 535–565. [CrossRef]
70. Rossi, V.; Ser-Giacomi, E.; López, C.; Hernández-García, E. Hydrodynamic provinces and oceanic connectivity from a transport network help designing marine reserves. *Geophys. Res. Lett.* **2014**, *41*, 2883–2891. [CrossRef]
71. Beron-Vera, F.J.; Wang, Y.; Olascoaga, M.J.; Goni, G.J.; Haller, G. Objective detection of oceanic eddies and the Agulhas leakage. *J. Phys. Oceanogr.* **2013**, *43*, 1426–1438. [CrossRef]
72. McWilliams, J.C. Submesoscale currents in the ocean. *Proc. R. Soc. A Math. Phys. Eng. Sci.* **2016**, *472*, 20160117. [CrossRef] [PubMed]
73. Mahadevan, A. The impact of submesoscale physics on primary productivity of plankton. *Annu. Rev. Mar. Sci.* **2016**, *8*, 161–184. [CrossRef] [PubMed]
74. Vignudelli, S.; Kostianoy, A.G.; Cipollini, P.; Benveniste, J. (Eds.) *Coastal Altimetry*; Springer Science and Business Media: Berlin/Heidelberg, Germany, 2011; pp. 3–50. [CrossRef]
75. Dibarboure, G.; Boy, F.; Desjonqueres, J.D.; Labroue, S.; Lasne, Y.; Picot, N.; Poisson, J.C.; Thibaut, P. Investigating short-wavelength correlated errors on low-resolution mode altimetry. *J. Atmos. Ocean. Technol.* **2014**, *31*, 1337–1362. [CrossRef]

76. Barceló-Llull, B.; Pascual, A. Recommendations for the design of in situ sampling strategies to reconstruct fine-scale ocean currents in the context of SWOT satellite mission. *Front. Mar. Sci.* **2023**, *10*, 1082978. [CrossRef]
77. Tintoré, J.; Pinardi, N.; Álvarez-Fanjul, E.; Aguiar, E.; Álvarez-Berastegui, D.; Bajo, M.; Balbin, R.; Bozzano, R.; Nardelli, B.B.; Cardin, V.; et al. Challenges for sustained observing and forecasting systems in the Mediterranean Sea. *Front. Mar. Sci.* **2019**, *6*, 568. [CrossRef]
78. Escudier, R.; Renault, L.; Pascual, A.; Brasseur, P.; Chelton, D.; Beuvier, J. Eddy properties in the Western Mediterranean Sea from satellite altimetry and a numerical simulation. *J. Geophys. Res. Ocean.* **2016**, *121*, 3990–4006. [CrossRef]
79. Mason, E.; Pascual, A. Multiscale variability in the Balearic Sea: An altimetric perspective. *J. Geophys. Res. Ocean.* **2013**, *118*, 3007–3025. [CrossRef]
80. Barceló-Llull, B.; Pascual, A.; Sánchez-Román, A.; Cutolo, E.; d'Ovidio, F.; Fifani, G.; Ser-Giacomi, E.; Ruiz, S.; Mason, E.; Cyr, F.; et al. Fine-scale ocean currents derived from in situ observations in anticipation of the upcoming SWOT altimetric mission. *Front. Mar. Sci.* **2021**, *8*, 679844. [CrossRef]
81. De Mey-Frémaux, P.; Ayoub, N.; Barth, A.; Brewin, R.; Charria, G.; Campuzano, F.; Ciavatta, S.; Cirano, M.; Edwards, C.A.; Federico, I.; et al. Model-observations synergy in the coastal ocean. *Front. Mar. Sci.* **2019**, *6*, 436. [CrossRef]

remote sensing

MDPI

Article

Retrieval of Internal Solitary Wave Parameters and Analysis of Their Spatial Variability in the Northern South China Sea Based on Continuous Satellite Imagery

Kexiao Lu [1], Tao Xu [1], Cun Jia [2,*], Xu Chen [1,3] and Xiao He [1]

1 Frontiers Science Center for Deep Ocean Multispheres and Earth System and Key Laboratory of Physical Oceanography, Ocean University of China, Qingdao 266100, China; lukexiao@stu.ouc.edu.cn (K.L.); xutao@stu.ouc.edu.cn (T.X.); chenxu001@ouc.edu.cn (X.C.); hexiao@stu.ouc.edu.cn (X.H.)

2 Ocean Dynamics Laboratory, Third Institute of Oceanography, Ministry of Natural Resources, Xiamen 361005, China

3 Laoshan Laboratory, Qingdao 266100, China

* Correspondence: jc@tio.org.cn

Abstract

The remote sensing inversion of internal solitary waves (ISWs) enables the retrieval of ISW parameters and facilitates the analysis of their spatial variability. In this study, we utilize continuous optical imagery from the FY-4B satellite to extract real-time ISW propagation speeds throughout their evolution from generation to shoaling. ISW parameters are retrieved in the northern South China Sea based on the quantitative relationship between sea surface current divergence and ISW surface features in optical imagery. The inversion method employs a fully nonlinear equation with continuous stratification to account for the strongly nonlinear nature of ISWs and uses the propagation speed extracted from continuous imagery as a constraint to determine a unique solution. The results show that as ISWs propagate from deep to shallow waters in the northern South China Sea, their statistically averaged amplitude initially increases and then decreases, while their propagation speed continuously decreases with decreasing depth. The inversion results are consistent with previous in situ observations. Furthermore, a three-day consecutive remote sensing tracking analysis of the same ISW revealed that the spatial variation in its parameters aligned well with the abovementioned statistical results. The findings provide an effective inversion approach and supporting datasets for extensive ISW monitoring.

Keywords: internal solitary waves; remote sensing inversion; continuous optical imagery; fully nonlinear equation; Northern South China Sea

1. Introduction

Internal solitary waves (ISWs) are isolated nonlinear dispersive waves that are widely present throughout the global ocean [1]. They typically exhibit large amplitudes [2,3], strong nonlinearity [4,5], long-range propagation [6,7], and stable waveforms [8]. As ISWs propagate, they are associated with significant vertical displacements [9,10] and horizontal shear [11,12], which, together, contribute to enhanced ocean mixing [13,14], the cross-isopycnal transport of momentum and tracers [15–18], and modifications to local acoustic environments and hydrodynamic loads on submerged maritime structures [19–24]. Therefore, accurately retrieving ISW parameters, such as the amplitude and propagation speed, is crucial for understanding their generation mechanisms, propagation dynamics, and dissipation processes.

In situ observation is one of the most effective approaches for measuring ISW parameters. The South China Sea (as shown in Figure 1) is an area of frequent ISW activity due to its strong stratification and the presence of steep bathymetric features. In the northern South China Sea, internal tides generated by the interaction between barotropic tides and the ridges of the Luzon Strait propagate westward and subsequently evolve into ISWs through instability or steepening [25]. These ISWs traverse the deep basin, then enter the continental slope, where their propagation paths diverge near Dongsha Island. One branch, influenced by topographic refraction, turns northwestward, while the other continues westward before gradually veering northwestward along the seafloor gradient. These waves propagate over 500 km and persist for more than four days before ultimately dissipating on the continental shelf [26].

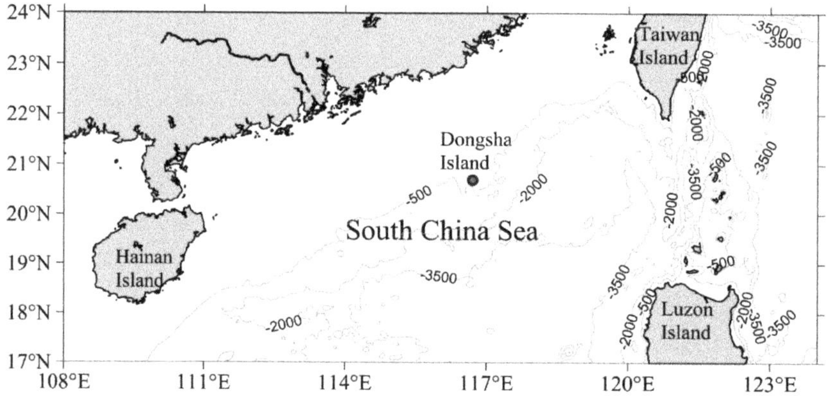

Figure 1. Geographical Map of the South China Sea.

In the northern South China Sea, a significant amount of research has been conducted, systematically measuring ISW parameters. Ramp et al. [27] conducted combined moored and shipboard observations near the continental shelf break between the southern tip of Taiwan and Dongsha Island. They observed ISWs with amplitudes ranging from 29 to 142 m and propagation speeds between 0.83 and 1.84 m/s. Yang et al. [28] deployed thermistor chain moorings on the continental slope northeast of Dongsha Island and reported ISWs with an average amplitude of 90 $\pm$ 15 m and an average propagation speed of 1.52 $\pm$ 0.04 m/s, estimated using the Korteweg–de Vries (KdV) equation. Klymak et al. [2], using sonar and a fast conductivity–temperature–depth (CTD) profiler, observed an ISW with an amplitude of 170 m and a propagation speed of 2.9 $\pm$ 0.1 m/s in the deep basin west of the Luzon Strait. Ramp et al. [29] deployed four oceanographic moorings along a transect from the Batanes Province, Philippines, in the Luzon Strait to just north of Dongsha Island on the Chinese continental slope. They observed ISWs with amplitudes ranging from 20 to 200 m and average propagation speeds of 3.23 $\pm$ 0.31 m/s in the deep basin and 2.22 $\pm$ 0.18 m/s over the continental slope. Lien et al. [30] observed five large-amplitude ISWs on the Dongsha slope during a spring tide using both shipboard and moored acoustic Doppler current profiler and CTD instruments. They reported amplitudes ranging from 106 to 173 m and propagation speeds between 1.6 and 1.72 m/s. Chen et al. [31] conducted year-long mooring observations west of Dongsha Island on the northern shelf slope, observing ISWs with an average amplitude of 43 $\pm$ 17 m and a propagation speed estimated by KdV of 1.38 $\pm$ 0.14 m/s. Chang et al. [32] deployed two mooring systems east of Dongsha Island and recorded ISWs with amplitudes ranging from 76 to 125 m and propagation speeds between 1 and 1.5 m/s. Ramp et al. [33] deployed an array of oceanographic moorings and a distributed temperature-sensing cable over the eastern slope of Dongsha Island, observing ISWs with amplitudes ranging from 20 to

180 m and propagation speeds between 1.1 and 2.6 m/s. In summary, in situ observations provide accurate measurements of ISW parameters. However, due to the high cost of deployment and maintenance, achieving extensive, continuous acquisition of ISW parameters remains challenging, thereby limiting the analysis of their spatial variability.

Recent advances in satellite remote sensing, particularly regarding spatial coverage, revisit frequency, and resolution, have enabled the systematic observation of ISWs over broad oceanic regions. Although remote sensing cannot directly measure ISW parameters such as amplitude and wavelength, relationships between surface features and wave parameters can be established through physical models, thereby enabling parameter retrieval through remote sensing inversion.

Zheng et al. [34] derived a theoretical model of ISW radar imagery based on the two-layer KdV equation and used synthetic aperture radar (SAR) imagery to retrieve the amplitudes of five ISWs over the Portuguese continental shelf and in the South China Sea. The retrieval results were consistent with the in situ measurements. Chen et al. [35] extended the KdV equation to the Benjamin–Ono (BO) equation and retrieved the amplitude of an ISW in the deep-water region of the northern South China Sea using optical imagery. The retrieved amplitude value was in reasonable agreement with historical in situ isotherm profile measurements. Huang and Zhao [36] applied a continuously stratified KdV model to retrieve ISW parameters from optical imagery in the deep-water region of the northern South China Sea, achieving good consistency with moored in situ observations. Zhang et al. [37] combined optical and SAR imagery with a corrected two-layer nonlinear Schrödinger (NLS) equation to retrieve the amplitudes of four ISWs in the Wenchang area, east of Hainan Island. The retrieved values closely matched the observed amplitudes. Jia et al. [38] utilized two consecutive SAR images and the extended KdV (eKdV) equation for a two-layer fluid to retrieve the amplitude of an ISW in shallow waters southeast of Hainan Island. The result was in close agreement with values derived using the classic KdV model under continuous stratification. Xie et al. [39] used both the KdV and NLS equations for a two-layer fluid, together with optical imagery, to retrieve the amplitudes of seven ISWs near Dongsha Island. The results aligned well with concurrent mooring observations. Chen et al. [40] applied the KdV, eKdV, and BO models for a two-layer fluid to SAR imagery to retrieve the amplitude of an ISW in the deep-water region of the northeastern South China Sea, yielding results consistent with historical mooring observations and with those from the continuously stratified KdV model.

Current remote sensing inversion studies of ISWs in the South China Sea have primarily focused on demonstrating the applicability of parameter retrieval methods in localized regions, with only a few case studies available. Systematic investigations into extensive ISW parameter retrieval and the spatial variability of these parameters remain lacking. In addition, most studies employ internal wave models based on two-layer fluid assumptions, without fully accounting for the continuous stratification commonly present in real ocean environments. By contrast, continuously stratified models better represent actual oceanic conditions [40] and yield more accurate inversion results than two-layer models [38]. Moreover, most existing inversion studies concentrate on retrieving the ISW amplitude, whereas propagation speed retrieval remains understudied, thereby limiting comprehensive insights into ISW dynamics. Most theoretical models used in remote sensing inversions are based on the weakly nonlinear assumption, which assumes that the ISW amplitude is small compared with their intrinsic vertical scale [41,42]. However, models based on the weakly nonlinear assumption are inadequate to describe strongly nonlinear ISWs [4,43], which account for a large proportion of ISWs in in situ oceanic observations [44–47]. Therefore, theoretical models with strong nonlinearity are needed when retrieving the parameters of large-amplitude ISWs from remote sensing imagery.

The Dubreil–Jacotin–Long (DJL) equation is a fully nonlinear model that makes no assumptions about wavelength or amplitude. Its applicability to strongly nonlinear ISWs has been validated in laboratory experiments and it has been successfully applied to retrieve ISW parameters [48]. In this study, the DJL equation under continuous stratification is employed in combination with continuous optical remote sensing imagery to retrieve ISW parameters over a broad region of the northern South China Sea. This study facilitates the analysis of the spatial variability of ISW parameters and reveals their distribution patterns in the northern South China Sea, thereby facilitating a more comprehensive understanding of ISW dynamics and providing essential data support for future investigations in the region.

The remainder of this study is organized as follows: In Section 2, we introduce the satellite remote sensing data, the theoretical equations of ISWs, and the methods used to extract ISW parameters. In Section 3, we present the inversion results and spatial variation characteristics of ISW parameters in the northern South China Sea. In Section 4, we discuss the comparison between the inversion results and in situ observations and analyze a case of propagation and evolution of a particular ISW. In Section 5, we summarize the main conclusions.

2. Data and Methods

2.1. Satellite Remote Sensing Data and ISW Parameter Extraction

The satellite imagery used in this study was obtained from the FY-4B Geo High-speed Imager (GHI), with a spatial resolution of 250 m and a swath width of 2000 km. FY-4B is the first operational satellite in China's new generation of geostationary meteorological satellites under the Fengyun-4 series. It was launched on 3 June 2021 and successfully positioned at 133°E on 11 April 2022. Operational data and application services began on 1 June 2022. Between 1 February and 5 March 2024, the satellite drifted from 133°E to 105°E, where it replaced FY-4A as the primary operational platform. Since March 5, FY-4B has resumed operational services at 105°E. The satellite's advanced GHI instrument captures imagery over a 2000 × 2000 km^2 area with spatial resolutions ranging from 250 to 2000 m and a temporal resolution of up to 1 min, enabling ISW propagation and evolution to be monitored in the northern South China Sea.

Figure 2a displays a 250 m resolution image from FY-4B GHI, captured at 04:02 UTC on 19 June 2023, following radiometric calibration, geometric correction, and cropping. The image clearly reveals the presence of an ISW. By analyzing the image intensity along the ISW crest line and its surroundings, the brightest points (indicated by the cyan dots in Figure 2a) are identified. A grayscale profile is then extracted perpendicular to the ISW crest (represented by the cyan oblique line in Figure 2a), as shown in the upper-left inset. In this profile, bright and dark stripes correspond to positive and negative peaks, respectively. The distance between these peaks, referred to as the peak-to-peak (PP) distance ($D_{p\text{-}p}$), can be measured from the grayscale profile. The PP distance is closely related to the characteristic half-width of ISWs [34] and serves as a key parameter for the retrieval of ISWs [49].

Since the observation locations of FY-4B GHI vary daily, a total of 46 days between June 2022 and July 2024 were selected, during which ISWs in the northern South China Sea were clearly observed, as listed in Table 1. The daily observation time of ISWs was mainly concentrated between 00:00 and 09:00 UTC. The northern South China Sea is divided into grids with a resolution of 0.25°, consistent with the grid structure of the 0.25° temperature and salinity fields in the World Ocean Atlas 2023 (WOA23) dataset [50]. Figure 2b shows the propagation and evolution of ISWs over a continuous 3 h period. The PP distance of the ISWs in each grid is defined as the average value of the PP distances extracted from multiple ISW crest lines within that grid (as indicated by the red and cyan dots in Figure 2b). Figure 2c shows the positions of the ISW crest lines on the left and right sides of a single grid cell. The ISW propagation speed in each grid is defined as the average speed of all of the brightest points propagating from the right to the left side within that grid. The retrieval of ISW parameters is based on the PP distances and the temperature and salinity data from the WOA23 within each grid.

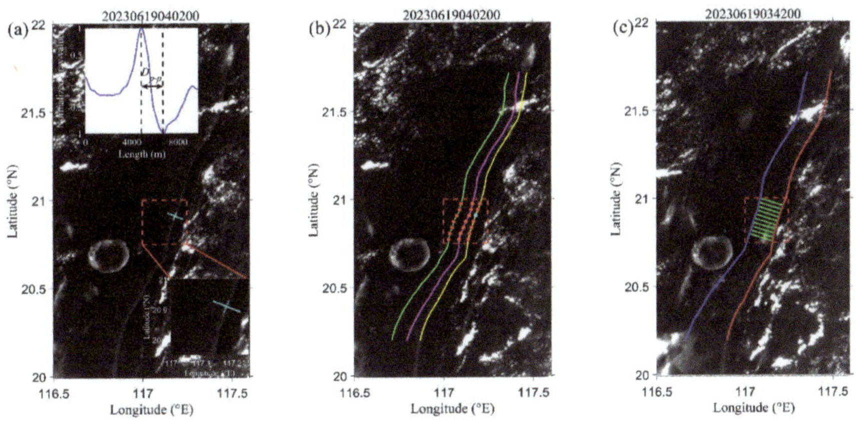

Figure 2. FY-4B GHI imagery of the northern South China Sea. (**a**,**b**) Images captured at 04:02:00 UTC on 19 June 2023; (**c**) images captured at 03:42:00 UTC on the same day. In (**a**), cyan dots mark the brightest points of the ISW, and the cyan oblique line indicates the profile extraction path, perpendicular to the crest and opposite to the propagation direction. The lower-right inset enlarges the dashed rectangle, while the upper-left inset shows the grayscale profile with black dashed lines marking the horizontal positions of the maximum and minimum gray values. In (**b**), red dots show where PP distances are extracted; cyan dots correspond to those in (**a**). Yellow, purple, and green lines indicate ISW crests at 04:02, 05:02, and 06:02 UTC, respectively. In (**c**), the red line shows the crest at 03:42 UTC, the blue line shows its position at 06:42 UTC, and the green line shows the propagation path. Red dashed boxes denote 0.25° × 0.25° WOA grid cells.

Table 1. Summary of remote sensing imagery for ISW observations in the northern South China Sea.

Index	Observation Period	ISW Packets	Index	Observation Period	ISW Packets
1	2 June 2022 03:01:00–09:01:00	5	24	19 June 2023 02:02:00–09:02:00	5
2	5 June 2022 02:01:00–09:01:00	4	25	20 June 2023 03:01:00–09:41:00	5
3	18 June 2022 02:01:00–06:31:00	5	26	21 June 2023 01:01:00–09:01:00	5
4	19 June 2022 02:01:00–05:01:00	3	27	22 June 2023 02:31:00–09:01:00	3
5	3 July 2022 03:02:00–09:02:00	3	28	24 June 2023 06:31:00–07:31:00	1
6	4 July 2022 05:31:00–08:31:00	1	29	25 June 2023 04:37:00–08:37:00	2
7	26 August 2022 03:01:00–09:01:00	2	30	26 June 2023 04:01:00–08:01:00	2
8	27 August 2022 01:12:00–08:02:00	1	31	20 July 2023 03:01:00–06:51:00	7
9	3 September 2022 00:01:00–08:01:00	4	32	22 July 2023 04:01:00–08:01:00	4
10	4 September 2022 02:31:00–06:31:00	3	33	23 July 2023 05:01:00–08:01:00	4
11	5 September 2022 03:01:00–08:01:00	3	34	24 July 2023 04:01:00–09:01:00	1
12	13 September 2022 01:01:00–09:01:00	5	35	15 August 2023 02:01:00–05:01:00	1
13	14 September 2022 03:01:00–06:01:00	4	36	16 August 2023 05:01:00–09:01:00	2
14	4 May 2023 03:01:00–08:01:00	2	37	17 August 2023 04:31:00–05:31:00	1
15	5 May 2023 03:01:00–08:01:00	5	38	22 August 2023 02:01:00–09:01:00	6
16	6 May 2023 03:01:00–09:01:00	2	39	20 September 2023 04:01:00–08:01:00	1
17	7 May 2023 03:01:00–07:01:00	2	40	30 April 2024 00:01:00–08:01:00	2
18	31 May 2023 03:01:00–08:01:00	1	41	1 June 2024 02:01:00–05:01:00	1
19	6 June 2023 05:01:00–07:01:00	1	42	10 June 2024 02:01:00–05:01:00	1
20	7 June 2023 03:01:00–08:01:00	5	43	21 June 2024 00:01:00–05:01:00	3
21	8 June 2023 00:31:00–08:31:00	4	44	9 July 2024 03:31:00–05:31:00	4
22	9 June 2023 00:01:00–09:01:00	5	45	10 July 2024 01:01:00–07:01:00	4
23	18 June 2023 01:01:00–09:01:00	2	46	11 July 2024 02:01:00–05:01:00	5

2.2. ISW Theoretical Equations and Parameter Extraction

Without making any assumptions about wavelength or amplitude, the DJL equation [51,52] was established and is expressed as follows:

$$\nabla^2\eta + \frac{N^2(z-\eta)}{c^2}\eta = 0, \tag{1}$$

where η represents the isopycnal displacement and c is the propagation speed of ISWs. The Laplacian operator is defined as $\nabla^2 = (\partial_x^2 + \partial_z^2)$, and x and z denote the zonal and vertical coordinates, respectively. In this study, the ISW amplitude (η_0) retrieved using the DJL equation refers to the maximum isopycnal displacement, defined as follows:

$$\eta_0 = \max(|\eta(x,z)|). \tag{2}$$

N in Equation (1) is the buoyancy frequency, given by the following:

$$N^2(z) = -\frac{g}{\rho_0}\frac{d\rho(z)}{dz}, \tag{3}$$

where g is the gravitational acceleration, $\rho(z)$ is the density profile, and ρ_0 is the reference density. The density profiles were derived from WOA23 temperature and salinity profiles using the equation of state proposed by Millero and Poisson [53]. The temperature and salinity inputs used here are the monthly and seasonal averages of seven decadal means (1955–2022), with a spatial resolution of $0.25° \times 0.25°$. Seasonal data are applied at depths greater than 1500 m, while monthly data are used for depths less than 1500 m. It should be noted that the DJL equation does not have explicit solutions, which can only be solved by the numerical method [52,54].

When retrieving ISW parameters from optical imagery, it is important to account for the relationship between sea surface current divergence and the surface features of ISWs. Lu et al. [55] investigated the relationship using both satellite remote sensing and laboratory experiments. By statistically analyzing the ISW PP distances in optical imagery and the PP distances of the surface current divergence for the same ISW, and performing linear fitting, they found that the PP distance of the sea surface current divergence is 0.83 times that of the optical imagery. The experimental and remote sensing results are consistent. Furthermore, the uncertainty of this relationship is derived as 0.03. Since the maximum PP distance extracted from remote sensing imagery in this study is 3207 m, the error introduced by this relationship is 96 m, which is smaller than the image resolution. The error in the inversion amplitude due to the PP distance is 1 m, and the error in propagation speed is 0.004 m/s. Therefore, the error in this relationship has no significant impact on the inversion results.

The PP distance of the sea surface current divergence is defined as follows:

$$D_{p\text{-}p} = \left| x_{|min(\frac{du}{dx})} - x_{|max(\frac{du}{dx})} \right|. \tag{4}$$

Therefore, the relationship between the ISW PP distance extracted from optical imagery ($(D_{p\text{-}p})_{\text{opt}}$) and the PP distance of the sea surface current divergence derived from the DJL theory ($(D_{p\text{-}p})_{\text{div}}$) can be given by the following:

$$(D_{p\text{-}p})_{\text{div}} = (D_{p\text{-}p})_{\text{opt}} \times 0.83 \tag{5}$$

3. Results

In this study, a total of 2067 ISW PP distances were extracted from 46 days of FY-4B GHI imagery that clearly captured ISWs in the northern South China Sea. The extraction locations are shown in Figure 3. The study area was divided into deep-water, transitional, and shallow-water regions based on the 500 m and 2000 m isobaths. Using the continuously stratified DJL equation and the extracted PP distances, ISW parameters in the northern South China Sea were retrieved, and their spatial variation characteristics were subsequently analyzed.

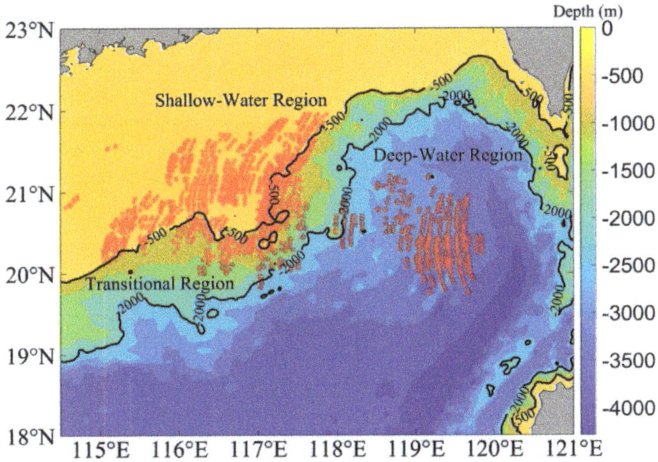

Figure 3. Spatial distribution of ISW PP distance extraction locations in the northern South China Sea. Red asterisks mark the extraction points. Based on the 500 m and 2000 m isobaths, the region is divided into deep-water, transitional, and shallow-water regions.

3.1. Retrieval of ISW Parameters in the Deep-Water Region

In the deep-water region, ISW parameters were retrieved using the continuously stratified DJL equation, driven by seasonal temperature and salinity data from WOA23. Figure 4a,b illustrate the relationships between the PP distance and amplitude and between the PP distance and propagation speed of ISWs, respectively, as derived from the DJL equation under different depth conditions. As shown, the relationship between the wave parameters and PP distance is no longer monotonic, and significant differences are exhibited when the amplitudes become large. Its typical feature is the existence of a turning point, which means that one PP distance will correspond to two parameters, that is, double solutions exist. Since it is difficult to distinguish between the two solutions using only the PP distance as a variable, Xue et al. [56] determined the double solutions based on the properties of wave packets in which the leading wave reaches the maximum amplitude. Instead, this study uses the ISW propagation speed estimated from continuous remote sensing imagery to more clearly determine whether to select the solution above or below the turning point. Once the propagation speed is determined, the corresponding amplitude is also fixed. By considering both the propagation speed and PP distance, the double solutions can be eliminated. Additionally, the selection of the double solutions can also be determined based on the historical measured data from the region. According to in situ observations by Huang et al. [57] in the deep-water region, the average ISW amplitude is typically less than 100 m. The turning point amplitude of the DJL solution in this region is always greater than 125 m. During their three-month continuous observations, a total of 177 ISWs were measured, of which only 4.5% had amplitudes exceeding 125 m. Therefore, the amplitudes below the turning point were selected, and the corresponding propagation speeds below the turning point were also chosen.

As shown in Figure 4a, the DJL model curves for the PP distance and amplitude exhibit distinct differences between the depth ranges of 2800–3400 m and 2200–2600 m. The ISWs in deeper waters (2800–3400 m) correspond to larger extracted PP distances than those in shallower waters (2200–2600 m). However, differences in the curves across the two depth ranges lead to only minor variations in the retrieved ISW amplitudes, despite the substantial differences in the PP distance. Figure 4b shows that the relationship between the PP distance and propagation speed displays only minor differences across different depths. Figure 4c compares the ISW propagation speed estimated from continuous satellite imagery with that retrieved using the DJL equation corresponding to solutions below the turning point. The results indicate that the remotely sensed propagation speed is generally greater than that retrieved from the DJL model. This discrepancy is similar to the in situ observations by Huang et al. [57], who found that the measured ISW propagation speeds in deep water were generally greater than the theoretical propagation speeds. This difference may be attributed to the proximity of the deep-water region to ISW generation sites. The surrounding area is an active region for background processes such as local tidal currents, internal tides, mesoscale eddies [47,57], and the Kuroshio [58]. As a result, ISWs in the deep-water region are more likely to be influenced by these processes. These background processes not only generate strong currents but also alter stratification, leading to deviations from the WOA climate mean profile. This study simulated the impact of stratification changes caused by an anticyclonic eddy with a 100 m amplitude on the ISW propagation speed retrieved using the DJL equation. The results showed that the stratification changes led to a 0.96 m/s increase in ISW propagation speed at a PP distance of 2500 m. The combined effects of background currents and stratification changes result in the difference between the ISW propagation speeds estimated from remote sensing and the theoretical propagation speeds in the deep-water region.

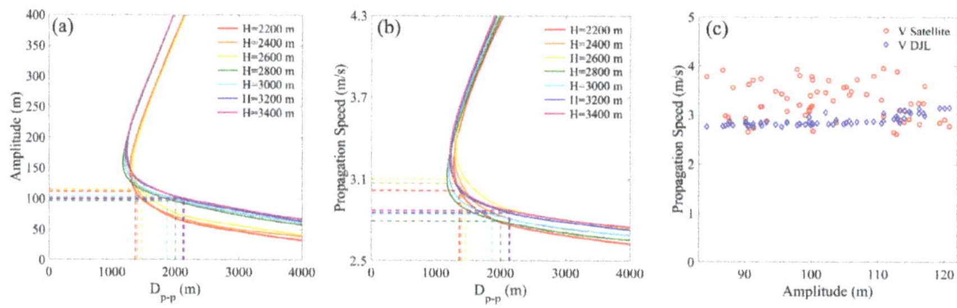

Figure 4. (**a**,**b**) Relationships between ISW PP distance and amplitude (**a**) and between PP distance and propagation speed (**b**), as derived from the DJL equation under different depth conditions in the deep-water region. In (**a**,**b**), the red, orange, yellow, green, cyan, blue, and purple lines correspond to DJL solutions at depths of 2200, 2400, 2600, 2800, 3000, 3200, and 3400 m (H represents the total depth), respectively. The dashed lines in corresponding colors indicate the ISW PP distances extracted and the amplitudes and propagation speeds retrieved at their respective depths. (**c**) Variation in ISW propagation speed with amplitude in the deep-water region. Red circles indicate propagation speeds estimated from continuous satellite remote sensing imagery, and blue diamonds represent those retrieved using the DJL equation.

In the deep-water region (depths ranging from 2200 to 3400 m), a total of 555 ISW PP distances were extracted from continuous optical imagery, and the corresponding amplitudes and propagation speeds were retrieved using the continuously stratified DJL equation. The spatial distribution characteristics of the PP distance, amplitude, and propagation speed of ISWs are presented in Figure 5a–c. As shown in Figure 5a, the PP distance of ISWs, extracted from consecutive optical imagery, ranges from 1440 to 3087 m. From east (120°E) to west (117°E), the PP distances gradually decrease with decreasing depth. Figure 5b shows that the ISW amplitude ranges from 89 to 121 m, increasing gradually from east to west. Figure 5c indicates that the ISW propagation speeds range from 2.6 to 3.7 m/s. Influenced by background currents, these speeds are unevenly distributed, with slightly greater values observed in the southern portion of the study area.

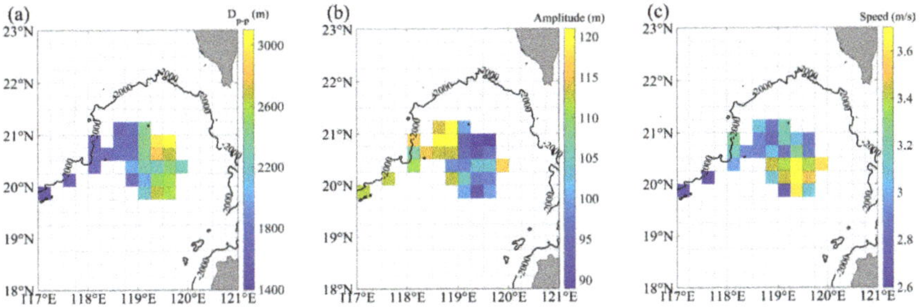

Figure 5. (**a**) Spatial distribution of ISW PP distances extracted from optical imagery, (**b**) ISW amplitudes retrieved using the DJL theory, and (**c**) ISW propagation speeds estimated from continuous remote sensing imagery, all in the deep-water region. The grid resolution for (**a**–**c**) is 0.25°.

3.2. Retrieval of ISW Parameters in the Transitional Region

In the transitional region, the monthly and seasonal WOA23 temperature and salinity profiles were used to derive the continuously stratified DJL equation, from which the relationships between the ISW PP distance and amplitude (Figure 6a) and between the PP distance and propagation speed (Figure 6b) were derived at different depths. Based on the ISW propagation speed estimated from continuous remote sensing imagery, the DJL solution above or below the turning point corresponding to the same PP distance was selected. The results indicated that most of the DJL solutions corresponded to parameters above the turning point. For a few cases in which parameters were initially matched below the turning point, resulting in unrealistic amplitude outliers, the solution above the turning point was instead selected to ensure physical consistency.

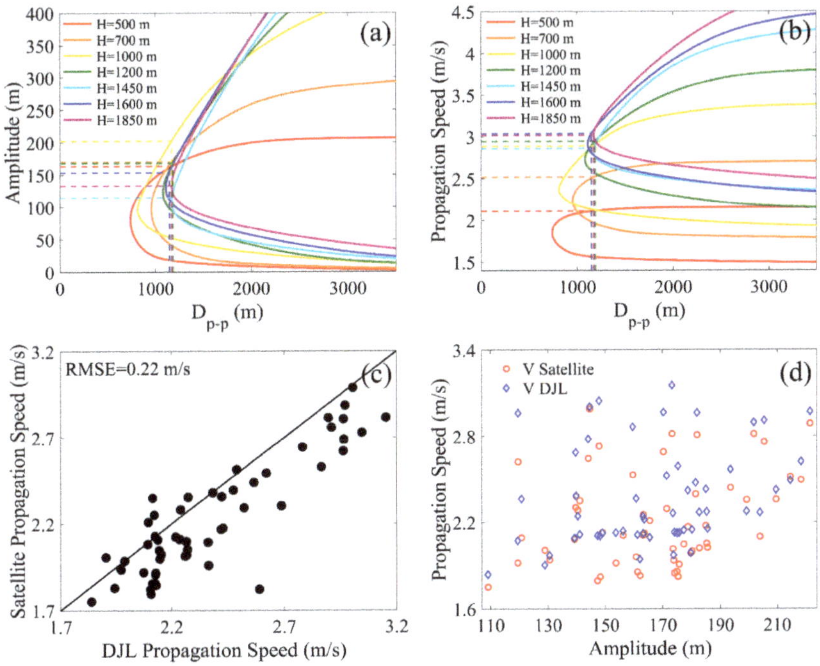

Figure 6. (**a**,**b**) Relationships between ISW PP distance and amplitude (**a**) and between PP distance and propagation speed (**b**), as derived from the DJL equation at different depths in the transitional

region. In (**a**,**b**), the red, orange, yellow, green, cyan, blue, and purple lines correspond to DJL solutions at depths of 2200, 2400, 2600, 2800, 3000, 3200, and 3400 m (H represents the total depth), respectively. The dashed lines in corresponding colors indicate the ISW PP distances extracted and the amplitudes and propagation speeds retrieved at their respective depths. (**c**) Comparison between ISW propagation speeds estimated from satellite remote sensing imagery and those retrieved using the DJL equation in the transitional region. RMSE denotes the root mean square error, and the solid black line represents the 1:1 reference. (**d**) Variations in ISW propagation speed with amplitude in the transitional region. Red circles indicate propagation speeds estimated from continuous satellite remote sensing imagery, and blue diamonds represent those retrieved using the DJL equation.

Figure 6a shows that the relationship between the ISW PP distance and amplitude, as derived from the DJL equation, exhibits significant variations at different depths. Although the PP distance of ISWs changes only slightly across this region, the retrieved amplitudes vary considerably. Figure 6b indicates that the curve of the propagation speed versus the PP distance derived from the DJL equation decreases as the depth becomes shallower. Given the relatively small variation in the PP distance, the propagation speed of ISWs progressively decreases as the water becomes shallower. As shown in Figure 6c, the ISW propagation speeds estimated from the continuous satellite remote sensing imagery closely match those retrieved from the DJL equation, with a root mean square error of 0.22 m/s. Figure 6d presents the variation in the ISW propagation speed with amplitude. The lack of a clear correlation suggests that depth may be the dominant factor influencing the ISW propagation speed in the transitional region.

In the transitional region, where depths range from 500 to 1850 m, a total of 486 ISW PP distances were extracted from continuous optical imagery. Using the continuously stratified DJL equation, the corresponding amplitudes and propagation speeds were retrieved, and the spatial distributions of the PP distance, amplitude, and propagation speed were further analyzed (Figure 7a–c). Figure 7a shows that the ISW PP distances range from 1182 to 1697 m, with less variability compared to those in the deep-water region. As shown in Figure 7b, the ISW amplitudes retrieved using the DJL model range from 120 to 210 m, exceeding those retrieved in the deep-water region. These amplitudes are strongly influenced by depth and density, leading to significant variations. Figure 7c indicates that the propagation speeds retrieved using the DJL equation range from 2.0 to 3.1 m/s, and thus are lower than those in the deep-water region. Overall, ISWs in the transitional region exhibit shorter PP distances, greater amplitudes, and lower propagation speeds compared to those in the deep-water region.

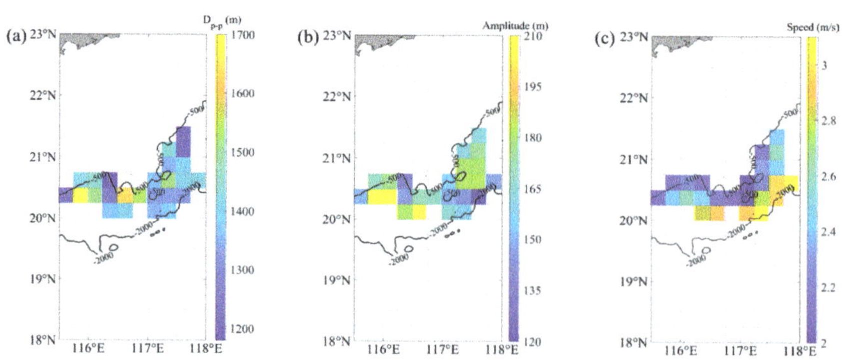

Figure 7. (**a**) Spatial distribution of ISW PP distances extracted from optical imagery, (**b**) ISW amplitudes retrieved using the DJL theory, and (**c**) ISW propagation speeds retrieved using the DJL theory, all in the transitional region. The grid resolution for (**a**–**c**) is 0.25°.

3.3. Retrieval of ISW Parameters in the Shallow-Water Region

In the shallow-water region, the DJL equation was solved using monthly WOA23 temperature and salinity profiles. Figure 8a,b illustrate the relationships between the ISW PP distance and amplitude, and between the PP distance and propagation speed, respectively, as obtained from the DJL equation under varying depth conditions. The DJL solutions were likewise selected based on the propagation speeds estimated using the continuous remote sensing imagery. Similarly to the transitional region, most cases corresponded to parameters above the turning point. For the few cases with initial mismatches, appropriate corrections were applied to ensure physical consistency.

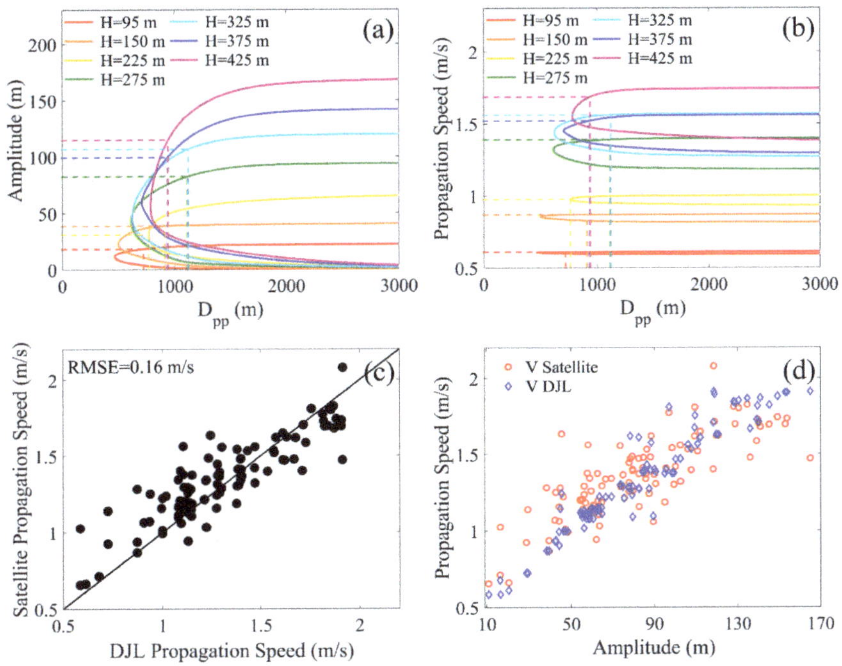

Figure 8. (**a**,**b**) Relationships between ISW PP distance and amplitude (**a**) between and PP distance and propagation speed (**b**), as derived from the DJL equation under different depth conditions in the shallow-water region. In (**a**,**b**), the red, orange, yellow, green, cyan, blue, and purple lines correspond to DJL solutions at depths of 2200, 2400, 2600, 2800, 3000, 3200, and 3400 m (H represents the total depth), respectively. The dashed lines in corresponding colors indicate the ISW PP distances extracted and the amplitudes and propagation speeds retrieved at their respective depths. (**c**) Comparison between ISW propagation speeds estimated from satellite remote sensing imagery and those retrieved using the DJL equation in the shallow-water region. RMSE denotes the root mean square error, and the solid black line represents the 1:1 reference. (**d**) Variation in ISW propagation speed with amplitude in the shallow-water region. Red circles indicate propagation speeds estimated using continuous satellite remote sensing imagery, and blue diamonds represent those retrieved using the DJL equation.

Figure 8a,b show that, as the depth gradually decreases, both the relationship between the PP distance and amplitude and that between the PP distance and propagation speed, as retrieved using the DJL equation, exhibit a decreasing trend. Although the ISW PP distances vary under different depth conditions, the retrieved amplitudes and propagation speeds generally decrease as the depth becomes shallower. This suggests that in the shallow-water region, the ISW amplitude and propagation speed are strongly influenced by depth, with both parameters tending to decline as the depth decreases. Figure 8c presents a comparison between the ISW propagation speeds estimated using the continuous satellite imagery and those retrieved using the DJL equation, showing a good overall agreement, with a root mean square error of 0.16 m/s. Figure 8d illustrates that the ISW propagation speed increases with amplitude, indicating a positive correlation between the two in shallow waters.

In the shallow-water region (depth range: 95–425 m), a total of 1026 ISW PP distances were extracted from the continuous optical imagery, and the corresponding amplitudes and propagation speeds were retrieved using the continuously stratified DJL equation. Figure 9a–c illustrate the spatial distributions of the ISW PP distance, amplitude, and propagation speed. As shown in Figure 9a, the ISW PP distances range from 876 to 1561 m, with only minor differences compared to those in the transitional region. Figure 9b shows that the ISW amplitudes, retrieved using the DJL model, range from 14 to 142 m, indicating a reduction relative to the transitional region. Figure 9c reveals that the ISW propagation speeds retrieved with the DJL equation range from 0.6 to 1.9 m/s, being lower than those in the transitional region. Overall, both the ISW amplitude and propagation speed decrease with decreasing depth, and all of the wave parameters are reduced compared to those in the transitional region.

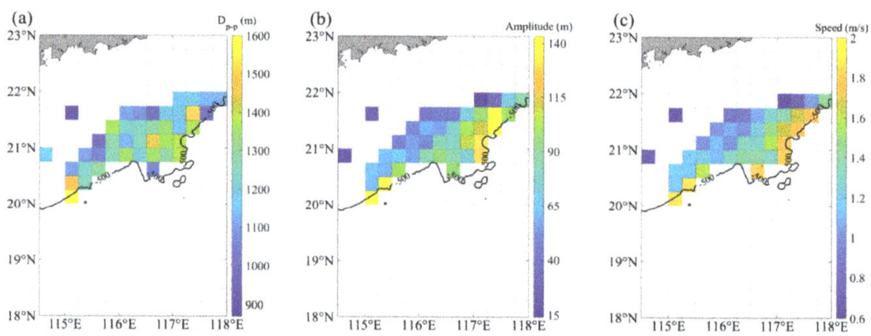

Figure 9. (**a**) Spatial distribution of ISW PP distances extracted from optical imagery, (**b**) ISW amplitudes retrieved using the DJL theory, and (**c**) ISW propagation speeds retrieved using the DJL theory, all in the shallow-water region. The grid resolution for (**a–c**) is 0.25°.

3.4. Spatial Variability of ISW Parameters in the Northern South China Sea

Using the continuously stratified DJL equation and continuous optical remote sensing imagery, the spatial distribution of ISW parameters across a depth range of 95–3400 m in the northern South China Sea was retrieved, as shown in Figure 10. Figure 10a illustrates that the ISW PP distances range from 876 to 3087 m, with the largest values observed in the deep-water region, where the average is 2134 m. As ISWs propagate westward, the PP distances decrease, averaging 1287 m. In the transitional and shallow-water regions, the PP distances show minimal variation. Figure 10b shows that the ISW amplitudes range from 14 to 210 m. In the deep-water region, the amplitudes remain relatively constant, averaging 102 m. As ISWs propagate westward into the transitional region, the intensified nonlinear effects lead to a notable increase in the amplitude. In the shallow-water region, the amplitudes gradually decrease with decreasing depth. Figure 10c indicates that the ISW propagation speeds range from 0.6 to 3.7 m/s, with the highest overall speeds in the deep-water region, where the average speed is 3.1 m/s. As ISWs propagate westward and the depth decreases, the propagation speeds decrease to an average of 2.3 m/s, and further reduce to 1.3 m/s in the shallow-water region.

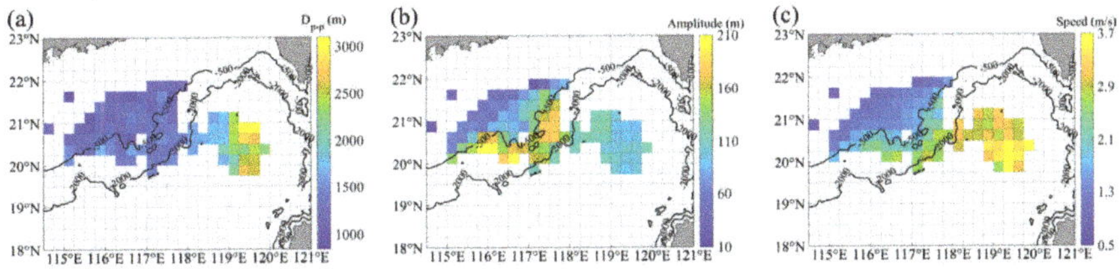

Figure 10. Spatial distributions of (**a**) ISW PP distances, (**b**) ISW amplitudes, and (**c**) ISW propagation speeds in the northern South China Sea, respectively. The grid resolution for (**a–c**) is 0.25°.

4. Discussion

4.1. Impact of Seasonal Stratification Variations on the Retrieval of ISW Parameters

This study examines the impact of seasonal stratification variations on the retrieval of ISW parameters based on the DJL equation. In the deep-water region, for ISW PP distances between 2500 and 3500, the retrieval amplitude was smallest in summer and largest in winter, with the difference between winter and summer being approximately 20.5%. The retrieval propagation speed was lowest in spring and highest in summer, with the difference between spring and summer being approximately 6.5%. In the transition region, for ISW PP distances between 1500 and 2500, the retrieval amplitude was smallest in winter and largest in spring, with the difference between spring and winter ranging from 20% to 34%. The retrieval propagation speed was lowest in winter and highest in summer, with the difference between summer and winter being approximately 25.5%. In the shallow-water region, for ISW PP distances between 1500 and 2500, the seasonal variations in the retrieved wave parameters were similar to those in the transition region. The difference between spring and winter in the retrieval amplitude ranged from 37% to 50%, and the difference in retrieval propagation speed between summer and winter was approximately 27.5%. In summary, the impact of seasonal stratification variations on the retrieval of ISW parameters was significant, with the seasonal effect increasing as the depth decreased.

4.2. Comparison Between Retrieved ISW Parameters and In Situ Observations

The ISW parameters in the northern South China Sea were retrieved by combining continuous optical remote sensing imagery with the continuously stratified DJL equation, and the results were compared with in situ observations. In the deep-water region, Huang et al. [57] reported average in situ ISW amplitudes below 100 m and propagation speeds ranging from 2.8 to 3.7 m/s. In comparison, the retrieval using the DJL equation produced an average amplitude of 102 m, and the propagation speeds estimated from the remote sensing imagery ranged from 2.6 to 3.7 m/s.

In the transitional region, Lien et al. [30] conducted mooring observations at a depth of 525 m (117.28°E, 21.07°N), reporting ISW amplitudes of 106–173 m and propagation speeds of 1.61–1.98 m/s. At the same location, the ISW amplitudes retrieved using the DJL equation ranged from 130 to 204 m, with propagation speeds between 1.97 and 2.27 m/s. Similarly, Ramp et al. [33] measured an ISW amplitude of 100 m and a propagation speed of 1.8 m/s at a depth of 505 m (117.08°E, 20.7°N), while the results retrieved from the DJL equation at the same site ranged from 109 to 176 m in amplitude and from 1.84 to 2.13 m/s in speed.

In the shallow-water region, Yang et al. [28] reported ISW amplitudes of 90 ± 15 m and propagation speeds of 1.52 ± 0.04 m/s, estimated using the KdV equation, at a depth of 426 m (117.22°E, 21.05°N). At a nearby location, Lien et al. [30] measured amplitudes of 106–173 m and speeds of 1.61–1.9 m/s. The amplitudes retrieved using the DJL equation at the same site ranged from 120 to 146 m, with propagation speeds of 1.63–1.84 m/s. Additionally, Chen et al. [31] observed amplitudes ranging from 10 to 100 m and estimated propagation speeds using the KdV equation as being between 0.74 and 1.88 m/s at a depth of 397 m (115.5°E, 20.52°N), while the DJL equation retrieved amplitudes of 89 to 112 m and speeds of 1.58 to 1.61 m/s.

Overall, the retrieved ISW parameters using the DJL model showed good agreement with in situ observations across various regions, supporting the applicability and reliability of the DJL equation for ISW parameter retrieval in the northern South China Sea.

4.3. An Example of the Propagation and Evolution of a Particular ISW in the Northern South China Sea

During the study period, the optical FY-4B GHI imagery featured the same ISW over three consecutive days. Using the continuously stratified DJL equation, the ISW's parameters were retrieved from different regions, facilitating a systematic analysis of its propagation and evolution, as shown in Figure 11 At 04:51 UTC on 18 June 2023, the ISW was observed in the deep-water region at 119.37°E, 20.63°N (point A1 in Figure 11a), with a depth of 2900 m. The extracted ISW PP distance was 3053 m, with a maximum amplitude of 88 m and a propagation speed of 2.8 m/s, as retrieved using the DJL equation (Figure 11c). Along its northwestward propagation path (shown by the black diagonal line in Figure 11a), the ISW propagated into progressively shallower waters (Figure 11e). At 04:52 UTC on 19 June 2023, the ISW was observed at the junction of the transitional and shallow-water regions (117.21°E, 21.12°N; point A2 in Figure 11a). The depth was 400 m, with a PP distance of 1450 m, a maximum amplitude of 148 m, and a propagation speed of 1.8 m/s. Compared to the ISW at point A1, the ISW at point A2 was located in shallower water and exhibited a shorter PP distance, larger amplitude, and lower propagation speed, consistent with parameter variations observed from the deep-water to the transitional region. From A1 to A2, the ISW propagated for 232 km, yielding an average speed of 2.7 m/s, in good agreement with the mean propagation speed (2.7 m/s) of ISWs along the black solid line between the two points.

On the second day, as the ISW propagated and encountered Dongsha Island, it underwent refraction and split into two branches (indicated by the green diagonal lines in Figure 11a,b). One branch propagated through the shallow waters above the Dongsha Island and was observed at location A3 (116.12°E, 21.38°N) at 04:51 UTC on 20 June 2023. At this site, the depth was 200 m, the extracted ISW PP distance was 1188 m, and the retrieved maximum amplitude and propagation speed were 60 m and 1.1 m/s, respectively. At A3, the ISW was located in shallower water than at A2, with the PP distance, amplitude, and propagation speed all decreasing with decreasing depth, in agreement with the parameter variation trend observed in shallow-water regions. The ISW propagated 116 km from A2 to A3 at an average speed of 1.3 m/s, which closely matched the mean propagation speed (1.3 m/s) along the black solid line between A2 and A3 in the shallow-water region.

Another ISW propagated beneath Dongsha Island. On the second day, it was observed at location B2 (116.89°E, 20.38°N), where the depth was 500 m. The extracted PP distance was 1669 m, the maximum amplitude was 181 m, and the propagation speed was 2.2 m/s. Although B2 and A2 correspond to the same ISW, differences in the local depth and density structure resulted in variations in the PP distance, amplitude, and propagation speed. On the third day, the ISW was observed at B3 (115.56°E, 20.88°N), where the depth was 225 m. The extracted PP distance was 1108 m, and the maximum amplitude and propagation speed retrieved using the DJL equation were 70 m and 1.2 m/s, respectively. At B3, compared to B2, the ISW was located in shallower water, with a decrease in the PP distance, propagation speed, and amplitude, consistent with the trend observed from the transitional to shallow-water regions. From B2 to B3, the ISW propagated 149 km at an average speed of 1.7 m/s, closely matching the mean speed (1.8 m/s) along the black solid line between B2 and B3. Although the travel time from A2 to A3 and from B2 to B3 was the same, the ISW from B2 to B3 covered a 33 km longer distance, resulting in a higher average propagation speed. This difference was attributed to a greater average depth along the B2–B3 path compared to A2–A3 (as shown in Figure 11e,f), since shallower depths correspond to slower ISW speeds. Although A3 and B3 are located in different regions, the ISW parameters differed little due to the similar depths, supporting the conclusion that depth is the primary factor influencing ISW parameters in shallow-water regions. In summary, variations in the parameters of the same ISW across different regions were consistent with the spatial variation trends analyzed in Section 3.

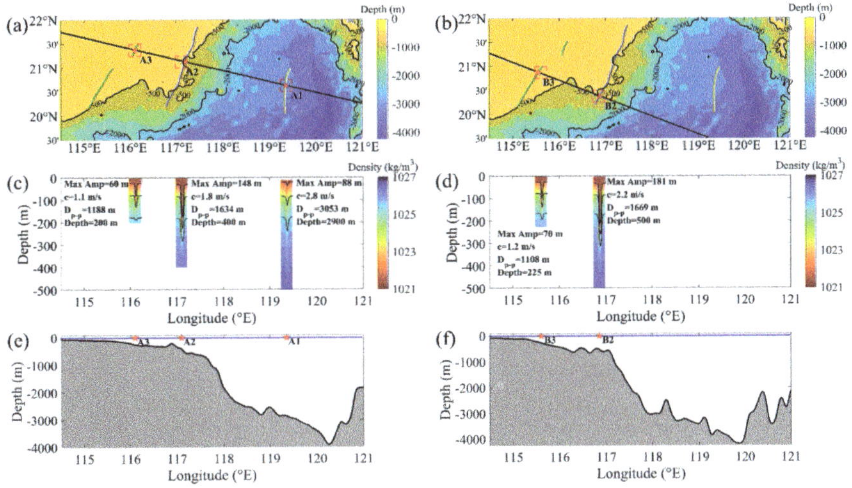

Figure 11. (**a**,**b**): Propagation of the same ISW in the northern South China Sea. The background shows the bathymetry, with yellow, blue, and green lines representing wave crests extracted at three different times: 04:51:00 UTC on 18 June 2023 (yellow), 04:52:00 UTC on 19 June 2023 (blue), and 04:51:00 UTC on 20 June 2023 (green), respectively. Black diagonal lines represent the ISW propagation directions, and red pentagrams mark the intersection points between black lines and the wave crests. Red dashed boxes denote 0.25° × 0.25° areas corresponding to grid cells in the WOA23 dataset. (**c**,**d**): Isopycnal displacements calculated using the DJL equation. Background shading represents density variation. Black contours indicate the displacements of isopycnals with densities of 1022, 1024, and 1025.5 kg·m^{-3}. (**e**,**f**): Bathymetric profiles along the propagation paths indicated by the black lines in panels (**a**,**b**), respectively. The blue line denotes the sea surface (0 m), and red pentagrams correspond to the same locations as in panels (**a**,**b**).

The ISW propagated from deep water (A1) to shallow water (A3), with its amplitude initially increasing and then decreasing, which aligned with the in situ observations of Alford et al. [6]. Lamb and Warn-Varnas [59], using a two-dimensional non-hydrostatic primitive equation model, also found that as the depth decreases from 3000 m, the ISW amplitude increases significantly, reaching a maximum between 300 and 600 m, after which it rapidly decreases. In the deep-water region, the ISW was generated through the nonlinear steepening of internal tides in the Luzon Strait. As the ISW propagated westward, the nonlinear effects intensified. In the transitional region, influenced by the seafloor topography, the ISW shoaled and underwent steepening deformation, resulting in an increase in amplitude. As the ISW propagated into the shallow-water region, its energy dissipated due to shear, instability, and breaking, resulting in a decrease in amplitude.

The ISW propagation speed decreased as the depth decreased (A1 to A3), consistent with the in situ observations of Alford et al. [6], who showed that the ISW propagation speed decreased from deep water (3668 m) to shallow water (320 m). According to the ISW theory equation [60], the ISW propagation speed consists of two components: nonlinear and linear phase speeds. The nonlinear phase speed is primarily influenced by the wave amplitude, while the linear phase speed dominates and is mainly determined by depth, with higher speeds in deeper water. Therefore, as the ISW propagated from deep to shallow water, its propagation speed decreased with the reduction in depth. In the shallow-water region, where the linear phase speed was lower, the relative contribution of nonlinear effects on wave speed became more prominent.

5. Conclusions

In this study, ISW parameters in the northern South China Sea were retrieved by integrating continuous optical FY-4B GHI imagery using the DJL equation under continuous stratification. The retrieved parameters were used to characterize spatial variations in ISWs across deep-water, transitional, and shallow-water regions. The results showed that in the deep-water region, ISWs exhibited the greatest PP distances and propagation speeds among all regions, with mean values of 2134 m and 3.1 m/s, respectively, while the amplitudes showed only limited variability, with an average of 102 m. As ISWs propagated westward into the transitional region, the depth decreased and nonlinearity intensified. As a result, the PP distance and propagation speed decreased to average values of 1389 m and 2.3 m/s, respectively, while the amplitude increased, ranging from 120 to 210 m. In the shallow-water region, ISWs propagated northwestward. As the depth continued to decrease, the PP distance remained relatively constant (averaging 1227 m), while both the amplitude (ranging from 14 to 142 m) and propagation speed (averaging 1.3 m/s) gradually decreased with depth. Overall, in the northern South China Sea, the ISW PP distances ranged from 876 to 3087 m, exhibiting a decreasing trend from east to west. The amplitudes ranged from 14 to 210 m, increasing initially and then decreasing as ISWs propagated westward. The propagation speeds ranged from 0.6 to 3.7 m/s and declined progressively with decreasing depth.

In the inversion process, selecting appropriate DJL solutions is critical. Appropriate DJL solutions for each area, except for the deep-water region, can be determined from an analysis of the ISW propagation speeds estimated from continuous remote sensing imagery. In both transitional and shallow-water regions, most of the DJL solutions corresponded to parameters above the turning point. Comparisons with in situ measurements confirmed that the retrieved amplitudes and speeds were consistent with the field observations. This supports the validity and applicability of the proposed inversion framework when it is applied in the northern South China Sea. Furthermore, a three-day consecutive remote sensing tracking analysis of the same ISW revealed that the spatial variation in its parameters was consistent with the statistical results presented earlier. This study provides essential data support for understanding the spatial variability and dynamic characteristics of ISWs in the northern South China Sea and offers an effective remote sensing inversion method for future extensive ISW research and monitoring.

Author Contributions: Conceptualization, K.L., T.X., C.J. and X.C.; methodology, K.L., T.X., C.J. and X.C.; software, K.L. and T.X.; writing—original draft preparation, K.L.; writing—review and editing, K.L., T.X., C.J., X.C. and X.H. All authors have read and agreed to the published version of the manuscript.

Funding: This work was supported by the National Key Research and Development Program of China (grant no. 2024YFC2815704) and the National Natural Science Foundation of China (grant no. 42476012).

Data Availability Statement: FY-4B GHI optical remote sensing imagery was downloaded from http://satellite.nsmc.org.cn/DataPortal/cn/home/index.html (accessed on 20 May 2025). The World Ocean Atlas 2023 is available at https://www.ncei.noaa.gov/access/world-ocean-atlas-2023/

(accessed on 20 May 2025). The results presented in this paper are available at https://zenodo.org/records/15474353 (accessed on 20 May 2025).

Acknowledgments: Figures were made with Matlab version 2021a (The MathWorks Inc., Natick, MA, USA, 2021), available at https://www.mathworks.com (accessed on 20 May 2025).

Conflicts of Interest: The authors declare no conflicts of interest.

References

1. Jackson, C.R. Internal wave detection using the Moderate Resolution Imaging Spectroradiometer (MODIS). *J. Geophys. Res. Ocean.* **2007**, *112*, C11012. [CrossRef]
2. Klymak, J.M.; Pinkel, R.; Liu, C.-T.; Liu, A.K.; David, L. Prototypical solitons in the South China Sea. *Geophys. Res. Lett.* **2006**, *33*, 11. [CrossRef]
3. Huang, X.; Chen, Z.; Zhao, W.; Zhang, Z.; Zhou, C.; Yang, Q.; Tian, J. An extreme internal solitary wave event observed in the northern South China Sea. *Sci. Rep.* **2016**, *6*, 30041. [CrossRef] [PubMed]
4. Grue, J.; Jensen, A.; Rusås, P.-O.; Sveen, J.K. Properties of large-amplitude internal waves. *J. Fluid Mech.* **1999**, *380*, 257–278. [CrossRef]
5. Lamb, K.G. A numerical investigation of solitary internal waves with trapped cores formed via shoaling. *J. Fluid Mech.* **2002**, *451*, 109–144. [CrossRef]
6. Alford, M.H.; Lien, R.-C.; Simmons, H.; Klymak, J.; Ramp, S.; Yang, Y.J.; Tang, D.; Chang, M.-H. Speed and evolution of nonlinear internal waves transiting the South China Sea. *J. Phys. Oceanogr.* **2010**, *40*, 1338–1355. [CrossRef]
7. da Silva, J.C.B.; New, A.L.; Magalhaes, J.M. On the structure and propagation of internal solitary waves generated at the Mascarene Plateau in the Indian Ocean. *Deep Sea Res. Part I Oceanogr. Res. Pap.* **2011**, *58*, 229–240. [CrossRef]
8. Osborne, A.R.; Burch, T.L. Internal solitons in the Andaman Sea. *Science* **1980**, *208*, 451–460. [CrossRef]
9. Xu, A.; Chen, X. A Strong Internal Solitary Wave with Extreme Velocity Captured Northeast of Dong-Sha Atoll in the Northern South China Sea. *J. Mar. Sci. Eng.* **2021**, *9*, 1277. [CrossRef]
10. Purwandana, A.; Cuypers, Y. Characteristics of internal solitary waves in the Maluku Sea, Indonesia. *Oceanologia* **2023**, *65*, 333–342. [CrossRef]
11. Tian, Z.; Jia, Y.; Du, Q.; Zhang, S.; Guo, X.; Tian, W.; Zhang, M.; Song, L. Shearing stress of shoaling internal solitary waves over the slope. *Ocean Eng.* **2021**, *241*, 110046. [CrossRef]
12. Huang, S.; Huang, X.; Zhao, W.; Chang, Z.; Xu, X.; Yang, Q.; Tian, J. Shear instability in internal solitary waves in the northern South China Sea induced by multiscale background processes. *J. Phys. Oceanogr.* **2022**, *52*, 2975–2994. [CrossRef]
13. Xu, J.; Xie, J.; Chen, Z.; Cai, S.; Long, X. Enhanced mixing induced by internal solitary waves in the South China Sea. *Cont. Shelf Res.* **2012**, *49*, 34–43. [CrossRef]
14. Bian, C.; Ruan, X.; Wang, H.; Jiang, W.; Liu, X.; Jia, Y. Internal solitary waves enhancing turbulent mixing in the bottom boundary layer of continental slope. *J. Mar. Syst.* **2022**, *236*, 103805. [CrossRef]
15. Chen, T.-Y.; Tai, J.-H.; Ko, C.-Y.; Hsieh, C.-H.; Chen, C.-C.; Jiao, N.; Liu, H.-B.; Shiah, F.-K. Nutrient pulses driven by internal solitary waves enhance heterotrophic bacterial growth in the South China Sea. *Environ. Microbiol.* **2016**, *18*, 4312–4323. [CrossRef] [PubMed]
16. Reid, E.C.; DeCarlo, T.M.; Cohen, A.L.; Wong, G.T.; Lentz, S.J.; Safaie, A.; Hall, A.; Davis, K.A. Internal waves influence the thermal and nutrient environment on a shallow coral reef. *Limnol. Oceanogr.* **2019**, *64*, 1949–1965. [CrossRef]
17. Dong, J.; Zhao, W.; Chen, H.; Meng, Z.; Shi, X.; Tian, J. Asymmetry of internal waves and its effects on the ecological environment observed in the northern South China Sea. *Deep Sea Res. Part I Oceanogr. Res. Pap.* **2015**, *98*, 94–101. [CrossRef]
18. Woodson, C.B. The fate and impact of internal waves in nearshore ecosystems. *Annu. Rev. Mar. Sci.* **2018**, *10*, 421–441. [CrossRef]
19. Badiey, M.; Wan, L.; Song, A. Three-dimensional mapping of evolving internal waves during the Shallow Water 2006 experiment. *J. Acoust. Soc. Am.* **2013**, *134*, EL7–EL13. [CrossRef]
20. Apel, J.R.; Ostrovsky, L.A.; Stepanyants, Y.A.; Lynch, J.F. Internal solitons in the ocean and their effect on underwater sound. *J. Acoust. Soc. Am.* **2007**, *121*, 695–722. [CrossRef]
21. Huang, M.; Gao, C.; Zhang, N. Numerical research on hydrodynamic characteristics and influence factors of underwater vehicle in internal solitary waves. *Int. J. Offshore Polar Eng.* **2023**, *33*, 27–35. [CrossRef]
22. Cheng, L.; Du, P.; Hu, H.; Xie, Z.; Yuan, Z.; Kaidi, S.; Chen, X.; Xie, L.; Huang, X.; Wen, J. Control of underwater suspended vehicle to avoid 'falling deep' under the influence of internal solitary waves. *Ships Offshore Struct.* **2024**, *19*, 1349–1367. [CrossRef]
23. Song, Z.J.; Teng, B.; Gou, Y.; Lu, L.; Shi, Z.M.; Xiao, Y.; Qu, Y. Comparisons of internal solitary wave and surface wave actions on marine structures and their responses. *Appl. Ocean Res.* **2011**, *33*, 120–129. [CrossRef]
24. Cui, J.; Dong, S.; Wang, Z.; Han, X.; Yu, M. Experimental research on internal solitary waves interacting with moored floating structures. *Mar. Struct.* **2019**, *67*, 102641. [CrossRef]

25. Lien, R.-C.; Tang, T.Y.; Chang, M.H.; D'Asaro, E.A. Energy of nonlinear internal waves in the South China Sea. *Geophys. Res. Lett.* **2005**, *32*, L05615. [CrossRef]
26. Simmons, H.; Chang, M.-H.; Chang, Y.-T.; Chao, S.-Y.; Fringer, O.; Jackson, C.R.; Ko, D.S. Modeling and prediction of internal waves in the South China Sea. *Oceanography* **2011**, *24*, 88–99. Available online: https://www.jstor.org/stable/24861123 (accessed on 20 May 2025). [CrossRef]
27. Ramp, S.R.; Tang, T.Y.; Duda, T.F.; Lynch, J.F.; Liu, A.K.; Chiu, C.-S.; Bahr, F.L.; Kim, H.R.; Yang, Y.J. Internal solitons in the northeastern South China Sea. Part I: Sources and deep water propagation. *IEEE J. Ocean. Eng.* **2004**, *29*, 1157–1181. [CrossRef]
28. Yang, Y.-J.; Tang, T.Y.; Chang, M.H.; Liu, A.K.; Hsu, M.-K.; Ramp, S.R. Solitons northeast of Tung-Sha Island during the ASIAEX pilot studies. *IEEE J. Ocean. Eng.* **2004**, *29*, 1182–1199. [CrossRef]
29. Ramp, S.R.; Yang, Y.J.; Bahr, F.L. Characterizing the nonlinear internal wave climate in the northeastern South China Sea. *Nonlinear Process. Geophys.* **2010**, *17*, 481–498. [CrossRef]
30. Lien, R.-C.; Henyey, F.; Ma, B.; Yang, Y.J. Large-amplitude internal solitary waves observed in the northern South China Sea: Properties and energetics. *J. Phys. Oceanogr.* **2014**, *44*, 1095–1115. [CrossRef]
31. Chen, L.; Zheng, Q.; Xiong, X.; Yuan, Y.; Xie, H.; Guo, Y.; Yu, L.; Yun, S. Dynamic and statistical features of internal solitary waves on the continental slope in the northern South China Sea derived from mooring observations. *J. Geophys. Res. Ocean.* **2019**, *124*, 4078–4097. [CrossRef]
32. Chang, M.-H.; Cheng, Y.-H.; Yang, Y.J.; Jan, S.; Ramp, S.R.; Reeder, D.B.; Hsieh, W.-T.; Ko, D.S.; Davis, K.A.; Shao, H.-J.; et al. Direct measurements reveal instabilities and turbulence within large amplitude internal solitary waves beneath the ocean. *Commun. Earth Environ.* **2021**, *2*, 15. [CrossRef]
33. Ramp, S.R.; Yang, Y.-J.; Jan, S.; Chang, M.-H.; Davis, K.A.; Sinnett, G.; Bahr, F.L.; Reeder, D.B.; Ko, D.S.; Pawlak, G. Solitary waves impinging on an isolated tropical reef: Arrival patterns and wave transformation under shoaling. *J. Geophys. Res. Ocean.* **2022**, *127*, e2021JC017781. [CrossRef]
34. Zheng, Q.; Yuan, Y.; Klemas, V.; Yan, X.-H. Theoretical expression for an ocean internal soliton synthetic aperture radar image and determination of the soliton characteristic half width. *J. Geophys. Res. Ocean.* **2001**, *106*, 31415–31423. [CrossRef]
35. Chen, G.-Y.; Su, F.-C.; Wang, C.-M.; Liu, C.-T.; Tseng, R.-S. Derivation of internal solitary wave amplitude in the South China Sea deep basin from satellite images. *J. Oceanogr.* **2011**, *67*, 689–697. [CrossRef]
36. Huang, X.; Zhao, W. Information of Internal Solitary Wave Extracted from MODIS Image: A Case in the Deep Water of Northern South China Sea. *Period. Ocean Univ. China* **2014**, *44*, 19–23.
37. Zhang, X.; Wang, J.; Sun, L.; Meng, J. Study on the amplitude inversion of internal waves at Wenchang area of the South China Sea. *Acta Oceanol. Sin.* **2016**, *35*, 14–19. [CrossRef]
38. Jia, T.; Liang, J.; Li, X.-M.; Fan, K. Retrieval of internal solitary wave amplitude in shallow water by tandem spaceborne SAR. *Remote Sens.* **2019**, *11*, 1706. [CrossRef]
39. Xie, H.; Xu, Q.; Zheng, Q.; Xiong, X.; Ye, X.; Cheng, Y. Assessment of theoretical approaches to derivation of internal solitary wave parameters from multi-satellite images near the Dongsha Atoll of the South China Sea. *Acta Oceanol. Sin.* **2022**, *41*, 137–145. [CrossRef]
40. Chen, H.; Wang, Z.; Cui, J.; Xia, H.; Guo, W. Application of different internal solitary wave theories for SAR remote sensing inversion in the northern South China Sea. *Ocean Eng.* **2023**, *283*, 115015. [CrossRef]
41. Choi, W.; Camassa, R. Fully nonlinear internal waves in a two-fluid system. *J. Fluid Mech.* **1999**, *396*, 1–36. [CrossRef]
42. Helfrich, K.R.; Melville, W.K. Long nonlinear internal waves. *Annu. Rev. Fluid Mech.* **2006**, *38*, 395–425. [CrossRef]
43. Grue, J.; Jensen, A.; Rusås, P.-O.; Sveen, J.K. Breaking and broadening of internal solitary waves. *J. Fluid Mech.* **2000**, *413*, 181–217. [CrossRef]
44. Stanton, T.P.; Ostrovsky, L.A. Observations of highly nonlinear internal solitons over the continental shelf. *Geophys. Res. Lett.* **1998**, *25*, 2695–2698. [CrossRef]
45. Chang, M.-H.; Lien, R.-C.; Lamb, K.G.; Diamessis, P.J. Long-term observations of shoaling internal solitary waves in the Northern South China Sea. *J. Geophys. Res. Ocean.* **2021**, *126*, e2020JC017129. [CrossRef]
46. Yang, Y.; Huang, X.; Zhao, W.; Zhou, C.; Huang, S.; Zhang, Z.; Tian, J. Internal solitary Waves in the Andaman Sea revealed by longterm mooring observations. *J. Phys. Oceanogr.* **2021**, *51*, 3609–3627. [CrossRef]
47. Huang, X.; Huang, S.; Zhao, W.; Zhang, Z.; Zhou, C.; Tian, J. Temporal variability of internal solitary waves in the northern South China Sea revealed by long-term mooring observations. *Prog. Oceanogr.* **2022**, *201*, 102716. [CrossRef]
48. Xu, T.; Chen, X.; Li, Q.; He, X.; Wang, J.; Meng, J. Strongly nonlinear effects on determining internal solitary wave parameters from surface signatures with potential for remote sensing applications. *Geophys. Res. Lett.* **2023**, *50*, e2023GL105814. [CrossRef]
49. Zhang, X.; Wang, H.; Wang, S.; Liu, Y.; Yu, W.; Wang, J.; Xu, Q.; Li, X. Oceanic internal wave amplitude retrieval from satellite images based on a data-driven transfer learning model. *Remote Sens. Environ.* **2022**, *272*, 112940. [CrossRef]

50. Reagan, J.R.; Boyer, T.P.; García, H.E.; Locarnini, R.A.; Baranova, O.K.; Bouchard, C.; Cross, S.L.; Mishonov, A.V.; Paver, C.R.; Seidov, D.; et al. *World Ocean Atlas 2023;* NOAA National Centers for Environmental Information: Silver Spring, MD, USA, 2024. Available online: https://www.ncei.noaa.gov/products/world-ocean-atlas (accessed on 20 May 2025).
51. Long, R.R. Some aspects of the flow of stratified fluids: I. A theoretical investigation. *Tellus* **1953**, *5*, 42–58. [CrossRef]
52. Stastna, M.; Lamb, K.G. Large fully nonlinear internal solitary waves: The effect of background current. *Phys. Fluids* **2002**, *14*, 2987–2999. [CrossRef]
53. Millero, F.J.; Poisson, A. International one-atmosphere equation of state of seawater. *Deep Sea Res. Part A Oceanogr. Res. Pap.* **1981**, *28*, 625–629. [CrossRef]
54. Dunphy, M.; Subich, C.; Stastna, M. Spectral methods for internal waves: Indistinguishable density profiles and double-humped solitary waves. *Nonlinear Process. Geophys.* **2011**, *18*, 351–358. [CrossRef]
55. Lu, K.; Xu, T.; Chen, X.; He, X.; Tan, J. Relationships between internal solitary wave surface features in optical and SAR satellite images: Insights from remote sensing and laboratory. *Ocean Eng.* **2024**, *309*, 118500. [CrossRef]
56. Xue, J.; Graber, H.C.; Lund, B.; Romeiser, R. Amplitudes estimation of large internal solitary waves in the Mid-Atlantic Bight using synthetic aperture radar and marine X-band radar images. *IEEE Trans. Geosci. Remote Sens.* **2012**, *51*, 3250–3258. [CrossRef]
57. Huang, X.; Zhang, Z.; Zhang, X.; Qian, H.; Zhao, W.; Tian, J. Impacts of a mesoscale eddy pair on internal solitary waves in the northern South China Sea revealed by mooring array observations. *J. Phys. Oceanogr.* **2017**, *47*, 1539–1554. [CrossRef]
58. Yu, Y.; Chen, X.; Cao, A.; Meng, J.; Yang, X.; Liu, T. Modulation of internal solitary waves by the Kuroshio in the northern South China Sea. *Sci. Rep.* **2023**, *13*, 6070. [CrossRef]
59. Lamb, K.G.; Warn-Varnas, A. Two-dimensional numerical simulations of shoaling internal solitary waves at the ASIAEX site in the South China Sea. *Nonlinear Process. Geophys.* **2015**, *22*, 289–312. [CrossRef]
60. Korteweg, D.J.; de Vries, G. On the change of form of long waves advancing in a rectangular canal, and on a new type of long stationary waves. *Lond. Edinb. Dublin Philos. Mag. J. Sci.* **1895**, *39*, 422–443. [CrossRef]

remote sensing

MDPI

Article

Automated Eddy Identification and Tracking in the Northwest Pacific Based on Conventional Altimeter and SWOT Data

Lan Zhang [1,2], Cheinway Hwang [3,*], Han-Yang Liu [3], Emmy T. Y. Chang [4] and Daocheng Yu [5]

1 Institute of Earthquake Forecasting, China Earthquake Administration, No. 63, Fuxing Road, Beijing 100036, China; zhanglan16@mails.ucas.ac.cn
2 Institute of Geodesy and Geoinformation, University of Bonn, 53115 Bonn, Germany
3 Department of Civil Engineering, National Yang Ming Chiao Tung University, 1001 Ta Hsueh Road, Hsinchu 300, Taiwan; hanyangliu.en10@nycu.edu.tw
4 Institute of Oceanography, National Taiwan University, No. 1, Sec. 4, Roosevelt Road, Taipei 106, Taiwan; etychang@ntu.edu.tw
5 School of Geomatics, Liaoning Technical University, Fuxin 123000, China; yudaocheng@lntu.edu.cn
* Correspondence: cheinway@nycu.edu.tw

Abstract: Eddy identification and tracking are essential for understanding ocean dynamics. This study employed the elliptical Gaussian function (EGF) simulations and the py-eddy-tracker (PET) algorithm, validated by Surface Velocity Program (SVP) drifter data, to track eddies in the western North Pacific Ocean. The PET method effectively identified large- and mesoscale eddies but struggled with submesoscale features, indicating areas for improvement. Simulated satellite altimetry by EGF, mirroring Surface Water and Ocean Topography (SWOT)'s high-resolution observations, confirmed PET's capability in processing fine-scale data, though accuracy declined for submesoscale eddies. Over 22 years, 1,188,649 eddies were identified, mainly concentrated east of Taiwan. Temporal analysis showed interannual variability, more cyclonic than anticyclonic eddies, and a seasonal peak in spring, likely influenced by marine conditions. Short-lived eddies were uniformly distributed, while long-lived ones followed major currents, validating PET's robustness with SVP drifters. The launch of the SWOT satellite in 2022 has enhanced fine-scale ocean studies, enabling the detection of submesoscale eddies previously unresolved by conventional altimetry. SWOT observations reveal intricate eddy structures, including small cyclonic features in the northwestern Pacific, demonstrating its potential for improving eddy tracking. Future work should refine the PET algorithm for SWOT's swath altimetry, addressing data gaps and unclosed contours. Integrating SWOT with in situ drifters, numerical models, and machine learning will further enhance eddy classification, benefiting ocean circulation studies and climate modeling.

Keywords: drifter; Kuroshio current; ocean eddy; Py_Eddy_Tracking algorithm; satellite altimetry; surface water and ocean topography (SWOT)

1. Introduction

Oceanic eddies, rotating fluid structures prevalent in the ocean's dynamic system, exhibit considerable variation across a broad spectrum of spatial and temporal scales. The spatial scales of the mesoscale eddies and submesoscale eddies are usually bounded by the Rossby deformation radius. The oceanic mesoscale can be characterized as consisting of variability with radius scales of 10–500 km [1]. Mesoscale eddies are primarily driven by geostrophic balance, significantly influencing large-scale ocean circulation and nutrient transport [2,3]. In contrast, submesoscale eddies are smaller (less than 10 km), transient (lasting days to weeks), and driven by frontal dynamics and mixed-layer instabilities, playing a crucial role in vertical exchanges of heat, momentum, and biogeochemical properties [3,4]. To more clearly delineate the scale range of eddies identified and tracked in this study, we classify oceanic eddies into three categories: large-scale eddies, with dimensions up to 100 km; mesoscale eddies, measuring between 10 and 100 km; and submesoscale or small-scale eddies, characterized by diameters less than 10 km. The small-scale eddies referred to in this study differ from those described by [1,2], whose work focused exclusively on mesoscale eddies with radii greater than approximately 40 km. The term 'small-scale eddies' in [1] refers to mesoscale eddies at the lower end of the mesoscale range, which still fall within the mesoscale regime. The temporal presence of these eddies can vary significantly, lasting from several days to multiple months, illustrating their dynamic nature [5,6].

The advent and subsequent advancement of satellite altimeter technology has made high-resolution altimeter satellites essential for studying ocean eddies. These satellites provide critical data on key oceanic parameters, such as sea level anomaly (SLA), absolute dynamic topography (ADT), and geostrophic currents, which are essential for robust identification, detection, and analysis of eddies. The formation and evolution of eddies have a substantial impact on ocean circulation, climate variability, marine biology, and other interconnected fields, underscoring the importance of their study [7–12]. The integration of data from multiple altimeter satellites for the identification and analysis of large-scale and mesoscale eddies has become a focal point in modern oceanographic research.

Despite these technological strides, the current literature reveals limitations in the comprehensive tracking and analysis of oceanic eddies, especially concerning mesoscale, submesoscale, and small-scaled formations. Early methods for using satellite altimeter data for eddy identification included manual visual analysis, which required researchers to identify oceanic eddies by visually analyzing SLA fields. This approach was inefficient and subjective, leading to inconsistent eddy identification and significant errors. The first automated procedure for detection and tracking of eddies in altimeter data were developed by [13], building upon techniques developed previously for turbulence studies from numerical model simulations [14–16]. The Okubo–Weiss threshold method often used simple thresholds to identify eddies, such as considering local maxima or minima on SLA images as eddy centers, requiring careful sensitivity analysis to avoid over-identification or under-identification of eddies [1]. Subsequent studies have fine-tuned the methodology to address various shortcomings [2,17,18]. The procedures for identifying eddies relate to geostrophic current velocity analysis [13,17–19], involving data smoothing and requiring the computation of SLA derivatives, which could reduce the resolution of the flow field and amplify any noise that is present in the SLA field [1,20]. These methods are mainly used for identifying large and mesoscale eddies.

The capacity to accurately detect and analyze submesoscale eddies remains a formidable challenge due to limitations in spatial resolution and the sensitivity of methodologies to inherent noise within satellite altimeter data. The nuanced dynamics of these submesoscale eddies are crucial for explaining the finer details and transient processes of ocean dynamics, especially in terms of vertical material exchange in the ocean [21,22], yet they are frequently underrepresented or insufficiently captured in existing studies [1,6,23]. Recent advancements, particularly the launch of the SWOT satellite in December 2022, have significantly enhanced the study of fine-scale oceanic features. SWOT offers new insights into ocean dynamics and the role of submesoscale eddies in climate processes (https://cnes.fr/en/press-releases/two-years-launch-exceptionally-precise-swot-data-reap-rewards-30-years-progress, accessed on 5 May 2025), providing an opportunity to overcome previous limitations in identifying submesoscale eddies [24]. For instance,

SWOT has successfully captured submesoscale cyclonic eddies in the northwestern Pacific, providing unprecedented detail on these dynamic structures [25].

In response to these challenges and opportunities, this paper aims to address the identified limitations in existing methodologies for eddy detection and analysis, with particular emphasis on identifying and tracking eddies across varying scales. By employing the Py_Eddy_Tracking (PET) algorithm developed by [26–28] based on the theoretical framework in [1], this study attempts to reconstruct the 22-year spatiotemporal variability of mesoscale eddies in the northwest Pacific and to evaluate the robustness of eddy tracking methodologies. Additionally, this research explores the potential of Surface Water and Ocean Topography (SWOT) sea level measurements in detecting ocean eddies. Launched on 16 December 2022, SWOT represents a breakthrough in identifying fine-scaled variations in ocean surface water [29]. Through the improvement of eddy tracking methodologies and validation using in situ data, this paper seeks to contribute to the oceanographic community's capacity to study and interpret the intricate and dynamic nature of oceanic eddies. We report on the short-term and long-term spatiotemporal evolution of ocean eddies in the northwest Pacific Ocean, validated by drifter data. For the first time, we demonstrate how SWOT altimeter data can be used in studying submesoscale eddies in the northwest Pacific Ocean and highlight SWOT's current limitations.

2. Study Area

The study area (Figure 1) is situated in the northwest Pacific Ocean, primarily south of Japan, encompassing the South China Sea (SCS), the waters off Taiwan, and the region east of Taiwan within the North Pacific subtropical countercurrent (STCC) zone. This region is characterized by dynamic oceanic conditions and complex circulatory patterns that play a critical role in the formation and behavior of ocean eddies.

The oceanography of this area is influenced by several major currents, most notably the Kuroshio Current, which flows northward along the eastern coast of Taiwan before turning eastward at the latitude of Japan. The interaction between the Kuroshio Current and the STCC creates a unique marine environment conducive to the generation of mesoscale and submesoscale eddies [30–32]. These eddies significantly influence heat, nutrient transport, and biological productivity across vast areas of the Pacific [7,33,34].

Furthermore, the Luzon Strait, situated south of Taiwan, acts as a key gateway for the exchange of water masses between the SCS and the Pacific Ocean. This strait is a region of intense eddy activity, influenced by the variable bathymetry and the interaction of the Kuroshio Current with local water masses [35,36]. The presence of these eddies contributes to complex circulation patterns that have significant implications for regional climate and marine ecosystems [37,38]. Recent studies highlight the prolific eddy activity in offshore eastern Taiwan, west of the STCC zone, where the convergence of different water masses and the influence of monsoonal winds exacerbate the conditions for eddy formation [39,40]. These eddies typically follow a west-northwestward trajectory, interacting with the Kuroshio and playing a crucial role in the lateral mixing of ocean waters and the distribution of biological and chemical constituents [12].

This study area was chosen due to its dynamic nature, providing an ideal setting for evaluating the capabilities of conventional nadir-looking altimeters and the wide-swath SWOT altimeter. The high-resolution data from these instruments enable a more detailed examination of mesoscale and submesoscale eddies, enhancing our understanding of their formation, evolution, and impacts on oceanic and atmospheric processes. The application of the PET algorithm and SWOT data in this region aims to address the current limitations in eddy detection and tracking, offering new insights into the intricate dynamics of ocean eddies.

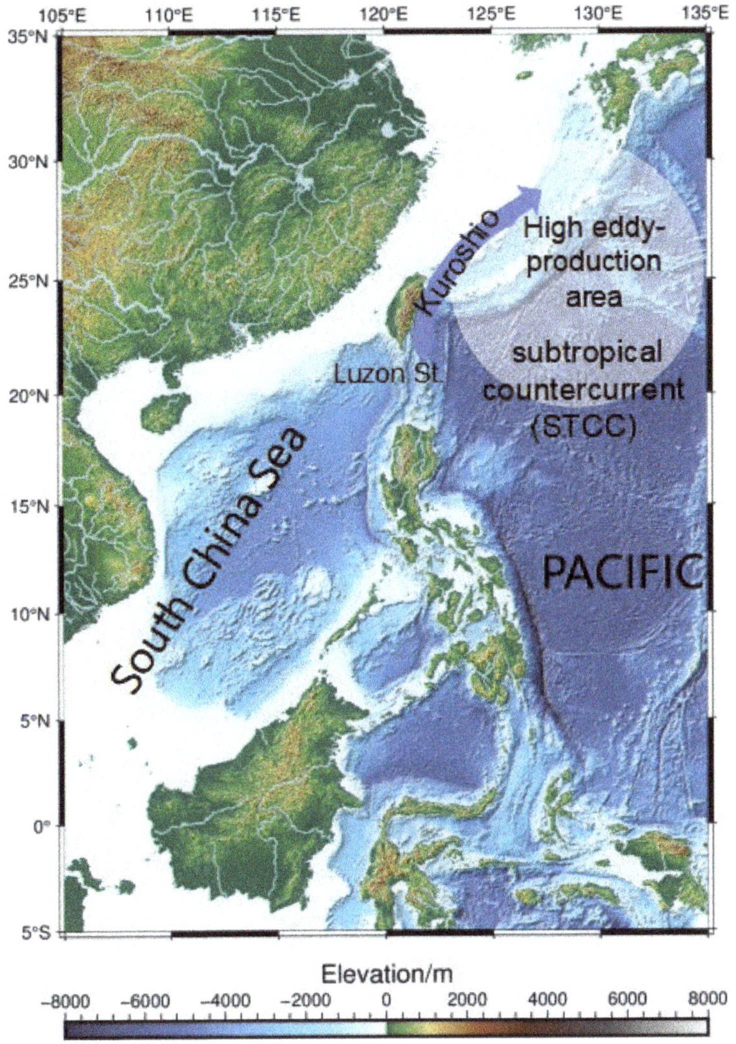

Figure 1. Seafloor and topographic map of the study area. The high-eddy-production area covers the North Pacific subtropical countercurrent (STCC). The bathymetric data from Generic Mapping Tools Version 6 @earth_relief are integrated from Shuttle Radar Topography Mission and other datasets.

3. Satellite Altimeter and In Situ Data

3.1. Altimeter Data

We use the daily global gridded SLA provided by the Copernicus Marine Environment Monitoring Service (CMEMS, marine.copernicus.eu), covering the period from 2000 to 2023. The SLA grids have a resolution of 0.25° × 0.25° and integrate observations from multiple altimetry satellites, including ERS-1/2, TOPEX/Poseidon (T/P), Jason-1/2/3, Cryosat-2, Saral/AltiKa, HY-2, and among others. SLA is calculated from the sea surface height (SSH) measured by satellite altimetry by subtracting the mean sea surface, which is the average SSH from 1993 to 2012 used in this study, and applying corrections for tides, atmospheric pressure, and electromagnetic biases. CMEMS datasets also provide geostrophic velocity components in eastward and northward directions, which are directly used in our tracking algorithm. In this paper, such SLA grids refer to grids from conventional altimetry.

The SWOT mission employs Ka-band Radar Interferometer (KaRIn), which makes measurements over a swath of 120 km (with a 20 km nadir gap that is sampled with coarse resolution along the centerline by a conventional altimeter), providing the first direct observations of ocean topography and land surface water in two dimensions [29]. One of its objectives is to measure SSH and terrestrial water heights. The KaRIn measurement technique reveals the cascading of ocean circulation processes from large scale to submesoscale, down to ~5 km in wavelength, using two synthetic aperture radar (SAR) antennas separated by a 10 m mast for interferometry in orbit [29]. We have used Level 3 expert products (L3_LR_SSH_Expert) from the SWOT mission, both in the CalVal orbit (pass 21 in the 1-day repeat/fast sampling orbit) from 29 March 2023 to 9 July 2023, and in the science orbit (21-day repeat; 4 cycles are used in this study) within the study area, from 7 September 2023, to 21 November 2023. The Level 3 products have employed a series of corrections (ocean tide from the FES22 model, solid earth tide, coherent internal tide, pole tide, and dynamic atmospheric corrections, removing spurious pixels) and mitigated random errors using an AI-based algorithm [41,42].

3.2. Drifter Data

To verify the eddy tracking results, we use drifter data from the Surface Velocity Program (SVP) [43]. A SVP drifter is a Lagrangian current-following device designed to track water currents at a depth of 15 m beneath the ocean surface. It is equipped with a buoy at the surface and a steel cable extending below, which holds multiple sensors. This setup allows the SVP drifter to collect data on ocean surface velocity by tracking the movement of the buoy and to gather temperature data via sensors on the cable. The collected information is transmitted to the surface buoy. This design allows effective measurement of the ocean's surface velocity and temperature over extended periods and across wide areas. We used SVP drifter data from 1 January 2004 to 1 January 2005, and from 1 January 2018 to 1 January 2019. The drifter data were obtained from the Global Drifter Program website (https://www.aoml.noaa.gov/phod/gdp/index.php, accessed on 5 May 2025).

4. Methods for Automatic Identification and Tracking of Eddies

4.1. Automated Identification and Tracking

We identified and tracked eddies by analyzing local extremum values in SLA and the velocity vectors of the eddies (geostrophic velocity) using an automated algorithm, PET (https://py-eddy-tracker.readthedocs.io/en/stable/, accessed on 5 May 2025), as described by [1,25]. Figure 2 shows the workflow for eddy identification. The process starts by applying a high-pass filter with a radius of 300–500 km to the daily SLA maps to emphasize significant SLA signals. These filtered SLA maps are then analyzed to find closed contours (isopleths) ranging from −100 to 100 cm, with a minimum separation of 0.4 cm between them. Isopleths that are not closed are disregarded. The algorithm then evaluates the containment relationships among polygons formed by closed isopleths. Polygons not enclosed by others are identified as eddy boundaries, while those not enclosing others are marked as eddy centers. Based on five closed-contour constraints [26] (Appendix A), eddies are classified as anticyclonic (center point higher than surroundings) or cyclonic (center point lower than surroundings).

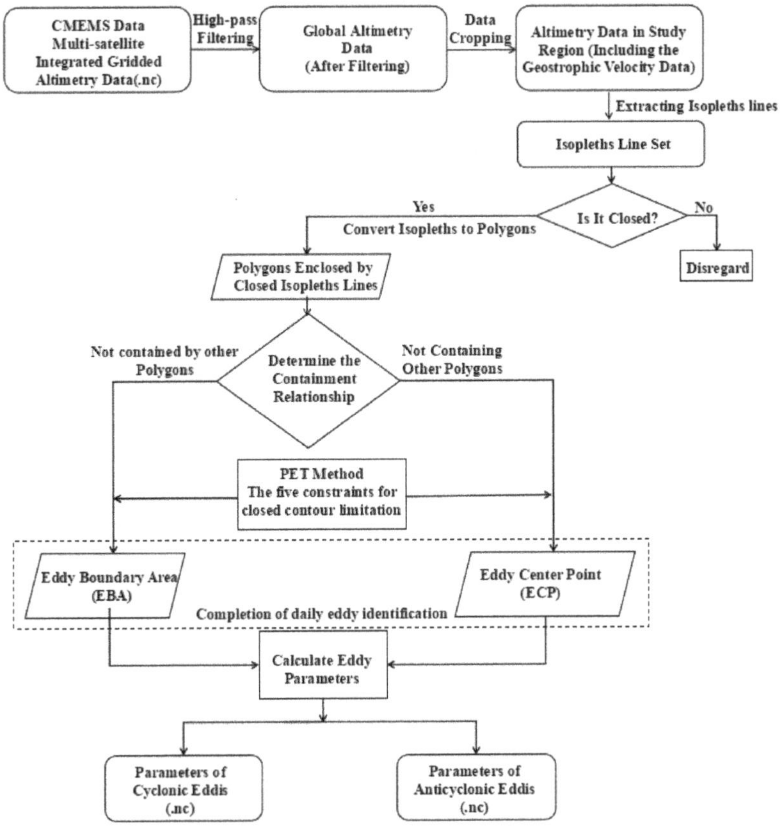

Figure 2. Workflow for eddy identification with PET.

Eddy tracking maps the movement of an eddy's center from formation to dissipation, conducted separately for anticyclonic and cyclonic types. The method identifies eddies that intersect across consecutive time steps. An eddy with no intersection is considered dissipated. If exactly two eddies intersect, they are treated as the same eddy. When more than two intersect, indicating potential merging or splitting, a cost function from [44] is applied to ensure continuity. Further details are provided in [26,44]. The five PET algorithm constraints for closed contours were also applied to higher-resolution altimeter datasets, such as SWOT SLA data. A minimum area of 8 pixels allows detection of the smallest submesoscale eddies, which typically exhibit SLA amplitudes of a few to several tens of centimeters. We use 0.4 cm contour intervals to meets the resolution needs for their identification. Details of the constraints are provided in Appendix A.

4.2. Validating PET Identification Using Simulated Eddies at Various Scales

Here, we validate eddy identification by the PET method using simulated eddies and explain the radii associated with eddy boundaries. We employ the elliptical Gaussian function (EGF) to simulate ocean eddies at the large, meso-, and sub-mesoscales to see how eddy sizes are related to identification errors. The expression for EGF is detailed in Appendix B. Ref. [45] noted that the most typical shape of an ocean eddy closely resembles that of a Gaussian curve. In this study, we simulated eddies on various spatial scales. The SLA values for these eddies are given on a $0.018° \times 0.018°$ grid within the study area

(Figure 1) to match the spatial resolution of SWOT. Table 1 shows the EGF parameters (Appendix B) for the simulated eddies.

Table 1. EGF parameters for the simulated eddies (see also Appendix B).

Eddy Scale	Eddy Center	Semi-Major Axis (km)	Semi-Minor Axis (km)	Amplitude (m)	Rotation Angle
Large-scale eddy	(125°E, 23.5°N)	125	93.75	1.0	30°
Mesoscale eddy	(124°E, 24°N)	50	37.5	0.3	45°
Submesoscale eddy	(123.5°E, 25°N)	5	3.75	0.1	60°

Figure 3 illustrates the simulated SLA and the corresponding closed contours of different scales. The results identified using the PET algorithm by inputting these simulated SLA data are shown in Figure 4. The identification of anticyclonic eddies is clearly demonstrated in all three examples, as indicated by the red contours in Figure 4, which correspond to the size and position settings in Figure 3. The red dashed line represents the identified eddies with the Effective Radius (ER) of the eddy, while the solid red line indicates the eddies with the Speed-based Radius (SR). ER represents the spatial extent or the average radius of an eddy, serving as a measure of the eddy's overall size. A larger effective radius suggests that an eddy can affect a more extensive area, potentially impacting a greater volume of water. Traditionally, the spatial extent of an eddy is determined by setting an SLA threshold, such as 2 mm [46]. In the PET algorithm, this could be adjusted based on the maximum pixel value. High-pass filtering tends to increase the area of the identified eddy. In our simulation, we assigned smaller maximum pixel values for eddies of various scales to reduce the size errors caused by filtering. The SR refers to the radius of a circle with the same area as that within the closed contours of the SLA, where the average geostrophic speed is maximized. SR helps in understanding the intensity and rotational dynamics of the eddy. Eddies with larger speed radii typically exhibit stronger rotational forces over a more extensive area. In the following sections, the closed contours corresponding to ER and SR are used to represent the shape and size of the identified eddy. We note from Figure 4c and Table 2 that for submesoscale eddies, the relative difference between ER and SR tends to decrease due to their compact structure and limited radial velocity gradient.

Table 2 shows the statistics for the parameters of identified eddies across various scales, including the ER, SR, the fitted circle radius (FCR, the squared root of the product of the set semi-major and semi-minor axes), and their associated errors. For eddies of all scales, the ER exceeds the FCR, a discrepancy attributed to high-pass filtering that enlarges the range of the eddies. With the spatial filtering applied in the objective analysis used to produce SLA gridded maps [1] (Appendix A), eddies having an amplitude weaker than ~1–2 cm or a radius smaller than ~40 km should probably be considered as noisy artifact structures on the daily SLA maps based on conventional altimeter data [44]. For large and mesoscale eddies, the ER closely matches the radii established in the simulated data, with relative errors for FCR of 5.2% and 10.2%, respectively; SR is only half of the ER. Ref. [26] pointed out that, compared to the CSS11 eddies datasets [1], where SR is about 70% of ER, the speed radius identified by the PET method is smaller. For submesoscale eddies, the relative error between ER and FCR significantly increases to 39%, and ER is the same size as SR. The results presented in Table 2 suggest that while the PET algorithm is effective, it can be further refined to improve the detection and characterization of submesoscale eddies.

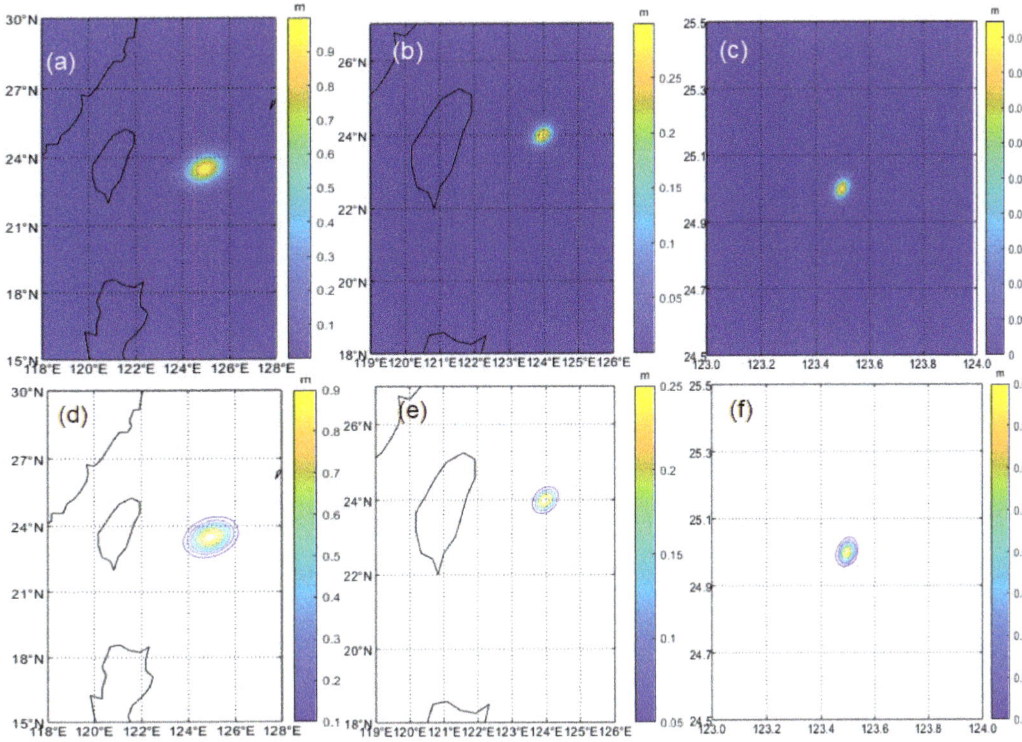

Figure 3. Simulated SLA with large-scale (**a**), mesoscale (**b**), and submesoscale (**c**) eddies. (**d**–**f**) are simulated close contours with the same scales as (**a**–**c**).

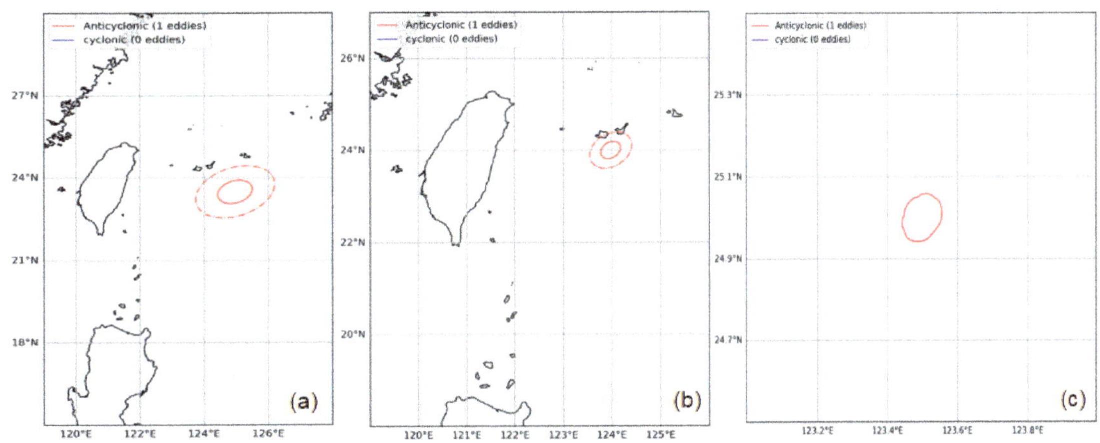

Figure 4. Identified eddies at (**a**) large-scale, (**b**) mesoscale, and (**c**) submesoscale using the PET method. The dashed red curves correspond to the eddy boundaries from ER, while the solid curves correspond to the eddy boundaries from SR.

Table 2. Parameters of eddy identification at various scales.

	ER (km)	SR (km)	FCR (km)	Relative Errors Between ER and FCR
Large-scale eddy	113.89	50.63	108.25	5.2%
Mesoscale eddy	47.73	27.57	43.30	10.2%
Submesoscale eddy	6.02	6.02	4.33	39.0%

5. Results and Discussions

5.1. Eddies from Conventional Altimeter Data

5.1.1. Snapshot of Ocean Eddies: A Case Study from 1 May 2023

As we transition into the results of our study, this section offers an analysis of the ocean eddies identified on a representative day in spring 2023. This initial exploration is important for understanding the complicated behaviors and characteristics of eddies within one day and for showing the capabilities of the PET algorithm in discerning subtle yet significant oceanic features for subsequent analyses.

We chose 1 May 2023 as the date for illustration. Figure 5a shows the spatial distribution of SLA following the application of a 500 km high-pass filter. Here, red denotes areas with higher SLA values, and blue signifies regions with lower SLA values, effectively highlighting large- to meso-scale sea surface height anomalies. The application of a high-pass filter helps to remove large-scale variability and emphasize mesoscale structures such as eddies. This method is particularly effective in regions with strong currents and complex topography [1]. Figure 5a shows noticeable eddy-like structures, particularly around the 20°N latitude.

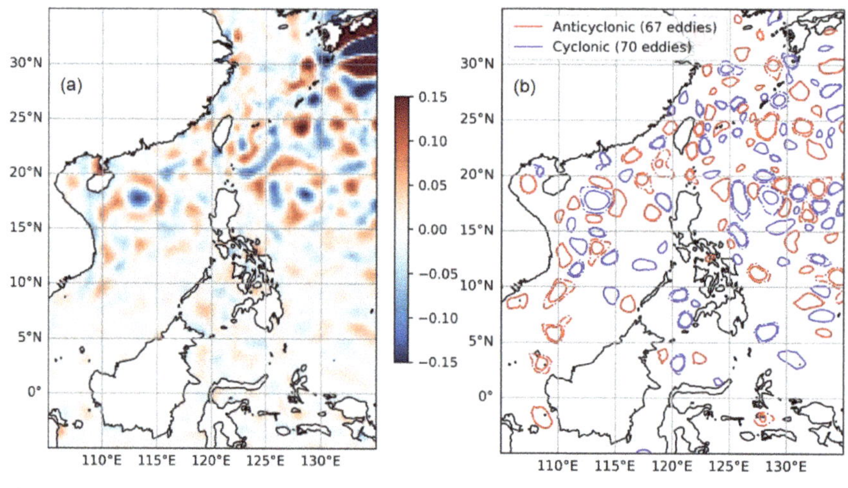

Figure 5. (**a**) SLA on 1 May 2023 by applying a 500 km high-pass filter; unit is m. (**b**) Identified CE (blue) and AE (red) obtained based on the contours extracted from Figure 5a and screened using the five contour constraints. The solid line and the dashed line represent the speed-based contour and the effective contour of the eddy, respectively.

Figure 5b shows that a total of 67 cyclonic and 70 anticyclonic eddies were identified from the analysis in Figure 5a and further refined through the PET algorithm using five contour constraints (Appendix A). The solid and dashed lines within Figure 5b delineate the effective and speed-related contours of these eddies, respectively. In the eastern region of Taiwan, there is a significant presence of highly active eddies originating from the distant

Pacific Ocean, which are pivotal in influencing the Kuroshio Current east of Taiwan. PET identification in the area east of Taiwan shows three adjacent anticyclonic eddies, consistent with the ocean current patterns associated with the Kuroshio Current east of Taiwan. In this vicinity, the ocean currents on both the east and west flanks of the Kuroshio's main stream are characterized by southward counterflows [37]. Furthermore, a considerable number of eddies are detected in the region spanning 15°S to 25°N and 105°E to 135°E, especially in the region east of the Luzon Strait. This area is known for its dynamic eddy activity, influenced by the interaction between the Kuroshio Current and the North Equatorial Current [47].

The interplay between eddy size, strength, and kinetic energy is vital for understanding ocean dynamics. Larger eddies with high eddy kinetic energy (EKE) can transport more energy and materials across long distances, influencing the ocean's thermohaline circulation and impacting weather and climate patterns. EKE is the quadratic sum of two geostrophic velocity components, which can depict the intensity of the flow, rotational patterns, fluid diffusion, and convergence, as well as the deformation behavior of the fluid under various mechanical effects [48]. Figure 6a,b display the ERs and SRs of the identified eddies, respectively. It is observed that both ERs and SRs correlate strongly with the eddies' size. In the region spanning 15°S to 20°N and 105°E to 130°E, particularly in the eastern part of Luzon Island, the area influenced by the eastward-flowing Kuroshio to the east of Japan, eddies exhibit larger ERs and VRs. The Kuroshio Current region is characterized by fast flow velocities and large flow rates, leading to an increase in energy. Under these conditions, larger eddies may be generated [49]. Eddies located near the coast are typically smaller, possibly because in the shallow waters near the coast, eddies cannot fully develop, resulting in smaller radii [11]. In Figure 6a,b, the ER of many eddies is observed to be much larger than the SR, consistent with our results in Section 4.2. The core of an eddy has high rotational speeds, represented by the SR, while the outer regions have lower speeds, contributing to a larger ER. Underwater topography, like continental shelves and ridges, can affect an eddy's size and intensity. For instance, an eddy encountering a slope might spread out horizontally, increasing the ER, but still maintain a high-speed core, as indicated by the SR [50].

Figure 6c,d present the EKE and mean EKE within each identified eddy contour calculated from SLA data. Regions characterized by large and dense eddy formations exhibit higher EKE levels, especially on the east side of Taiwan Island and Japan. This is linked to the Kuroshio Extension. Ref. [51] pointed out that the Kuroshio leaves the east coast of Japan, forming an eastward unstable jet, making the Kuroshio Extension the most active eddy region and the peak area of eddy kinetic energy in the North Pacific. A higher mean EKE does not always correspond to a larger ER. For instance, an eddy located east of Taiwan shows a relatively small ER but has a high mean EKE, as indicated by the darker color within the eddy contour in Figure 6d. Both the underwater topography and different stages of the eddy's lifespan could cause this situation. Notably, due to the Rossby parameter nearing zero in the region of 5°S~5°N, which results in an infinitely large geostrophic flow speed, the EKE values in this area should not be considered for reference because the equatorial region's dynamics are dominated by other forces, such as wind-driven currents and equatorial waves, which are not accurately represented by geostrophic calculations [52].

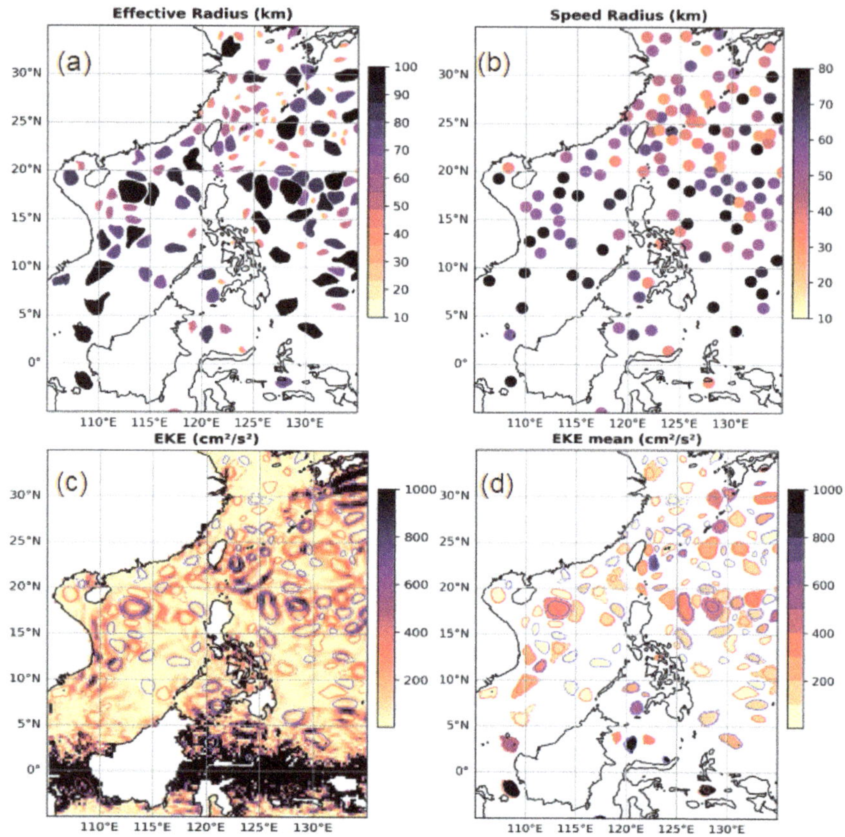

Figure 6. (**a**) ER, (**b**) SR, (**c**) EKE, and (**d**) EKE mean of identified eddies on 1 May 2023. The EKE mean refers to the filling of EKE within each identified eddy contour.

5.1.2. Long-Term Spatiotemporal Characteristics of the Detected Eddies

After examining the detailed characteristics of ocean eddies on a single spring day in 2023, we now situate these observations within longer-term trends. This section presents an analysis of the spatiotemporal patterns of eddy occurrences from 2000 to 2022, spanning more than two decades. Over the study period, the total occurrences of eddies identified using the PET method were aggregated. In our study area, a total of 1,188,649 eddies were identified during these 22 years, consisting of 600,398 cyclonic eddies and 588,251 anticyclonic eddies. This extensive dataset allows for a comprehensive analysis of eddy dynamics, providing insights into their spatial distribution and temporal variability, which are crucial for understanding oceanic processes and their impacts on climate and marine ecosystems.

Figure 7a shows the spatial distribution of all identified cyclonic eddies between 1 January 2000 and 31 December 2022 on a $1° \times 1°$ grid. The occurrences of cyclonic and anticyclonic eddies have similar spatial distributions. Significant peaks in eddy occurrences are noted in the northern SCS, the northeastern region of Taiwan adjacent to the Kuroshio. These regions are known for their high eddy activity due to the interaction of various ocean currents and the complex topography [53,54]. The highest number of cyclonic and anticyclonic eddies in a single grid ($1° \times 1°$) reached 2188 and 2380, respectively. To explore a detailed understanding of the distribution patterns and identify regions with significant eddy activity, we conducted a statistical analysis of cyclonic and anticyclonic eddy counts along both longitude and latitude, compiling counts at 0.5° intervals. The results are shown with line charts in Figure 7a,b.

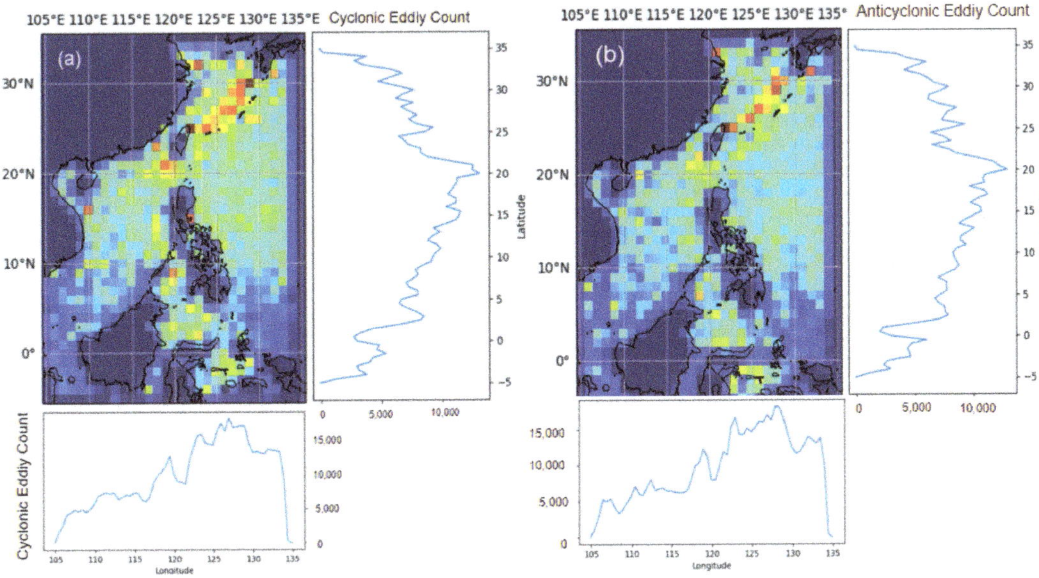

Figure 7. (**a**) The spatial distribution of identified cyclonic eddies by the PET method from 1 January 2000 to 31 December 2022 on a 1° × 1° grid. The curves on the right and bottom represent the eddy counts along longitude and latitude, respectively. (**b**) Same as (**a**) but for anticyclonic eddies.

The meridional analysis highlights a significant peak in eddy occurrences in the eastern region of Taiwan Island (123°E–130°E). Eddies in the ocean generally propagate westward due to the influence of the Coriolis force [55]. In the northwest Pacific, westward-propagating eddies are blocked by the Kuroshio, leading to the accumulation of eddy signals in the area east of Taiwan. This is reflected in Figure 7, where the number of eddies in this region is significantly greater than at other longitudes. The zonal eddies count statistics plot shows a notably higher count within the latitude range of approximately 15°N to 25°N compared to other latitude regions. This finding is consistent with the research by [7] which identified 18°N–25°N as a frequent eddy occurrence zone, and with the strong EKE observed in the 18°N–25°N, 125°E–140°E region in the 8-year satellite altimetry reanalysis by [56]. The higher eddy count in this latitude range can be attributed to the dynamic interactions between the North Equatorial Current, the Kuroshio, and the regional wind patterns, which create favorable conditions for eddy formation and propagation [49].

Figure 8 shows the mean amplitudes (peak SLA) of eddies within 1° × 1° grid cells during the period from 1 January 2000 to 31 December 2022. The amplitude of an eddy is a crucial parameter as it reflects the strength of the sea level anomaly associated with the eddy, which in turn indicates the eddy's potential impact on oceanic mixing and nutrient transport [1]. Regions with higher amplitudes still appear in the northeastern waters of Taiwan and the northern region of the SCS, consistent with statistics on effective radius and velocity radius, indicating that larger eddies experience stronger amplitude variations in SLA. This consistency suggests a robust relationship between the size of the eddy and its amplitude, which can influence the energy transfer processes within the ocean [10].

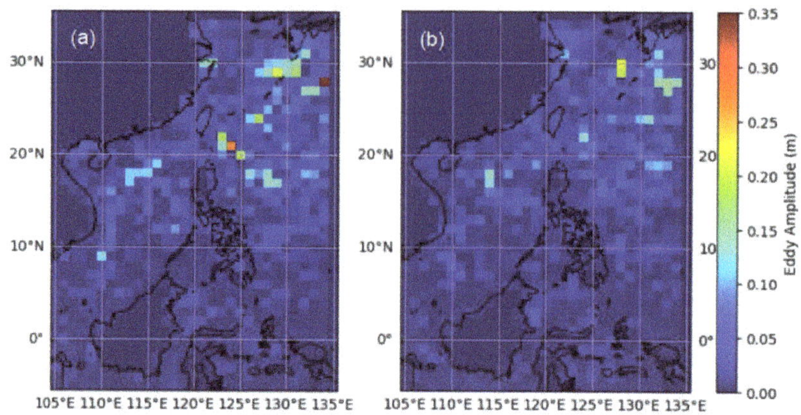

Figure 8. The spatial distribution of amplitudes of cyclonic (**a**) and anticyclonic (**b**) eddy was identified from 1 January 2000 to 31 December 2022 on a $1^{\circ} \times 1^{\circ}$ grid.

The maximum amplitudes of cyclonic and anticyclonic eddies reach 0.036 m and 0.022 m, respectively. Cyclonic eddies, characterized by their counterclockwise rotation in the northern hemisphere, typically show higher amplitudes due to the upwelling of deeper, colder water to the surface, leading to greater sea level anomalies [57]. Compared to anticyclonic eddies, regions with higher amplitudes of cyclonic eddies show a wider range and relatively larger values. The broader range of cyclonic eddies may be attributed to the more dynamic and variable nature of these features, allowing them to propagate over larger distances and interact with various oceanic currents and topographical features [58]. Eddies with high amplitudes are generated southeast of Japan. In the spatial distribution of EKE on the same day, as shown in Figure 6c also exhibits extreme values in this area, indicating a correlation between the amplitude of eddy and locally oceanic hydrodynamics.

Figure 9a illustrates the interannual variations in the number of cyclonic and anticyclonic eddies from 2000 to 2022, indicating that the number of both cyclonic and anticyclonic eddies exhibits significant interannual variability without a clear long-term increasing or decreasing trend. The occurrence of eddies in this region is influenced by a combination of factors, including oceanic currents, wind patterns, and possibly climate phenomena such as El Niño and La Niña [1]. As shown in Figure 9a, among the nine moderate-to-strong La Niña events, seven corresponded to peaks in eddy counts, especially for anticyclonic eddies. Conversely, during the three moderate-to-strong El Niño events, the counts of both cyclonic and anticyclonic eddies reached their lowest values. This finding is consistent with the analysis by [59], who investigated the influence of El Niño–Southern Oscillation (ENSO) on eddy variability in the northwest Pacific Ocean and found that the total number of eddies tends to decrease during El Niño and increase during La Niña events. Ref. [60] investigated the variability of EKE in the South China Sea and found that the EKE levels were lower than normal during El Niño events but higher during La Niña events. They suggested that El Niño (La Niña) events induce anomalous anticyclonic (cyclonic) wind stress curl over the South China Sea, which weakens (strengthens) the background cyclonic circulation and consequently leads to suppressed (enhanced) eddy activity during El Niño (La Niña) periods. The number of cyclonic and anticyclonic eddies in the region exhibits similar interannual variability, but the peak number of cyclonic eddies surpasses that of anticyclonic eddies. This is consistent with the global statistics reported by [61] and the finding of [62] in the Southwestern Taiwan Strait.

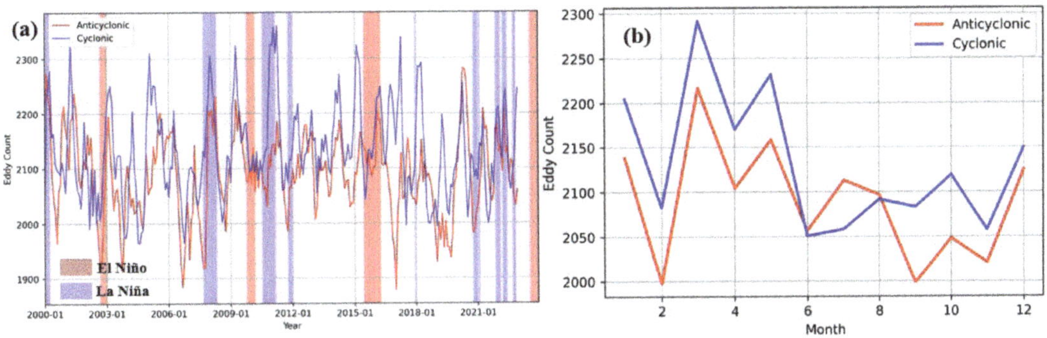

Figure 9. (**a**) Statistical analysis of cyclonic and anticyclonic eddy counts for each month in the northwest Pacific Ocean, south of Japan, with a three-month moving average. ENSO events are highlighted whenever the ONI.v5 exceeds a threshold of 1 (moderate to strong El Niño, red) or −1 (moderate to strong La Niña, blue). (**b**) The climatologically monthly mean of the number of cyclonic and anticyclonic eddies during 2000–2022.

Figure 9b shows a 22-year statistical analysis of cyclonic and anticyclonic eddy counts for each month, revealing a similar change pattern for both types of eddies. Both cyclonic and anticyclonic eddy counts peak around February and June, with a subpeak for cyclonic eddies occurring in August. The seasonal variation suggests that certain periods of the year are more conducive to eddy formation, possibly due to seasonal changes in wind patterns, ocean stratification, and the strength of the Kuroshio Current [49]. The sharp increase in eddy numbers from February to March can be attributed to seasonal variations in oceanic conditions, which is consistent with the findings of [63], who reported a notable peak in the number of cyclonic eddies between January and April in this region, such as changes in wind patterns and ocean currents, which significantly influence eddy formation and intensity [1,7,64]. The identified eddies show a relative decrease in their number during the second half of the year, particularly in October and November. This decrease could be linked to more stable oceanic conditions and dynamics that are less favorable for eddy formation during these months.

5.1.3. Eddy Validation Using SVP Drifters

We employed the PET method to track eddies within our study area. Figure 10 shows eddy trajectories with lifespans exceeding 8, 16, 32, and 48 weeks from 1 January 2000 to 5 June 2023, respectively. Figure 10a shows a total of 3875 trajectories, with roughly equal numbers of cyclones and anticyclones. Eddies with lifespans exceeding 8 weeks are widely distributed across the study area, displaying high density and complex, interwoven tracks. The region also exhibits frequent short-lived eddies, suggesting that shorter-lived eddies can form and dissipate in a variety of environments. This reflects the dynamic and variable nature of mesoscale oceanic processes [1]. Additionally, high-density eddy trajectories can be observed east and north of Taiwan and in the northern SCS regions, showing spatial characteristics consistent with the distribution of major ocean currents. Figure 10b shows the eddy trajectories exceeding 16 weeks, which are significantly fewer compared to those in Figure 10a. The number of trajectories shows that eddies with lifespans between 8 and 16 weeks account for 80% of eddies lasting longer than 8 weeks. The number of cyclonic trajectories slightly exceeds that of anticyclonic ones. Ref. [16] found that cyclones with lifespans greater than 16 weeks outnumber anticyclones by 6% globally. Tracks of eddies with lifespans over 16 weeks cover a wide area, with particularly high density in the Kuroshio region and the northern SCS. In contrast, there are almost no eddy trajectories along the coast and in enclosed sea areas formed by islands, suggesting that few eddies survive beyond 16 weeks in these regions. This is consistent with [11], who noted that eddies cannot fully develop in the shallow waters near the coast, resulting in shorter lifespans.

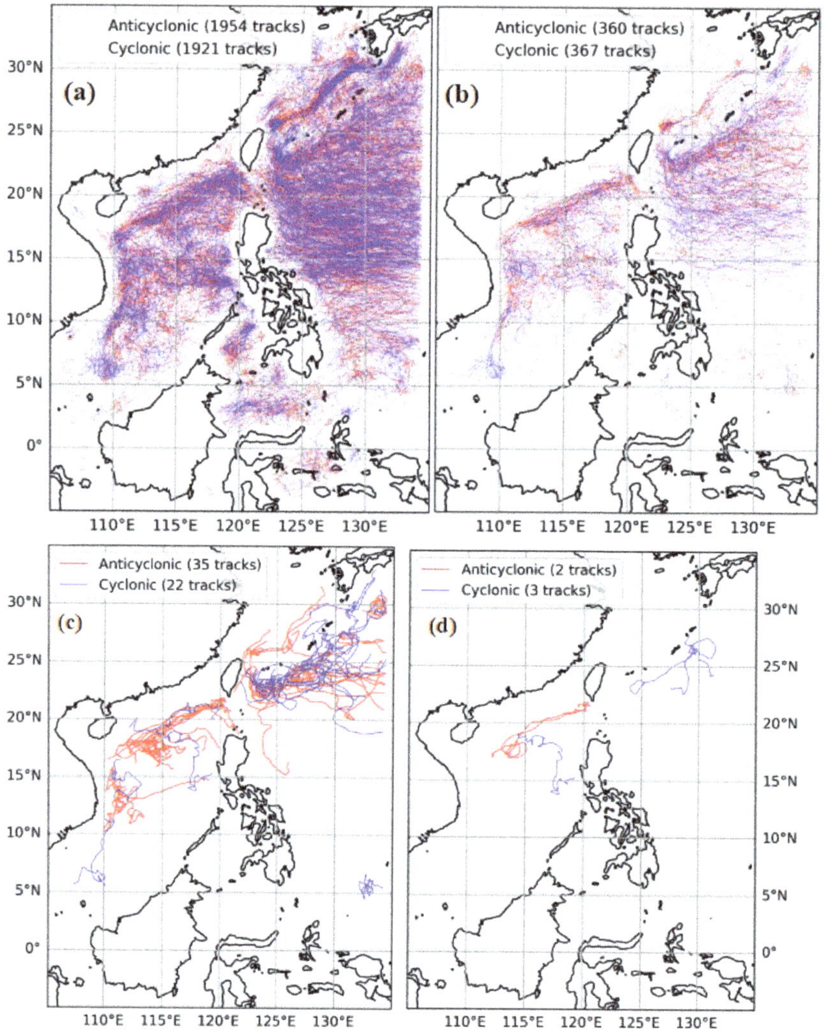

Figure 10. (**a**) Eddy trajectories with a lifespan greater than 8 (**a**), 16 (**b**), 32 (**c**), and 48 (**d**) weeks tracked from 1 January 2000 to 5 June 2023. Here, the eddy lifetime refers to the duration of the eddy trajectories within the study area.

Figure 10c,d show the eddy trajectories with a lifespan longer than 32 weeks and 48 weeks tracked from 1 January 2000 to 5 June 2023, respectively. Only 35 anticyclonic and 22 cyclonic eddies exhibit lifespans longer than 32 weeks, indicating that while many eddies are generated, only a few persist for extended periods. This is consistent with [16], who found that globally, the eddies with the longest lifetimes and longest propagation distances are predominantly anticyclonic. It can be observed that eddies with lifespans exceeding 32 weeks are mainly concentrated in the northeastern waters of Taiwan and the northern part of the SCS, showing certain regularities and following major current systems closely in their movement. Around 25°N, eddies form within the shear zone between the eastward Subtropical Countercurrent (18–25°N) and the westward North Equatorial Current [7]. They propagate westward at several centimeters per second, evolve during transit, and eventually dissipate upon encountering the Kuroshio Current [64]. In the northern South China Sea, eddies typically form near strait exits and dissipate in shallow coastal waters. This concentration highlights the significant role of major current systems and regional topography in sustaining long-lived eddies [10]. Only five eddies exceed the lifespan of 48 weeks. Although their number is very small, they have a significant impact on SSH. According to [16], SSH is typically influenced by four to six eddies per year within eddy-rich regions.

These long-lived eddies can be categorized into two types of eddy trajectories based on their propagation paths. The first type propagates westward along the Kuroshio Current, reaching the eastern waters of Taiwan and dissipating after encountering the Ryukyu Islands of Japan. The second type originates near the northern Philippines in the Luzon Strait, extends into the SCS, and further stretches towards Hainan Island. This classification is consistent with the study by [65], who tracked eddy trajectories in southern and eastern Taiwan using ECCO2-SSEDG data from 2008 to 2015. The former trajectory type accounts for 96% of the total tracked trajectories, indicating a dominant pattern influenced by the strong, persistent flow of the Kuroshio Current.

Figure 11 shows the percentages of cyclonic eddies and anticyclonic eddies with various speed radii and amplitudes, categorized by the lifespan of the eddies from 1 January 2000 to 5 June 2023. The distribution patterns for both cyclonic eddies and anticyclonic eddies are similar, suggesting that the processes governing the formation, growth, and dissipation of these eddies are comparable. The majority of the total number of cyclonic eddies and anticyclonic eddies have lifespans ranging from 3 to 48 weeks, which are critical for mesoscale ocean dynamics and play a substantial role in energy transfer and material transport within the ocean [17].

Figure 11a,c show the percentage distribution of anticyclonic and cyclonic eddies by speed radius, grouped by different lifespans. Eddies with shorter lifespans (less than 3 weeks) tend to have smaller speed radii, indicating that these eddies are generally less stable and dissipate more quickly. In contrast, eddies with longer lifespans (over 48 weeks) exhibit a broader range of speed radii, suggesting that larger eddies are more stable and can persist for extended periods. This pattern corresponds with the understanding that larger eddies possess more energy and structural coherence, allowing them to maintain their integrity over longer durations [1].

Figure 11b,d show the percentage distribution of anticyclonic and cyclonic eddies by amplitude, categorized by different lifespans. Similarly to the speed radius distribution, shorter-lived eddies typically have lower amplitudes, reflecting their weaker intensity and shorter persistence. Longer-lived eddies, however, can have a wide range of amplitudes, with many showing higher values. Short-lived eddies are likely influenced by transient atmospheric conditions and local topographic features, while long-lived eddies are more affected by large-scale oceanic currents and climatic patterns [49].

For our analysis, we selected three distinct regions—offshore eastern Taiwan, the area encompassing southern Taiwan and the Luzon Strait, and the northern SCS region—to use SVP float data for validating the eddy trajectories identified in this study. Figure 12a presents the validation of an eddy trajectory in the eastern region of Taiwan using SVP float data from 2018 to 2019, specifically selecting the SVP float with ID 63995470. The blue contours in Figure 12 shows the closed contours of this eddy at 10-day intervals. By comparing the eddy trajectories with the float data based on their temporal and spatial alignment, we identified a cyclonic eddy characterized by clockwise rotation. The SVP floats' trajectories closely follow the eddy contours, demonstrating a coincident path between the observed drifter paths and the identified eddy movements. This indicates the effectiveness of the eddy tracking method.

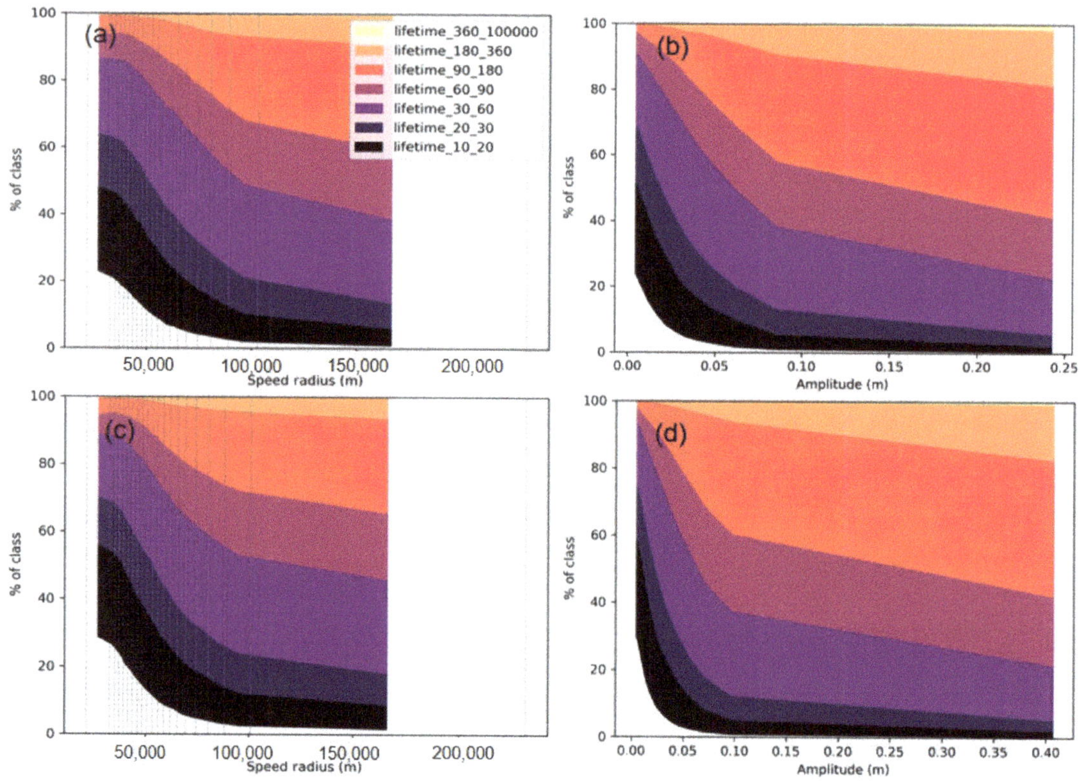

Figure 11. (**a**,**b**) are statistics on the percentage of anticyclonic eddies by speed radius and amplitude, grouped by different lifespans; (**c**,**d**) are same as (**a**,**b**) but for cyclonic eddies.

Figure 12b,d validate two eddy trajectories in the region south of Taiwan for the period 2004–2005, using SVP float trajectories with IDs 41489 and 41482. The ocean currents in this region are significantly influenced by monsoons, resulting in pronounced seasonal variations. During the winter, the northeast monsoon alters the flow dynamics, causing the main stream of the Kuroshio Current to split at the southernmost tip of Taiwan's Hengchun Peninsula, forming a branch. This branch partly intrudes into the northern SCS and partly merges with cyclonic eddies southwest of Taiwan, aiding in the formation of the Loop Current [37]. The distinct differences between the two trajectories in the southern region, as shown in Figure 12b,c, are influenced by the Kuroshio Current, its branches, and the loop currents south of Taiwan [66]. This complex interaction highlights the dynamic nature of the region, where monsoonal winds and major ocean currents create varying eddy patterns and behaviors [49]. The consistency between the SVP float and the eddy contours indicates that the PET method can capture the complex movements of eddies in regions with strong currents and undulating topography.

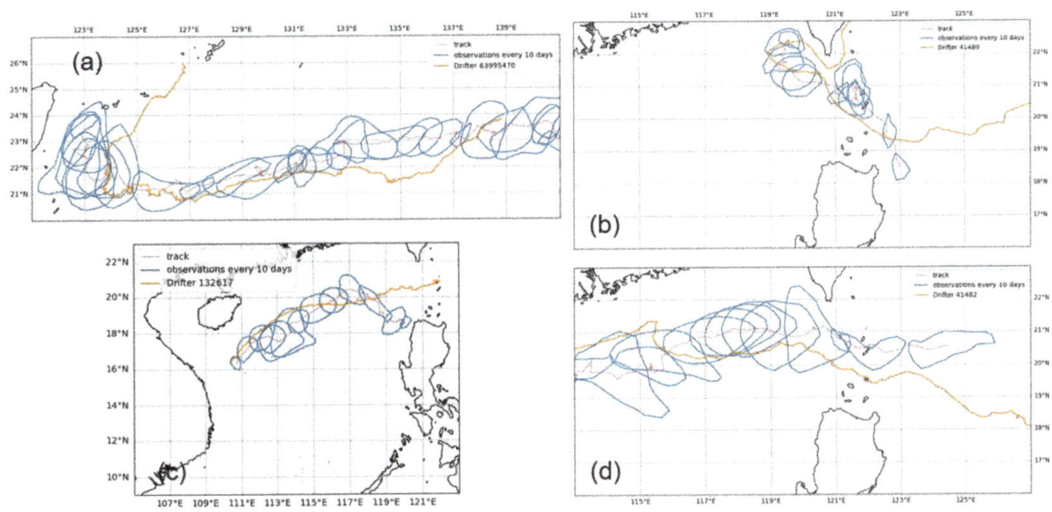

Figure 12. Spatial overlay verification of cyclonic eddy trajectories (in red) and SVP floats (in orange). The blue contours represent the closed contours of the eddy with a 10-day interval. (**a**) Eastern region of Taiwan; (**b**,**d**) southern region of Taiwan; (**c**) SCS region. The arrows indicate the travel direction of the eddies. The dates are shown in the text.

The SCS region, as validated in Figure 12c, is known for its complex circulation patterns influenced by both the monsoon system and regional currents. Using the SVP float trajectory with ID 132617, we validated the eddy trajectories identified across all four regions. These trajectories are consistent with the motion of the SVP floats, providing evidence for the accuracy of our eddy tracking method. The consistency between the SVP drifters and the eddy contours over time shows the robustness of the PET method for tracking eddies.

5.2. A Preliminary Assessment of SWOT Observations for Eddy Detection

Building on the results of mesoscale eddy detection and tracking using conventional altimeter data in Section 5.1, we further explore the potential of high-resolution satellite observations for identifying smaller-scale features. Given the limitations of traditional altimetry in resolving submesoscale dynamics, the newly available SWOT data provide an opportunity to examine submesoscale eddies with unprecedented spatial detail. The following section presents a preliminary investigation into the identification of submesoscale eddies based on SWOT observations.

5.2.1. 'Eddy' Detection Using Observations from SWOT's One-Day Orbit

Since early 2023, we have joined the SWOT CalVal team to collect sea surface height measurements by Global Navigation Satellite System (GNSS) on a research vessel from Taiwan in June 2023. This effort aims to validate the measurements of SWOT under Pass 21 of SWOT's one-day repeat mission in the region shown in Figure 13. The GNSS validation result is still under investigation.

Figure 13 displays the SLA on 1 May 2023, obtained from conventional altimeter satellites and the SWOT satellite, both demonstrating the basic characteristics of an anticyclonic eddy. A comparison between SWOT and conventional AVISO altimeter observations indicates that SWOT and conventional altimeter data generally agree on the position and magnitudes of this anticyclonic eddy. Additionally, SWOT exhibits the capability to discern the frontal regions of eddies, as also noted by [29]. The detailed observation capabilities of

SWOT, as shown in Figure 13, highlight its potential for revolutionizing our understanding of ocean surface topography. By providing higher spatial resolution data, SWOT may reveal the structure and dynamics of eddies with unprecedented clarity, which is essential for studying their role in ocean mixing, heat transport, and biogeochemical cycles [26,67].

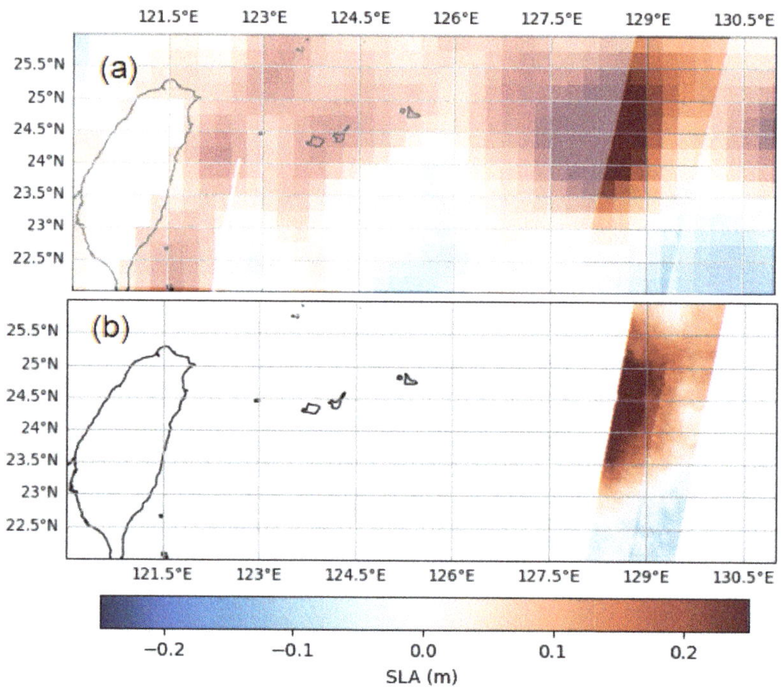

Figure 13. (**a**) SLA observations from conventional altimetry and (**b**) SWOT's one-day orbit on 1 May 2023.

Figure 14 compares the cyclonic and anticyclonic eddies identified using the PET method from conventional altimeter and small structures identified from SWOT data. It is noted that the PET algorithm did not identify the largest anticyclonic eddy with the SWOT data, which is shown in Figure 14a, with an effective radius of approximately 83 km. Instead, the algorithm detected dozens of submesoscale features in Figure 14b. It is observed that such features primarily occur within and around the peripheries of two large-scale eddies, labeled as 3 and 4. This suggests a complex interplay between mesoscale and submesoscale processes that could be influenced by the dynamics of larger eddies [6]. The edges of large-scale and mesoscale eddies intensify gradients at submesoscale, leading to mesoscale strain-induced frontogenesis. These fronts, with km-scale gradients, generate smaller-scale instabilities [68] (SWOT Science Team Meeting, 2024). These fine-scale features are recognized as a key structuring regime for biogeochemistry and ecology [69]. We cannot confirm whether the submesoscale features identified in Figure 14b around the large- and mesoscale eddies are submesoscale eddies, nor can we confirm their key structure and role in the dynamics. Further repeated measurements by SWOT may improve this understanding.

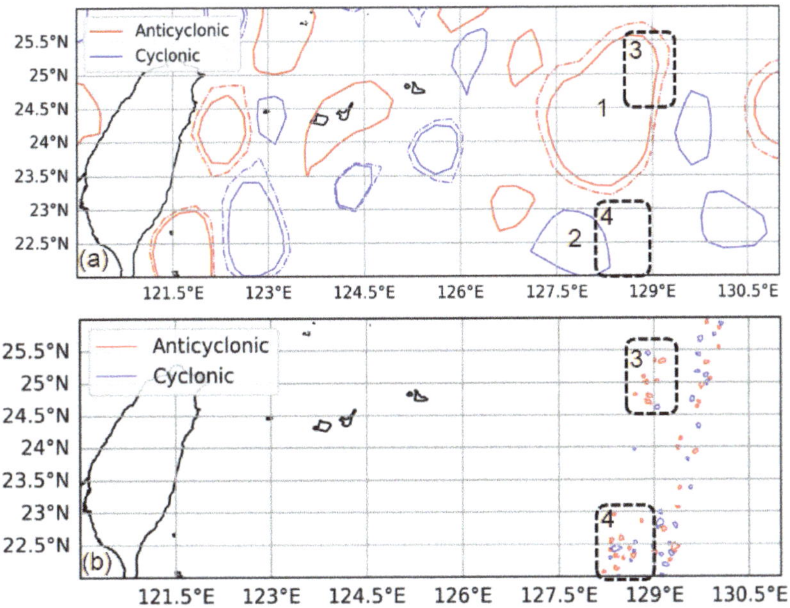

Figure 14. (**a**) Cyclonic and anticyclonic submesoscale eddies detected from conventional altimetry, and (**b**) from SWOT's one-day orbit on 1 May 2023. See the text for the numbers in the dashed rectangles.

Figure 15 shows the submesoscale structures identified from three consecutive days of SWOT satellite observations in regions labeled 3 and 4 in Figure 14. We observed that these structures identified at the same location for two or three consecutive days are considered the same submesoscale features. The emergence and dissipation of these small-scale structures within the three-day period are apparent. Few structures persist for the entire duration of three days, except for the three eddies depicted in Figure 15a, which is located within the edge of the large-scale eddy labeled 1 in Figure 14a. It is hypothesized that the lifespan of these small-scale structures is correlated with their positions within large-scale eddies. Large-scale eddies can influence the formation and stability of smaller eddies, providing a favorable environment for their persistence [6]. Statistical analysis in Figure 15d reveals that the SRs and ERs of the identified submesoscale features fall within the range of 2 to 8 km. We also provided the SRs and ERs of large and mesoscale eddies identified in Figure 15c for comparison. The discrepancy between SR and ER in Figure 15c suggests that large-scale and mesoscale eddies have more pronounced structural differences. Compared to large and mesoscale eddies, the ERs and SRs of submesoscale features lie closer to the line of equality. This observation is consistent with our simulation result of the submesoscale eddies in Section 4.2. This highlights the capability of SWOT to capture finer details in smaller structures.

5.2.2. Eddy Detection Using Observations from SWOT's 21-Day Orbit

Figure 16 compares the SLA values from conventional altimeter data and SWOT's 21-day orbit data from 7 September 2023 to 21 September 2023, as well as the identified eddies over the study area. The SWOT 21-day orbit basically covers the entire study area, despite small gaps due to the inherent gap problem in the 21-day orbit data. Due to SWOT's orbital revolution, the eddies observed in different swaths have shifted over time. However, large-scale eddies have lifespans of several months while undergoing slight changes in shapes and positions within two weeks. Therefore, in comparison with the SLA observed from conventional altimeter data, we found that SWOT can effectively capture large-scale and mesoscale eddies. The sizes and positions of these eddies from SWOT data are generally consistent with those from conventional altimeter observations, as shown in Figure 16a,b. Besides detailing the frontal edges of eddies, SWOT also reveals intricate details in non-eddy areas, which are regions where submesoscale eddies are generated.

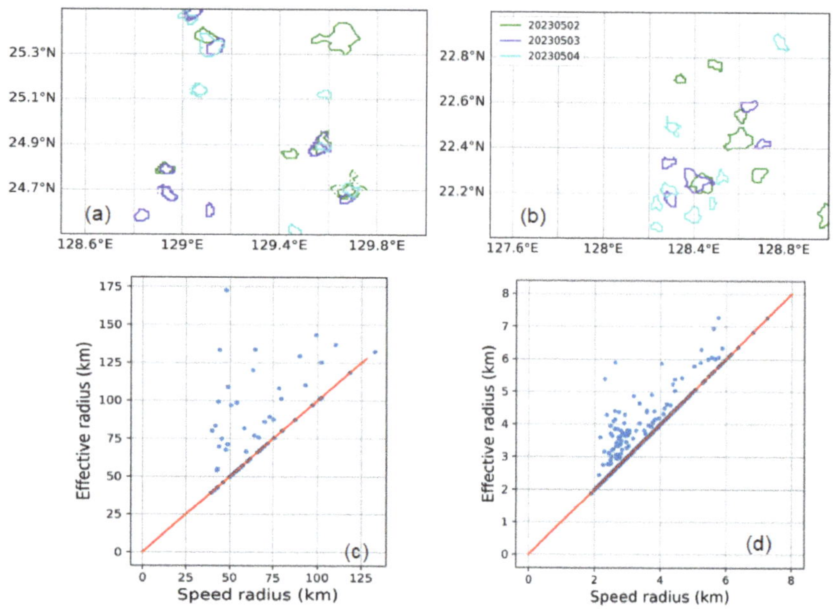

Figure 15. (**a**,**b**) Submesoscale features identified from three-day observations of SWOT satellites in two sub-regions labeled 3 and 4 in Figure 14. (**c**,**d**) SRs and ERs based on data from conventional altimeter data (Figure 5b) and SWOT data (pass 21) on 1 May 2023.

Using the PET method, we also identified submesoscale features in SWOT's 21-day data; the large-scale eddies cannot be identified solely through SWOT data because the boundaries of the swaths disrupt closed contour lines. The spatial distribution of these submesoscale features is widespread throughout the study area, with a higher density near large and medium-scale eddies and in the enclosed sea areas formed by numerous islands. The causes and spatial distribution of submesoscale eddies have not been fully studied yet. Ref. [70] noted that eddies smaller than 50 km in scale can be generated by several processes, such as interactions between large-scale ocean currents and bottom topography, islands, or headlands, atmospheric forcing, and barotropic or baroclinic instabilities in currents and fronts. Ref. [71] analyzed submesoscale eddy activity in the Baltic, Black, and Caspian Seas during 2009–2010, using over 2000 radar images from European Remote Satellites ERS-1, and ERS-2, and Envisat. This analysis identified more than 14,000 eddy structures with diameters ranging from 1 to 75 km, of which 95% were between 1 and 15 km. This indicates a widespread presence of submesoscale eddies, under 15 km, in these regions.

Figure 17 presents the SLA observations over the study area, obtained from two cycles of SWOT's 21-day repeat orbit, effectively covering the entire marine region depicted in Figure 5. The SLA data reveals detailed spatial patterns, including prominent ripple-like features in the SCS region west of the Luzon Strait, which are shown in greater detail in Figure 18. These features suggest the presence of internal waves, which are typically undetectable by conventional altimetry. Internal waves, which oscillate within the ocean's interior rather than on the surface, play a significant role in ocean mixing and energy transfer processes. Detecting these features is crucial for an improved understanding of ocean dynamics [72]. The study of internal waves is beyond the scope of this paper.

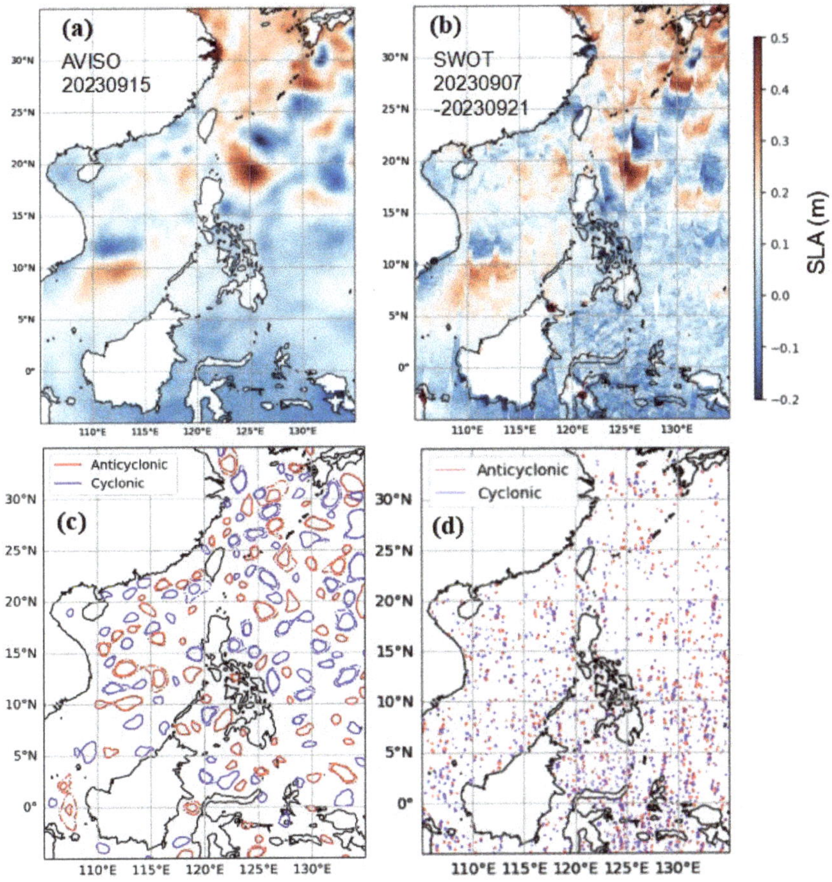

Figure 16. (**a**) SLA from conventional altimetry, (**b**) from SWOT (one cycle of 21-day science orbit), (**c**) eddies identified from the SLA in (**a**), and (**d**) eddies identified the SLA in (**b**). In (**d**), only eddy centers (circles) are shown without contours like in (**c**).

Figure 19 illustrates the temporal evolution of two large-scale eddies shown in Figure 16a. These datasets, obtained through the combination of conventional altimetry and SWOT's KaRIn data, cover the period from 7 September to 8 October in 2023. Due to the temporal discontinuity of SWOT data, only snapshots of the SWOT orbit passing through this area are shown in the subfigures of Figure 19. The combination of data from conventional altimeters and the SWOT altimeter provides a more complete picture of these eddies' formation, maturation, and dissipation, revealing the complex interactions between cyclonic and anticyclonic eddies.

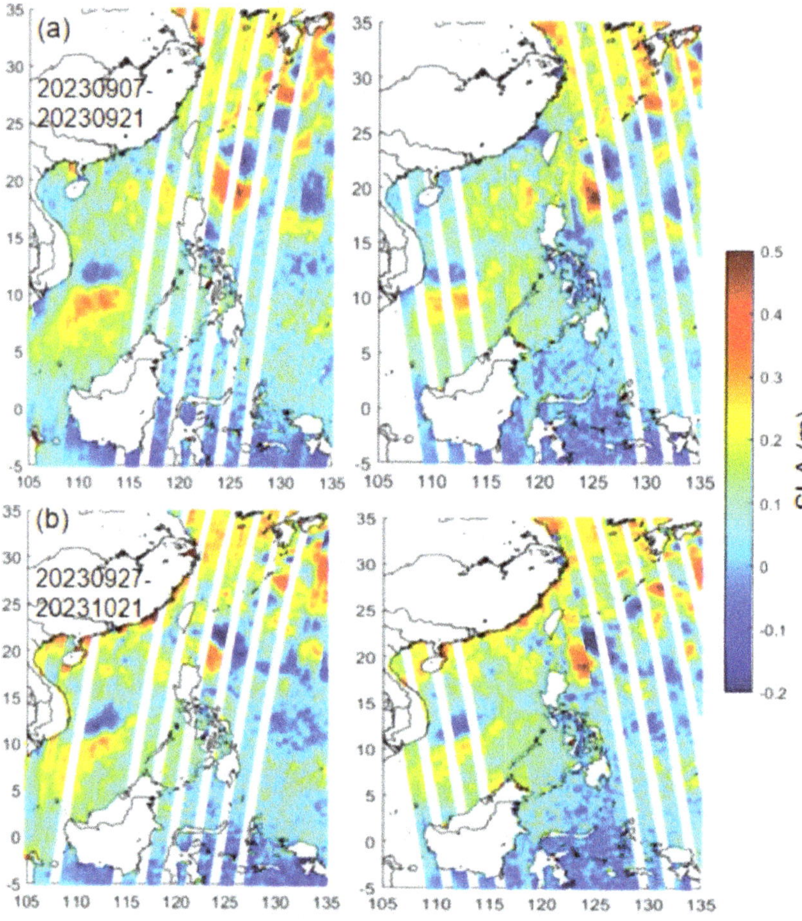

Figure 17. (**a**,**b**) SLA observations from two cycles of ascending and descending SWOT's 21-day orbit. The dates of the cycles are shown in the figures.

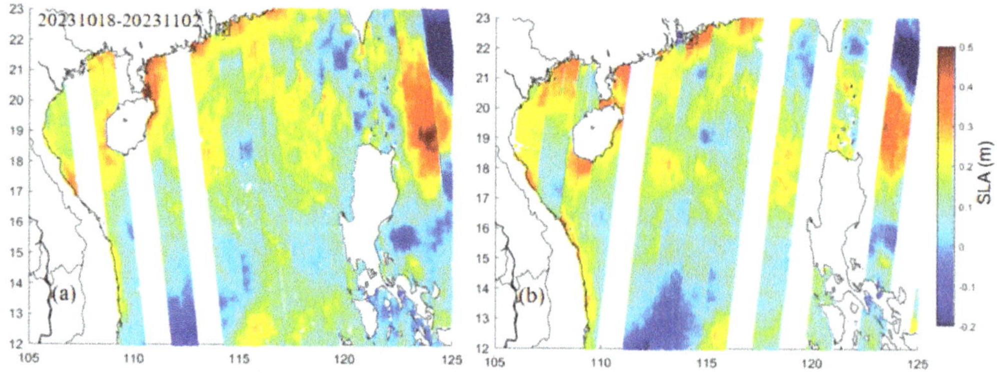

Figure 18. (**a**,**b**) Ripple-like SLA features from one cycle of SWOT's 21-day ascending and descending orbit on the west side of the Luzon Strait. The dates are shown in the figures.

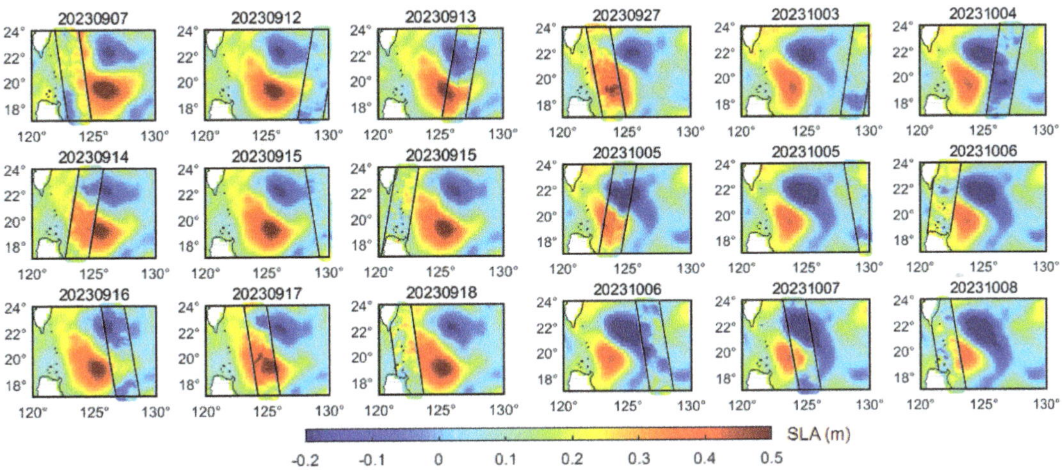

Figure 19. Evolutions of two large-scale eddies shown in Figure 16a obtained by the combination of the radar altimeters and KaRIn in the days that SWOT orbit during the period from 7 September to 8 October 2023.

6. Discussion

6.1. Errors Induced by High-Pass Filter

High-pass filtering introduces potential errors in eddy identification, as it tends to increase the detected size of eddies. As discussed in Section 4.2, applying a high-pass filter at different radii (20–300 km) resulted in significant variations in eddy count and spatial distribution. Ref. [28] pointed out that using ADT instead of SLA maps as input can improve eddy identification in regions with strong sea surface height (SSH) gradients. However, larger filter radii may cause signal leakage in strong gradient regions, leading to the loss of physical content after filtering. Therefore, we ultimately chose SLA maps as input and applied a 300 km high-pass filter. Comparing our results with the Mesoscale Eddy Trajectory Atlas Product (META3.1, [28]) from 1 January 2000 to 31 December 2020, META3.1 identified 448,912 cyclonic and 441,442 anticyclonic eddies in the study area, while our results show 516,747 and 505,112, respectively, surpassing META3.1 by over 100,000 eddies. The use of a deep learning-based approach, such as that proposed by [73,74], may optimize the choice of filter length. Other detection algorithms, such as TOEddies, do not require spatial filtering and operate directly on the ADT field. In global eddy tracking statistics, this approach yields approximately 2% more eddies compared to META3.2 [75].

6.2. Limitations of PET in Identifying Submesoscale Eddies

Submesoscale eddies may not strictly adhere to geostrophic balance [67] and may not always form closed structures [76]. As a result, it remains uncertain whether the small-scale features identified using SWOT data genuinely represent submesoscale eddies or are merely artifacts caused by data noise. Further verification is required to confirm the validity of these detected structures. Moreover, in our efforts to identify Submesoscale eddies, we utilized SLA data input while applying a smaller filtering radius and lower pixel threshold. This methodological choice, although intended to enhance the detection of finer-scale features, may have inadvertently introduced additional errors. Specifically, the reduced filtering radius can exaggerate noise, making it difficult to distinguish actual submesoscale eddies from transient fluctuations in the SLA field. Similarly, the lower pixel threshold may contribute to the over-identification of weak or short-lived eddies, many of which could be spurious.

Addressing these issues requires further refinement of the PET algorithm, particularly in terms of optimizing filter settings to strike a balance between detecting genuine submesoscale eddies and minimizing false detections. Incorporating complementary datasets, such as in situ drifter measurements and high-resolution ocean model outputs, may provide an additional means of verifying the existence and properties of these small-scale features. Future studies should focus on refining detection criteria and testing alternative filtering techniques to improve the reliability of submesoscale eddy identification.

7. Conclusions

In this study, we employed EGF simulations and the PET algorithm, validated by SVP drifter data, to identify and track eddies in the northwest Pacific Ocean. The PET method demonstrated high accuracy in detecting large- and mesoscale eddies, with the long-lived eddies associated with major currents, notably abundant around 20°N latitude, with peaks in EKE, eddy radius, and amplitude along the Kuroshio Current and northern South China Sea. Over 22 years, we identified 1,188,649 eddies, with a seasonal increase in spring linked to marine environmental factors. Simulated satellite altimetry data, mirroring SWOT's high-resolution capabilities, confirmed the PET method's effectiveness in processing detailed datasets, and revealed eddies as small as a few km in diameter, despite a noted decline in accuracy for submesoscale eddies. Future work should enhance the PET algorithm for SWOT's swath altimetry, addressing data fragmentation and unclosed contours. Integrating SWOT with in situ drifters and models will refine eddy tracking, while machine learning can further improve classification, benefiting ocean circulation and climate studies.

Author Contributions: C.H., L.Z. and H.-Y.L. conceptualized the initial idea and experimental design. L.Z. and H.-Y.L. made all computations and wrote the first draft, with C.H. and L.Z. assessing the SWOT observations. E.T.Y.C. and D.Y. commented on the draft. All authors have read and agreed to the published version of the manuscript.

Funding: This study is supported by the National Science and Technology Council, Taiwan, under grant numbers 113-2611-M-A49-001 and 112-2221-E-A49-025-MY3. This study is also supported by the Central Public-interest Scientific Institution Basal Research Fund (CEAIEF20240305); the NNSFC (National Natural Science Foundation of China) (4230040788), and State Key Laboratory of Geodesy and Earth's Dynamics, Innovation Academy for Precision Measurement Science and Technology, Chinese Academy of Sciences (SKLGED2023-2-1).

Data Availability Statement: The SWOT_L3_SSH product, derived from the L2 SWOT KaRIn low-rate ocean data products (NASA/JPL and CNES), is produced and made freely available by AVISO and DUACS teams as part of the DESMOS Science Team project. AVISO/DUACS, 2023. SWOT Level 3 SSH Expert (v0.3). CNES, https://doi.org/10.24400/527896/A01-2023.018, accessed on 5 May 2025. The drifter data are provided by the Surface Velocity Program and available on the Global Drifter Program website (https://www.aoml.noaa.gov/phod/gdp/index.php, accessed on 5 May 2025). The code for the Py_Eddy_Tracking (PET) algorithm is provided by [26–28].

Acknowledgments: The PET eddy automatic identification method is based on [1,26–28].

Conflicts of Interest: The authors declare no conflicts of interest.

Appendix A. Five Constraints for Closed Polygons

The five constraints on the closed contours of eddies used in this study considered the recommendations of Mason et al., (2014) [26], but we also made some improvements. First, whereas the SLA interval used in their article was 1 cm, this study has adjusted it to 0.4 cm, meaning there is a contour for every 0.4 cm interval of SLA. This change was made because this study aims to capture small-scale oceanic dynamical phenomena more accurately. Second, this study has increased the shape error from the original 55% to 70%. The shape error is the difference in area between the outermost closed contour of the original eddy and its best-fit circle, as a proportion of the best-fit circle's area. The reason for loosening the restriction to 70% is to ensure detection of elongated eddies, which are often seen in dynamically active regions [28]. The other constraints follow the recommendation of PET as follows:

1. An eddy's closed contour includes only one eddy center SLA extremum, with anticyclonic eddies containing only one SLA maximum value and cyclonic eddies containing only one SLA minimum value, differing from the constraint of [1].
2. The area of the region enclosed by the closed SLA isopleth lines is between 8 pixels and 1000 pixels. For submesoscale eddies, which are generally smaller than 10 km in radius, an excessive number of pixels may fail to accurately capture their compact structure and can lead to an overestimation of their ER during detection. Therefore, we set the maximum pixel value to 150 pixels.
3. The eddy amplitude (Amplitude, A) is between 1 cm and 150 cm. A = | SLA_center—SLA_contour |, where SLA_center is the SLA at the center of the eddy within the closed SLA isopleth, and SLA_contour is the average SLA on that closed isopleth.
4. For anticyclonic eddies, the formed eddy area only includes those pixels where the SLA value is greater than the current set SLA interval value; for cyclonic eddies, the formed eddy area only includes those pixels where the SLA value is less than the current set SLA interval value, with the interval value set at 0.4 cm in this study.
5. Passing the shape test with Error_Shape $\leq$ 70%, where Error_Shape = Area_deviation/ Area_(p_eff), with Area_(p_eff) being the area of the green best-fit circle shown in Figures 3 and 4, which has the same area as the red closed contour, and Area_deviation being the area enclosed outside the green best-fit circle and within the red closed contour.

Appendix B. Elliptical Gaussian Functions for Simulated Eddies

This paper uses the elliptical Gaussian function (EGF) to simulate the SLA values across an ocean eddy. With a simulated eddy, the SLA value is expressed as proposed by Nobach and Honkanen (https://en.wikipedia.org/wiki/Gaussian_function, accessed on 5 May 2025):

$$f(x,y) = A\ \exp(-(a(x - x_0)^2 + 2b(x - x_0)(y - y_0) + c(y - y_0)^2)) \quad (A1)$$

where

$$a = \frac{cos^2\theta}{2\sigma_x^2} + \frac{sin^2\theta}{2\sigma_y^2} \quad (A2)$$

$$b = -\frac{sin2\theta}{4\sigma_x^2} + \frac{sin2\theta}{4\sigma_y^2} \quad (A3)$$

$$c = \frac{sin^2\theta}{2\sigma_x^2} + \frac{cos^2\theta}{2\sigma_y^2} \quad (A4)$$

where $f(x,y)$ is SLA at a location (x,y) and (x_0, y_0) is the eddy center of the EGF with the maximum or minimum SLA value. σ_x, σ_y are the standard deviations along the x-axis (east) and y-axis (north), respectively, which determine the spread of the Gaussian function along these axes. A is the amplitude of the eddy, i.e., the height difference between the eddy center and the eddy edge. θ is the rotation angle of the ellipse relative to the x-axis, which indicates how the major and minor axes of the ellipse are oriented. In this paper, the semi-major axis and semi-minor axis of an EGF are related to σ_x, σ_y as

$$\sigma_x = \frac{R_{major}}{\sqrt{2}} \tag{A5}$$

$$\sigma_y = \frac{R_{minor}}{\sqrt{2}} \tag{A6}$$

Equations (A5) and (A6) imply that the spread of SLA within an eddy is smoother for larger major and minor axes compared to eddies with smaller axes. The choice in Equations (A5) and (A6) are realistic but not mandatory.

References

1. Chelton, D.B.; Schlax, M.G.; Samelson, R.M. Global observations of nonlinear mesoscale eddies. *Prog. Oceanogr.* **2011**, *91*, 167–216. [CrossRef]
2. Chelton, D.B.; Schlax, M.G.; Samelson, R.M.; de Szoeke, R.A. Global observations of large oceanic eddies. *Geophys. Res. Lett.* **2007**, *34*, L15606. [CrossRef]
3. Garabato, A.C.N.; Yu, X.; Callies, J.; Barkan, R.; Polzin, K.L.; Frajka-Williams, E.E.; Griffies, S.M. Kinetic energy transfers between mesoscale and submesoscale motions in the open ocean's upper layers. *J. Phys. Oceanogr.* **2022**, *52*, 75–97. [CrossRef]
4. Thomas, L.N.; Tandon, A.; Mahadevan, A. Submesoscale processes and dynamics. *Ocean Model. Eddying Regime* **2008**, *177*, 17–38.
5. McWilliam, J.C. Fluid dynamics at the margin of rotational control. Environ. *Fluid Mech.* **2008**, *8*, 441–449. [CrossRef]
6. McWilliams, J.C. Submesoscale currents in the ocean. *Proc. R. Soc. A Math. Phys. Eng. Sci.* **2016**, *472*, 20160117. [CrossRef]
7. Qiu, B. Seasonal eddy field modulation of the North Pacific Subtropical Countercurrent: TOPEX/Poseidon observations and theory. *J. Phys. Oceanogr.* **1999**, *29*, 2471–2486. [CrossRef]
8. Moum, J.N.; Nash, J.D. Mixing measurements on an equatorial ocean mooring. *J. Atmos. Oceanic Technol.* **2009**, *26*, 317–336. [CrossRef]
9. Stammer, D.; Ray, R.D.; Andersen, O.B.; Arbic, B.K.; Bosch, W.; Carrère, L.; Thomas, M. Accuracy assessment of global barotropic ocean tide models. *Rev. Geophys.* **2014**, *52*, 243–282. [CrossRef]
10. Zhang, Z.; Wang, W.; Qiu, B. Oceanic mass transport by mesoscale eddies. *Science* **2014**, *345*, 322–324. [CrossRef]
11. Dong, C.; McWilliams, J.C.; Liu, Y.; Chen, D. Global heat and salt transports by eddy movement. *Nat. Commun.* **2014**, *5*, 3294. [CrossRef] [PubMed]
12. Mensah, V.; Jan, S.; Andres, M.; Chang, M.H. Response of the Kuroshio east of Taiwan to mesoscale eddies and upstream variations. *J. Oceanogr.* **2020**, *76*, 271–288. [CrossRef]
13. Isern-Fontanet, J.; García-Ladona, E.; Font, J. Identification of marine eddies from altimetric maps. *J. Atmos. Oceanic Technol.* **2003**, *20*, 772–778. [CrossRef]
14. Okubo, A. Horizontal dispersion of floatable particles in the vicinity of velocity singularities such as convergences. *Deep Sea Res. Oceanogr. Abstr.* **1970**, *17*, 445–454. [CrossRef]
15. Weiss, J. The dynamics of entropy transfer in two-dimensional hydrodynamics. *Phys. D Nonlinear Phenom.* **1991**, *48*, 273–294. [CrossRef]
16. Fu, L.L.; Chelton, D.B.; Le Traon, P.Y.; Morrow, R. Eddy dynamics from satellite altimetry. *Oceanography* **2010**, *23*, 14–25. [CrossRef]
17. Morrow, R.; Birol, F.; Griffin, D.; Sudre, J. Divergent pathways of cyclonic and anticyclonic ocean eddies. *Geophys. Res. Lett.* **2004**, *31*, 24. [CrossRef]
18. Chaigneau, S.; Gizolme, A.; Grados, C. Mesoscale eddies of Peru in altimeter records: Identification algorithms and eddy spatio-temporal patterns. *Prog. Oceanogr.* **2008**, *79*, 106–119. [CrossRef]
19. Hwang, C.; Chen, S.A. Circulations and eddies over the South China Sea derived from TOPEX/Poseidon altimetry. *J. Geophys. Res. Ocean.* **2000**, *105*, 23943–23965. [CrossRef]
20. Barabinot, Y.; Speich, S.; Carton, X. Defining mesoscale eddies boundaries from in-situ data and a theoretical framework. *J. Geophys. Res. Ocean.* **2024**, *129*, e2023JC020422. [CrossRef]
21. Lapeyre, G.; Klein, P. Impact of the small-scale elongated filaments on the oceanic vertical pump. *J. Mar. Res.* **2006**, *64*, 835–851. [CrossRef]
22. Hewitt, H.; Fox-Kemper, B.; Pearson, B. The small scales of the ocean may hold the key to surprises. *Nat. Clim. Change* **2022**, *12*, 496–499. [CrossRef]
23. Qiu, B.; Chen, S.; Klein, P.; Sasaki, H.; Sasai, Y. Seasonal mesoscale and submesoscale eddy variability along the North Pacific Subtropical Countercurrent. *J. Phys. Oceanogr.* **2014**, *44*, 3079–3098. [CrossRef]

24. Michael, D.; Fu, L.L.; Lettenmaier, D.P.; Alsdorf, D.E.; Rodriguez, E.; Esteban-Fernandez, D. The Surface Water and Ocean Topography Mission: Observing Terrestrial Surface Water and Oceanic Submesoscale Eddies. *Proc. IEEE* **2010**, *98*, 766–779.
25. Zhang, Z.; Miao, M.; Qiu, B.; Tian, J.; Jing, Z.; Chen, G.; Chen, Z.; Zhao, W. Submesoscale eddies detected by SWOT and moored observations in the Northwestern Pacific. *Geophys. Res. Lett.* **2024**, *51*, e2024GL110000. [CrossRef]
26. Mason, E.; Pascual, A.; McWilliams, J.C. A new sea surface height–based code for oceanic mesoscale eddy tracking. *J. Atmos. Ocean. Technol.* **2014**, *31*, 1181–1188. [CrossRef]
27. Mason, E.; Pascual, A.; Gaube, P.; Ruiz, S.; Pelegrí, J.L.; Delepoulle, A. Subregional characterization of mesoscale eddies across the Brazil-Malvinas Confluence. *J. Geophys. Res. Oceans* **2017**, *122*, 3329–3357. [CrossRef]
28. Pegliasco, C.; Delepoulle, A.; Mason, E.; Morrow, R.; Faugère, Y.; Dibarboure, G. META3.1exp: A new global mesoscale eddy trajectory atlas derived from altimetry. *Earth Syst. Sci. Data* **2022**, *14*, 1087–1107. [CrossRef]
29. Fu, L.L.; Pavelsky, T.; Cretaux, J.F.; Morrow, R.; Farrar, J.T.; Vaze, P.; Dibarboure, G. The Surface Water and Ocean Topography Mission: A breakthrough in radar remote sensing of the ocean and land surface water. *Geophys. Res. Lett.* **2024**, *51*, e2023GL107652. [CrossRef]
30. Metzger, E.J.; Hurlburt, H.E. The nondeterministic nature of Kuroshio penetration and eddy shedding in the South China Sea. *J. Phys. Oceanogr.* **2001**, *31*, 1712–1732. [CrossRef]
31. Yuan, D.; Han, W.; Hu, D. Anti-cyclonic eddies northwest of Luzon in summer-fall observed by satellite altimeters. *Geophys. Res. Lett.* **2007**, *34*, L13610. [CrossRef]
32. Chen, G.; Hou, Y.; Chu, X. Mesoscale eddies in the South China Sea: Mean properties, spatiotemporal variability, and impact on thermohaline structure. *J. Geophys. Res. Ocean.* **2011**, *116*, C06018. [CrossRef]
33. Xiu, P.; Chai, F.; Shi, L.; Xue, H.; Chao, Y. A census of eddy activities in the South China Sea during 1993–2007. *J. Geophys. Res. Oceans* **2010**, *115*, C03012. [CrossRef]
34. He, Y.; Xie, J.; Cai, S. Interannual variability of winter eddy patterns in the eastern South China Sea. *Geophys. Res. Lett.* **2016**, *43*, 5185–5193. [CrossRef]
35. Yuan, D.; Han, W.; Hu, D. Surface Kuroshio path in the Luzon Strait area derived from satellite remote sensing data. *J. Geophys. Res. Ocean.* **2006**, *111*, C11007. [CrossRef]
36. Caruso, M.J.; Gawarkiewicz, G.G.; Beardsley, R. Interannual variability of the Kuroshio intrusion in the South China Sea. *J. Oceanogr.* **2006**, *62*, 559–575. [CrossRef]
37. Jan, S.; Mensah, V.; Andres, M.; Chang, M.H.; Yang, Y.J. Eddy-Kuroshio interactions: Local and remote effects. *J. Geophys. Res. Ocean.* **2017**, *122*, 9744–9764. [CrossRef]
38. Jan, S.; Chern, C.S.; Wang, J.; Chiou, M.D. Generation and propagation of baroclinic tides modified by the Kuroshio in the Luzon Strait. *J. Geophys. Res. Ocean.* **2012**, *117*, C2. [CrossRef]
39. Zhang, Y.; Zhang, Z.; Chen, D.; Qiu, B.; Wang, W. Strengthening of the Kuroshio current by intensifying tropical cyclones. *Science* **2020**, *368*, 988–993. [CrossRef]
40. Ren, Q.; Yu, F.; Nan, F.; Li, Y.; Wang, J.; Liu, Y.; Chen, Z. Effects of mesoscale eddies on intraseasonal variability of intermediate water east of Taiwan. *Sci. Rep.* **2022**, *12*, 9182. [CrossRef]
41. Gómez-Navarro, L.; Cosme, E.; Sommer, J.L.; Papadakis, N.; Pascual, A. Development of an image de-noising method in preparation for the Surface Water and Ocean Topography Satellite Mission. *Remote Sens.* **2020**, *12*, 734. [CrossRef]
42. Tréboutte, A.; Carli, E.; Ballarotta, M.; Carpentier, B.; Faugère, Y.; Dibarboure, G. KaRIn noise reduction using a convolutional neural network for the SWOT Ocean Products. *Remote Sens.* **2023**, *15*, 2183. [CrossRef]
43. Lumpkin, R.; Pazos, M. Measuring surface currents with Surface Velocity Program drifters: The instrument, its data, and some recent results. Lagrangian Anal. *Predict. Coast. Ocean Dyn.* **2007**, *39*, 67.
44. Pegliasco, C.; Chaigneau, A.; Morrow, R. Main eddy vertical structures observed in the four major Eastern Boundary Upwelling Systems. *J. Geophys. Res. Oceans* **2015**, *120*, 6008–6033. [CrossRef]
45. Wang, Z.; Li, Q.; Sun, L.; Li, S.; Yang, Y.; Liu, S. The most typical shape of oceanic mesoscale eddies from global satellite sea level observations. *Front. Earth Sci.* **2015**, *9*, 202–208. [CrossRef]
46. Hwang, C.; Wu, C.R.; Kao, R. TOPEX/Poseidon observations of mesoscale eddies over the Subtropical Countercurrent: Kinematic characteristics of an anticyclonic eddy and a cyclonic eddy. *J. Geophys. Res. Ocean.* **2004**, *109*, C08013. [CrossRef]
47. Qiu, B.; Lukas, R. Seasonal and interannual variability of the North Equatorial Current, the Mindanao Current, and the Kuroshio along the Pacific western boundary. *J. Geophys. Res. Oceans* **1996**, *101*, 12315–12330. [CrossRef]
48. Ferrari, R.; Wunsch, C. The distribution of eddy kinetic and potential energies in the global ocean. *Tellus A Dyn. Meteorol. Oceanogr.* **2010**, *62*, 92–108. [CrossRef]
49. Qiu, B.; Chen, S. Variability of the Kuroshio extension jet, recirculation gyre, and mesoscale eddies on decadal time scales. *J. Phys. Oceanogr.* **2005**, *35*, 2090–2103. [CrossRef]
50. Kamenkovich, V.M.; Leonov, Y.P.; Nechaev, D.A.; Byrne, D.A.; Gordon, A.L. On the influence of bottom topography on the Agulhas eddy. *J. Phys. Oceanogr.* **1996**, *26*, 892–912. [CrossRef]

51. Qiu, B. *Encyclopedia of Ocean Sciences*; Academic Press: Cambridge, MA, USA, 2001; pp. 1413–1426.
52. Chelton, D.B.; Schlax, M.G. Global observations of oceanic Rossby Waves. *Science* **1996**, *272*, 234–238. [CrossRef]
53. Zhang, Z.; Zhao, W.; Qiu, B.; Tian, J. Anticyclonic eddy sheddings from Kuroshio loop and the accompanying cyclonic eddy in the northeastern South China Sea. *J. Phys. Oceanogr.* **2017**, *47*, 1243–1259. [CrossRef]
54. Shi, Y.; Liu, X.; Liu, T.; Chen, D. Characteristics of mesoscale eddies in the vicinity of the Kuroshio: Statistics from satellite altimeter observations and OFES model data. *J. Mar. Sci. Eng.* **2022**, *10*, 1975. [CrossRef]
55. Kamenkovich, I.; Berloff, P.; Pedlosky, J. Role of eddy forcing in the dynamics of multiple zonal jets in a model of the North Atlantic. *J. Phys. Oceanogr.* **2009**, *39*, 1361–1379. [CrossRef]
56. Lee, I.H.; Ko, D.S.; Wang, Y.H.; Centurioni, L.; Wang, D.P. The mesoscale eddies and Kuroshio transport in the western North Pacific east of Taiwan from 8-year (2003–2010) model reanalysis. *Ocean Dyn.* **2013**, *63*, 1027–1040. [CrossRef]
57. Chow, C.H.; Shih, Y.Y.; Chien, Y.T.; Chen, J.Y.; Fan, N.; Wu, W.C.; Hung, C.C. The wind effect on biogeochemistry in eddy cores in the Northern South China Sea. *Front. Mar. Sci.* **2021**, *8*, 717576. [CrossRef]
58. Chu, F.; Si, Z.; Yan, X.; Liu, Z.; Yu, J.; Pang, C. Physical structure and evolution of a cyclonic eddy in the Northern South China sea. *Deep Sea Res. Part I Oceanogr. Res. Pap.* **2022**, *189*, 103876. [CrossRef]
59. Simanungkalit, Y.A.; Pranowo, W.; Purba, N.; Riyantini, I.; Nurrahman, Y. Influence of El Nino Southern Oscillation (ENSO) phenomena on eddies variability in the northwest Pacific Ocean. *IOP Conf. Ser. Earth Environ. Sci.* **2018**, *176*, 012002. [CrossRef]
60. Cheng, X.; Qi, Y. Variations of eddy kinetic energy in the South China Sea. *J. Oceanogr.* **2010**, *66*, 85–94. [CrossRef]
61. Zhao, H.; Yue, X.; Wang, L.; Wu, X.; Chen, Z. Submesoscale short-lived eddies in the southwestern Taiwan Strait observed by high-frequency surface-wave radars. *Remote Sens.* **2024**, *16*, 589. [CrossRef]
62. Yin, Y.; Lin, X.; Hou, Y. Seasonality of the Kuroshio intensity east of Taiwan modulated by mesoscale eddies. *J. Mar. Syst.* **2019**, *193*, 84–93. [CrossRef]
63. Jia, F.; Wu, L.; Qiu, B. Seasonal modulation of eddy kinetic energy and its formation mechanism in the southeast Indian Ocean. *J. Phys. Oceanogr.* **2011**, *41*, 657–665. [CrossRef]
64. Qiu, B.; Chen, S. Eddy-mean flow interaction in the decadally-modulating Kuroshio Extension system. *Deep Sea Res. II* **2010**, *57*, 1098–1110. [CrossRef]
65. Yang, S.; Xing, J.; Sheng, J.; Chen, S.; Chen, D. A process study of interactions between a warm eddy and the Kuroshio Current in Luzon Strait: The fate of eddies. *J. Mar. Syst.* **2019**, *194*, 66–80. [CrossRef]
66. Chen, C.T.A.; Jan, S.; Huang, T.H.; Tseng, Y.H. Spring of no Kuroshio intrusion in the southern Taiwan Strait. *J. Geophys. Res. Ocean.* **2010**, *115*, C8. [CrossRef]
67. Fu, L.L.; Ferrari, R. Observing oceanic submesoscale processes from space. *Eos* **2008**, *89*, 488. [CrossRef]
68. Tom, F.; d'Ovidio, F.; Gerald, D. Ocean Finescale Dynamics—SWOT Science Team Meeting 2024. Available online: https://swot.jpl.nasa.gov/events/63/2024-swot-science-team-meeting/ (accessed on 5 May 2025).
69. Lévy, M.; Franks, P.J.S.; Smith, K.S. The role of submesoscale currents in structuring marine ecosystems. *Nat. Commun.* **2018**, *9*, 4758. [CrossRef]
70. Alpers, W.; Brandt, P.; Lazar, A.; Dagorne, D.; Sow, B.; Faye, S.; Hansen, M.W.; Rubino, A.; Poulain, P.M.; Brehmer, P. A small-scale oceanic eddy off the coast of West Africa studied by multi-sensor satellite and surface drifter data. *Remote Sens. Environ.* **2013**, *129*, 132–143. [CrossRef]
71. Karimova, S. Spiral eddies in the Baltic, Black and Caspian seas as seen by satellite radar data. *Adv. Space Res.* **2012**, *50*, 1107–1124. [CrossRef]
72. Whalen, C.B.; De Lavergne, C.; Naveira Garabato, A.C.; Klymak, J.M.; MacKinnon, J.A.; Sheen, K.L. Internal wave-driven mixing: Governing processes and consequences for climate. *Nat. Rev. Earth Environ.* **2020**, *1*, 606–621. [CrossRef]
73. Duo, Z.; Wang, W.; Wang, H. Oceanic mesoscale eddy detection method based on deep learning. *Remote Sens.* **2019**, *11*, 1921. [CrossRef]
74. Sun, H.; Li, H.; Xu, M.; Yang, F.; Zhao, Q.; Li, C. A lightweight deep learning model for ocean eddy detection. *Front. Mar. Sci.* **2023**, *10*, 1266452. [CrossRef]
75. Ioannou, A.; Guez, L.; Laxenaire, R.; Speich, S. Global Assessment of Mesoscale Eddies with TOEddies: Comparison Between Multiple Datasets and Colocation with In Situ Measurements. *Remote Sens.* **2024**, *16*, 4336. [CrossRef]
76. Ni, Q.; Zhai, X.; Wilson, C.; Chen, C.; Chen, D. Submesoscale eddies in the South China Sea. *Geophys. Res. Lett.* **2021**, *48*, e2020GL091555. [CrossRef]

remote sensing

MDPI

Article

A Comparative Evaluation of Two Bias Correction Approaches for SST Forecasting: Data Assimilation Versus Deep Learning Strategies

Wanqiu Dong [1], Guijun Han [1], Wei Li [1], Haowen Wu [1,*], Qingyu Zheng [1], Xiaobo Wu [2], Mengmeng Zhang [1], Lige Cao [1] and Zenghua Ji [1]

1 Tianjin Key Laboratory for Marine Environmental Research and Service, School of Marine Science and Technology, Tianjin University, Tianjin 300072, China; wanqiu@tju.edu.cn (W.D.); guijun_han@tju.edu.cn (G.H.); liwei1978@tju.edu.cn (W.L.); zqywork1998@tju.edu.cn (Q.Z.); zhangmengmeng@tju.edu.cn (M.Z.); clg@tju.edu.cn (L.C.); jzh1999@tju.edu.cn (Z.J.)

2 National Marine Environmental Forecasting Center, Beijing 100080, China; xb_wu@tju.edu.cn

* Correspondence: whw@tju.edu.cn

Abstract: This study introduces two distinct post-processing strategies to address systematic biases in sea surface temperature (SST) numerical forecasts, thereby enhancing SST predictive accuracy. The first strategy implements a spatiotemporal four-dimensional multi-grid analysis (4D-MGA) scheme within a three-dimensional variational (3D-Var) data assimilation framework. The second strategy establishes a hybrid deep learning architecture integrating empirical orthogonal function (EOF) analysis, empirical mode decomposition (EMD), and a backpropagation (BP) neural network (designated as EE–BP). The 4D-MGA strategy dynamically corrects systematic biases through a temporally coherent extrapolation of analysis increments, leveraging its inherent capability to characterize intrinsic temporal correlations in model error evolution. In contrast, the EE–BP strategy develops a bias correction model by learning the systematic biases of the SST numerical forecasts. Utilizing a satellite fusion SST dataset, this study conducted bias correction experiments that specifically addressed the daily SST numerical forecasts with 7-day lead times in the Kuroshio region south of Japan during 2017, systematically quantifying the respective error reduction potentials of both strategies. Quantitative verification reveals that EE–BP delivers enhanced predictive skill across all forecast horizons, achieving 18.1–22.7% root–mean–square error reduction compared to 1.2–9.1% attained by 4D-MGA. This demonstrates deep learning's unique advantage in capturing nonlinear bias evolution patterns.

Keywords: bias correction; SST numerical forecasts; 4D-MGA; EOF; EMD; BP neural network

1. Introduction

Sea surface temperature (SST) plays a significant role in both climate and weather systems, as well as within ecosystems. The accurate prediction of SST provides a key source of information for various applications. Current SST forecasting methods primarily encompass three approaches: a numerical model, empirical statistics, and machine learning [1]. Numerical model methods simulate ocean dynamics by solving coupled hydrodynamic–thermodynamic equations to forecast SST [2]. Empirical statistical methods leverage historical data to build predictive relationships but often fail to resolve complex nonlinear processes [3]. Machine learning, particularly deep learning, offers advantages in capturing complex spatiotemporal patterns directly from the data [4]. As of now,

SST forecasting remains predominantly reliant on numerical methods in the operational oceanography community. However, SST forecasts generated through numerical modeling frequently demonstrate systematic deviations, primarily stemming from three fundamental limitations: structural deficiencies inherent in computational frameworks, inadequately specified initial conditions, and errors in physical parameterization schemes [5]. Biases in SST forecasts can lead to significant inaccuracies in weather predictions, climate projections, and oceanographic studies. Therefore, bias correction is essential to improve the reliability and accuracy of SST numerical forecasts.

Statistical post-processing methods have become a mainstream solution for mitigating systematic errors in numerical weather prediction, primarily due to their operational affordability and demonstrated skill in refining forecast precision [6]. The emergence of machine learning techniques, however, is reshaping this paradigm by introducing data-driven post-processing frameworks capable of addressing inherent biases in both weather forecasts and climate projection models [7,8]. Researchers have also pioneered machine learning explorations for SST forecast bias correction in recent studies. For instance, Han et al. proposed a hybrid algorithm integrating empirical orthogonal function (EOF) analysis with a backpropagation (BP) neural network, achieving a 64% reduction in root–mean–square error (RMSE) for 1-day lead SST forecasts in the South China Sea [9]. Similarly, Fei et al. employed a convolutional long short-term memory (ConvLSTM) network enhanced with multi-attention mechanisms, reducing the RMSE for 1-day lead SST forecasts by 41% in the same region, but their approach overlooked the modulation effects of seasonal variations on bias characteristics [10]. For another sea area, Liu et al. utilized a LSTM neural network to decrease the mean absolute error of SST forecasts by 70%, yet their single-point training paradigm disregarded spatial correlations, necessitating point-wise modeling for regional applications and compromising computational efficiency [11]. Yuan et al. focused on global marine regions using a generative adversarial network (GAN)-based integrated model, which reduced forecast mean squared error (MSE) by 90.3% [12]. However, their monthly-averaged data framework is inadequate for operational rapid-response requirements. These studies reveal two limitations: (1) the existing methods mostly focus on bias correction for 1-day lead time forecasts, and there is a lack of research and discussion on the bias correction for multi-day continuous forecasts; (2) despite the achievements in the application of AI-based methods for SST bias correction, the requirements of bias correction in operational forecasting have not been considered. Additionally, the performance of neural network methods in handling different scales, such as interannual, seasonal, and daily scales, has not been systematically diagnosed.

In this study, we propose two distinct post-processing strategies to address systematic biases in SST numerical forecasts: (1) a data assimilation-based strategy implementing the spatiotemporal four-dimensional multi-grid analysis scheme, denoted as 4D-MGA [13], and (2) a hybrid deep learning-based strategy integrating EOF analysis, empirical mode decomposition (EMD), and a BP neural network, formally designated as EE–BP [14]. Specifically, the 4D-MGA framework is fundamentally grounded in a three-dimensional variational (3D-Var) assimilation scheme [15], with explicit temporal dimension integration. This spatiotemporal formulation establishes the methodology as a cost-effective post-processing mechanism for numerical forecast bias mitigation. It is worth noting that this critical distinction sets 4D-MGA apart from conventional data assimilation approaches, which are primarily designed for computationally expensive online bias correction systems [5]. By leveraging satellite fusion SST data, bias correction experiments are conducted targeting daily 7-day SST forecasts in 2017 in the Kuroshio region south of Japan, followed by systematic validation and comparative analysis to evaluate the efficacy of these two strategies.

The remainder of this paper is organized as follows. In Section 2, the data and methods used in this study are introduced. In Section 3, we present and analyze the results of the bias correction experiments. Finally, conclusions are given in Section 4.

2. Materials and Methods

2.1. Data

This study comparatively evaluates these two novel post-processing strategies for bias correction in SST forecasts. The target area is the dynamically complex Kuroshio region in the south of Japan (28–36°N, 128–142°E; Figure 1). As the most powerful western boundary current in the North Pacific, the Kuroshio region exhibits strong nonlinear dynamics that generate pronounced SST variability through intense air–sea interactions [16,17]. This establishes the study area as an ideal validation testbed for assessing bias correction methodologies under challenging forecasting conditions.

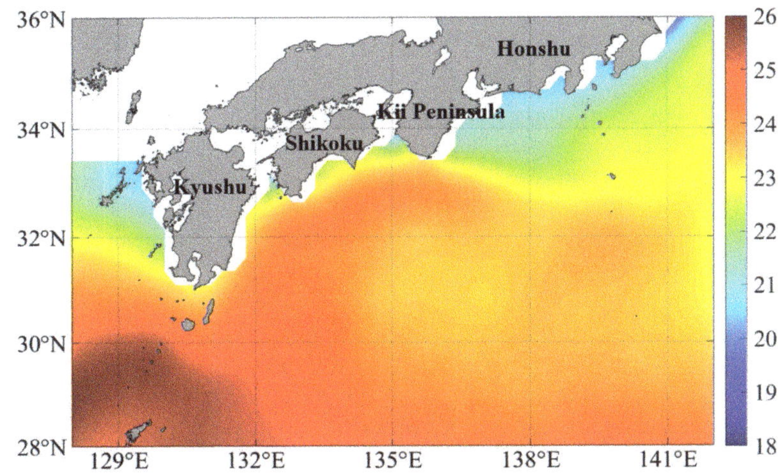

Figure 1. Maps of annual mean SST (°C) from the OISST in 2017 in the Kuroshio region in the south of Japan (28–36°N, 128–142°E).

The 7-day daily SST numerical forecasts used are generated by an ocean model custom-developed to simulate the 2017 Kuroshio large meander event [18]. The modeling framework utilizes the Princeton Ocean Model based on a generalized coordinate system (POMgcs) [19], driven by the ERA5 reanalysis data [20]. The initialization protocol employs a hybrid adaptive approach, synergistically integrating the ensemble adjustment Kalman filter with multi-grid variational analysis [21]. This ensures the effective assimilation of multi-source observations, including in situ temperature/salinity profiles, satellite-retrieved SSTs, and altimeter sea surface height anomalies. The configured model produced three-dimensional daily mean ocean state forecasts at 1/24° spatial resolution from 2016 to 2017, establishing a high-fidelity baseline for subsequent bias correction analysis.

For bias correction validation and performance evaluation, we employ the daily Optimum Interpolation SST (OISST, https://www.ncei.noaa.gov/products/optimum-interpolation-sst, accessed on 5 January 2025) v2.1 from the National Oceanic and Atmospheric Administration (NOAA) as the reference data [22]. The OISST dataset employs Advanced Very High-Resolution Radiometer (AVHRR) satellite data as its primary input, complemented by in situ measurements (ships, buoys, and Argo floats), with a horizontal resolution of $1/4° \times 1/4°$. Validation studies show that the RMSE of OISST in the Northwest Pacific is about 0.3–0.8 °C compared to independent buoy measurements [23].

Forecast biases are computed as the difference between the SST numerical forecasts and the OISST data. To ensure consistency in spatial resolution, the SST numerical forecasts are gridded to match the OISST grid resolution ($1/4° \times 1/4°$) through spatial averaging.

2.2. Data Assimilation-Based Strategy for Bias Correction

The data assimilation-based strategy for bias correction leverages the temporal correlation property of 4D-MGA, which enables it to fit historical bias trends. By extrapolating the trends to the forecast period, the bias correction of the SST forecasts can be achieved.

2.2.1. Principle of 4D-MGA Method

The 4D-MGA method introduces the temporal and spatial correlations of the increments of the background fields due to the observations (called analysis increments) by incorporating the Laplace operator. The method employs a dichotomy-based multi-grid analysis (MGA) [24,25], enabling variational analysis from coarse to fine grids (from large scales to small scales). The 4D-MGA uses bilinear interpolation to calculate observation increments (biases). This interpolation scheme is applied to grids of different levels. On coarser grids, the interpolation spans larger spatial scales, capturing broad-scale error trends. On finer grids, the interpolation resolves localized biases with higher precision. This hierarchical mechanism ensures the robust spatiotemporal extrapolation of analysis increments. For the nth level of the grid, the number of nodes formed via dichotomy is $2^{(n-1)}+1$ for each individual dimension (3D spatial + 1D temporal), and the cost function for 4D-MGA is:

$$J^{(n)}(\overset{\sim(n)}{\mathbf{X}}) = \lambda \overset{\sim(n)}{\mathbf{X}}^{T} \mathbf{S}^{(n)} \overset{\sim(n)}{\mathbf{X}} + \frac{1}{2}[\mathbf{H}^{(n)}\overset{\sim(n)}{\mathbf{X}} - \overset{\sim(n)}{\mathbf{Y}}]^{T}\mathbf{O}^{(n)^{-1}}[\mathbf{H}^{(n)}\overset{\sim(n)}{\mathbf{X}} - \overset{\sim(n)}{\mathbf{Y}}](n = 1, \cdots N), \tag{1}$$

where N is the last level of the grid; $\overset{\sim(n)}{\mathbf{X}}$ is the analysis increment at the nth level, that is, the bias correction; $\overset{\sim(n)}{\mathbf{Y}}$ is the increment of the observation relative to the background field (called observation increment) at the nth level; $\mathbf{H}$ denotes the linear interpolation operator from the background field to the observation point; $\mathbf{O}$ is the observation error covariance matrix; $\mathbf{S}$ denotes the Laplace operator; $\lambda \overset{\sim(n)}{\mathbf{X}}^{T} \mathbf{S}^{(n)} \overset{\sim(n)}{\mathbf{X}}$ is the smoothing term controlling the smoothing scale of the analysis field [25]; and λ is the smoothing coefficient governing the weight of the smoothing term.

Equation (1) is solved iteratively by the minimization algorithm to obtain $\overset{\sim(n)}{\mathbf{X}}$. The final results of 4D-MGA can be obtained by the superposition of the results from each level:

$$\mathbf{X}^{a} = \mathbf{X}^{b} + \sum_{n=1}^{N} \overset{\sim(n)}{\mathbf{X}}, \tag{2}$$

where $\mathbf{X}^{a}$ is the analysis results and $\mathbf{X}^{b}$ is the background field used in the analysis—in our case, the SST numerical forecasts of POMgcs.

We utilize the analysis increments of the analysis period (where observations are present) to extrapolate the bias trend via the smoothing term. For a one-dimensional grid with M uniformly distributed grid points, the smoothing term can be described as:

$$\lambda \tilde{\mathbf{X}}^{T} \mathbf{S} \tilde{\mathbf{X}} = \lambda \sum_{i=2}^{M-1} \left(\tilde{\mathbf{X}}_{i-1} - 2 \cdot \tilde{\mathbf{X}}_{i} + \tilde{\mathbf{X}}_{i+1} \right)^{2}. \tag{3}$$

Here, the Laplace operator serves to define the smoothness, or the spatiotemporal correlation of $\widetilde{\mathbf{X}}$. When a grid point, such as $\widetilde{\mathbf{X}}_i$, is adjusted under the influence of observations, all the other analysis increments will be constrained by the Laplace operator and adjusted following the variation trend of $\widetilde{\mathbf{X}}_i$.

In this study, the 4D-MGA strategy employs the Laplace operator to enforce smoothness constraints, allowing for the linear extrapolation of bias trends from the analysis period (with observations) to the forecast period (without observations). During the analysis phase, increments are computed to minimize mismatches between forecasts and observations; these increments are then extended into the future under the assumption of gradual temporal evolution. Thereby, we can obtain the bias corrections for the forecast period.

2.2.2. Workflow of 4D-MGA

The spatiotemporal scale of MGA mainly depends on the smoothing coefficient and the space–time grid resolution. Through trial and error, the smoothing coefficient λ is set to 200. A large smoothing coefficient can better filter out the random errors and instantaneous signals in the analysis increments, thus retaining the large-scale temporal signals and enabling their effective extrapolation. In order to achieve better performance and lower computational cost, again, via trial-based calibration, the grid levels of the three dimensions used in this study are 7, 6, and 7 (longitude, latitude, and time dimensions), respectively. The finest grid comprises $65 \times 33 \times 65$ nodes, with a horizontal resolution of $0.22° \times 0.25°$ and a temporal grid step of 1 day.

The 4D-MGA primarily estimates bias by fitting temporal trends, including the following four steps:

1. Background field generation: The background fields for each day within the time window (65 days) are set as $\mathbf{X}_1^b \cdots \mathbf{X}_{65}^b$. Taking the forecast start time (day 59) as the splitting point, days 1 to 58 are the analysis period, and days 59 to 65 are the forecast period. The background fields for the analysis period ($\mathbf{X}_1^b \cdots \mathbf{X}_{58}^b$) are derived from the day 1 output of the 7-day SST forecasts, while the background fields for the forecast period ($\mathbf{X}_{59}^b \cdots \mathbf{X}_{65}^b$) are the 7-day SST forecasts corresponding to $\mathbf{X}_{59}^b$.
2. Observation increment calculation: For the analysis period, the observation increments ($\widetilde{\mathbf{Y}}_1 \cdots \widetilde{\mathbf{Y}}_{58}$) are calculated as the difference between the OISST data ($\mathbf{Y}_1^{obs} \cdots \mathbf{Y}_{58}^{obs}$) and the background fields ($\mathbf{X}_1^b \cdots \mathbf{X}_{58}^b$) for the corresponding dates. These observation increments are equal to the negative of the SST forecast biases.
3. Data assimilation: Based on the smoothing term, 4D-MGA fits the observation increments ($\widetilde{\mathbf{Y}}_1 \cdots \widetilde{\mathbf{Y}}_{58}$) to generate smoothed analysis increments ($\widetilde{\mathbf{X}}_1 \cdots \widetilde{\mathbf{X}}_{58}$) for the forecast period, which are extrapolated to obtain the analysis increments ($\widetilde{\mathbf{X}}_{59} \cdots \widetilde{\mathbf{X}}_{65}$) for the analysis period.
4. Bias correction: Add the analysis increments ($\widetilde{\mathbf{X}}_{59} \cdots \widetilde{\mathbf{X}}_{65}$) for the forecast period to the corresponding background fields ($\mathbf{X}_{59}^b \cdots \mathbf{X}_{65}^b$), producing the bias-corrected SST analysis fields ($\mathbf{X}_{59}^a \cdots \mathbf{X}_{65}^a$).

It is worth noting that the 4D-MGA bias correction strategy (the same as the following EE–BP strategy) is applied as a post-processing step to the SST numerical forecasts. Corrected outputs do not feed back into the subsequent forecast cycles, ensuring computational tractability and methodological isolation.

2.3. Deep Learning-Based Strategy for Bias Correction

The deep learning-based strategy (named EE–BP) constructs SST bias forecast models through learning systematic biases between the numerical model outputs and the OISST data. By integrating the spatiotemporal decomposition capability of EOF analysis, the multi-scale signal-processing capability of EMD, and the nonlinear learning capability of BP neural networks, the hybrid method becomes adept at capturing nonlinear signals (see Section 3).

2.3.1. Principle of EE–BP Method

The EOF analysis decomposes a data matrix into spatial modes (called EOFs) and their corresponding time series (called principal components, PCs) [26]. These EOFs and PCs capture the spatial distribution and temporal evolution characteristics of the data, respectively. The proportion of explained variance for each mode reflects the concentration of information, allowing for data simplification by selecting EOFs and PCs with high variance contributions. In this study, EOF analysis is applied to the SST forecast bias data. To balance information preservation and computational efficiency, we retain EOFs and PCs explaining 99.9% cumulative variance. This threshold is validated through comparative experiments (see Table S1 in the Supplementary Material). When the threshold rises, the increase in the correction effect diminishes, and the computational cost increases significantly. During the training of the bias correction model, the PCs are taken as the predicted values. With the predicted PCs, the SST biases can be reconstructed by combining them with the EOFs. The decomposition strategy reduces the computational complexity of neural network training by focusing on the temporal dimension. Compared with single-point neural network prediction, combining EOF analysis with the neural network improves the training efficiency while preserving the spatial correlations.

The EMD method is an effective signal-processing technique [27] that has been widely used in marine data processing and forecasting [28–32]. It adaptively decomposes complex and non-stationary signals into a series of intrinsic mode functions (IMFs) and a residual component. This decomposition smooths the signal and reduces its nonlinearity for further analysis. In this study, EMD is applied to each PC derived from EOF analysis, decomposing them into multiple IMFs. These IMFs represent different frequency components within the PCs, providing finer input features for neural network training. By trial and error, each PC is decomposed into three IMFs and one residual component, capturing the primary frequency components while avoiding over-decomposition and information redundancy.

The BP neural network, a multi-layer feedforward network trained via error back-propagation, excels in modeling nonlinear relationships through iterative weight optimization [4,33,34]. Its capability to approximate complex mappings between input–output variables proves advantageous for marine data characterized by strong nonlinearity, such as SST. Through extensive training, the BP network learns the weight relationships between input and output data, enabling accurate predictions even for new input data. Therefore, it is an ideal tool for bias correction and forecast accuracy improvement. In this study, we build three-layer (input–hidden–output) BP deep neural network models to conduct bias correction experiments for SST forecasts. The model performance is optimized by minimizing the RMSE between the predicted values and the ground truth values, using gradient descent. Predictions are generated through forward propagation, and weights are adjusted via backward error propagation to reduce errors iteratively. Additionally, the adaptive moment estimation (Adam) optimizer is employed to solve sparse gradients and noise problems. The rectified linear unit (ReLU) activation function is chosen to prevent the gradient vanishing problem and improve computational efficiency.

In this study, EOF analysis first decomposes the SST biases of the training set into EOFs and PCs, reducing dimensionality while preserving spatiotemporal correlations. EMD further decomposes each PC into multi-scale components (IMFs), isolating nonlinear and non-stationary signals. A BP neural network model is then trained on these IMFs. Leveraging the model, we can predict the future IMFs of the test set by inputting IMFs in the past few days. By reconstructing predicted IMFs into PCs and combining them with precomputed EOFs, EE–BP generates bias-corrected SST forecasts.

2.3.2. Workflow of EE–BP

The EE–BP combines EOF analysis, EMD analysis, and BP neural networks to build seven bias correction models for each forecast lead time. Figure 2 shows the flowchart of the EE–BP on day m ($m = 1, 2 \cdots 7$), which includes the following five steps:

1. EOF analysis: The 2016 SST forecast bias data are decomposed by EOF analysis to obtain EOFs and PCs. EOFs and PCs accounting for 99.90% of the cumulative variance are selected. The 2017 SST forecast bias data are projected onto the 2016 EOFs to derive the corresponding time series, called pseudo-PCs.
2. EMD analysis: Each PC is decomposed into three IMFs and one residual component (called derived PCs) using EMD analysis. For each derived PC, the 2016 data are used as the training set, and the 2017 data are used as the test set.
3. Model training: The BP neural network is constructed and trained on the training set. The size of the time window used to predict the derived PCs is set to m, which means that we use the preceding m-day-derived PC data to predict m-step-derived PCs.
4. Model validation: based on the trained BP neural network, we predict the derived PCs in 2017, which are compared with the test set to evaluate model accuracy.
5. Bias correction: The predicted derived PCs are reconstructed into PCs, which are then combined with EOFs to generate SST bias forecasts. By combining the biases and SST forecasts, the corrected SST can be obtained.

2.4. Evaluation Metrics

To evaluate the performance of the two bias correction strategies, we employ time-averaged and space-averaged bias, RMSE, and the spatial anomaly correlation coefficient (ACC) as the evaluation criteria. These calculation formulas are defined as follows:

$$bias_i = \frac{1}{N}\sum_{t=1}^{N}\left(x_{i,t}^{P} - x_{i,t}^{T}\right), \tag{4}$$

$$bias_t = \frac{1}{M}\sum_{i=1}^{M}\left(x_{i,t}^{P} - x_{i,t}^{T}\right), \tag{5}$$

$$RMSE_i = \frac{1}{N}\sqrt{\sum_{t=1}^{N}\left(x_{i,t}^{P} - x_{i,t}^{T}\right)^2}, \tag{6}$$

$$ACC = \frac{\sum_{i=1}^{M}\left(y_{i,t}^{P} - \overline{y}_t^{P}\right)\left(y_{i,t}^{T} - \overline{y}_t^{T}\right)}{\sqrt{\sum_{i=1}^{M}\left(y_{i,t}^{P} - \overline{y}_t^{P}\right)^2\left(y_{i,t}^{T} - \overline{y}_t^{T}\right)^2}}, \tag{7}$$

where $x_{i,t}^{P}$ and $x_{i,t}^{T}$ are the predictive and true SST of the ith grid point on the tth day, respectively; M is the number of spatial grid points; and N is the number of days of testing data. $Bias_i$ is the mean error of the ith grid point and $bias_t$ is the mean error on the tth day. Their joint use ensures a balanced evaluation of correction performance across temporal and spatial scales. $RMSE_i$ is the root–mean–square error of the ith grid point. $y_{i,t}^{P} = x_{i,t}^{P} - c_i$ and $y_{i,t}^{T} = x_{i,t}^{T} - c_i$ are the predictive and satellite SST anomalies (SSTA) of the ith grid point on the tth day, respectively; c_i is the climatological SST of the OISST during from 1991 to

2020 of the ith grid point; and $\overline{y}_i^P = \frac{1}{M} y_{i,t}^P$ and $\overline{y}_i^T = \frac{1}{M} y_{i,t}^T$ are mean values of SSTA from prediction and satellite on the tth day, respectively. In this study, all evaluation metrics calculations use the OISST dataset as ground truth.

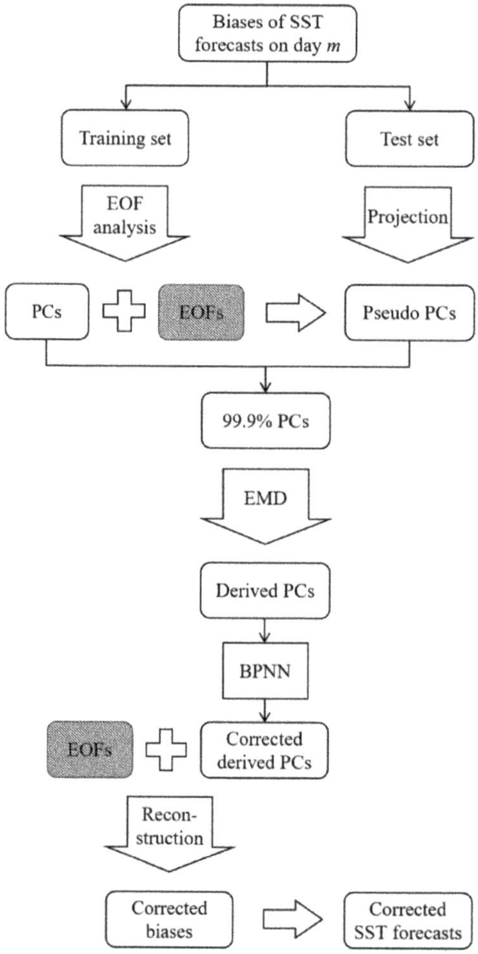

Figure 2. Flowchart of the EE–BP on day *m*.

3. Results

3.1. Overall Performance Evaluation

In this section, the SST forecast results in 2017 are analyzed to assess the overall performance of the two bias correction strategies (4D-MGA and EE–BP). Specifically, the means of RMSEs and spatial ACCs are calculated and compared in Figure 3. As shown in Figure 3b, the uncorrected 7-day forecasts exhibit ACC values consistently exceeding 0.6, confirming the reliability of the SST numerical forecasts [35]. Compared to the uncorrected forecasts (blue bars), both 4D-MGA (orange bars) and EE–BP (red bars) strategies demonstrate measurable improvements. 4D-MGA shows enhanced performance during days 1–3, reducing RMSEs by 4.5–9.1% while improving ACC values. However, its corrective effectiveness progressively declines with an increasing forecast horizon. In terms of RMSE, improvement in the forecast accuracy at day 7 is only 1.2%. Such results reflect the inherent defects of the 4D-MGA strategy—the bias correction of 4D-MGA is based on linear extrapolation. Therefore, the 4D-MGA strategy naturally lacks the capability to properly capture non-linear signals. Moreover, to eliminate the random errors and instantaneous signals in the analysis increments, 4D-MGA has to employ a relatively large smoothing coefficient. This also leads to the fact that 4D-MGA is unable to capture the complex changes in temporal trends, which is also reflected in the discussion about the seasonal biases after correction in Section 3.2.3. Consequently, its performance in the aspect of bias correction for a long lead time is unsatisfactory. In contrast, the EE–BP maintains stable corrective performance across all seven forecast lead times. After bias correction by the EE–BP, the RMSEs decrease from 0.44–0.83°C to 0.34–0.6 8°C (18.1–22.7% reduction), with ACC values improving to above 0.79. These results indicate that while both strategies enhance SST forecast accuracy, EE–BP exhibits superior correction magnitude and temporal robustness throughout the 7-day lead times.

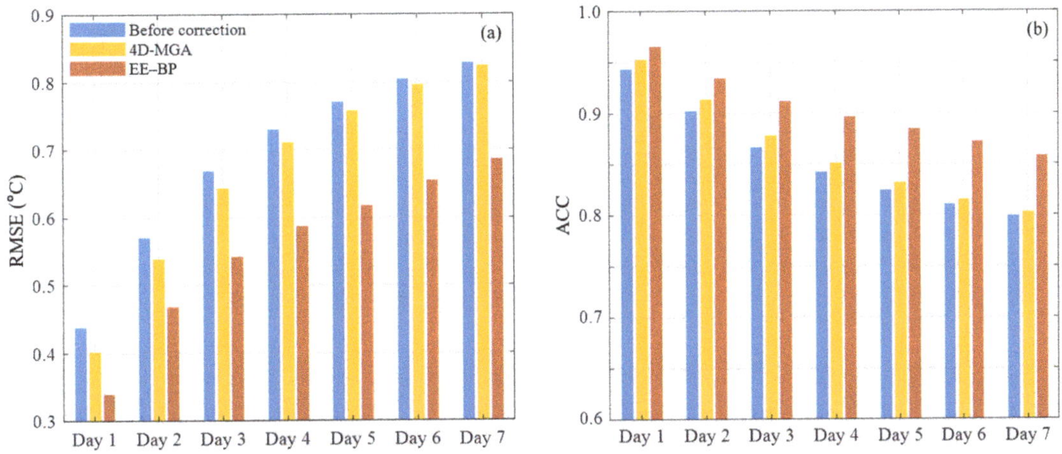

Figure 3. Means of RMSEs (a) and spatial ACCs (b) for the SST forecasts before (blue bars) and after bias correction by 4D-MGA (orange bars) and EE–BP (red bars).

3.2. *Skill Comparison of Two Strategies*

To comprehensively compare the correction skills of 4D-MGA and EE–BP, we conduct a systematic evaluation from the perspective of multiple temporal scales: annual, seasonal, monthly, and daily.

3.2.1. Annual Mean Bias Correction

Figure 4 illustrates the spatial distribution of annual mean SST forecast biases before and after correction. Previous studies show that satellite fusion SST data in the East Asian marginal seas exhibit errors ranging from 0.5 °C to 0.7 °C [36–38], which is consistent with the error of OISST in the Northwest Pacific. Therefore, we define ±0.50 °C as the threshold for assessing the validity of bias correction (black contours in Figure 4). It is a clear standard to judge the correction effect in different regions by analyzing the spatial distribution of SST biases. As shown in Figure 4a, the spatial distribution of SST forecast biases before correction reveals a clear trend of increasing biases with forecast horizon. Additionally, significant warm and cold biases exceeding ±0.5 °C initially appear near Shikoku Island and the Kii Peninsula during days 1–2, then expand spatially and intensify from day 3 onward. This trend reflects the error accumulation in SST numerical forecasts, which amplifies with lead time. Notably, regions with large biases are predominantly concentrated along and around the Kuroshio axis. This is attributed to the highly nonlinear spatiotemporal evolution of SST driven by the multiscale dynamical processes related to the Kuroshio. Numerical models face limitations in accurately forecasting such complex dynamics, particularly in strong mixing zones, resulting in relatively large forecast biases [39,40].

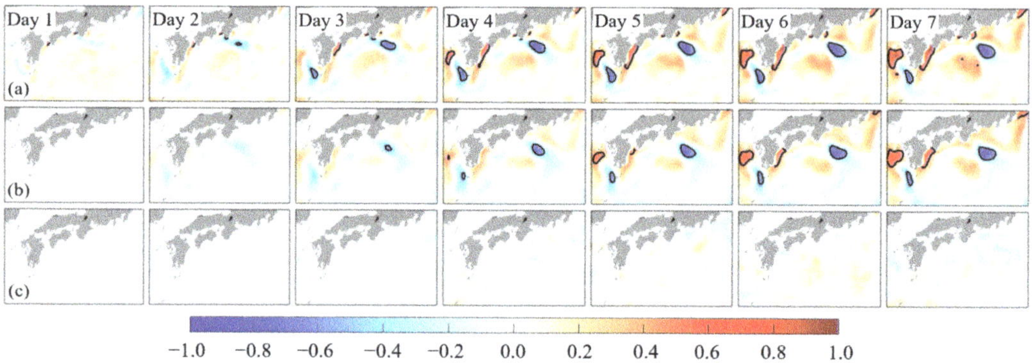

Figure 4. Maps of annual mean biases (°C) of SST forecasts before (**a**) and after bias correction by 4D-MGA (**b**) and EE–BP (**c**). The black contours indicate the error range of ±0.50 °C.

To quantitatively evaluate the bias correction performance of 4D-MGA and EE–BP, we calculate the RMSEs and biases at all model grid points within the study region before and after the bias correction, respectively. Subsequently, the proportions of grid points with improved bias or RMSE for the 7-day SST forecast correction are calculated and presented in Table 1.

Table 1. Proportions (%) of improved grid points for bias correction in SST forecasts.

Metric	Method	Day 1	Day 2	Day 3	Day 4	Day 5	Day 6	Day 7
RMSE	4D-MGA	85.51	59.82	49.62	43.10	39.11	36.20	35.20
	EE–BP	100.00	98.01	97.16	94.71	93.10	86.89	82.36
Bias	4D-MGA	69.56	55.75	46.78	41.87	38.80	37.19	37.04
	EE–BP	99.62	96.32	94.71	91.95	91.72	85.12	79.22

The results show that the annual mean SST forecast bias corrected by 4D-MGA (Figure 4b) is significantly reduced during days 1–2, with biases effectively constrained within ±0.50 °C. On day 1, the proportions of improved grid points for 4D-MGA reach 85.5% for RMSE and 69.6% for bias. However, its correction performance diminishes with increasing forecast horizon, particularly in regions with extensive and intense warm or cold biases. In contrast, EE–BP (Figure 4c) demonstrates more stable correction performance, maintaining annual mean SST forecast biases within ±0.50 °C throughout the 7-day forecast horizons. In terms of improved grid points (Table 1), EE–BP consistently achieves proportions above 82.4% for RMSE and 79.2% for bias across all forecast horizons. Notably, while EE–BP outperforms 4D-MGA in the proportions of improved grid points on day 1, the annual mean biases corrected by 4D-MGA are generally lower in most regions. This discrepancy arises from the offsetting effects of positive and negative biases in the annual mean calculation. 4D-MGA, which primarily captures large-scale bias characteristics, benefits from this offsetting effect.

In summary, the 4D-MGA method demonstrates effective bias correction capabilities only during short forecast lead times, whereas EE–BP delivers stable and significant corrections across the entire 7-day forecast horizons. The performance divergence stems from fundamental methodological differences underlying the two strategies. 4D-MGA relies on linear extrapolation, providing accurate short-term alignment but progressively diverging from reality as nonlinear dynamics dominate long lead-time forecasts. Conversely, EE–BP leverages its strong nonlinear learning capabilities to effectively address the nonlinear signals of SST forecast biases in the Kuroshio region, providing consistent and robust corrections.

It is worth noting that the EE–BP method can more effectively extract long-term SST bias signals because of the capability of fitting the nonlinear bias changes in the training data. In contrast, the 4D-MGA method conducts a real-time linear fitting of the large-scale time-varying characteristics of the bias based on the smoothing term. Under the constraint of the Laplace operator, biases with closer time distances have a greater correlation with each other, and the characteristics of biases with larger time distances are unlikely to have a significant impact on the forecast period.

3.2.2. Seasonal Bias Correction

Figure 5 presents the proportions of improved grid points (with reduced RMSE) for 4D-MGA and EE–BP across four seasons, with blue and orange lines representing their respective daily trends during days 1–7. Both strategies exhibit consistent seasonal performance: summer achieves the highest correction efficacy, followed by spring, while autumn and winter demonstrate relatively weaker improvements. Except for a slight increase for 4D-MGA during days 3–7 in summer, the correction efficacy of both strategies generally declines with increasing forecast horizon and peaks on day 1. For 4D-MGA, the proportions of improved grid points range from 29.1% to 60.7%. Except in summer, 4D-MGA maintains >50% improved grid points only during days 1–3, reflecting its instability of correction performance. In contrast, EE–BP outperforms 4D-MGA across all seasons, with improved grid points ranging from 50.2% to 96.8%. Although its performance also declines marginally with increasing forecast horizon, EE–BP sustains 50% effectiveness even on day 7. In summary, both strategies effectively correct seasonal SST forecast biases, with optimal performance in summer. However, EE–BP exhibits greater stability across all seasons, particularly for forecasts exceeding a 3-day lead time.

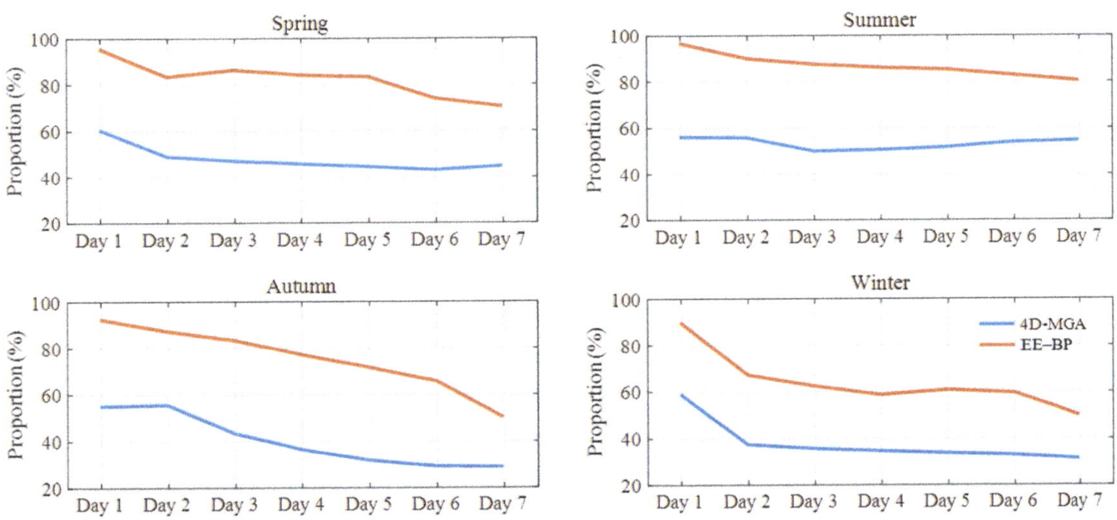

Figure 5. Maps of the proportions (%) of improved grid points (with reduced RMSE) after bias correction by 4D-MGA and EE–BP in four seasons.

To further explore the spatial scale bias correction effects of the two strategies in four seasons, we plotted the maps of seasonal SST forecast bias distributions before and after correction (Figures 6–9). As revealed in Figures 6a, 7a, 8a and 9a, the uncorrected SST forecast biases exhibit distinct seasonal characteristics: significant warm and cold biases dominate in spring and winter, warm biases prevail in summer, and autumn shows relatively smaller biases compared to other seasons. Additionally, all four seasons exhibit large biases in the northern part of the study area, closely aligned with the Kuroshio axis, consistent with the annual mean bias distribution described in Section 3.2.1.

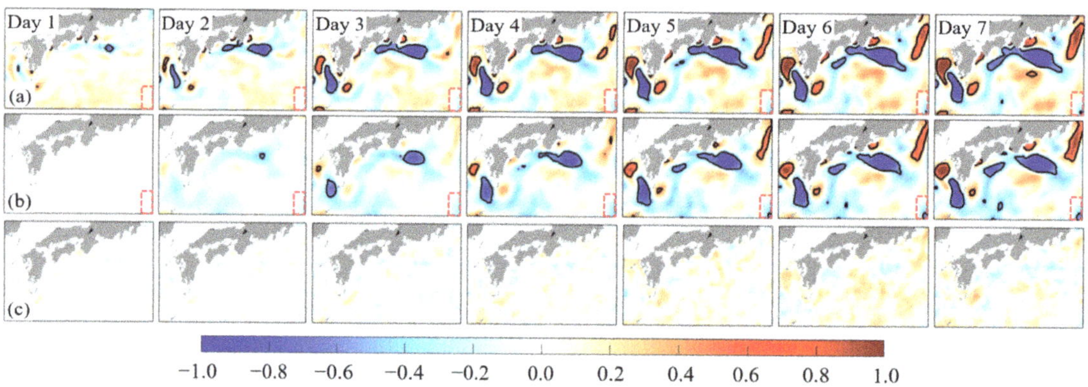

Figure 6. Maps of the spring biases (°C) of SST forecasts before (**a**) and after bias correction by 4D-MGA (**b**) and EE–BP (**c**). The black contours indicate the error range of ±0.50 °C. The pink dashed boxes delineate regions where bias growth direction reverses during the forecast period.

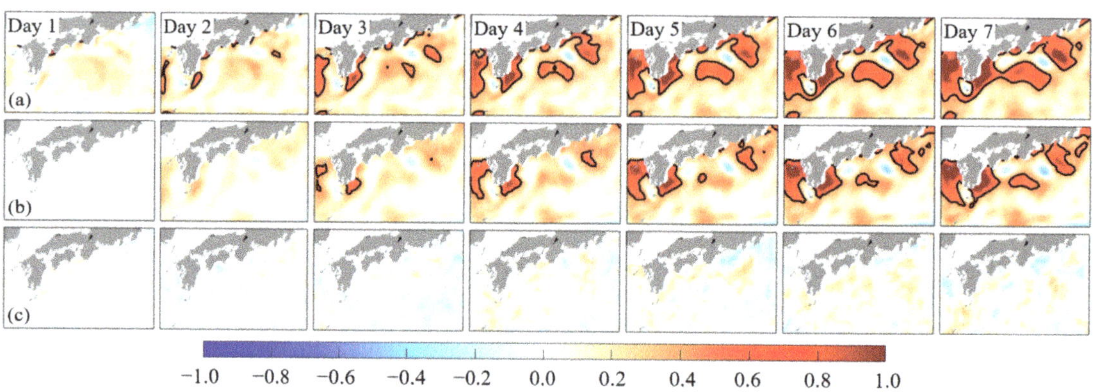

Figure 7. Maps of the summer biases (°C) of SST forecasts before (**a**) and after bias correction by 4D-MGA (**b**) and EE–BP (**c**). The black contours indicate the error range of ±0.50 °C.

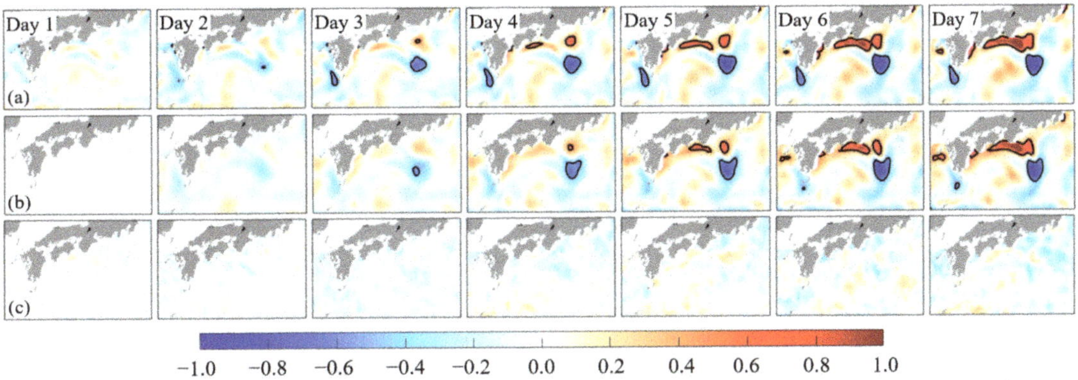

Figure 8. Maps of the autumn biases (°C) of SST forecasts before (**a**) and after bias correction by 4D-MGA (**b**) and EE–BP (**c**). The black contours indicate the error range of ±0.50 °C.

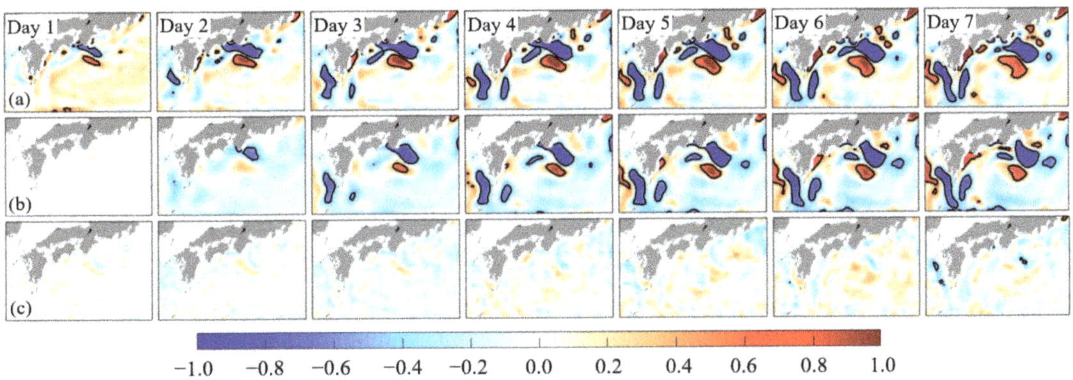

Figure 9. Maps of the winter biases (°C) of SST forecasts before (**a**) and after bias correction by 4D-MGA (**b**) and EE–BP (**c**). The black contours indicate the error range of ±0.50 °C.

By comparing with Figure 4, we find that the performance of 4D-MGA (Figures 6b, 7b, 8b and 9b) and EE–BP (Figures 6c, 7c, 8c and 9c) for seasonal bias correction aligns with the annual mean bias correction results. Specifically, 4D-MGA demonstrates significant correction efficacy on the first forecast day across all seasons but its performance degrades as the forecast horizon increases. 4D-MGA predicts the trend of bias changes based on the analysis increments in the analysis period, and therefore, the error accumulation of the 7-day SST forecasts is ignored. As the lead time increases, the accumulated forecasting errors gradually dominate the SST biases. As a result, the forecasting errors corrected by 4D-MGA will continuously increase and finally resemble those before correction in all seasons. This also explains the significant seasonal characteristics of the spatial bias distribution in Figures 6b, 7b, 8b and 9b: the numerical model's overly conservative dynamic interpolation leads to significant warm biases in summer and cold biases in winter. In contrast, while EE–BP is affected by seasonal biases, it consistently provides significant seasonal bias corrections and is less influenced by seasonal variations and error accumulations (Figures 6c, 7c, 8c and 9c). By generating bias correction models independently, EE–BP can consider the model error of each forecast lead time, which leads to robust correction performance across all forecast horizons.

Despite relatively large seasonal biases, 4D-MGA exhibits the best and most stable improvement ratios of the proportions of improved grid points in summer (Figure 5). As shown in Figure 7a, the biases are warm biases that grow unidirectionally throughout the forecast horizons in most areas. While 4D-MGA can smoothly extrapolate the unidirectional growth trend from the analysis increments, for regions where the bias growth direction changes during the forecast period (as shown by the pink dashed boxes in Figure 6), it is difficult to make inferences relying on linear extrapolation. In fact, this may even lead to an increase in biases. On the other hand, EE–BP can better perceive the complex temporal trends of biases. As shown in Figures 6c, 7c, 8c and 9c, the bias trend corrected by EE–BP has no obvious correspondence with that before correction.

In summary, although both bias correction strategies are influenced by seasonal biases in numerical forecasts, the extent of this influence varies significantly. For 7-day SST forecasts, EE–BP demonstrates a clear advantage, effectively addressing model error accumulation and accurately capturing the impact of seasonal forecast biases, outperforming 4D-MGA in seasonal bias correction capability.

3.2.3. Monthly and Daily Mean Bias Correction

In this section, we compare their correction performance at monthly and daily mean time scales to further investigate the sensitivity of the two strategies to seasonal biases. Figure 10 presents histograms of daily mean biases and line graphs of monthly mean biases for SST forecasts before and after correction. As shown in Figure 10a, the biases in the 2016 SST forecasts reveal seasonal variation patterns similar to those observed in 2017, with both years showing pronounced warm biases from April to August. As the forecast horizon increases, the seasonal biases become more pronounced, reaching 1 °C on day 7, reflecting the model error accumulation. It is critically important to emphasize that this inherent similarity in systematic biases exhibited by the 2016 and 2017 annual cycles enables the EE–BP model, trained solely on the 2016 SST forecast bias sequence, to effectively correct the 2017 SST forecast biases. The seasonal biases and error accumulation remain apparent after being corrected by 4D-MGA. The residual seasonal signals on day 3 are nearly identical to those before correction. In contrast, EE–BP effectively mitigates (and in some cases eliminates) seasonal biases in SST forecasts (Figure 10b). Its bias correction performance remains largely unaffected by seasonal biases across the 7-day forecast horizons, exhibiting more stable characteristics on monthly and daily scales.

The trends of daily mean biases from day 1 to day 7 further show that the limitations of 4D-MGA are manifested in two ways. Firstly, 4D-MGA struggles to identify the alteration of the bias growth direction. At the inflections of the monthly mean biases (represented by the black curve), the extrapolation direction of 4D-MGA deviates from the real bias evolution, resulting in the increase of biases. As aforementioned, 4D-MGA only extracts large-scale linear signals. Throughout the entire forecast horizon, the bias correction amount increases linearly following the trend of the analysis periods. The deficiency of linear extrapolation mechanisms renders 4D-MGA incapable of identifying the nonlinear bias signals. 4D-MGA achieves better correction efficacy in spring and summer (from March to August) where biases exhibit temporally stable and unidirectional trends but performs less effectively in autumn and winter (from September to February) due to frequent bias sign reversals and nonlinear variability. This seasonal contrast (also seen in Figure 5) is consistent with its linear extrapolation mechanism. Secondly, as the forecast horizon extends, the cumulative model errors keep growing, and the correction efficacy of 4D-MGA gradually deteriorates. Consequently, residual seasonal biases persist in 4D-MGA corrections, particularly manifesting pronounced biases during days 3–7.

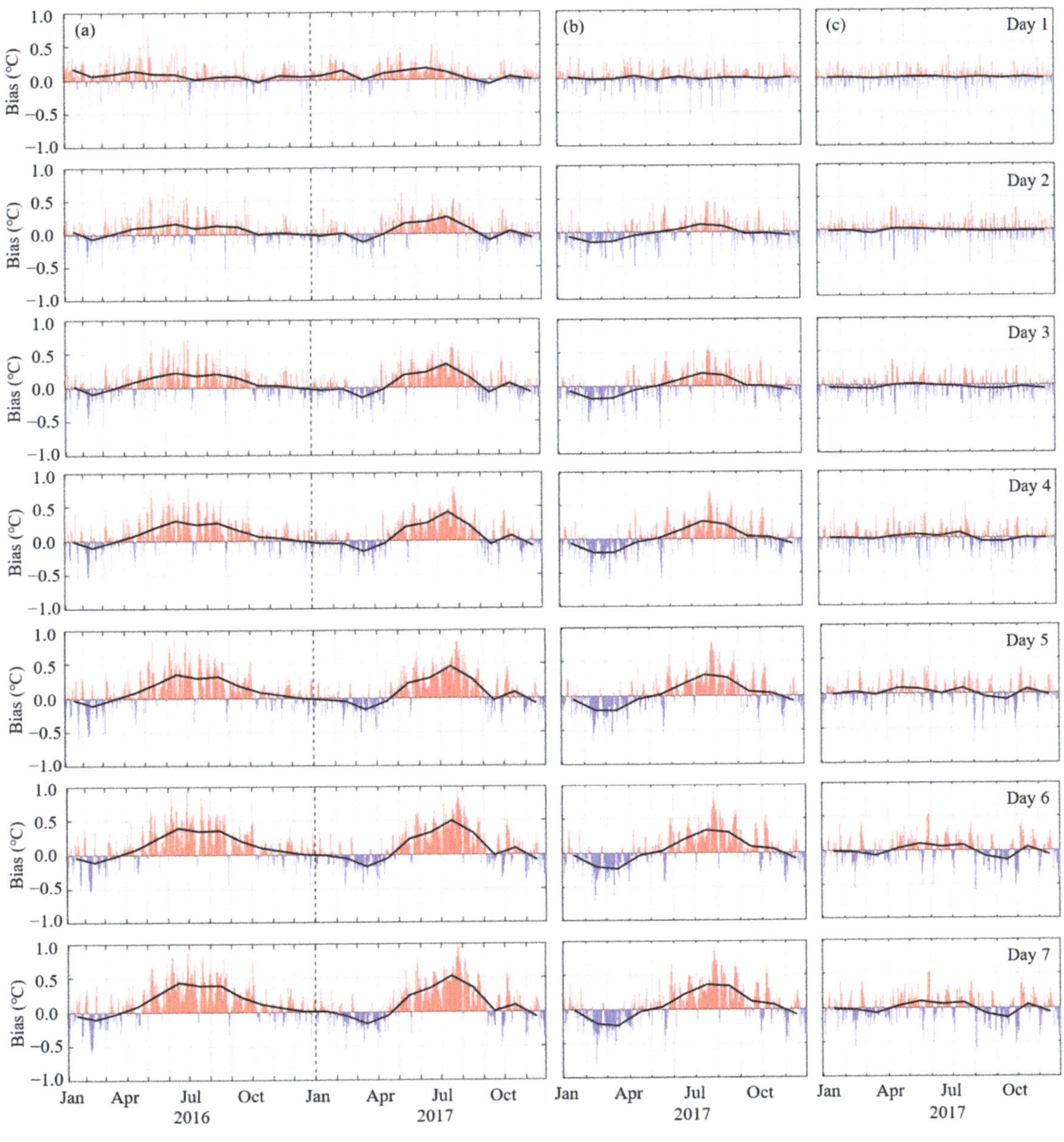

Figure 10. Monthly mean (black curve) and daily mean (red and blue bars) time series of the biases in SST forecasts before (**a**) and after bias correction by 4D-MGA (**b**) and EE–BP (**c**).

In comparison, EE–BP shows better identification of bias inflections and superior stability throughout the entire forecast horizon. The daily mean biases display minimal seasonal characteristics. Due to its strong nonlinear learning capability, EE–BP can precisely understand the consistent bias variation trends inherent in the training data, allowing it to effectively capture and correct significant seasonal signals. Additionally, the independent modeling strategy ensures that EE–BP can focus on the bias changes of individual forecast days, avoiding the impact of model error accumulation. As a result, it becomes more robust in handling seasonal biases and improves the accuracy of SST forecasts.

Figure 11 presents the daily ACC distributions for 7-day lead-time SST forecasts in 2017, where the blank areas indicate that the effective threshold of forecast skill is not reached (ACC < 0.6) [35]. The uncorrected forecasts (Figure 11a) exhibit substantial skill-deficient areas (blank areas) accounting for 1.9%, while the 7-day sustained effective forecast rate, with all 7-day ACCs > 0.6, is only 94.0%. Notably, there are more forecast days in June and October when the threshold is not reached, which corroborates the systematic peaks of biases in Figure 10a. After applying 4D-MGA (Figure 11b), the percentage of skill-deficient areas increases to 2.1%, although the horizon of the high-skill areas (ACC > 0.9, red areas) is extended. The 7-day sustained effective forecast rate only improves by 0.3 percentage points to 94.3%, which indicates that 4D-MGA has a limited ability to extend the effective lead time of the forecast. Conversely, EE–BP (Figure 11c) achieves remarkable progress: the percentage of skill-deficient areas plunges from 1.9% to 0.1%, with only a very small number of skill-deficient areas remaining, mainly in the summer. Notably, the 7-day sustained effective forecast rate improves by 5.7 percentage points to 99.7%, accompanied by systematic ACC enhancement, as evidenced by the expansion of red coverage (high-skill region expansion). In conclusion, EE–BP is effective in both extending the effective forecast lead time and expanding the horizon of the high-skill areas, further confirming the accuracy and stability of EE–BP in 7-day SST forecasts.

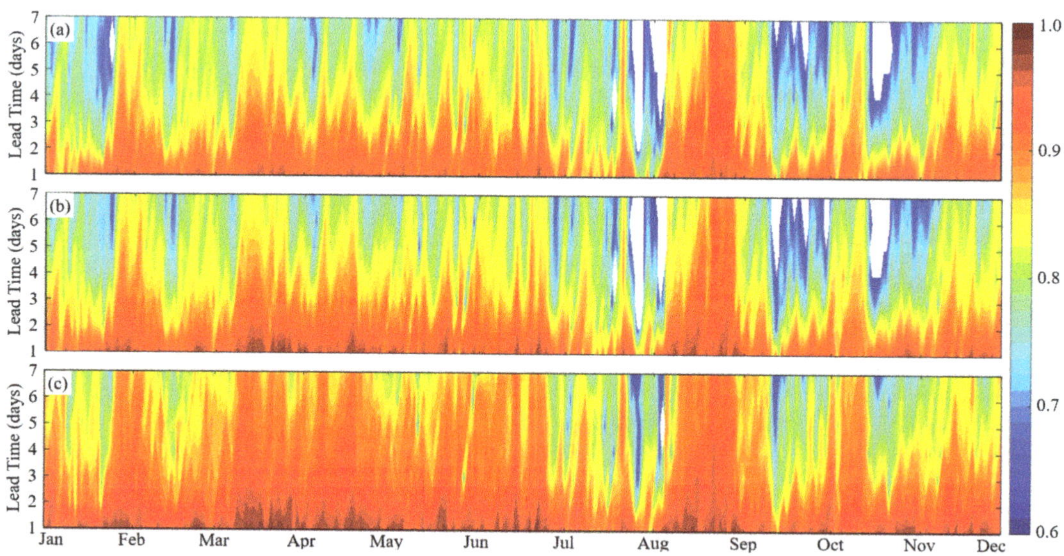

Figure 11. Daily ACCs of the SST forecasts before (**a**) and after bias correction by 4D-MGA (**b**) and EE–BP (**c**) at 1–7 day lead times in 2017. The blank areas indicate ACCs smaller than the threshold of 0.6.

4. Discussion

In this study, two bias correction strategies, 4D-MGA and EE–BP, are proposed for the bias correction in 7-day SST forecasts. Their performance and skills are systematically assessed across multiple temporal scales (annual, seasonal, monthly, and daily). The results show that both strategies can improve the accuracy of SST numerical forecasts, and EE–BP exhibits superior performance in correcting long-term forecast biases and handling complex seasonal biases. Specifically, 4D-MGA excels in short forecast lead times (particularly on day

1), improving forecast accuracy (RMSE) by approximately 9.1%. However, its correction efficacy diminishes with increasing forecast horizon, showing limited effectiveness in regions with high-intensity warm or cold biases and pronounced seasonal variations. These limitations stem from the model error accumulation and inherent challenges of 4D-MGA in addressing the complex nonlinear dynamics in the Kuroshio region. In contrast, the EE–BP strategy delivers stable and significant corrections throughout the 7-day forecast horizons with consistently improving forecast accuracy (RMSE) by over 18.1%. As a robust nonlinear deep learning method, EE–BP effectively captures nonlinear features in SST forecast biases, providing consistent and reliable corrections for 7-day forecasts. EE–BP outperforms 4D-MGA across all temporal scales, particularly in mitigating seasonal biases, demonstrating strong robustness.

While the EE–BP strategy demonstrates superior bias correction performance, its practical implementation faces two interrelated limitations. Firstly, the method's reliance on a large amount of historical data introduces scalability challenges. Although one year of SST training data achieves satisfactory results in this study, its data demand for training escalates significantly when applied to scenarios with more predicted variables, such as three-dimensional corrections. Secondly, EE–BP requires significant preprocessing and model training time. For example, it takes approximately 56 h on a portable workstation for the training process and the SST correction for the year 2017, whereas 4D-MGA (without training process) completes the correction in 1.5 h. 4D-MGA provides a lightweight alternative, requiring only 58-day historical data and minimal processing time, making it ideal for real-time applications with limited infrastructure.

5. Conclusions

In conclusion, 4D-MGA is particularly advantageous for short-term forecasts, especially when model biases exhibit regular, linear trends. Its low application cost and lack of reliance on model training make it suitable for rapid, short-term ocean model forecasts. Conversely, the EE–BP strategy offers higher correction accuracy and longer effective correction horizons. Its ability to handle complex nonlinear problems ensures robust performance in correcting 7-day forecast biases, particularly in addressing seasonal biases. Furthermore, there remains room for improvement in its short-term forecast bias correction capabilities. Future research could explore enhanced neural network training schemes to address this limitation. Additionally, a synergistic application of both strategies could be developed, leveraging their complementary strengths to create customized bias correction models tailored to specific regional and seasonal characteristics. Such advancements would further enhance the accuracy of SST numerical forecasts.

Supplementary Materials: The following supporting information can be downloaded at: https://www.mdpi.com/article/10.3390/rs17091602/s1, Table S1: Effect of selecting different cumulative variances on EE–BP on day 1; Figure S1: Means of RMSEs (**a**) and spatial ACCs (**b**) for the SST forecasts before (blue bars) and after bias correction by EOF–BP (orange bars) and EE–BP (red bars); Table S2: RMSEs (°C) and ACCs for the SST forecasts before and after bias correction by EOF–BP and EE–BP.

Author Contributions: Conceptualization, G.H. and W.L.; methodology, W.D., G.H., W.L., H.W. and X.W.; software, W.D., H.W. and X.W.; validation, W.D., G.H., X.W., H.W., Q.Z., L.C., M.Z. and Z.J.; formal analysis, W.D., G.H., H.W. and X.W.; investigation, W.D., G.H., X.W., H.W., L.C., M.Z. and Z.J.; resources, W.D., G.H., W.L., H.W. and X.W.; data curation, W.D., H.W. and X.W.; writing—original draft preparation, W.D., G.H. and H.W.; writing—review and editing, W.D., G.H. and Q.Z.; visualization, W.D.; supervision, G.H.; project administration, G.H.; funding acquisition, G.H. and W.L. All authors have read and agreed to the published version of the manuscript.

Funding: This research was supported in part by the National Key Research and Development Program under Grant 2023YFC3107800 and in part by the National Natural Science Foundation under Grants 42376190 and 41876014.

Data Availability Statement: The NOAA OISST v2.1 dataset is publicly accessible via the National Centers for Environmental Information (NCEI) at https://www.ncei.noaa.gov/products/optimum-interpolation-sst (accessed on 5 January 2025). The daily SST numerical forecasts generated by the POMgcs ocean model are available from the corresponding author (whw@tju.edu.cn) upon reasonable request.

Acknowledgments: The authors thank the following data and tool providers: NOAA for OISST-V2.1-AVHRR data, available at https://www.ncei.noaa.gov/products/optimum-interpolation-sst (accessed on 5 January 2025), Google for machine learning-related open-source software including TensorFlow, available at https://tensorflow.google.cn/ (accessed on 5 January 2025), Keras, available at https://keras.io/ (accessed on 5 January 2025), and Scikit-learn, available at http://scikit-learn.org/stable/ (accessed on 5 January 2025).

Conflicts of Interest: The authors declare no conflicts of interest.

References

1. Hou, S.; Li, W.; Liu, T.; Zhou, S.; Guan, J.; Qin, R.; Wang, Z. MIMO: A unified spatio-temporal model for multi-scale sea surface temperature prediction. *Remote Sens.* **2022**, *14*, 2371. [CrossRef]
2. Krishnamurti, T.N.; Chakraborty, A.; Krishnamurti, R.; Dewar, W.K.; Clayson, C.A. Seasonal prediction of sea surface temperature anomalies using a suite of 13 coupled atmosphere–ocean models. *J. Clim.* **2006**, *19*, 6069–6088. [CrossRef]
3. Lins, I.D.; Araujo, M.; das Chagas Moura, M.; Silva, M.A.; Droguett, E.L. Prediction of sea surface temperature in the tropical Atlantic by support vector machines. *Comput. Stat. Data Anal.* **2013**, *61*, 187–198. [CrossRef]
4. Wei, L.; Guan, L.; Qu, L. Prediction of sea surface temperature in the South China Sea by artificial neural networks. *IEEE Geosci. Remote Sens. Lett.* **2019**, *17*, 558–562. [CrossRef]
5. Chepurin, G.A.; Carton, J.A.; Dee, D. Forecast model bias correction in ocean data assimilation. *Mon. Weather Rev.* **2005**, *133*, 1328–1342. [CrossRef]
6. Hewson, T.D.; Pillosu, F.M. A low-cost post-processing technique improves weather forecasts around the world. *Commun. Earth Environ.* **2021**, *2*, 132. [CrossRef]
7. Kim, H.; Ham, Y.G.; Joo, Y.S.; Son, S.W. Deep learning for bias correction of MJO prediction. *Nat. Commun.* **2021**, *12*, 3087. [CrossRef]
8. Frnda, J.; Durica, M.; Rozhon, J.; Vojtekova, M.; Nedoma, J.; Martinek, R. ECMWF short-term prediction accuracy improvement by deep learning. *Sci. Rep.* **2022**, *12*, 7898. [CrossRef]
9. Han, G.; Zhou, J.; Shao, Q.; Li, W.; Li, C.; Wu, X.; Zhou, G. Bias correction of sea surface temperature retrospective forecasts in the South China Sea. *Acta Oceanol. Sin.* **2022**, *41*, 41–50. [CrossRef]
10. Fei, T.; Huang, B.; Wang, X.; Zhu, J.; Chen, Y.; Wang, H.; Zhang, W. A hybrid deep learning model for the bias correction of sst numerical forecast products using satellite data. *Remote Sens.* **2022**, *14*, 1339. [CrossRef]
11. Liu, B.; Xie, B.; Huang, B.; Yin, X.; Wang, Z.; Yang, Y. Deviation correction of the SST prediction in global high resolution ocean prediction system. *Adv. Mar. Sci.* **2023**, *41*, 444–455. [CrossRef]
12. Yuan, S.; Feng, X.; Mu, B.; Qin, B.; Wang, X.; Chen, Y. A generative adversarial network–based unified model integrating bias correction and downscaling for global SST. *Atmos. Ocean. Sci. Lett.* **2024**, *17*, 100407. [CrossRef]
13. Zhou, G.; Han, G.; Li, W.; Wang, X.; Wu, X.; Cao, L.; Li, C. High-resolution gridded temperature and salinity fields from Argo floats based on a spatiotemporal four-dimensional multigrid analysis method. *J. Geophys. Res. Ocean.* **2023**, *128*, e2022JC019386. [CrossRef]
14. Zhang, M.; Han, G.; Wu, X.; Li, C.; Shao, Q.; Li, W.; Cao, L.; Wang, X.; Dong, W.; Ji, Z. SST forecast skills based on hybrid deep learning models: With applications to the South China Sea. *Remote Sens.* **2024**, *16*, 1034. [CrossRef]
15. Lorenc, A.C. Analysis methods for numerical weather prediction. *Q. J. R. Meteorol. Soc.* **1986**, *112*, 1177–1194. [CrossRef]
16. Hsueh, Y. The kuroshio in the east China sea. *J. Mar. Syst.* **2000**, *24*, 131–139. [CrossRef]
17. Tao, L.; Sun, X.; Yang, X.Q.; Fang, J.; Cai, D.; Zhou, B.; Chen, H. Cross-season effect of spring Kuroshio-Oyashio extension SST anomalies on following summer atmospheric circulation. *Geophys. Res. Lett.* **2024**, *51*, e2024GL108750. [CrossRef]
18. Wu, X. Analysis and Prediction of the Kuroshio Path South of Japan. Ph.D. Thesis, Tianjin University, Tianjin, China, 2024.
19. Mellor, G.L.; Häkkinen, S.M.; Ezer, T.; Patchen, R.C. A generalization of a sigma coordinate ocean model and an intercomparison of model vertical grids. In *Ocean Forecasting*; Springer: Berlin/Heidelberg, Germany, 2002; pp. 55–72. [CrossRef]

20. Hersbach, H.; Bell, B.; Berrisford, P.; Hirahara, S.; Horányi, A.; Muñoz-Sabater, J.; Nicolas, J.; Peubey, C.; Radu, R.; Schepers, D.; et al. The ERA5 global reanalysis. *Q. J. R. Meteorol. Soc.* **2020**, *146*, 1999–2049. [CrossRef]
21. Cao, L.; Wu, X.; Han, G.; Li, W.; Wu, X.; Wu, H.; Li, C.; Li, Y.; Zhou, G. EAKF-based parameter optimization using a hybrid adaptive method. *Mon. Weather Rev.* **2022**, *150*, 3065–3079. [CrossRef]
22. Reynolds, R.W.; Smith, T.M.; Liu, C.; Chelton, D.B.; Casey, K.S.; Schlax, M.G. Daily high-resolution-blended analyses for sea surface temperature. *J. Clim.* **2007**, *20*, 5473–5496. [CrossRef]
23. Huang, B.; Liu, C.; Banzon, V.; Freeman, E.; Graham, G.; Hankins, B.; Smith, T.; Zhang, H.M. Improvements of the daily optimum interpolation sea surface temperature (DOISST) version 2.1. *J. Clim.* **2021**, *34*, 2923–2939. [CrossRef]
24. Li, W.; Xie, Y.; He, Z.; Han, G.; Liu, K.; Ma, J.; Li, D. Application of the multigrid data assimilation scheme to the China Seas' temperature forecast. *J. Atmos. Ocean. Technol.* **2008**, *25*, 2106–2116. [CrossRef]
25. Li, W.; Xie, Y.; Han, G. A theoretical study of the multigrid three-dimensional variational data assimilation scheme using a simple bilinear interpolation algorithm. *Acta Oceanol. Sin.* **2013**, *32*, 80–87. [CrossRef]
26. Hannachi, A.; Jolliffe, I.T.; Stephenson, D.B. Empirical orthogonal functions and related techniques in atmospheric science: A review. *Int. J. Climatol.* **2007**, *27*, 1119–1152. [CrossRef]
27. Huang, N.E.; Shen, Z.; Long, S.R.; Wu, M.C.; Shih, H.H.; Zheng, Q.; Yen, N.C.; Tung, C.C.; Liu, H.H. The empirical mode decomposition and the Hilbert spectrum for nonlinear and non-stationary time series analysis. *Proc. R. Soc. Lond. Ser. A Math. Phys. Eng. Sci.* **1998**, *454*, 903–995. [CrossRef]
28. Shao, Q.; Hou, G.; Li, W.; Han, G.; Liang, K.; Bai, Y. Ocean reanalysis data-driven deep learning forecast for sea surface multivariate in the South China Sea. *Earth Space Sci.* **2021**, *8*, e2020EA001558. [CrossRef]
29. Shao, Q.; Li, W.; Hou, G.; Han, G.; Wu, X. Mid-term simultaneous spatiotemporal prediction of sea surface height anomaly and sea surface temperature using satellite data in the South China Sea. *IEEE Geosci. Remote Sens. Lett.* **2022**, *19*, 1501705. [CrossRef]
30. Liu, X.; Li, N.; Guo, J.; Fan, Z.; Lu, X.; Liu, W.; Liu, B. Multistep-ahead prediction of ocean SSTA based on hybrid empirical mode decomposition and gated recurrent unit model. *IEEE J. Sel. Top. Appl. Earth Obs. Remote Sens.* **2022**, *15*, 7525–7538. [CrossRef]
31. Wu, X.; Han, G.; Li, W.; Ji, Z.; Cao, L.; Dong, W. A hybrid deep learning model for predicting the Kuroshio path south of Japan. *Front. Mar. Sci.* **2023**, *10*, 1112336. [CrossRef]
32. Wang, L.; Cao, Y.; Deng, X.; Liu, H.; Dong, C. Significant wave height forecasts integrating ensemble empirical mode decomposition with sequence-to-sequence model. *Acta Oceanol. Sin.* **2023**, *42*, 54–66. [CrossRef]
33. Aparna, S.G.; D'souza, S.; Arjun, N.B. Prediction of daily sea surface temperature using artificial neural networks. *Int. J. Remote Sens.* **2018**, *39*, 4214–4231. [CrossRef]
34. Bai, Y.; Li, W.; Shao, Q. A prediction model of Sea Surface Height Anomaly based on Empirical Orthogonal Function and machine learning. *Mar. Sci. Bull.* **2020**, *39*, 678–688.
35. Pendlebury, S.F.; Adams, N.D.; Hart, T.L.; Turner, J. Numerical weather prediction model performance over high southern latitudes. *Mon. Weather Rev.* **2003**, *131*, 335–353. [CrossRef]
36. Sakaida, F.; Kudoh, J.I.; Kawamura, H. A-HIGHERS—The system to produce the high spatial resolution sea surface temperature maps of the western North Pacific using the AVHRR/NOAA. *J. Oceanogr.* **2000**, *56*, 707–716. [CrossRef]
37. Lee, M.; Chang, Y.; Sakaida, F.; Kawamura, H.; Cheng, C.H.; Chan, J.; Huang, I. Validation of satellite-derived sea surface temperatures for waters around Taiwan. *TAO Terr. Atmos. Ocean. Sci.* **2005**, *16*, 1189–1204. [CrossRef]
38. Qiu, C.; Wang, D.; Kawamura, H.; Guan, L.; Qin, H. Validation of AVHRR and TMI-derived sea surface temperature in the northern South China Sea. *Cont. Shelf Res.* **2009**, *29*, 2358–2366. [CrossRef]
39. Kagimoto, T.; Miyazawa, Y.; Guo, X.; Kawajiri, H. High resolution Kuroshio forecast system: Description and its applications. In *High Resolution Numerical Modelling of the Atmosphere and Ocean*; Hamilton, K., Ohfuchi, W., Eds.; Springer: New York, NY, USA, 2008; pp. 209–239. [CrossRef]
40. Wang, Z.; Liu, G.; Li, W.; Wang, H.; Wang, D. Development of the operational oceanography forecasting system in the Northwest Pacific. *J. Phys. Conf. Ser.* **2023**, *2486*, 012032. [CrossRef]

remote sensing

MDPI

Article

Deriving Coastal Sea Surface Current by Integrating a Tide Model and Hourly Ocean Color Satellite Data

Songyu Chen, Fang Shen *, Renhu Li, Yuan Zhang and Zhaoxin Li

State Key Laboratory of Estuarine and Coastal Research, East China Normal University, Shanghai 200241, China; 51253904023@stu.ecnu.edu.cn (S.C.); 52203904014@stu.ecnu.edu.cn (R.L.); 52213904008@stu.ecnu.edu.cn (Y.Z.); zxli@sklec.ecnu.edu.cn (Z.L.)

* Correspondence: fshen@sklec.ecnu.edu.cn

Abstract: Sea surface currents (SSCs) play a pivotal role in material transport, energy exchange, and ecosystem dynamics in coastal marine environments. While traditional methods to obtain wide-range SSCs, such as satellite altimetry, often struggle with limited performance in coastal regions due to waveform contamination, deriving SSCs from sequential ocean color data using maximum cross-correlation (MCC) has emerged as a promising approach. In this study, we proposed a novel SSC estimation method, called tide-restricted maximum cross-correlation (TRMCC), and implemented it on hourly ocean color data obtained from the Geostationary Ocean Color Imager II (GOCI-II) and the global tide model FES2014 to derive SSCs in coastal seas and turbid estuaries. Cross-comparison over three years with buoy data, high-frequency radar, and numerical model products shows that TRMCC is capable of obtaining high-resolution SSCs with good accuracy in coastal and estuarine areas. Both large-scale ocean circulation patterns in seas and fine-scale surface current structures in estuaries can be effectively captured. The deriving accuracy, especially in coastal and estuarine areas, can be significantly improved by integrating tidal current data into the MCC workflow, and the influence of invalid data can be minimized by using a flexible reference window size and normalized cross-correlation in the Fourier domain technique. Seasonal SSC structure in the Bohai Sea and diurnal SSC variation in the Yangtze River Estuary were depicted via the satellite method, for the first time. Our study highlights the vast potential of TRMCC to improve the understanding of current dynamics in complex coastal regions.

Keywords: sea surface current; coastal waters; Yangtze River Estuary; ocean color; GOCI-II

1. Introduction

As a medium of material transportation, sea surface currents (SSCs) serve as vital conduits for materials exchange within marine ecosystems [1]. They mediate the transfer of heat and moisture between the ocean and the atmosphere, thereby influencing weather patterns [2]. SSCs also play a critical role in coastal areas by shaping coastal landscapes [3], influencing larval dispersion [4] and pollutant distribution [5], and affecting maritime activities, such as shipping and human engineering [6], etc. Accurately deriving SSCs is crucial, both for better understanding global climate change and optimizing coastal management strategies.

Although traditional methods, such as the use of surface drifters [7] and acoustic Doppler current profilers (ADCPs) [8], provide reliable SSC measurements, their observation range is often limited. High-frequency (HF) radars can effectively monitor coastal SSCs by measuring backscattered radar signals [9,10]; a global HF radar network was established to improve coastal management [11], but its mapping area is still inadequate to monitor the global coastline, since its deployment and maintenance can be difficult in remote and extreme environments [12]. Satellite altimetry has also been widely utilized to calculate wide-range ocean currents based on geostrophic balance [13,14]; however, it can only retrieve geostrophic currents [15], while realistic ocean surface currents are also affected by tidal currents, wind-driven currents (Ekman current), Stokes drift, and other ageostrophic components. Moreover, currents retrieved from an altimeter are rather unreliable due to the distortion and contamination inherent in altimeter waveforms in coastal regions [15–17].

The movement of ocean surface features such as chlorophyll *a* (Chl-*a*), total suspended matter (TSM), and sea surface temperature (SST) can effectively indicate the SSC structure [18], making feature tracking an effective method to retrieve SSCs using sequential thermal or ocean color images. From the time that Emery et al. [19] first applied the maximum cross-correlation (MCC) approach, which is commonly used in image registration and cloud motion detection, to the advent of advanced, very high-resolution radiometer (AVHRR) images and extracted SSCs in the coastal ocean of Colombia, MCC has been widely used to retrieve SSCs in various sea areas using different satellite images. Taniguchi et al. [20] derived the surface velocity in the Lombok Strait using hourly Himawari-8 SST data. Liu et al. [12] investigated SSCs along the California coast by merging MCC results from different surface tracers, including SST, Chl-*a* concentration, and remote sensing reflectance (R_{rs}). Yang et al. [21] derived SSC in the U.S. East and Gulf coasts by utilizing the overlapping area of a visible infrared imaging radiometer (VIIRS) and compared SSC products generated by different tracer images. Ren et al. [22] applied MCC to X-band synthetic aperture radar (SAR) images to derive high resolution tidal currents in Hangzhou Bay and around Amrum Island. Apart from MCC, various other approaches have been implemented on sequential satellite images to obtain SSCs, including optical flow [23,24], Doppler shift [25,26], inter-band time-lag [27], displaced frame central difference equation [28], etc.

During the process of feature tracking, inconsistencies in spatial resolution and band settings is inevitable for multi-sensor images. However, a geostationary satellite can avoid these issues and stably generate consecutive images. The Korean Geostationary Ocean Color Imager (GOCI) provides hourly images in the time range of UTC 0:00–7:00, making it one of the most commonly used sensors in SSC extraction. Yang et al. [29] analyzed the diurnal variation in SSCs and the eddy structures on the west coast and East Sea of Korea, using GOCI-derived TSM and Chl-*a* as tracers for the MCC technique. Based on GOCI images, Jiang et al. [30] comprehensively tested different parameters to determine the best performance of MCC in the Bohai Sea, validated with ADCP, and HF radar and the numerical model both showed good results. Hu et al. [31] tested the particle image velocimetry (PIV) method on GOCI-derived TSM and Chl-*a* to derive SSCs in the East China Sea and the Sea of Japan, achieving solid performance, especially in eddy areas with strong rotation and deformation. Using five years of GOCI Level-2 (L2) products and optimal interpolation, Liu et al. [32] extracted SSCs in the entire GOCI observation range and analyzed annual and seasonal mean flows.

The aforementioned studies mainly focus on sea areas where surface tracers display a distinct gradient and stable structure. In shallow nearshore areas, such as the Yangtze River Estuary (YRE) and the Korean Coastal Sea (KCS), land boundaries are much more intricate; horizontal transport and vertical mixing are jointly influenced by river discharge, tidal current, and wind, leading to a more complex and dynamic current structure and surface tracers [33,34], causing obstacles to precise high-resolution SSC estimation. However, due the tide-dominance characteristic in these areas [35], surface total current direction can be highly consistent with tidal current during the flooding and ebbing phases, when the tidal current magnitude is large [30,36].

The objective of our study is to fill the gaps in SSC estimation in coastal and estuary areas. An improved MCC workflow, called tide restricted maximum cross-correlation (TRMCC), is developed by integrating the FES2014 tide model and applying it to GOCI-II ocean color data from 2021 to 2023. Cross-comparisons are performed between SSC estimation results, buoy data, HF radar, and a numerical model. High-resolution SSC structures in multiple regions are revealed. Our study provides new insight into SSC estimation in coastal areas via ocean color remote sensing.

The structure of this article is organized as follows: Section 2 introduces the data used in this study. Section 3 explains the methodology of our research. Section 4 presents the SSC estimation performance in both the long and short term, with cross-comparisons with SSC data obtained from the GOCI-II L2 product, a numerical model product, HF radar, and a buoy. TRMCC's ability to reveal high-resolution SSC structures in turbid coastal areas at different time scales is also tested. Finally, discussions and conclusions are presented in Section 5.

2. Materials and Methods

2.1. Materials

2.1.1. GOCI-II Data

Sequential images, with both high spatial resolution and short time intervals, are the cornerstone of retrieving SSCs from ocean color data. The GOCI-II (Geostationary Ocean Color Imager II) was mounted on Korean satellite GEO-KOMPSAT-2B, launched in February 2020; it has a spatial resolution of 250 m, with 12 bands in the visible and near-infrared range, covering most of the sea areas of Korea, Japan, and China. It is capable of conducting 10 observations per day, from UTC 23:30 to 8:30, with an interval of 1 h, perfect for the extraction of SSCs. GOCI-II data can be obtained from the Korea Ocean Satellite Center (KOSC) website (https://kosc.kiost.ac.kr/, accessed on 6 January 2024). Since a full scene of GOCI-II image consists of 12 sub-images obtained by the step and stare method, data from two different slots were used in this study: Slot S007, which covers the Korean Sea, and Slot S010, covering the East China Sea (Figure 1a).

For the case in the Korean Sea, previous studies have proved that Chl-*a* can be a suitable tracer in sea areas with relatively low turbidity, such as along the California Coast and in the Korean Sea [29,31,37,38]. Therefore, the GOCI-II L2 Chl-*a* product of slot S007 (green square in Figure 1a) was selected as the tracer data to derive SSCs in the sea area of the Korean Peninsula. This product is generated based on the L2 atmospheric corrected R_{rs} product using the ocean color index (OCI) algorithm [39]. Full data from 2021 to 2023 were used, adding up to a total of 10,407 scenes.

For the case in the East China Sea, due to the shallow topography and abundant terrestrial sediment input from the Yangtze River and the Yellow River, the optical properties in the East China Sea are strongly influenced by suspended sediment, creating obstacles for accurately deriving Chl-*a* [40,41]. Therefore, in these area, TSM has been widely used in SSC estimation [31,42–44]. However, in extremely turbid nearshore and estuary areas such as the YRE, the GOCI-II official atmospheric correction method performs poorly, resulting in large blank area in the ocean color product of slot S010. Therefore, the high-quality Level 1B top-of-atmosphere radiance product of slot S010 (orange square in Figure 1a) from 2021 to 2023 was manually filtered (Figure 1b), adding up to a total of 5445 scenes in 573 days. Rayleigh correction was performed with an atmospheric vector radiative transfer model [45,46]. Cloud areas were then masked, based on an improved cloud masking method [47]. Atmospheric correction was performed with XGB-CW [48], a new atmospheric correction algorithm especially designed for highly turbid waters which combines a coupled ocean–atmosphere model with the eXtreme Gradient Boosting (XGBoost). After R_{rs} was obtained, TSM was derived via a semi-empirical radiative transfer model [49] and was selected as tracer data to derive SSCs in the East China Sea areas.

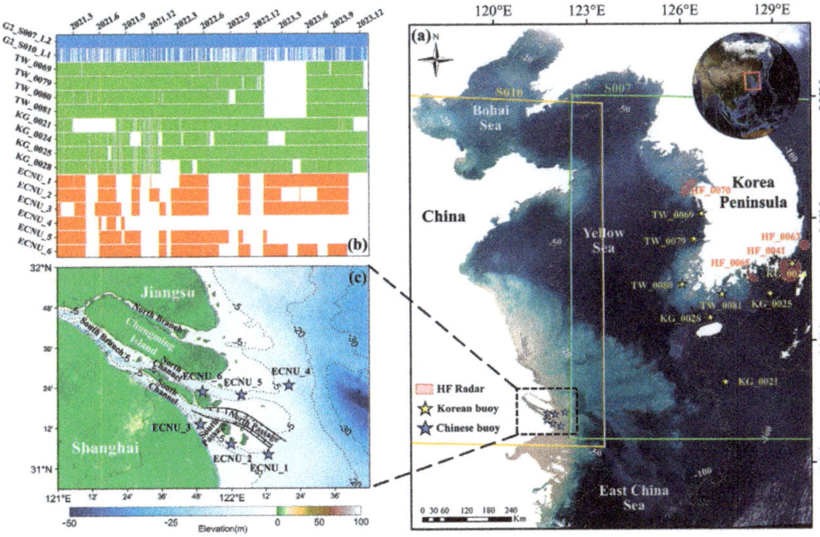

Figure 1. (**a**) Area of interest (AOI) of this study. Orange and green squares represent the observation range of GOCI- II data in two different slots (named by S010 and S007). The white dashed line shows the isobath in the AOI. The stars and red-dashed border represent the location of buoys and HF radar. (**b**) Valid data time range of ocean color data (blue), Korean buoy (green), and Chinese buoy (orange). (**c**) Underwater bottom topography of the YRE; black solid line represents the location of the Deep-Water Channel Project.

Moreover, GOCI-II official L2 SSC data products for slot S007 and S010 from 2021 to 2023 were also downloaded for cross-comparison. These SSC data were developed by using the L2 Chl-*a* product as the tracer image, along with a primitive MCC algorithm based on the work of Barton et al. [50] and Emery et al. [19]. This product has been subjected to a quality control process using global mean ocean surface velocities calculated from 30 years of satellite-tracked surface drifter data.

2.1.2. Buoy Data

In order to exam the accuracy of retrieved SSC, hourly SSC data of 14 buoys from 2021 to 2023 was collected to be used as ground truth (GT), 8 of which are located in KCS (as shown in Figure 1a) and 6 of which are located in the YRE, China (as shown in Figure 1c). Korean buoy data, which includes surface current speed and direction, were downloaded from the Korea Hydrographic and Oceanographic Agency website (http://www.khoa.go.kr/, accessed on 10 February 2024); four of the buoys were located in the outer sea (KG_0021, KG0024, KG0025, KG0028), and four of them were located in the coastal sea, with lower water depth and more intricate land boundaries (TW_0069, TW_0079, TW_0080, TW_0081). Moreover, six buoys were deployed in the YRE, located in the North Channel, North Passage, and South Passage (as shown in Figure 1c), capable of providing the current speed and the direction of the vertical water column. In order to reduce noise, the current data from the depth of 0 to 2 m were averaged and used as SSCs. The available time range of 14 buoys can be seen in Figure 1b.

2.1.3. High Frequency Radar Data

The data from four high frequency (HF) radar sites (as shown in Figure 1a) were also downloaded from the Korea Hydrographic and Oceanographic Agency website (http://www.khoa.go.kr/, accessed on 12 February 2024) to perform an accuracy verification of the retrieved SSCs. These HF radar sites are located in different geographical environments (e.g., shallow coast, water bay, deep sea), with resolutions ranging from 0.75 to 3 km, an operating frequency ranging from 13 to 25 MHz, and band width ranging from 50 to 200 kHz [51]. Additionally, buoy KG_0024 is located in the observation range of HF radar site HF_0041, which makes it possible to make a cross-comparison between SSCs retrieved by different methods.

2.1.4. Numerical Model Data

CMEMS (Copernicus Marine Environment Monitoring Service) products have been widely used to study coastal ocean dynamics [52,53]. In this study, SSC data from the CMEMS product "GLOBAL_ANALYSISFORECAST_PHY_001_024" was used for crosscomparison. The product is based on version 3.6 of the NEMO (Nucleus for European Modelling of the Ocean) ocean model, coupled with new data assimilation method and observation data [54], and is capable of providing hourly SSCs, with a resolution of 1/12°. This product is available at the Copernicus Marine Data Store (https://data.marine.copernicus.eu/products, accessed on 6 February 2024).

2.1.5. Tide Model

In this study, a finite element solution (FES) tide model was utilized to simulate the surface tidal current and tidal height. As the latest version of FES tide model, FES2014 has a higher horizontal resolution of 1/16°, a more accurate bathymetry and shoreline grid in shallow water areas, an expanded altimetry and tidal gauges assimilation database, and a more advanced data assimilation method [55]. The FES2014 tide model can be downloaded from the Archiving, Validation, and Interpretation of Satellite Oceanographic Satellite Data (Aviso) website (https://www.aviso.altimetry.fr/en/data/products/auxiliary-products/global-tide-fes.html, accessed on 4 December 2023). Previous studies have pointed out that FES2014 provides a more reliable result than other ocean tide models in both the East China Sea [56,57] and the coastal sea of the Korean Peninsula [58].

2.2. Methods

Considering the above mentioned obstacles in SSC estimation and the notable contribution of tidal currents in coastal seas, we proposed an improved MCC workflow, called tide restricted maximum cross-correlation (TRMCC). It is capable of producing high-resolution and high-accuracy SSC results, especially in coastal sea areas. The workflow diagram of the TRMCC can be seen in Figure 2, and its main improvements are described in the following section.

2.2.1. Normalized Cross-Correlation in Fourier Domain

Invalid data in the data window can greatly influence the SSC-deriving result. In order to minimize the influence caused by land or cloud cover, a masked normalized cross-correlation (NCC) technique [59] was adopted to evaluate the cross-correlation between the reference window and the suspected target window. This is an image registration method originally developed for medical imaging. By introducing a masked area into the Fourier–Merlin algorithm, it can effectively avoid the influence of the masked area, which in our case, is invalid data caused by land and cloud cover. The NCC can be calculated as follows:

$$NCC = \frac{f^{-1}(F_1 \cdot F_2^*) - \frac{f^{-1}(F_1 \cdot M_2^*) \cdot f^{-1}(M_1 \cdot F_2^*)}{f^{-1}(M_1 \cdot M_2^*)}}{\sqrt{f^{-1}(f(W_1 \cdot W_1) \cdot M_2^*) - \frac{(f^{-1}(F_1 \cdot M_2^*))^2}{f^{-1}(M_1 \cdot M_2^*)}} \cdot \sqrt{f^{-1}(M_1 \cdot f(W_2' \cdot W_2')) - \frac{(f^{-1}(M_1 \cdot F_2^*))^2}{f^{-1}(M_1 \cdot M_2^*)}}}, \quad (1)$$

where NCC denotes the normalized cross-correlation; f and f^{-1} denote the Fourier and inverse Fourier transform, respectively; W_1 and W_2 are the two image windows; F_1 and F_2 are the Fourier transformed image windows; and M_1 and M_2 are the Fourier transformed masks of the two images. More details can be found in Ref. [59].

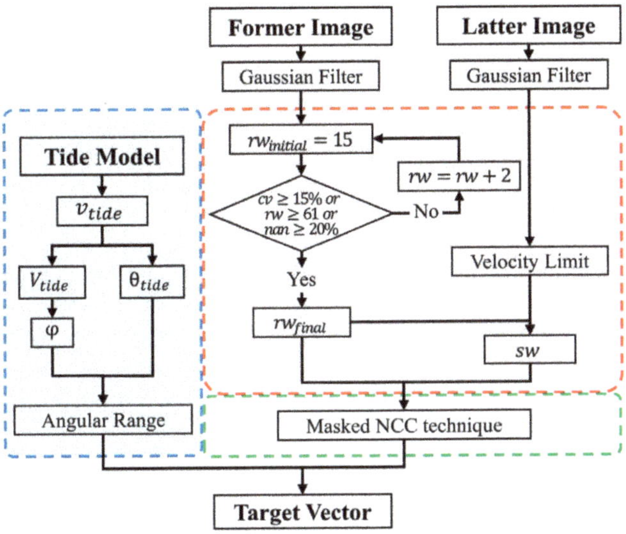

Figure 2. Workflow diagram of TRMCC. Former and latter images represent an image pair, with a time interval. the blue area represents the process of tide-restriction. The red area shows the circulation used to determine the reference window size (rw) and the search window size (sw) by evaluating the feature in the reference window. The green area shows the masked normalized cross-correlation technique.

2.2.2. Restricted Angular Search Range

In coastal seas, the tidal current, rather than geostrophic current, the wind-driven current, etc., can be the dominant current component. The genuine current's direction can be significantly affected by the tidal current. As shown in the density plot (Figure 3) of over 630,000 data pairs between the simulated tidal current and the genuine current observed by buoys, a clear pattern between the tidal current magnitude (V_{tide}) and the angular difference of genuine current and tidal current ($|\theta_{buoy} - \theta_{tide}|$) can be concluded. The larger the V_{tide}, the smaller the angular difference between the genuine current and the tidal current. A total of 63.3% of V_{tide} distributes in the range of 5 to 75 cm/s, while the angular difference distributes in the range of 0 to 30. Therefore, in this study, we defined a variable φ, which stands for the angular difference between the simulated tidal current direction and the genuine current direction, and assume that φ has a functional relationship with V_{tide} that can be expressed as a linear combination of exponential functions, as follows:

$$\varphi = a \cdot e^{b \cdot V_{tide}} + c \cdot e^{d \cdot V_{tide}} \quad (2)$$

where φ is the angular difference; V_{tide} is the tidal current magnitude; and a, b, c, and d are undetermined coefficients.

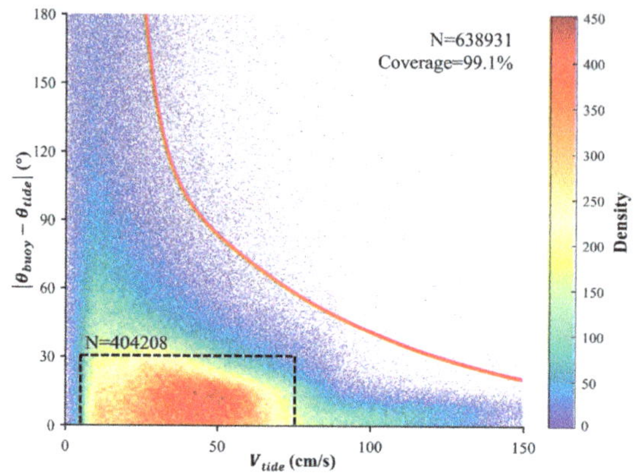

Figure 3. Scatter density plot showing the tidal current magnitude and the angular difference between the genuine current and the tidal current. The red line indicates the curve of $\varphi = f(V_{tide})$.

Based on a vast number of data pairs between the simulated tidal current and the genuine current, four coefficients in Equation (2) were obtained using Gaussian kernel density estimation (GKDE) to model the distribution of φ and V_{tide}, and an optimization algorithm was employed to ensure that the resulting function can cover at least 99% of the data pairs, while minimizing the angular difference. Finally, the functional relationship between φ and V_{tide} can be expressed as follows:

$$\varphi = 9036.18 \cdot e^{-0.19 \cdot V_{tide}} + 168.24 \cdot e^{-0.014 \cdot V_{tide}} \tag{3}$$

where φ is the angular difference, and V_{tide} is the tidal current magnitude.

During the workflow of the TRMCC (as shown in Figure 4), the tidal current magnitude (V_{tide}) and direction (θ_{tide}) between the time range of two sequential images was generated using tide model; the mean value ($\overline{V_{tide}}$ and $\overline{\theta_{tide}}$) will be seen as the tidal current vector corresponding to the estimation time. φ was determined by $\overline{V_{tide}}$, based on Equation (3), and the angular restriction range can be defined as $[\overline{\theta_{tide}} - \varphi\ ,\ \overline{\theta_{tide}} + \varphi]$. As shown in Figure 4, taking two Chl-*a* images from 8:30 to 9:30 (UTC + 9) on 25 April 2021 as an example, the suspected vector (gray arrow) that is outside the restriction range was discarded, even though it has a higher NCC.

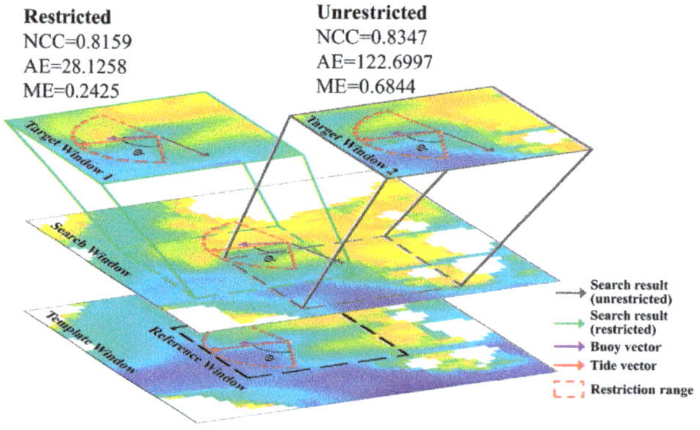

Figure 4. Schematic illustration of angular restriction range, taking Chl-*a* images from 8:30 to 9:30 (UTC + 9) on 25 April 2021 as an example. The vector and window borders in green and gray represent the search result after and before tide-restriction.

2.2.3. Adaptive Reference Window Size

In the workflow of the conventional MCC approach, the size of the reference window (rw) significantly influences the accuracy of SSC estimation and can vary depending on the research area. For example, in a previous study that applied MCC on GOCI data to derive SSCs, rw varied from 10 to 22 km [29,32,38,44,60]. However, in coastal sea area such as the YRE, SSCs show a much finer structure and can be dissimilar, even in an area of 10×10 km, not to mention in the inner estuary, where the width of the river channel can be less than 6 km. When implementing MCC in different areas, rw should be decided based on the feature scale and signal abundance of the surface tracer, since an excessively large window size cannot reflect the fine structure of SSCs [22].

Therefore, we brought the coefficient of variation (cv) into the workflow (as shown in Figure 2) to evaluate the signal abundance. The cv of an image window can be computed as follows:

$$cv = \frac{SD}{MN} \times 100\% \tag{4}$$

where SD and MN are the standard deviation and mean value of a window.

With an initial rw of 15 pixels (3.75 km), if cv is less than a threshold of 15% and the invalid data percentage is less than 20%, rw will be expanded for 2 pixels, and this process will be repeated until cv reaches the threshold or rw exceeds 61 pixels (15.25 km). After this cycle, an optimal rw can be determined, ensuring a suitable tracer signal amount, while assuring an acceptable invalid data percentage. Then, the search window size (sw) can be determined based on velocity limit, which is set to 2 m/s, according to the in situ observation of the buoys.

2.2.4. Accuracy Evaluation Metrics

Following the methods of previous research [28,61,62], average angular error (AAE) and average magnitude error (AME) were used to evaluate the performance of SSC estimation. AAE, AE, AME, and ME can be calculated as follows:

$$\{AAE, AME\} = \{\overline{AE}, \overline{ME}\} = \left\{\overline{\Delta\theta}, \overline{\frac{\Delta V}{V}}\right\} = \frac{1}{N}\sum_{i,j}\left\{\arccos\left(\frac{\boldsymbol{V}\cdot\boldsymbol{V}_{gt}}{|\boldsymbol{V}||\boldsymbol{V}_{gt}|}\right), \frac{|\boldsymbol{V}-\boldsymbol{V}_{gt}|}{|\boldsymbol{V}_{gt}|}\right\} \tag{5}$$

where $\boldsymbol{V}$ is the current vector being evaluated, and $\boldsymbol{V}_{gt}$ is the ground truth (buoy) vector.

3. Results

3.1. Validation Using Buoy Data

To thoroughly verify TRMCCs in different sea areas, SSCs in 14 buoy locations from 2021 to 2023 were derived, based on GOCI-II data. In order to demonstrate the effect of tide-restriction, a reference data group, without undergoing the process of tide-restriction, is also generated. The performance of tide-restricted method (TRMCC), the unrestricted method, and the GOCI-II official method are shown in Table 1 (eight Korean buoys) and Table 2 (six Chinese buoys).

In the offshore area of the Korean Sea, although the accuracy of the official SSC product is only slightly lower than that of our method, it yields only about one-third the amount of valid data. Considering that all three results were based on the same L2 Chl-*a* data, it is possible that the production process of the SSC product cannot deal with invalid data caused by cloud cover. The difference between tide-restricted method and unrestricted method is very small, since V_{tide} in the offshore area is relatively smaller than in the nearshore area. According to Equation (3), V_{tide} needs to be greater than 26.1 cm/s, or the φ will be over 180°; under this circumstance, the angular search range is still [0, 360], which means that the tide-restriction did not take effect.

Table 1. Comparison of different SSC results in the Korean Coastal Sea from 2021 to 2023.

Type	Buoy ID	Tide-Restricted			Unrestricted			Official SSC Product		
		AME	AAE (°)	Count	AME	AAE (°)	Count	AME	AAE (°)	Count
Offshore	KG_0021	0.481	35.472	340	0.48	35.605	340	0.584	38.937	146
	KG_0024	0.429	39.562	1326	0.429	39.353	1329	0.528	55.835	361
	KG_0025	0.46	34.45	1340	0.46	34.512	1340	0.519	55.747	334
	KG_0028	0.571	33.56	960	0.573	34.764	964	0.504	53.879	396
	Mean/Sum	**0.48525**	**35.761**	**3966**	**0.4855**	**36.0585**	**3973**	**0.53375**	**51.0995**	**1237**
Nearshore	TW_0069	0.546	29.375	819	0.554	38.126	842	0.835	80.53	41
	TW_0079	0.394	26.072	1132	0.396	28.543	1144	0.806	75.479	37
	TW_0080	0.544	33.447	1056	0.534	40.52	1078	0.868	101.846	36
	TW_0081	0.409	33.347	1226	0.412	35.144	1235	0.795	101.245	54
	Mean/Sum	**0.47325**	**30.5603**	**4233**	**0.474**	**35.5833**	**4299**	**0.826**	**89.775**	**168**

Table 2. Comparison of different SSC results in the YRE from 2021 to 2023.

Buoy ID	Tide-Restricted			Unrestricted		
	AME	AAE (°)	Count	AME	AAE (°)	Count
ECNU_01	0.612	41.801	134	0.646	44.105	134
ECNU_02	0.647	23.607	562	0.746	39.458	743
ECNU_03	0.563	34.114	578	0.857	79.771	710
ECNU_04	0.528	35.473	464	0.587	78.665	547
ECNU_05	0.651	22.195	307	0.695	67.113	513
ECNU_06	0.437	33.672	477	0.858	89.789	566
Mean/Sum	**0.573**	**31.810333**	**2522**	**0.7315**	**66.4835**	**3213**

However, in the nearshore area, the advantages of our method began to take shape. While the official SSC product shows poor accuracy and few valid data counts, the unrestricted results, which utilize an adaptive reference window size and a masked *NCC* technique, achieved valid observations numbering 4299, with a good accuracy of 35.6° in regards to angular error and a relative magnitude error of 0.47. The AAE was further improved by 14.1%, on average, under the effect of tide-restriction; this improvement varies depending on the location of the buoy, with the greatest improvement of 23.0% at TW_0069, and the least improvement of 5.1% at TW_0081. However, the improvement in AME is relatively small.

As shown in Table 2, in the YRE, where water is much shallower than that in the KCS, tide-restriction yielded huge improvements in both AME and AEE. On average, AME improved by 21.7%, and AAE improved by 52.2%. This improvement is most obvious in ECNU_03 and ECNU_06, as each are improved by 34.3% and 49.1% for AME and 57.2% and 62.5% for AAE, respectively. However, the valid vector count dropped by 21.5% on average, indicating a failure in finding a target vector that matches the correlation limit, while staying in the angular restriction range. Moreover, comparisons between the official SSC product and the buoy data are not presented, as the official SSC product provided nearly no valid data in the YRE, since the official atmospheric correction method scarcely works in the highly turbid water of the YRE, and the official SSC extraction method is unable to deal with the invalid data caused by land boundaries.

3.2. Cross-Comparison of SSC Derived from Satellite, Numerical Model, and HF Radar

To further investigate the SSC pattern in a larger spatial scale, the mean SSC of the CMEMS product and the TRMCC results from 2021 to 2023 can be seen in Figure 5. Overall, both results show a similar current structure. After the Kuroshio Current (black arrow) arrived at the Jeju Island, it was divided into the Yellow Sea Warm Current (YSWC, yellow arrow) and the East Korea Warm Current (EKWC, red arrow). The EKWC was bifurcated by the Korean Strait (KS), flowing into the Sea of Japan. However, differences exist in the distribution of SSC magnitude. The CMEMS product shows a higher magnitude in the EKWC than in the TRMCC results, while in the YSWC and the area of the Changjiang (Yangtze) Diluted Water (CDW), the TRMCC results show a higher magnitude.

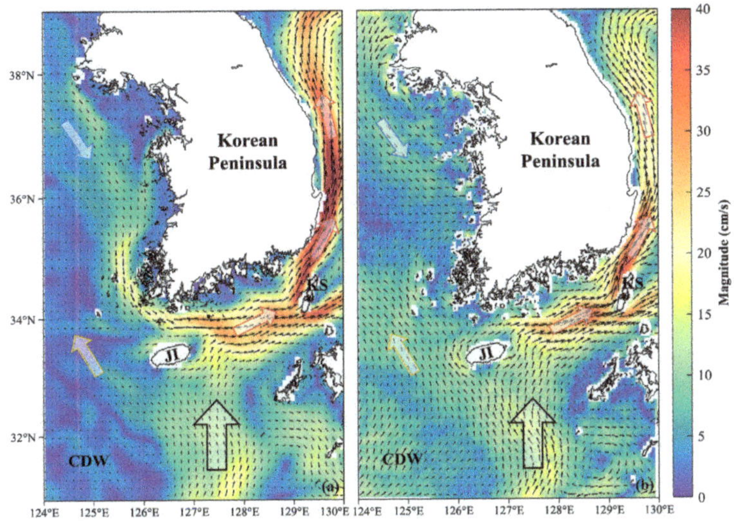

Figure 5. Mean SCC obtained by CMEMS product (**a**) and TRMCC (**b**) from 2021 to 2023 in the sea area of the Korean Peninsula, overlayed with the average current magnitude. Arrows with black, red, blue, and yellow borders represents the Kuroshio, East Korea Warm Current, Korean Coastal Current, and Yellow Sea Warm Current. KS: Korean Strait. JI: Jeju Island. CDW: Changjiang (Yangtze) Diluted Water.

In order to examine the differences between the CMEMS product and the TRMCC results, three representative buoys were selected, and their magnitude differences and the direction distribution can be seen in Figure 6. For buoy KG_0021 (115 m in depth), located in the outer sea, as shown in Figure 6a,d, both results match the actual situation, where the SSC mainly flows to the north and northwest, while the TRMCC results yields a relatively higher accuracy in regards to magnitude. For buoy TW_0079 (25 m in depth), located on the west coast of Korea, as shown in Figure 6a,e, the SSC mainly flows to the southwest; both results slightly overestimated the flow in the northeast, but the TRMCC results still show a higher accuracy in regards to magnitude. However, for ECNU_05 (5 m in depth), located in the shallow YRE, as shown in Figure 6c,f, the CMEMS product shows a rather poor performance in regards to both magnitude and direction. The SSCs in the YRE were confined by the river channel and were highly influenced by runoff and tidal currents, resulting in a reciprocal flow pattern in a west and east direction. The TRMCC results correspond well with the buoy data, while preserving a relatively good accuracy in regards to magnitude.

Furthermore, data from three HF radar sites located in different water depths were selected to demonstrate TRMCC's performance in the costal sea and the influence caused by invalid data in sequential images. The TRMCC results were resampled into a corresponding resolution of HF radar, and the results are presented in Figure 7. It can be seen that the TRMCC results generally agree well with the SSCs observed by HF radar in regard to both direction and velocity. However, the extremely fine scale structure of SSCs in coastal areas can be neglected, for example, in the northwest part of HF_0065 (Figure 7c) and the west part of HF_0070 (Figure 7e), where HF radar depicts a more detailed current structure. Moreover, since the GOCI-II L2 Chl-*a* product was used as a tracer image, invalid data exist in the L2 product due to the failure of atmospheric correction in turbid water (e.g., Type 1 in Figure 7a,b and Type 3 in Figure 7e,f) and cloud cover (e.g., Type 2 in Figure 7c,d). Although the TRMCC is able to ignore a certain amount of invalid data using the masked NCC technique, information loss caused by a huge blank area can still inevitably lead to failure in the TRMCC process.

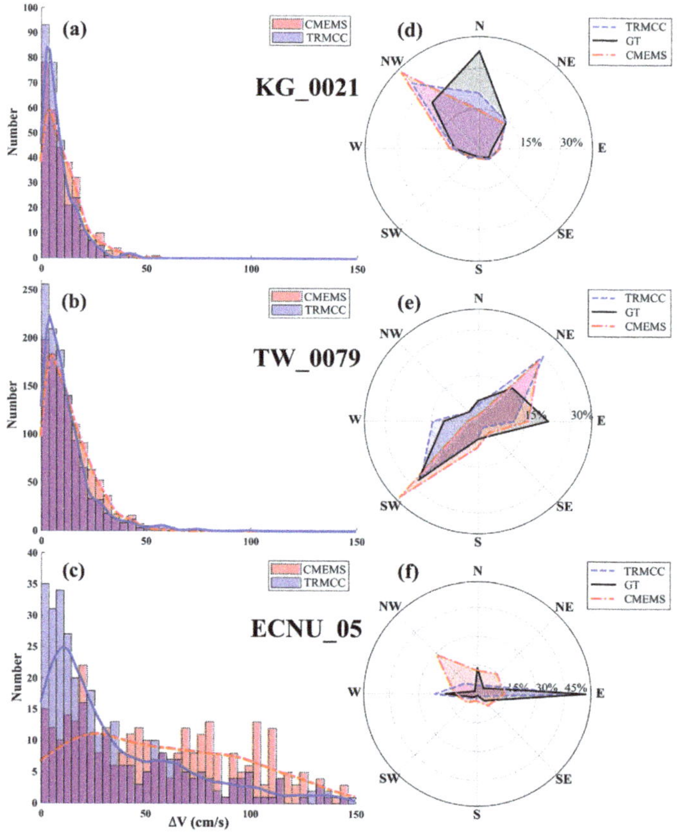

Figure 6. (**a–c**) Distribution of magnitude differences between TRMCC results, CMEMS product, and buoy data. (**d–f**) Distribution of velocity direction between TRMCC results, CMEMS product, and buoy data.

To thoroughly investigate the performance of different SSC results, data from a buoy located in the Korean Strait were regard as the ground truth. SSCs from HF radar, the CMEMS product, and TRMCC results were resampled into the same grid, with a resolution of 4 km; the daily average SSC from different data can be seen in Figure 8a. In the northwest part, where water depth is around 50 m, three SSC data correspond well in regards to direction, with the CMEMS product overestimating the velocity magnitude. However, in the south part and the area with high water depth, the TRMCC results agree well with the CMEMS product, showing that the SSC mainly flows in a northeasterly direction, while HF radar depicts a completely opposite direction. Furthermore, the eastward and northward component of the flow vector from 9:00 to 16:00 (UTC + 9) in three different datasets was compared with the buoy data. It can be seen in Figure 8b that although the vector direction from the CMEMS product agrees well with the ground truth, with an AAE of 30.431°, it overestimated the velocity, resulting in an AME of 1.6. The vector derived by HF radar and TRMCC, on the other hand, shows a better accuracy, in regards to both velocity and direction, with an AME of 0.326 and an AAE of 22.229° for HF radar, exhibiting even better results for TRMCC, with the AME and AEE being 0.151 and 17.878°, respectively.

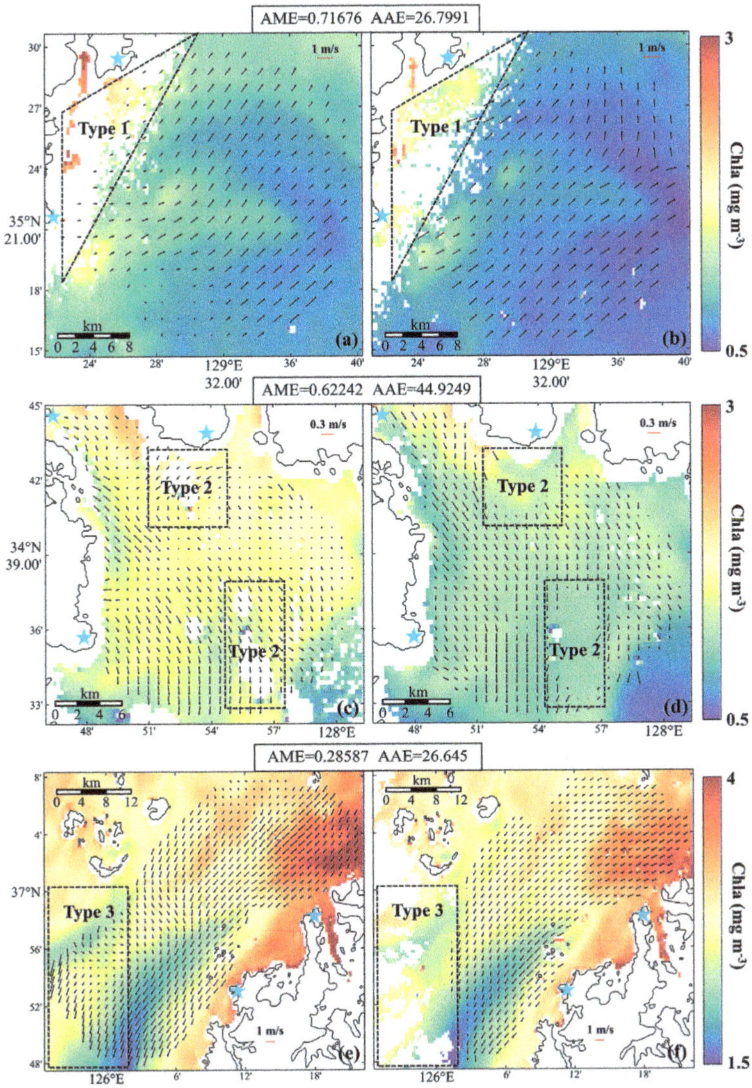

Figure 7. SSC derived by HF radar (**a**,**c**,**e**) and TRMCC (**b**,**d**,**f**) at 9:00 (UTC+9) on 24 Oct 2022, overlayed with the GOCI-II Chl-a product at 8:30 (left) and 9:30 (right). (**a**) HF_0063. (**c**) HF_0065. (**e**) HF_0070. Blue stars represent the location of radar sites. SCC derived by TRMCC outside the observation range of HF radar were removed for better visualization. Type 1, 2 and 3 denote in-valid data area caused by various reasons (e.g., atmospheric correction, cloud cover, etc.).

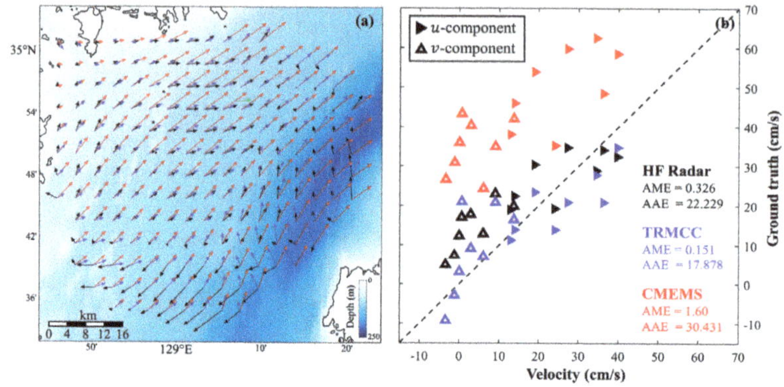

Figure 8. (**a**) Average SSC in Korean Strait from 9:00 to 16:00 (UTC + 9) on 5 May 2022. Green, black, blue, and red vectors represents currents derived by buoys, HF radar, TRMCC, and CMEMS, respectively. SCCs derived by TRMCC and CMEMS that are outside the observation range of HF radar were removed for better visualization. (**b**) Comparison between HF radar (black), TRMCC (blue), CMEMS (red), and ground truth (buoy). Solid triangle represents an eastward component (u), and hollow triangle represents a northward component (v).

3.3. Fine-Scale SSC Seasonal Variation in Bohai Sea

As one of the biggest shelf seas in the world, the East China Sea is known for its shallow water depth and abundant terrestrial sediment input, making it hard for altimeter and numerical models to retrieve SSCs. However, sediment floating on the sea surface can also be a perfect tracer, facilitating the extraction of SSC by feature tracking.

Taking the Bohai Sea (BS) as an example, the seasonal mean SSC was derived by the TRMCC, based on TSM images from 2021 to 2023, and the results can be seen in Figure 9. Due to the dominant southeast and northwest winds in the summer and winter, it can be seen that in the Bohai Strait, the SSC flows from the Yellow Sea (YS) to the BS in summer and the opposite in winter, with a mean velocity of around 30 cm/s. SSCs around the Yellow River Estuary show a consistent outward diffusion pattern in every season due to the runoff of the Yellow River and its convergence into the Laizhou Bay in autumn, under the influence of the north wind. The number of valid vectors reach up to around 500 for each season in the inner Bohai Sea, where the water is rich in sediment. But the number is generally lower in summer due to a higher rate of cloud cover.

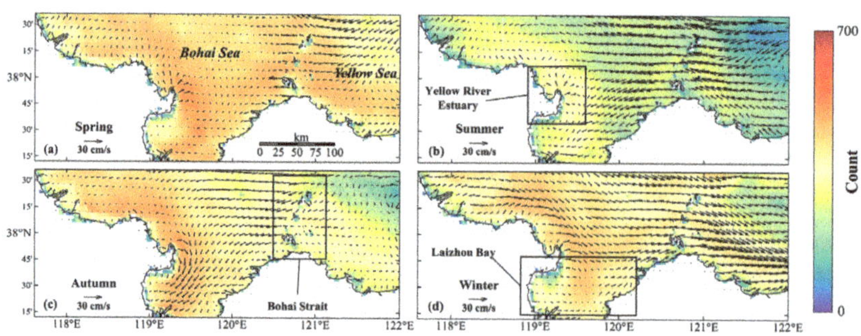

Figure 9. Seasonal mean SSC from 2021 to 2023 in the Bohai Sea, derived from TSM images, overlayed with valid vector data count. (**a**) Spring. (**b**) Summer. (**c**) Autumn. (**d**) Winter.

3.4. Fine-Scale SSC Diurnal Variation in Yangtze River Estuary

Another example of TRMCC's ability to retrieve high-resolution SSCs is given in Figure 10. The hydrological condition in the YRE is very complex under the joint influence of runoff, tidal current, wind and human activities [34], making it hard to accurately retrieve SSCs. Since in situ measurement methods, such as buoys and ADCP, have a limited observation range, previous research that studies SSCs in the YRE generally utilize a numerical model that requires reliable boundary conditions and initial fields [63,64]. Therefore, by applying the TRMCC to GOCI-II TSM images, we retrieved the SSC in the YRE via satellite images, for the first time. Since the nature of the TRMCC is feature-tracking, surface flow trajectories can be observed in the Lagrange method rather than using a fixed-point observation, such as fixed-buoy, as in the Euler method. Firstly, the tidal height in the YRE was simulated by a tide model and compared with the tidal height observed by three tide-gauge stations; the result can be seen in Figure 10b. The simulated tidal height agrees well with the observed data, and based on these, the trajectories consist of nine consequential vectors in various locations of the YRE, as shown in Figure 10a. Each vector was colored in red (flood period) and blue (ebb period), according to local tidal phase, and the SSC characteristics in the YRE can be vividly depicted.

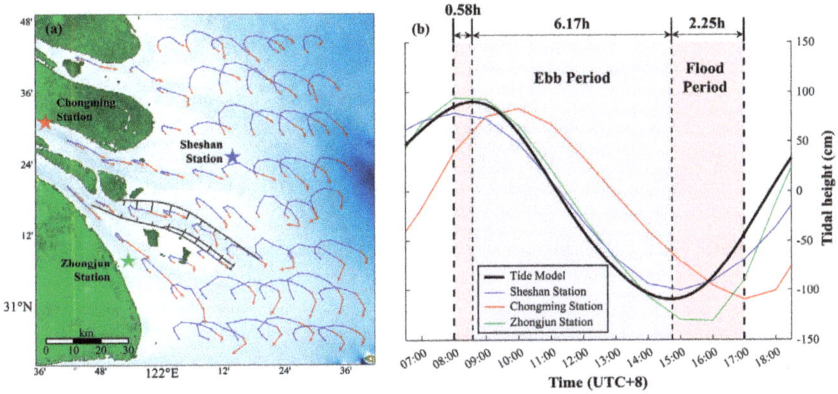

Figure 10. (**a**) SSC trajectories in YRE from 8:00 to 17:00 (UTC + 8) on 6 June 2021. The color of the trajectories represents the concurrent tidal phase: red and blue represent the flood and ebb periods, respectively. Stars represents the location of tide-gauge stations. (**b**) Tidal height in the YRE on 6 June 2021. The black line represents average tidal height simulated by tide model; blue, red, and green lines represent tidal height measured by the Sheshan, Chongming, and Zhongjun tide-gauge stations. Red and blue backgrounds represent the flood and ebb period determined by simulated tidal height.

It is indicated in Figure 10a that the SSCs have shown a rotational flow style, with an elliptical trajectory in the outer estuary, gradually transforming into a reciprocal flow style. Moreover, since the observation time of GOCI-II encompassed half of a tide period in the YRE, the responding mechanism between SSCs and the tide can also be interpreted based on these trajectories. In the inner estuary, from 8:00 to 11:00, the northwesterly current that flows into the YRE slowly decelerates and turns to the southeast, flowing to the outer sea; although the tidal current already began ebbing at 9:00, it takes 2 h for the actual SSC to change its direction. From 11:00 to 15:00, a southeasterly current gradually accelerates during the ebb phase and keeps accelerating from 15:00 to 17:00, when the tide begins to rise. Overall, the current in the estuary that is strongly influenced by runoff often stalls in response to the variation in the tidal current, while in the outer estuary, where the water is deeper, tidal variation mostly influences the current direction instead of its velocity. Over the whole observation period, the current direction gradually changes from northwest to southeast.

4. Discussion

In this study, an improved MCC workflow, called TRMCC, is proposed. By assimilating tidal current data, adaptive reference window size, and the masked normalized cross-correlation technique, TRMCC is able to retrieve high-resolution SSCs in the coastal sea area with a good accuracy, providing a new insight for SSC estimation in coastal areas. The performance of TRMCC was thoroughly examined by cross-comparison with SSC data obtained by multiple methods in various spatial and temporal scales. On a long timescale, TRMCC retains a high accuracy in a total of 10,721 data-pairs from 14 buoys located in the Korean seas and the YRE, with an average AME of 0.51 and AAE of 32.58°, on average (Tables 1 and 2). Surface current systems such as the Kuroshio Current, the East Korea Warm Current, and the Korean Coastal Current can be depicted with a higher accuracy than that of the CMEMS product (Figures 5 and 6), and the seasonal pattern of SSCs were analyzed in the Bohai Sea (Figure 9). On a short timescale, TRMCC is capable of revealing SSC trajectories in a shallow estuary like the YRE during GOCI-II's observation time (Figure 10), The SSCs' response to tidal currents can also be analyzed by combining SSC results and simulated tidal height. In general, TRMCC exhibits a promising result for SSC estimation in coastal seas and estuaries, but room for improvement still exists in various aspects.

This study utilized multiple SSC data to investigate the accuracy of TRMCC in various spatial-temporal scales; however, the discrepancy between various SSC data can potentially influence the cross-comparison results. For example, different SSC measurements represent currents at different depths. Buoys measure currents at the ocean surface; SSCs from different numerical datasets may differ in depth, e.g., 0–30 m for OSCAR (Ocean Surface Current Analysis Real-Time), 5 m for SODA (Simple Ocean Data Assimilation), and 0.5 m for the CMEMS product used in this study. The equivalent integration depth of HF radar depends on the wavelength, according to the Bragg scattering theory [65], but the HF radar data in this study mainly represent currents within 0.44–1.44 m. Ocean color data can only reflect current information from the surface to the optical depth [38], which depends on the optical properties of the water, including absorption and scattering. Moreover, in order to quantitatively compare different SSC data, we resampled them into the same spatial resolution (as shown in Figure 8), despite the huge differences among them. Other influencing factors, such as wind, ocean waves, attenuation of radar signals, etc., are not considered in this study.

In the process of tide-restriction, the equation to calculate the angular difference φ using the tidal current magnitude was determined by statistical analysis, ensuring that 99% of the genuine current can be covered in the angular restriction range. The results show that tide-restriction significantly improves the accuracy of SSC estimation in coastal seas, especially in the tide-dominated YRE, where AME was improved by 21.7% and AAE by 52.2%, on average. Tidal current data were independently pre-generated by the FES2014 tide model, facilitating the process of the TRMCC workflow. However, although FES2014 has a proven accuracy in coastal seas due to its high horizontal resolution and fine shoreline grid, for estuaries with more complex land boundaries, errors still exist in simulated tidal currents. Therefore, in these areas, the regional tidal models which can produce more accurate tide data can be considered to replace FES2014 in the TRMCC workflow.

From SSC estimation methods, previous studies have shown that erroneous vectors exist in the SSCs derived, which can be identified and eliminated by applying additional filter methods such as the reciprocal filter [50] and the nearest neighbor filter [12,66]. Moreover, the performance of derived SSCs can be further improved by applying optimization method such as optimal interpolation (OI) [32,67]. Since our research mainly focus on exploiting the potential of integrating tidal currents into the process of SSC estimation, these filtering and reconstructing methods are not considered in this study.

When it comes to the selection of tracer images, GOCI-II can be the perfect candidate due to its temporal resolution time of one hour, its spatial resolution of 250 m, and its capability to acquire 10 images per day. But apart from GOCI-II, TRMCC's application for other satellite images is not yet tested. Satellites with a better temporal resolution, such as Himawari-8 (20 min) and GF-4 (up to 20 s), can further reduce the influence caused by the vertical exchange and mixing of surface tracers, thus providing a better SSC estimation. Moreover, as a sensor running in the geostationary orbit, GOCI-II's observation range is limited to the western Pacific. In other regions, as long as two images have a short time interval, such as MODIS and VIIRS [12], Landsat 8 and Sentinel 2 [23], two different bands of Sentinel-2 that have an inter-band time lag [27], can also be selected as tracer images. Furthermore, tracer images are not limited to ocean color data, which can be influenced by cloud cover. As Liu et al. [32] pointed out, the valid data percentage of the ocean color product obtained by GOCI is only around 15~40%; this deficiency in data volume caused by cloud cover may compromise the authenticity of long-term SSC observation results. Therefore, applying TRMCC on other tracer data that is not affected by cloud cover, such as SAR images [68], model simulated sea surface temperature data [67], etc., can be a potential solution.

The SSC is an important physical property for understanding material exchange, biogeochemical processes, and ocean–atmosphere interactions in estuaries and coastal areas. Conventional methods for deriving coastal SSCs have their own limitations, such as in situ observation methods, e.g., fixed-buoy and ADCP have limited observation ranges, satellite altimeters are unreliable due to waveform contamination, and numerical models require complex boundary conditions and initial fields. Deriving SSCs from sequential ocean color images offers a new approach that offers lower computational costs and higher timeliness. Previous studies that utilized ocean color images, such as Chl-*a*, TSM, and R_{rs} obtained by GOCI, VIIRS, and AVHRR, to derive SSCs rarely involved estuary and coastal areas, due to obstacles posed by intricate land boundaries and complex dynamic environments.

5. Conclusions

In this study, we propose a novel SSC estimation method, called TRMCC, that can overcome the aforementioned obstacles and implemented it on ocean color data obtained from GOCI-II. Validation using in situ current data obtained from 14 buoys located in different water areas from 2021 to 2023 showed that TRMCC is capable of deriving high-accuracy SSCs, even in turbid coastal areas, with an average AME of 0.51 and an AAE of 32.58°. Additionally, it was shown that estimation accuracy can be significantly improved by integrating tidal current data into the MCC process. Cross-comparisons were made between TRMCC results, HF radar observations, CMEMS numerical model products, and buoy observations, showing that both large-scale surface circulation patterns and fine-scale current structures can be extracted by TRMCC. Seasonal SSC structures in the Bohai Sea and diurnal SSC variation in the YRE were depicted via satellite methods, for the first time. By combining hourly ocean color data from GOCI-II, TRMCC can provide new insights into studying sea surface currents in highly turbid water areas with intricate land boundaries, expanding our understanding of complex physical processes in these regions.

Author Contributions: Conceptualization, S.C., Z.L. and F.S.; methodology, S.C. and F.S.; software, S.C. and R.L.; validation, S.C., R.L. and Y.Z.; formal analysis, S.C. and R.L.; investigation, S.C.; resources, S.C. and R.L.; data curation, S.C. and R.L.; writing—original draft preparation, S.C. and F.S.; writing—review and editing, S.C. and F.S.; visualization, S.C. and R.L.; supervision, F.S.; project administration, F.S.; funding acquisition, F.S. All authors have read and agreed to the published version of the manuscript.

Funding: This study was funded by the National Natural Science Foundation of China [Grant No. 42271348].

Data Availability Statement: The raw data supporting the conclusions of this article will be made available by the authors on request.

Acknowledgments: We would like to thank the Korea Ocean Satellite Center for providing satellite data, the Korea Hydrographic and Oceanographic Agency for providing buoy and high-frequency radar data, the Archiving, Validation, and Interpretation of Satellite Oceanographic Satellite Data for providing the FES2014 tide model, and the Copernicus Marine Environment Monitoring Service for providing the numerical model product.

Conflicts of Interest: The authors declare no conflicts of interest.

References

1. Portela, E.; Kolodziejczyk, N.; Gorgues, T.; Zika, J.; Perruche, C.; Mignot, A. The Ocean's Meridional Oxygen Transport. *J. Geophys. Res. Oceans* **2024**, *129*, e2023JC020259. [CrossRef]
2. Oldenburg, D.; Kwon, Y.; Frankignoul, C.; Danabasoglu, G.; Yeager, S.; Kim, W.M. The Respective Roles of Ocean Heat Transport and Surface Heat Fluxes in Driving Arctic Ocean Warming and Sea Ice Decline. *J. Clim.* **2024**, *37*, 1431–1448. [CrossRef]
3. Ma, S.; Wang, N.; Zhou, L.; Yu, J.; Chen, X.; Chen, Y. Inversion of Tidal Flat Topography Based on the Optimised Inundation Frequency Method—A Case Study of Intertidal Zone in Haizhou Bay, China. *Remote Sens.* **2024**, *16*, 685. [CrossRef]
4. Ospina-Alvarez, A.; Weidberg, N.; Aiken, C.; Navarrete, S. Larval transport in the upwelling ecosystem of central Chile: The effects of vertical migration, developmental time and coastal topography on recruitment. *Prog. Oceanogr.* **2018**, *168*, 82–99. [CrossRef]
5. Mazoyer, C.; Vanneste, H.; Dufresne, C.; Ourmières, Y.; Magaldi, M.G.; Molcard, A. Impact of wind-driven circulation on contaminant dispersion in a semi-enclosed bay. *Estuar. Coast. Shelf Sci.* **2020**, *233*, 106529. [CrossRef]
6. Kuang, C.; Chen, W.; Gu, J.; He, L. Comprehensive analysis on the sediment siltation in the upper reach of the deepwater navigation channel in the Yangtze Estuary. *J. Hydrodyn.* **2014**, *26*, 299–308. [CrossRef]
7. Fitzenreiter, K.; Mao, M.; Xia, M. Characteristics of Surface Currents in a Shallow Lagoon–Inlet–Coastal Ocean System Revealed by Surface Drifter Observations. *Estuaries Coasts.* **2022**, *45*, 2327–2344. [CrossRef]
8. Rypina, I.; Kirincich, A.; Peacock, T. Horizontal and vertical spreading of dye in the coastal ocean of the northern Mid-Atlantic bight. *Cont. Shelf Res.* **2021**, *230*, 104567. [CrossRef]
9. Barrick, D.; Evans, M.; Weber, B. Ocean Surface Currents Mapped by Radar. *Science* **1977**, *198*, 138–144. [CrossRef]
10. Georges, T.M.; Harlan, J.A.; Lematta, R.A. Large-scale mapping of ocean surface currents with dual over-the-horizon radars. *Nature* **1996**, *379*, 434–436. [CrossRef]
11. Roarty, H.; Cook, T.; Hazard, L.; George, D.; Harlan, J.; Cosoli, S.; Wyatt, L.; Alvarez Fanjul, E.; Terrill, E.; Otero, M.; et al. The Global High Frequency Radar Network. *Front. Mar. Sci.* **2019**, *6*, 164. [CrossRef]
12. Liu, J.; Emery, W.; Wu, X.; Li, M.; Li, C.; Zhang, L. Computing Coastal Ocean Surface Currents from MODIS and VIIRS Satellite Imagery. *Remote Sens.* **2017**, *9*, 1083. [CrossRef]
13. Carret, A.; Birol, F.; Estournès, C.; Zakardjian, B. Assessing the capability of three different altimetry satellite missions to observe the Northern Current by using a high-resolution model. *Ocean Sci.* **2023**, *19*, 903–921. [CrossRef]
14. Emery, W.; Baldwin, D.G.; Matthews, D. Sampling the mesoscale ocean surface currents with various satellite altimeter configurations. *Geosci. Remote Sens. IEEE Trans. On.* **2004**, *42*, 795–803. [CrossRef]
15. Rio, M.; Santoleri, R.; Bourdalle-Badie, R.; Griffa, A.; Piterbarg, L.; Taburet, G. Improving the Altimeter-Derived Surface Currents Using High-Resolution Sea Surface Temperature Data: A Feasability Study Based on Model Outputs. *J. Atmos. Ocean. Technol.* **2016**, *33*, 2769–2784. [CrossRef]
16. Andersen, O.B.; Scharroo, R. Range and Geophysical Corrections in Coastal Regions: And Implications for Mean Sea Surface Determination. In *Coastal Altimetry*; Vignudelli, S., Kostianoy, A.G., Cipollini, P., Benveniste, J., Eds.; Springer: Berlin/Heidelberg, Germany, 2011; pp. 103–145.
17. Peng, F.; Deng, X.; Shen, Y. Assessment of Sentinel-6 SAR mode and reprocessed Jason-3 sea level measurements over global coastal oceans. *Remote Sens. Environ.* **2024**, *311*, 114287. [CrossRef]
18. Kelly, K.A. The influence of winds and topography on the sea surface temperature patterns over the northern California slope. *J. Geophys. Res. Ocean* **1985**, *90*, 11783–11798. [CrossRef]
19. Emery, W.J.; Thomas, A.C.; Collins, M.J.; Crawford, W.R.; Mackas, D.L. An objective method for computing advective surface velocities from sequential infrared satellite images. *J. Geophys. Res. Ocean* **1986**, *91*, 12865–12878. [CrossRef]

20. Taniguchi, N.; Kida, S.; Sakuno, Y.; Mutsuda, H.; Syamsudin, F. Short-Term Variation of the Surface Flow Pattern South of Lombok Strait Observed from the Himawari-8 Sea Surface Temperature. *Remote Sens.* **2019**, *11*, 1491. [CrossRef]
21. Yang, H.; Arnone, R.; Jolliff, J. Estimating advective near-surface currents from ocean color satellite images. *Remote Sens. Environ.* **2015**, *158*, 1–14. [CrossRef]
22. Ren, Y.; Li, X.; Gao, G.; Busche, T. Derivation of Sea Surface Tidal Current From Spaceborne SAR Constellation Data. *IEEE Trans. Geosci. Remote Sens.* **2017**, *55*, 3236–3247. [CrossRef]
23. Osadchiev, A.; Sedakov, R. Spreading dynamics of small river plumes off the northeastern coast of the Black Sea observed by Landsat 8 and Sentinel-2. *Remote Sens. Environ.* **2019**, *221*, 522–533. [CrossRef]
24. Volkov, D.L.; Negahdaripour, S. Implementation of the Optical Flow to Estimate the Propagation of Eddies in the South Atlantic Ocean. *Remote Sens.* **2023**, *15*, 3894. [CrossRef]
25. Moiseev, A.; Johnsen, H.; Hansen, M.W.; Johannessen, J.A. Evaluation of Radial Ocean Surface Currents Derived From Sentinel-1 IW Doppler Shift Using Coastal Radar and Lagrangian Surface Drifter Observations. *J. Geophys. Res. Oceans* **2020**, *125*, e2019JC015743. [CrossRef]
26. Moiseev, A.; Johnsen, H.; Johannessen, J.A.; Collard, F.; Guitton, G. On Removal of Sea State Contribution to Sentinel-1 Doppler Shift for Retrieving Reliable Ocean Surface Current. *J. Geophys. Res. Oceans* **2020**, *125*, e2020JC016288. [CrossRef]
27. Yurovskaya, M.; Kudryavtsev, V.; Chapron, B.; Collard, F. Ocean surface current retrieval from space: The Sentinel-2 multispectral capabilities. *Remote Sens. Environ.* **2019**, *234*, 111468. [CrossRef]
28. Chen, W. Surface Velocity Estimation From Satellite Imagery Using Displaced Frame Central Difference Equation. *IEEE Trans. Geosci. Remote Sens.* **2012**, *50*, 2791–2801. [CrossRef]
29. Yang, H.; Choi, J.; Park, Y.; Han, H.; Ryu, J. Application of the Geostationary Ocean Color Imager (GOCI) to estimates of ocean surface currents. *J. Geophys. Res. Oceans* **2014**, *119*, 3988–4000. [CrossRef]
30. Jiang, L.; Wang, M. Diurnal Currents in the Bohai Sea Derived From the Korean Geostationary Ocean Color Imager. *IEEE Trans. Geosci. Remote Sens.* **2017**, *55*, 1437–1450. [CrossRef]
31. Hu, Z.; Zhang, H.; Wang, D. A Novel Approach for Estimating Sea Surface Currents From Numerical Models and Satellite Images: Validation and Application. *IEEE Trans. Geosci. Remote Sens.* **2024**, *62*, 4203708. [CrossRef]
32. Liu, J.; Emery, W.J.; Wu, X.; Li, M.; Li, C.; Zhang, L. Computing Ocean Surface Currents From GOCI Ocean Color Satellite Imagery. *IEEE Trans. Geosci. Remote Sens.* **2017**, *55*, 7113–7125. [CrossRef]
33. Liu, H.; He, Q.; Wang, Z.; Weltje, G.J.; Zhang, J. Dynamics and spatial variability of near-bottom sediment exchange in the Yangtze Estuary, China. *Estuar. Coast. Shelf Sci.* **2010**, *86*, 322–330. [CrossRef]
34. Cai, L.; Chen, S.; Yan, X.; Bai, Y.; Bu, J. Study on High-Resolution Suspended Sediment Distribution under the Influence of Coastal Zone Engineering in the Yangtze River Mouth, China. *Remote Sens.* **2022**, *14*, 486. [CrossRef]
35. Byun, D.; Hart, D.E. Tidal current classification insights for search, rescue and recovery operations in the Yellow and East China Seas and Korea Strait. *Cont. Shelf Res.* **2022**, *232*, 104632. [CrossRef]
36. Zhang, E.; Savenije, H.H.G.; Wu, H.; Kong, Y.; Zhu, J. Analytical solution for salt intrusion in the Yangtze Estuary, China. *Estuar. Coast. Shelf Sci.* **2011**, *91*, 492–501. [CrossRef]
37. Crocker, R.I.; Matthews, D.K.; Emery, W.J.; Baldwin, D.G. Computing Coastal Ocean Surface Currents From Infrared and Ocean Color Satellite Imagery. *IEEE Trans. Geosci. Remote Sens.* **2007**, *45*, 435–447. [CrossRef]
38. Warren, M.A.; Quartly, G.D.; Shutler, J.D.; Miller, P.I.; Yoshikawa, Y. Estimation of ocean surface currents from maximum cross correlation applied to GOCI geostationary satellite remote sensing data over the Tsushima (Korea) Straits. *J. Geophys. Res. Oceans* **2016**, *121*, 6993–7009. [CrossRef]
39. Hu, C.; Lee, Z.; Franz, B. Chlorophyll aalgorithms for oligotrophic oceans: A novel approach based on three-band reflectance difference. *J. Geophys. Res. Oceans* **2012**, *117*, 5–9. [CrossRef]
40. Shen, F.; Zhou, Y.; Li, D.; Zhu, W.; Suhyb Sal-ama, M. Medium resolution imaging spectrometer (MERIS) estimation of chlorophyll-a concentration in the turbid sediment-laden waters of the Changjiang (Yangtze) Estuary. *Int. J. Remote Sens.* **2010**, *31*, 4635–4650. [CrossRef]
41. Cai, L.; Yu, M.; Yan, X.; Zhou, Y.; Chen, S. HY-1C/D Reveals the Chlorophyll-a Concentration Distribution Details in the Intensive Islands' Waters and Its Consistency with the Distribution of Fish Spawning Ground. *Remote Sens.* **2022**, *14*, 4270. [CrossRef]
42. Hu, Z.; Pan, D.; He, X.; Bai, Y. Diurnal Variability of Turbidity Fronts Observed by Geostationary Satellite Ocean Color Remote Sensing. *Remote Sens.* **2016**, *8*, 147. [CrossRef]
43. Hu, Z.; Wang, D.P.; Pan, D.; He, X.; Miyazawa, Y.; Bai, Y.; Wang, D.; Gong, F. Mapping surface tidal currents and Changjiang plume in the East China Sea from Geostationary Ocean Color Imager. *J. Geophys. Res. Oceans* **2016**, *121*, 1563–1572. [CrossRef]
44. Ma, Y.; Yin, W.; Guo, Z.; Xuan, J. The Ocean Surface Current in the East China Sea Computed by the Geostationary Ocean Color Imager Satellite. *Remote Sens.* **2023**, *15*, 2210. [CrossRef]
45. He, X.; Pan, D.; Bai, Y.; Gong, F. A general purpose exact Rayleigh scattering look-up table for ocean color remote sensing. *Acta Oceanol. Sin.* **2006**, *25*, 48–56.

46. He, X.; Bai, Y.; Zhu, Q.; Gong, F. A vector radiative transfer model of coupled ocean–atmosphere system using matrix-operator method for rough sea-surface. *J. Quant. Spectrosc. Radiat. Transf.* **2010**, *111*, 1426–1448. [CrossRef]
47. Lu, S.; He, M.; He, S.; He, S.; Pan, Y.; Wenbin, Y.; Li, P. An Improved Cloud Masking Method for GOCI Data over Turbid Coastal Waters. *Remote Sens.* **2021**, *13*, 2722. [CrossRef]
48. Li, R.; Shen, F.; Zhang, Y.; Li, Z.; Chen, S. Identifying algal bloom types and analyzing their diurnal variations using GOCI-II data. *Int. J. Appl. Earth Obs. Geoinf.* **2025**, *136*, 104377. [CrossRef]
49. Shen, F.; Verhoef, W.; Zhou, Y.; Salama, M.; Liu, X. Satellite Estimates of Wide-Range Suspended Sediment Concentrations in Changjiang (Yangtze) Estuary Using MERIS Data. *Estuaries Coasts* **2010**, *33*, 1420–1429. [CrossRef]
50. Barton, I.J. Ocean Currents from Successive Satellite Images: The Reciprocal Filtering Technique. *J. Atmos. Ocean. Technol.* **2002**, *19*, 1677–1689. [CrossRef]
51. Fujii, S.; Heron, M.L.; Kim, K.; Lai, J.; Lee, S.; Wu, X.; Wu, X.; Wyatt, L.R.; Yang, W. An overview of developments and applications of oceanographic radar networks in Asia and Oceania countries. *Ocean Sci. J.* **2013**, *48*, 69–97. [CrossRef]
52. Sanchez-Arcilla, A.; Staneva, J.; Cavaleri, L.; Badger, M.; Bidlot, J.; Sorensen, J.T.; Hansen, L.B.; Martin, A.; Saulter, A.; Espino, M.; et al. CMEMS-Based Coastal Analyses: Conditioning, Coupling and Limits for Applications. *Front. Mar. Sci.* **2021**, *8*, 604741. [CrossRef]
53. Pirooznia, M.; Raoofian Naeeni, M.; Tourian, M.J. Modeling total surface current in the Persian Gulf and the Oman Sea by combination of geodetic and hydrographic observations and assimilation with in situ current meter data. *Acta Geophys.* **2023**, *71*, 2839–2863. [CrossRef]
54. Lellouche, J.; Greiner, E.; Le Galloudec, O.; Garric, G.; Regnier, C.; Drevillon, M.; Mounir, B.; Testut, C.; Bourdalle-Badie, R.; Gasparin, F.; et al. Recent updates on the Copernicus Marine Service global ocean monitoring and forecasting real-time 1/12° high resolution system. *Ocean Sci.* **2018**, *14*, 1093–1126. [CrossRef]
55. Lyard, F.H.; Allain, D.J.; Cancet, M.; Carrere, L.; Picot, N. FES2014 global ocean tide atlas: Design and performance. *Ocean Sci.* **2021**, *17*, 615–649. [CrossRef]
56. Fu, Y.; Feng, Y.; Zhou, D.; Zhou, X.; Li, J.; Tang, Q. Accuracy assessment of global ocean tide models in the South China Sea using satellite altimeter and tide gauge data. *Acta Oceanol. Sin.* **2020**, *39*, 1–10. [CrossRef]
57. Wang, N.; Gao, Y.; Guo, H.; Lu, J. Analysis of Characteristics of Tide and Tidal Current in the east China Seas. *J. Phys. Conf. Ser.* **2023**, *2486*, 12039. [CrossRef]
58. Nguyen, V.T.; Lee, M. Effect of Open Boundary Conditions and Bottom Roughness on Tidal Modeling around the West Coast of Korea. *Water* **2020**, *12*, 1706. [CrossRef]
59. Padfield, D. Masked Object Registration in the Fourier Domain. *IEEE Trans. Image Process.* **2011**, *21*, 2706–2718. [CrossRef]
60. Hu, Z.; Pan, D.; He, X.; Song, D.; Huang, N.; Bai, Y.; Xu, Y.; Wang, X.; Zhang, L.; Gong, F. Assessment of the MCC method to estimate sea surface currents in highly turbid coastal waters from GOCI. *Int. J. Remote Sens.* **2017**, *38*, 572–597. [CrossRef]
61. Chen, W.; Mied, R.P.; Gao, B.; Wagner, E. Surface Velocities From Multiple-Tracer Image Sequences. *IEEE Geosci. Remote Sens. Lett.* **2012**, *9*, 769–773. [CrossRef]
62. Cui, H.; Chen, J.; Cao, Z.; Huang, H.; Gong, F. A Novel Multi-Candidate Multi-Correlation Coefficient Algorithm for GOCI-Derived Sea-Surface Current Vector with OSU Tidal Model. *Remote Sens.* **2022**, *14*, 4625. [CrossRef]
63. Ge, J.; Ding, P.; Chen, C.; Hu, S.; Fu, G.; Wu, L. An integrated East China Sea–Changjiang Estuary model system with aim at resolving multi-scale regional–shelf–estuarine dynamics. *Ocean Dyn.* **2013**, *63*, 881–900. [CrossRef]
64. Shi, S.; Xu, Y.; Li, W.; Ge, J. Long-term response of an estuarine ecosystem to drastic nutrients changes in the Changjiang River during the last 59 years: A modeling perspective. *Front. Mar. Sci.* **2022**, *9*, 1012127. [CrossRef]
65. Rubio, A.; Mader, J.; Corgnati, L.; Mantovani, C.; Griffa, A.; Novellino, A.; Quentin, C.; Wyatt, L.; Schulz-Stellenfleth, J.; Horstmann, J.; et al. HF Radar Activity in European Coastal Seas: Next Steps toward a Pan-European HF Radar Network. *Front. Mar. Sci.* **2017**, *4*, 8. [CrossRef]
66. Notarstefano, G.; Poulain, P.; Mauri, E. Estimation of Surface Currents in the Adriatic Sea from Sequential Infrared Satellite Images. *J. Atmos. Ocean. Technol.* **2008**, *25*, 271–285. [CrossRef]
67. Ye, F.; Hao, Z.; Pan, D. An Optimization Method Based on Decorrelation Scales Analysis for Improving Surface Currents Retrieval From Sea Surface Temperature. *IEEE Trans. Geosci. Remote Sens.* **2024**, *62*, 4202717. [CrossRef]
68. Qazi, W.; Emery, W.; Fox-Kemper, B. Computing Ocean Surface Currents Over the Coastal California Current System Using 30-Min-Lag Sequential SAR Images. *IEEE Trans. Geosci. Remote Sens.* **2013**, *52*, 7559–7580. [CrossRef]

remote sensing

MDPI

Article

Potential Impact of Sea Surface Temperature Variability on the 2007 Sudden Bloom of *Ulva prolifera* in the Southern Yellow Sea

Yufeng Pan [1,2], Pin Li [3,*], Jiaxuan Sun [4], Siyu Liu [1,5], Lvyang Xing [1,2], Di Yu [1,2] and Qi Feng [1,2]

1 Key Laboratory of Submarine Sciences and Prospecting Techniques, Ministry of Education of the People's Republic of China, Ocean University of China, Qingdao 266100, China; panyufeng@stu.ouc.edu.cn (Y.P.); 11220411020@stu.ouc.edu.cn (S.L.); xinglvyang@stu.ouc.edu.cn (L.X.); yuayu@stu.ouc.edu.cn (D.Y.); 21210411048@stu.ouc.edu.cn (Q.F.)
2 College of Marine Geoscience, Ocean University of China, Qingdao 266100, China
3 First Institute of Oceanography, MNR, Qingdao 266100, China
4 Qingdao West Coast New Area Marine Development Bureau, Qingdao 266000, China; sjx1251@stu.ouc.edu.cn
5 Academy of the Future Ocean, Ocean University of China, Qingdao 266100, China
* Correspondence: pinli@fio.org.cn

Abstract: Since 2007, *Ulva prolifera* (*U. prolifera*) originating in northern Jiangsu (NJ) has consistently expanded to the southern coast of the Shandong Peninsula. However, the underlying reasons for the 2007 sudden bloom of *U. prolifera* on a large scale remain unknown. This study uses remote sensing data from MODIS/AQUA spanning the period 2003–2022 to investigate the sea surface temperature (SST) structure changes in the southern Yellow Sea (SYS) over the past 20 years. The results demonstrate the following. (1) Since 2007, the NJ northward current and the Yangtze estuary warm current have exhibited higher temperatures, earlier northward intrusions, and larger influence areas, leading to a faster warming rate in NJ before mid-May. This rapid increase in SST to a level suitable for early *U. prolifera* growth triggers large-scale blooms. (2) The change in temperature structure is primarily induced by a prolonged and intense La Niña event in 2007–2008. However, since 2016, under stable global climate conditions, the temperature structure of the SYS has returned to the pre-2007 state, corresponding to a decrease in the scale of *U. prolifera* blooms.

Keywords: *Ulva prolifera*; sea surface temperature; Northern Jiangsu northward current; warm current across the Yangtze River Estuary; MODIS/AQUA

1. Introduction

The green tide has emerged as a significant marine disaster impacting coastal regions worldwide. Investigating the underlying mechanisms contributing to its onset holds crucial practical implications for predicting and mitigating large-scale disasters. Since 2007, the proliferation of *Enteromorpha* in the western region of the southern Yellow Sea (SYS) has resulted in large-scale blooms. This phenomenon has had severe impacts on coastal fisheries and tourism in Jiangsu and Shandong provinces, making it one of the most severe marine disasters in China. The tongue-shaped topography (Figure 1, yellow dashed square) off the coast of northern Jiangsu (NJ) is characterized by extensive shallow waters (particularly the NJ shallow, with depths of less than 20 m), located between the abandoned Yellow River Estuary and the Yangtze Estuary where radial sand ridges spread, creating a significant breeding base for *Porphyra yezoensis*. Previous studies using remote sensing monitoring and homology analysis have indicated that *Enteromorpha* blooms in the SYS predominantly originate along the *Porphyra* aquaculture rafts coast of Jiangsu province [1–3]. Morphological and molecular analyses have revealed five to seven species of attached algae from *Enteromorpha*, including *Ulva Prolifera* (*U. Prolifera*), *Ulva linza*, *Ulva compressa*, *Ulva intestinalis*, *Ulva flexuosa*, and *Ulva clathrata* [4,5]. The dominant species responsible for these blooms is *U. prolifera* [6,7].

The occurrence of green tide events closely relates to multiple factors [8] including sea surface temperature (SST), nutrient concentrations, and light intensity. In previous research, the primary focus has been the overall oceanographic conditions and their impact on *U. prolifera* within a single year or a relatively short timescale in the SYS [9–14]. Despite the eutrophication status of the SYS, the nutrient concentration remains sufficient for *U. prolifera* to bloom, despite reduced nutrient levels in springtime [15–17]. Analysis of local aquaculture and environmental data strongly suggests that the *U. prolifera* in both the SYS and East China Sea, including massive blooms in the summers of 2008 and 2009, is unlikely to be solely caused by coastal eutrophication [3,15,18]. Therefore, a suitable SST has emerged as the critical factor in the large-scale bloom of *U. prolifera*, with temperatures of 15–25 °C being most favorable for its growth [19]. Laboratory studies have demonstrated that the early growth of *U. prolifera* is more substantial than in other species when the SST reaches 10–20 °C [20]. As for chlorophyll-a concentration, this has a negative correlation with the occurrence of *U. prolifera* in the SYS [21]. Furthermore, the prevailing southeast summer monsoon in the SYS acts as the primary driving force for the northward drift and diffusion of *U. prolifera* [3,22,23]. Hu et al. established a robust relationship to link biomass per area to the reflectance-based floating algae index (FAI) [24]. Since 2007, *U. prolifera* has originated near the NJ shallow from April to May each year, drifting northward to reach its maximum coverage and distribution area in June. Subsequently, the bloom gradually diminishes from July to August along the southern coast of the Shandong Peninsula.

Existing studies have overlooked the relationship between long-term changes in SST structure and the occurrence of *U. prolifera* blooms. Although *U. prolifera* was observed almost every year from 2000–2009, the extent of its coverage before 2008 was considerably smaller than in 2008 and 2009 [18], and there are notable interannual variations in its occurrence. The reasons for the 2007 sudden large-scale blooms have not been thoroughly investigated. Therefore, investigating the underlying mechanisms contributing to its onset holds crucial practical implications for predicting and mitigating large-scale disasters. Our research suggests that, since 2007, *U. prolifera* has originated in NJ during April and May, leading to widespread northward dispersion driven by wind drift and the increasing SST in the SYS [25]. To further explore these changes, remote sensing data from MODIS/AQUA spanning the period from 2003–2022 are used to compare the SST structure before and since 2007. By conducting correlation analyses between SST and the coverage area of *U. prolifera*, this study investigates the effects of temperature structure changes influenced by the current system on the sudden bloom of *U. prolifera* and its intensity.

2. Geological Setting

The study area (Figure 1) is a typical plain siltation coast. In the central and northern expanses of the SYS, the sea plain is wide and shallow, producing a gently contoured sector. This zone is a stable accumulation landform, marked by sediments of a notably fine granularity [26]. The central and southern sectors of the western coastline of the SYS are distinguished by an array of emanating tidal sand ridges, colloquially referred to as the NJ shallow, which extend across an aggregate area of roughly 35,000 square kilometers.

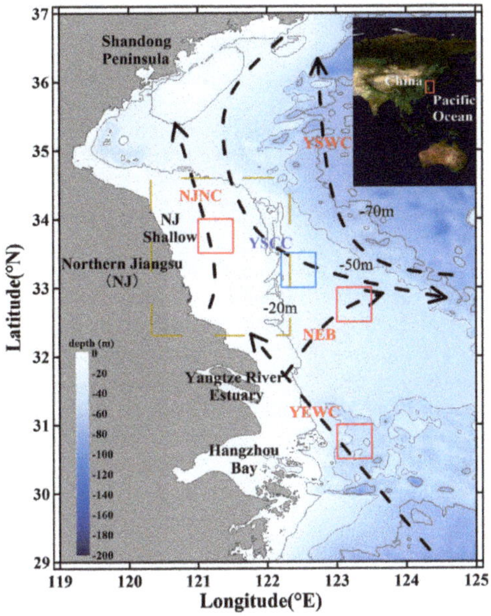

Figure 1. Schematic map of the study area. The black single-arrow dashed curves depict the primary flow paths, as primarily determined by previous studies [27–29] and the temperature structure in Chapter 4.2. The red and blue squares indicate the characteristic regions associated with each respective current. The yellow dashed square is the tongue-shaped topography area where *Ulva prolifera* (*U. prolifera*) originates. The visual representation is rendered using the Surfer16 software.

In the western region of the SYS, where *U. prolifera* blooms occur, there is a marked coastal current. Traditionally, it was believed that the Yellow Sea coastal current (YSCC) originated near Haizhou Bay and flowed southward to the north of the Yangtze River Estuary. The YSCC is formed by the runoff from the Guan River and Sheyang River in NJ, which mixes with coastal seawater. The SST in this region is approximately 28 °C, and the salinity ranges from 28–29 psu [30,31]. However, recent studies based on Argo data and numerical simulations suggest that the YSCC exhibits a northward flow during the summer months, specifically along the NJ northward current (NJNC) region. This northward current aligns with the drift path of *U. prolifera* from the NJ shallow to the southern coast of the Shandong Peninsula [32,33]. Furthermore, preliminary research indicates that the warm current across the Yangtze River Estuary (YEWC), of which the Taiwan warm current is the main component, influences the SST of the SYS during summer [25].

3. Materials and Methods

3.1. Materials

The remote sensing data used in this research consist of SST Level 2 data obtained from the MODIS/AQUA satellite. These data were acquired through NASA's Ocean-Color Web platform (https://oceancolor.gsfc.nasa.gov/, Accessed on 17 October 2022) and were made available with the support of the Ocean Biology Processing Group (OBPG) at NASA's Goddard Space Flight Center. The SST data exhibit a temporal resolution of 1 day and a spatial resolution of 1 km. To assess the reliability and accuracy of the measurements, each pixel in the data was assigned a numeric quality level: 0-BEST, 1-GOOD, 2-QUESTIONABLE, 3-BAD, and 4-NOT PROCESSED.

The nutrient concentrations were taken from the Global Ocean Biogeochemistry Analysis and Forecast (GOBAF) dataset, provided by the EU Copernicus Marine Service Information (https://doi.org/10.48670/moi-00015, Accessed on 15 September 2023). This dataset has a temporal resolution of 1 day and a spatial resolution of 0.25°.

The in-situ SST data used in this study were obtained from multiple sources and were collected using conductivity–temperature–depth measurements with Sea-bird911 instruments. These measurements were taken at depths ranging from 0–5 m. Specifically, a total of 677 in-situ data points were obtained in the SYS region from 2006–2019, as illustrated in Figure 2a.

The supplementary data pertaining to the impacts of *U. prolifera* in this paper were sourced from the *Bulletin of China Marine Disaster* (*BCMD*) [9,34,35]. The *BCMD* is an official publication issued annually by the Ministry of Natural Resources of the People's Republic of China, typically released in March or April each year. It provides comprehensive and detailed records of various aspects related to *U. prolifera* blooms, including the earliest discovery location, maximum distribution date and area (MDA), maximum coverage date and area (MCA), and other relevant parameters. Prior to 2021, the MDA and MCA metrics were identified and monitored using satellite imagery with a resolution of 250 meters. From 2021 onwards, these parameters have been tracked using satellite imagery with an enhanced resolution of 50 meters. These records have been consistently documented since 2009. Additionally, this study references pertinent information from the *Green Tide Remote Sensing Monitoring Bulletin*, published by the National Satellite Marine Application Centre [36].

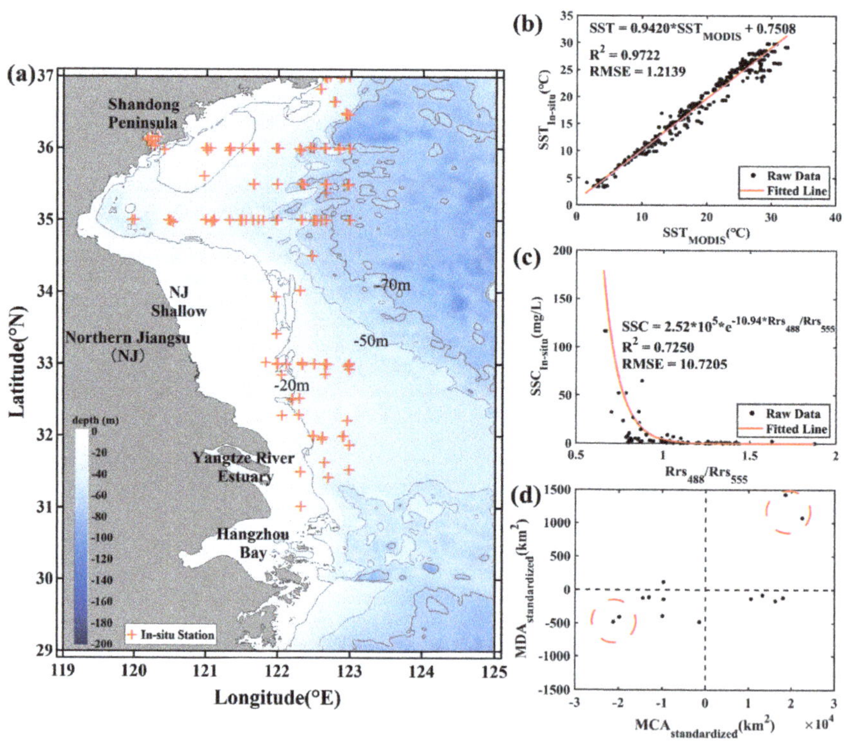

Figure 2. (**a**) Distribution of in-situ data survey stations. (**b**) Correlation analysis between the sea surface temperature (SST) derived from MODIS/AQUA satellite imagery and in-situ data. (**c**) Correlation analysis between the suspended sediment concentration (SSC) derived from MODIS/AQUA satellite imagery and in-situ data. (**d**) Standardized calculation for the classification of the *U. prolifera*.

3.2. Methods

The MODIS/AQUA remote sensing data used in this study were subjected to a series of preprocessing and inversion procedures, involving both data formatting and the utilization of in-situ data for calibration. Poor quality data points (indicated by a quality level greater than 2) were excluded. The inversion process primarily relies on establishing a correlation between the values obtained from in-situ data and the corresponding remote sensing data. By comparing the time, longitude, and latitude of the survey stations, this study identified matching points in the remote sensing data that align with the same time and location, allowing for the retrieval of the corresponding values. This study collected data from a total of 677 in-situ data stations spanning the period from 2006–2019. Among these, 222 stations were successfully matched to the remote sensing data, resulting in the acquisition of 540 valid remote sensing values. When identifying matching points, the observed remote sensing data for the day were used within a circular area with a 1 km radius around the corresponding in-situ station. The high coefficient of the SST inversion formula (Table 1, Equation (1)) determination (R^2) of 0.9722, as shown in Figure 2b, signifies a strong correlation between the in-situ data and the MODIS/AQUA data.

Table 1. Methods and formulas.

Datasets		Methods	Formulas	Parameters	Remarks
MODIS/AQUA data	SST	Inversion	$SST = 0.942 * SST_{MODIS} + 0.7508$ (1)		Obtained the spatial pattern of SST during *U. prolifera* bloom events.
		Warming rate (WR)	$WR = \left(X^T X\right)^{-1} X^T t$ (2)	X is the SST of a specific pixel in different time periods; t is the different time periods.	Obtained the speed of SST changes over different time periods.
		Inversion of the suspended sediment concentration (SSC)	$WRA = WR_{mid-May\ to\ late-June} - WR_{early-April\ to\ mid-May}$ (3)		Highlighted the difference in SST changes between the origin and outbreak stage of the *U. prolifera* event.
		The monthly mean value of nutrient concentrations	$SSC = 2.52 * 10^5 * e^{-10.94*(\frac{Rrs_{488}}{Rrs_{555}})}$ (4) [37,38]	Rrs_{488} and Rrs_{555} are 488 nm and 555 nm bands, respectively.	Obtained the SSC value and its spatial pattern during *U. prolifera* bloom events.
GOBAF dataset		The monthly mean value of nutrient concentrations	/		Obtained nutrient concentrations including the mole concentrations of nitrate (NO_3), phosphate (PO_4), and silicate (Si) in seawater.
Bulletin of China Marine Disaster (BCMD)		Standardized calculation for the classification of the *U. prolifera*	$S = X - \overline{X}$ (5)	S represents the standardized value; X represents the MCA or MDA value of each year; and $\overline{X}$ represents the mean value of the MCA or MDA from 2008–2022.	

Table 1. *Cont.*

Datasets	Methods	Formulas	Parameters	Remarks
/	Pearson product-moment correlation coefficient (PPMCC)	$r(X,Y) = \frac{n\sum_{i=1}^{n} x_i y_i - \left(\sum_{i=1}^{n} x_i\right)\left(\sum_{i=1}^{n} y_i\right)}{\sqrt{n\sum_{i=1}^{n} x_i^2 - \left(\sum_{i=1}^{n} x_i\right)^2}\sqrt{n\sum_{i=1}^{n} y_i^2 - \left(\sum_{i=1}^{n} y_i\right)^2}}$ (6)	X is the SST of a specific pixel and time period, but from different years; Y is the coverage area of *U. prolifera* for the same time period, but from different years (within 2010–2019)	Obtained the spatial pattern of the PPMCC between SST and the coverage area of *U. prolifera* and assessed the extent to which SST impacts *U. prolifera* in different periods and locations.

To obtain the spatial pattern of SST during *U. prolifera* bloom events (from April to August), the inverted SST data were processed using a 10-day average. However, for months containing 31 days, the average for the 11th day was used as the late average for that month. This approach allows for the characterization of the SST spatial patterns in the early, middle, and late periods of each month.

Additionally, the SST for different time series was calculated using the least-squares method to analyze the SST warming rate (WR) (Table 1, Equation (2)). To further analyze the WR difference between different time periods, the WR anomaly (WRA) (Table 1, Equation (3)) was computed by subtracting the WR of the early April to mid-May period (the origin stage of the *U. prolifera* event) from the WR of the mid-May to late June period (the outbreak stage of the *U. prolifera* event). To assess the extent to which the SST impacts *U. prolifera* in different periods and locations, the spatial pattern of the Pearson product-moment correlation coefficient (PPMCC) (Table 1, Equation (5)) between the SST and the coverage area of *U. prolifera* was derived using MODIS/AQUA data and the coverage area data from 2010–2019 [36].

In addition, previous studies [37,38] have demonstrated that the suspended sediment concentration (SSC) can be estimated using the 488 nm (Rrs_{488}) and 555 nm (Rrs_{555}) bands of MODIS/AQUA satellite data. This study collected data from a total of 289 in-situ data stations spanning the period from 2006–2019. Among these, 46 stations were successfully matched to the remote sensing data, resulting in the acquisition of 77 valid remote sensing values. The R^2 of the SSC inversion formula reaches 0.7250, as shown in Figure 2c. The nutrient concentrations considered in this study include the mole concentrations of nitrate (NO_3), phosphate (PO_4), and silicate (Si) in seawater. The data span the period 2003–2020, and the monthly mean values for the study area were calculated. The processing of all nutrient concentration datasets was conducted using MATLAB/2023a.

4. Results

4.1. Impact Factors: Nutrient Concentration, SSC, and SST

4.1.1. Nutrient Concentration and SSC

By processing the GOBAF dataset and inverting the SSC, the monthly spatial averages for the study area were derived (Figure 3). From April to June, the nutrient concentrations remain relatively consistent both before and since the years 2007–2008. The mean and standard deviation ($x \pm s$) before and since 2007 are presented in Table 2. There are only minor changes in nutrient concentrations during these years (Figure 3, blue box). The most substantial alteration pertains to the concentration of PO_4 in the month of June, which increases by 45.28%. Following this increase, the PPMCC between the concentration of PO_4 and the MCA of *U. prolifera* reaches a value of 0.19.

Turbidity refers to the degree of cloudiness or haziness in water caused by suspended particles, which directly influences the availability of light to *U. prolifera* and subsequently impacts its growth [25]. In the study area, the average SSC values for June in the years 2007–2008 are lower than in other periods (Figure 3). However, there is an increasing trend in SSC both before and since 2007–2008. This enhanced turbidity could potentially attenuate the sunlight intensity, thereby limiting the growth of *U. prolifera*. The PPMCC values between the SSC for April, May, and June and the MCA of *U. prolifera* in different years are −0.05, −0.16, and −0.05, respectively.

Table 2. Statistics of nutrient concentration.

		NO_3 * (mmol/m^3)	PO_4 * (mmol/m^3)	Si * (mmol/m^3)
April	2003–2006	5.73 ± 0.31	0.0093 ± 0.0023	12.43 ± 0.58
	2007–2020	5.64 ± 0.48	0.0093 ± 0.0024	12.91 ± 0.63
May	2003–2006	7.63 ± 0.72	0.0059 ± 0.0009	12.61 ± 0.61
	2007–2020	7.57 ± 0.71	0.0058 ± 0.0014	13.22 ± 0.76
June	2003–2006	9.73 ± 0.57	0.0053 ± 0.0012	13.20 ± 0.42
	2007–2020	9.53 ± 0.95	0.0077 ± 0.0018	13.50 ± 0.86

* ($x \pm s$): x represents the mean value and s represents the standard deviation.

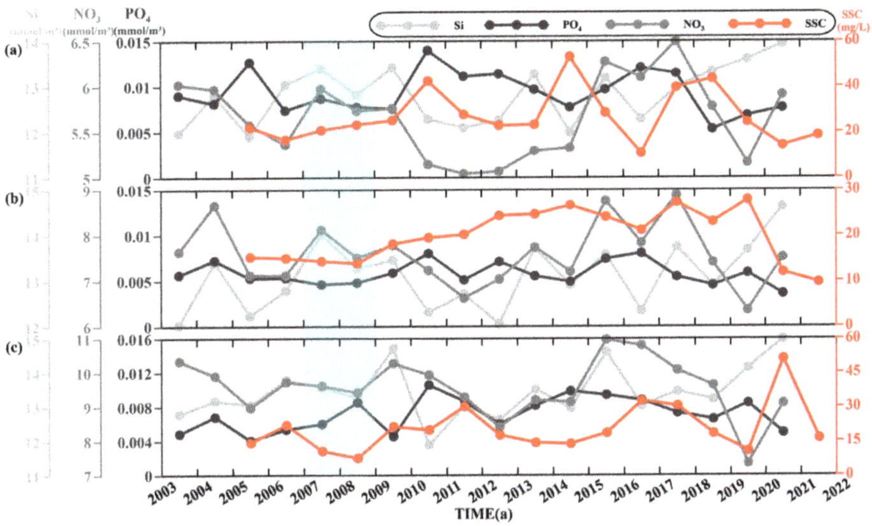

Figure 3. Diagram illustrating nutrient concentration and Suspended Sediment Concentration (SSC) from different months. Blue box indicates the period of the sudden large-scale bloom of *U. prolifera* in 2007–2008. (**a**) April. (**b**) May. (**c**) June.

4.1.2. SST

Following the processing of MODIS/AQUA data, the inversion of the SST, and the computation of PPMCC values, the spatial patterns of PPMCC between SST and *U. prolifera* in different time periods were derived (Figure 4). Prior to early May, *U. prolifera* occurrences are sporadic, making it challenging to calculate its coverage area accurately. By mid-May, the scale of *U. prolifera* starts to become more pronounced, although the PPMCC values for its coverage area are consistently negative. However, a certain positive correlation is observed in the area influenced by the YEWC. In late May, the PPMCC values increase markedly, particularly in the western SYS, reaching 0.5 in most areas. Coastal regions in NJ and the NJNC area, where *U. prolifera* originates, exhibit PPMCC values exceeding 0.75. By early June, there is a positive correlation between SST and the coverage area of *U. prolifera* throughout the SYS. However, from mid-June to late June, when *U. prolifera* generally reaches its MCA, the overall PPMCC values drop below 0.5. The maximum PPMCC value in the central SYS during this period is 0.618. In July, *U. prolifera* gradually disappears from the southern coast of the Shandong Peninsula, resulting in a predominantly negative correlation with SST. Therefore, from mid-May to late May, the YEWC and the NJNC contribute to the warming of the source area of *U. prolifera*, creating favorable SST conditions for its early growth.

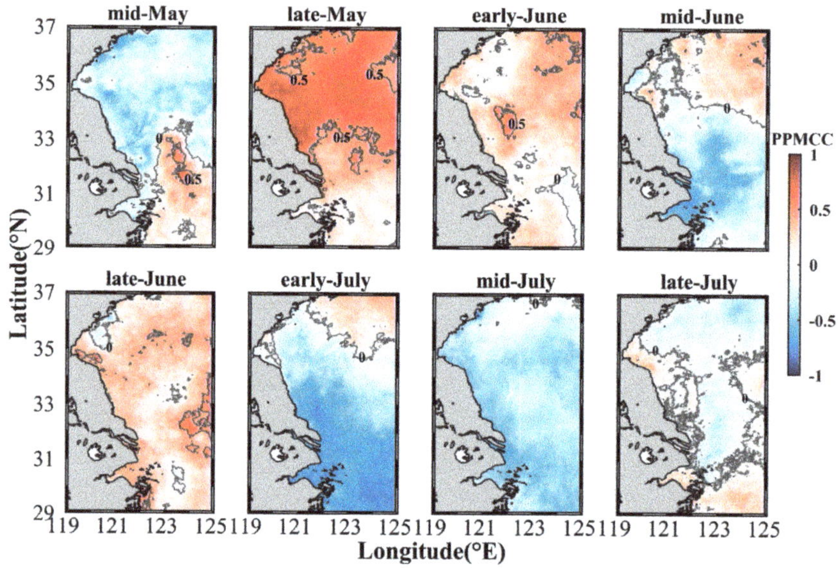

Figure 4. Spatial pattern of the Pearson product-moment correlation coefficient (PPMCC) between SST and the coverage area of *U. prolifera* from mid-May to late July.

4.2. SST Structure Derived from MODIS/AQUA

4.2.1. SST Structure Changes Before and Since 2007

Based on the processing of MODIS/AQUA data and the inversion of the SST, it is possible to derive the spatial pattern of the mean SST structure for different time periods (Figure 5), specifically 2003–2006 and 2007–2022. The main current systems in the SYS (black single-arrow dashed curves in Figure 1) were primarily determined based on previous studies [27–29] and the temperature structure (Figure 5). This study does not consider the SST in August as it exceeds the suitable temperature range for the growth of *U. prolifera* in the entire SYS region.

Figure 5 indicates that the SST structure in the SYS is primarily influenced by the YEWC, NJNC, YSCC, and Yellow Sea warm current (YSWC). From 2003–2006, the middle region of the SYS is strongly influenced by the YSCC, which hinders the northward extension of the YEWC and the NJNC from early April to early May. However, from mid-May onward, as the YSCC and YSWC weaken, the southeast extension of the YSCC retreats toward the coastline. Consequently, the influence of the YEWC and NJNC increases markedly, leading to a rapid warming of the overall SYS, with temperatures exceeding 15 °C. During this period, the YEWC mainly extends northwestward in the northern region of the Yangtze River Estuary. Additionally, the north-east branch (NEB) of the YEWC emerges following the intensification of the Yangtze River Plume in summer [39]. By early June, the YSCC has dissipated, and the SST in the SYS is primarily governed by the NJNC and YEWC. Notably, the influence area of the NJNC gradually extends northward, reaching 36°N by mid-July.

In the period from 2007–2022, the SST during all periods is higher than in the period 2003–2006 (Figure 5). NJNC experiences a notable enhancement in this period, reaching as far north as 36°N in early April. The retraction of the YSWC and YSCC toward the shore occurs earlier, specifically in early May, weakening their blocking effect on the northward intrusion of YEWC and allowing for its enhanced northwestward extension. From June to July, the YSCC disappears, and the SST in the SYS is primarily influenced by the NJNC and YEWC, similar to the conditions observed during the 2003–2006 period. Additionally, there has been a notable change in the path of the YSCC since 2007, with the main axis shifting eastward. This corresponds to the strengthening of the NJNC, further influencing the SST distribution in April.

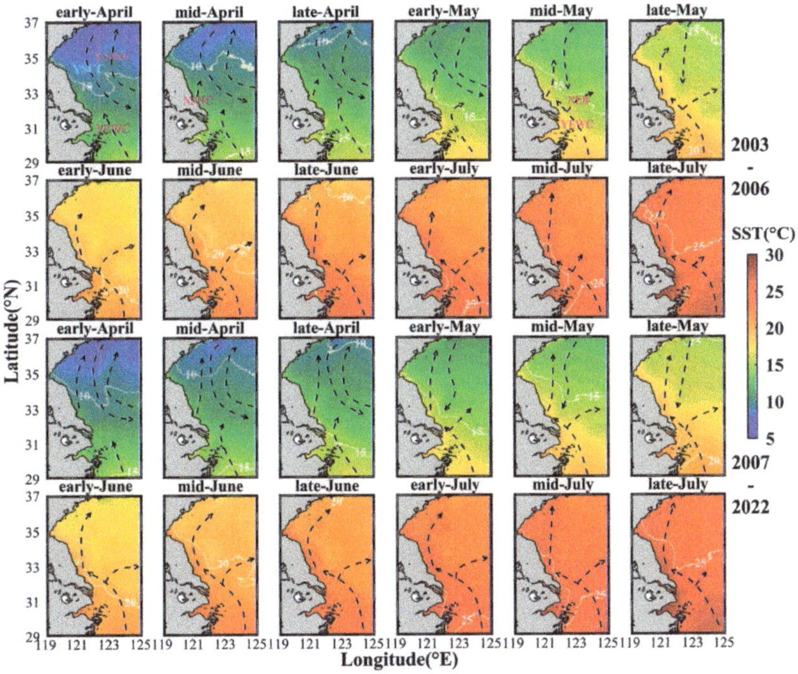

Figure 5. Spatial pattern of SST structure in the Southern Yellow Sea (SYS) from April–July. The black single-arrow dashed curves represent the primary current paths within the SST structure spatial pattern.

The results demonstrate the differences in the SST structure of the SYS from 2003–2006 and from 2007–2022. Following 2007, the increased YEWC and NJNC, characterized by an earlier northward invasion and larger impact area, contribute to a rapid expansion of the *U. prolifera* source area in NJ before mid-May. This region experiences suitable SST conditions for the early development and rapid growth of *U. prolifera*, acting as the key trigger for its large-scale bloom.

4.2.2. Changes in Different *U. prolifera* Bloom Cases

Initially, both the MCA and MDA were subjected to standardized calculations, followed by the application of quadrant segmentation to the functions derived from the MCA and MDA (Figure 2c). The standardized value was calculated using Equation (5) in Table 1.

The *U. prolifera* event exhibits notable interannual variations in bloom time and scale, which can be attributed to environmental changes and human interventions [40]. Thus, it is imperative to classify the intensity of the *U. prolifera* blooms. It is evident that 2009 and 2021 fall within the first quadrant, characterized by substantial-scale blooms. Correspondingly, in the fourth quadrant, two points with the maximum distance from the two points in the first quadrant, i.e., the points exhibiting the highest dissimilarity, can be identified, i.e., 2012 and 2020 are characterized by minor-scale blooms. Therefore, three typical bloom cases are delineated (Table 3), including the years devoid of *U. prolifera* occurrences from 2003–2006.

Table 3. Typical bloom cases of *U. prolifera*.

Case 1: No *U. prolifera*	Case 2: Minor-Scale (MDA Is Less Than 20,000 km^2 and MCA Is Less Than 500 km^2)	Case 3: Substantial-Scale (MDA Is Greater Than 50,000 km^2 and MCA Is Greater Than 1500 km^2)
2003–2006	2012, 2020	2009, 2021

Based on the average values during the same time periods across years within the same case, the spatial patterns of WR from early April to mid-May and from mid-May to late June of different *U. prolifera* bloom cases are obtained (Figure 6). During the early April to mid-May period, the WRs of the three typical bloom cases exhibit higher values near the coast than in the open sea, primarily influenced by the YSCC. The YEWC and NJNC also display high WR values. However, from mid-May to late June, the WR near the coast becomes lower than that in the open sea.

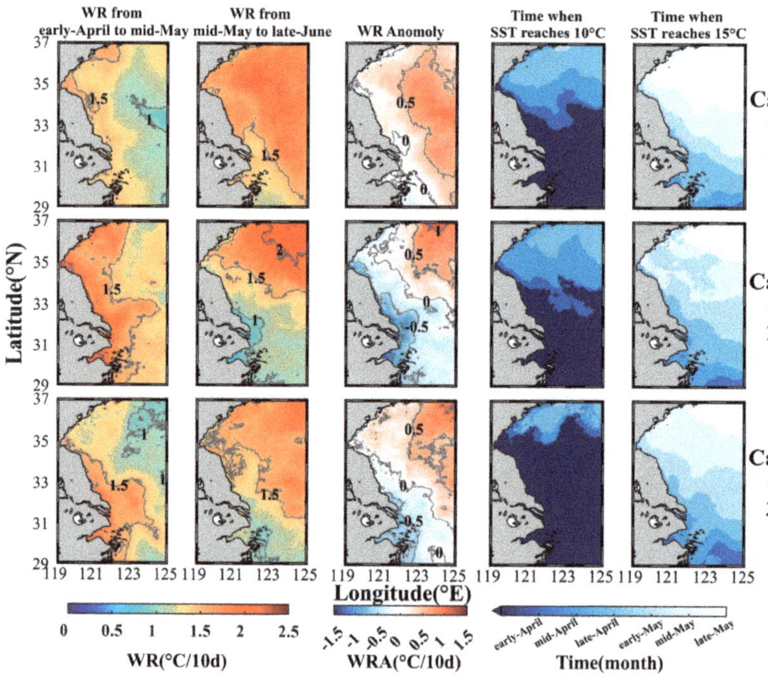

Figure 6. Spatial pattern of SST structure characteristics. Case 1: Period from 2003–2006, characterized by the absence of *U. prolifera* blooms. Case 2: The years of 2012 and 2020, characterized by small-scale blooms of *U. prolifera*. Case 3: The years of 2009 and 2021, characterized by large-scale blooms of *U. prolifera*. The warming rate anomaly (WRA) is calculated by subtracting the warming rate (WR) index from early April to mid-May from that of mid-May to late June for each respective case.

Notably, there is a difference in WR before and after mid-May when comparing Case 1 with Case 2 and Case 3. Therefore, the WRA is calculated in the different areas. Affected by the YEWC and NJNC, Case 1 exhibits a small positive anomaly. In contrast, Case 2 and Case 3 display clear negative anomalies, particularly Case 2, where the anomaly exceeds −0.5 °C/10 d at the Yangtze River estuary and the NJ shallow.

However, when calculating the SST WR, the initial SST (in early April) influences the results. For Case 2, despite having a smaller scale *U. prolifera* bloom than Case 3, the WR is higher before mid-May. Moreover, the area over which WR exceeds 1.5 °C/10 d extends northward to the southern coast of the Shandong Peninsula, rather than being limited to the NJ shallow. Previous studies have shown that early development of *U. prolifera* is initiated when SST reaches 10 °C and large-scale reproduction occurs when SST reaches 15 °C [19]. From the spatial patterns of the time when SST reaches 10 °C and 15 °C for each case (Figure 6), it is evident that SST reaches 10 °C notably earlier in Case 3 than in Case 1 and Case 2. At the time when SST reaches 10 °C, both the YSCC and YSWC play marked roles. For Case 1, in particular, the influence area of the YSCC is between 35°N and 36°N, closer to the shore than in Case 2 and Case 3, resulting in a stronger effect. In contrast, the NJNC only exhibits a noticeable influence area in Case 2 and Case 3. The SST reaches 15 °C at similar times in Case 2 and Case 3, and is notably earlier than in Case 1.

4.2.3. Multi-Year SST Along the Main Current Systems

To investigate the timing and causes of these SST structure changes, square areas with a spatial resolution of 0.5° × 0.5° were selected along the main axes of YEWC, NEB, NJNC, and YSCC (Figures 1 and 5, and Table 4). This study investigated the annual distribution of SST structures and the influence extents of current systems since 2003. This was undertaken to ensure that the selected regions effectively encapsulate the distinctive variations characterizing each current system. By calculating the WRA from early April to mid-May and from mid-May to late June, we obtained a measure of the WR difference during these two time intervals (Figure 7).

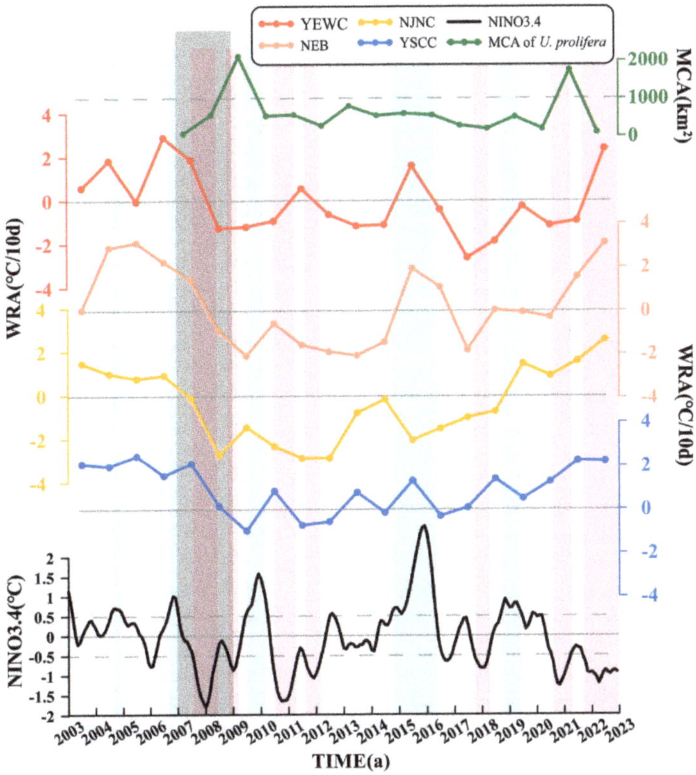

Figure 7. Black curve represents the continuous change in the three-month moving average of the NINO3.4 index from 2003–2022, which is an indicator of El Niño-Southern Oscillation (ENSO) activity.

Green curve represents the maximum coverage area (MCA) of U. prolifera, indicating the extent of U. prolifera blooms. Other curves represent the WRA from early April to mid-May and from mid-May to late June of each year, respectively, as calculated using (3). Blue box indicates an El Niño event, red box indicates a La Niña event.

Table 4. Square area of the main axis of the current system.

Current System	Latitude Range (°N)	Longitude Range (°E)
Warm current across the Yangtze River estuary (YEWC)	30.5–31	123–123.5
Northeast branch (NEB) of YEWC	32.5–33	123–123.5
Northern Jiangsu northward current (NJNC)	33.5–34	121–121.5
Yellow Sea coastal current (YSCC)	33–33.5	122.2–122.7

According to the curves in Figure 7, the WRA values for all current systems undergo a transition from positive to negative during 2007–2008 (gray box in Figure 7). The WRA of the YEWC exhibits a marked change around 2007–2008. Prior to 2008, the SST warms up faster from mid-May to late June (indicating a higher WR). However, after 2008, there is a shift, and the SST starts warming faster from early April to mid-May. This change suggests that, since 2008, the YEWC has contributed to an overall earlier increase in SST in the SYS region. The WRA of the NEB also changes from positive to negative in 2008, indicating a shift in the WR. After 2008, the WR becomes higher from early April to mid-May. This suggests that, although the spatial appearance of NEB is influenced by the intensification of the Yangtze River Plume in June [39], its SST structure is still primarily influenced by the main body of the YEWC. In summary, the changes observed in both the YEWC and NEB indicate that the SST structure in these regions is mainly influenced by the YEWC, with a notable shift occurring around 2007–2008. Additionally, starting from 2007, the WRA of the NJNC from early April to mid-May surpasses that from mid-May to late June. This characteristic remains consistent throughout the period 2007–2018. Both the YEWC and its NEB experience a resurgence in 2015. Furthermore, the influence area of the NJNC corresponds to the source region of *U. prolifera*, directly affecting its early growth. The WRA changes in the NJNC exhibit a closer correspondence to the MCA than the other currents. In terms of the YSCC, this generally indicates a higher WR from mid-May to late June. However, there is a decrease in WRA to a low value in 2008, suggesting that the YSCC plays a stronger role in slowing WR before mid-May than after mid-May. From June onward, the YSCC gradually weakens and eventually disappears, making it difficult to impede the northward intrusion of the YEWC and NJNC. As a result, the overall SYS experiences a rapid warming in SST.

5. Discussion

While the occurrence of green tide events closely relates to multiple factors [8], previous studies [15–17] and the results presented above indicate that nutrient concentration and SSC exhibit a minor variation around 2007–2008 and their correlations with *U. prolifera* are negligible in the SYS. Moreover, the significant increase in PO_4 levels in June primarily observed since 2010 further supports the conclusion that it was not the cause of the 2007 sudden large-scale bloom of *U. prolifera*.

Conversely, the intensity of *U. prolifera* blooms is highly correlated with SST, especially marked by a discernible shift in 2007–2008. Therefore, in this Chapter, we focus on discussing the impact of SST on the 2007 sudden large-scale bloom of *U. prolifera*.

5.1. Response of U. prolifera to SST Variations in the SYS Under Global Changes

Previous studies have indicated that the western Pacific Ocean is particularly susceptible to La Niña events, during which the region experiences elevated temperatures [41,42]. In consideration of the potential influence of global warming on the observed upward trend in SST within the current systems of the SYS over the past two decades, the NINO3.4

index provides a standardized measure to characterize global changes [43]. The NINO3.4 index represents the regional mean SST anomaly within the NINO3.4 region (170–120°W, 5°S–5°N). Its mean value was calculated based on data collected from 1981–2010. To identify El Niño and La Niña events, the *Identification method for El Nino/La Nina events* (GB/T33666-2017, https://openstd.samr.gov.cn/bzgk/gb/index, Accessed on 25 February 2023) was employed [44,45]. According to this method, an El Niño event is determined when the 3-month moving average of the NINO3.4 index exceeds 0.5 °C for a minimum duration of five months. Conversely, a La Niña event is identified when the 3-month moving average falls below −0.5 °C. Analyzing the frequency of El Niño/La Niña events since 1982 (Table 5), it is evident that the occurrence of La Niña events increases sharply from 0.32 to 0.44 between the years 2007 and 2022.

Table 5. Statistics of El Niño/La Niña events.

	No Events (Frequency)	El Niño (Frequency)	La Niña (Frequency)
1982–2006	7 (0.28)	10 (0.40)	8 (0.32)
2007–2022	4 (0.25)	5 (0.31)	7 (0.44)

On a larger timescale, the two notable temperature structure variations that occurred after 2007 are closely associated with El Niño/La Niña events, specifically during the periods 2007–2011 and 2015–2016 (Figure 7). In the 2007–2011 period, the NINO3.4 index exhibits substantial fluctuations, with a notably high occurrence of strong events, leading to prolonged global climate impacts [44]. In particular, the occurrence of the exceptional La Niña event in 2008 corresponds to changes in the SST structure within the SYS during 2007–2008. This event corresponds to the rapid warming stage of each current system shifting from after mid-May to before mid-May. Between 2012 and 2014, no El Niño/La Niña events were observed, resulting in a renewed elevation of WRA in the NJNC. During the super El Niño event in 2015, the WRA of the YEWC and its NEB experienced a renewed elevation.

Following 2016, the NINO3.4 index exhibits a notable decrease in volatility, and the frequency of El Niño/La Niña events returns to a more typical pattern. Concurrently, the WRAs within each current system demonstrate an upward trend, resembling the conditions observed prior to 2007. This trend is particularly evident for the YEWC, its NEB, and the NJNC. Hence, both the MCA and MDA of *U. prolifera* display a declining trend; the exception is the extensive *U. prolifera* bloom in 2021, which was triggered by a typhoon in May that led to a large influx of breeding rafts into the open sea [46]. The MCA notably decreases to less than 510 km^2, in contrast to the 2100 km^2 observed in 2009. This observation further underscores the impact and role of global changes in *U. prolifera* events.

5.2. Impact Factors of the 2007 Sudden Bloom of *U. prolifera*

The cultivation of *Porphyra yezoensis* along the NJ shallow dates back to the 1990s, with a sharp increase in its area observed since 2003. This increase creates favorable conditions for the attachment of *U. prolifera* to *Porphyra yezoensis* rafts [7]. However, there were no notable blooms or large-scale spread observed before 2007. Since its initial discovery in the waters of the SYS in 2007, *U. prolifera* has bloomed consecutively for 16 years [8]. The characteristic area parameter known as MCA (Figure 7, green curve), demonstrates substantial peaks in bloom intensity in the years 2009 and 2021, each exceeding 1000 km^2. Notably, since 2010, the MCA has exhibited a certain periodicity with a cycle of 3–4 years. While the *U. prolifera* biomass has been taken into account [24], it presents some differences from the bloom intensity as represented by the MCA. However, the available data in this paper are currently insufficient to facilitate a comprehensive discussion on these discrepancies.

Nevertheless, the reasons behind the 2007 sudden bloom and subsequent large-scale blooms in 2008 were previously unexplained. Based on the results and discussion above, the occurrence of *U. prolifera* events can be attributed to changes in the temperature structure within the SYS region under the influence of global climate change. The WRA values for all current systems undergo a transition from positive to negative during 2007–2008 (gray box in Figure 7). When there are sharp fluctuations in the global climate, particularly with a high frequency of strong El Niño/La Niña events, the SST structure within the SYS undergoes changes [3,47]. This leads to the early appearance of the YEWC and NJNC, occurring before mid-May and impacting a wider area. Consequently, the WR accelerates from early April to mid-May, surpassing the temperature rise observed from mid-May to late June (Figure 6). These conditions provide a more favorable SST environment for the initiation and early growth of *U. prolifera* and are the key factor in the 2007 sudden bloom of *U. prolifera* and its subsequent large-scale blooming since 2008.

The intense and prolonged La Niña event in 2007–2008 obviously influenced the temperature structure changes [48]. Conversely, when the global climate is relatively stable, the temperature structure within the SYS tends to return to its pre-2007 state (Figure 7). The northward intrusion of YEWC and NJNC strengthens after mid-May, which has a negative impact on the scale of *U. prolifera* blooms. Both the 2009 and 2021 *U. prolifera* bloom events in Case 3 occurred after La Niña events. Even the strongest El Niño event in 2015 had only a minor correlation to the scale of the *U. prolifera* event. This suggests that large-scale blooms of *U. prolifera* are more influenced by La Niña events, which increase the likelihood of such blooms occurring. Since the end of 2021, the global climate has been affected by a prolonged La Niña event, although its intensity is not as strong as that of the 2007–2008 event (Figure 7). It has triggered large-scale *U. prolifera* blooms in 2023, with the MCA of 998 km^2, which is the great evidence of the correlation between *U. prolifera* and global climate change.

6. Conclusions

Using MODIS/AQUA data and the GOBAF dataset covering the timeframe from 2003–2022, this study has analyzed the impact of SST, nutrient concentration, and SSC on *U. prolifera*. By integrating the influence of global changes, an investigation into the reasons behind the 2007 sudden bloom has been undertaken. The following conclusions can be drawn.

(1) Nutrient concentration and light intensity have only a minor correlation with *U. prolifera* in the SYS, with the maximum PPMCC values being 0.19 and −0.16, respectively. Instead, the occurrence and intensity of *U. prolifera* events in the SYS are primarily determined by the SST structure, as influenced by the current systems. Coastal regions in NJ and the NJNC area, where *U. prolifera* originates, exhibit PPMCC values exceeding 0.75. Due to the limited availability of in-situ data, further studies may require more extensive discussion of these factors.

(2) Changes in the SST structure serve as a pivotal trigger for the subsequent large-scale bloom of *U. prolifera* after 2007. Notable differences in the SST structure in the SYS exist between the periods 2003–2006 and 2007–2022. During the latter period, there is a marked increase in the intensity of the YEWC and NJNC. These current systems exhibit greater magnitudes and earlier northward invasion times, and cover larger areas. As a result, the WR in the *U. prolifera* source area accelerates before mid-May, particularly in NJ, where the SST becomes conducive to the early development and rapid growth of *U. prolifera*. Since 2008, the time at which the SST reaches 10 °C, a temperature suitable for *U. prolifera* to bloom, has emerged as a critical factor influencing the final scale of *U. prolifera* events. The earlier the SST reaches this threshold, the larger the eventual scale of the *U. prolifera* bloom tends to be.

(3) Since 2007, the changes in the SST structure in the SYS (where *U. prolifera* originates) can be attributed to the impact of global climate change. During the period 2007–2011, there were obvious fluctuations in the global climate, characterized by a high frequency of

intense El Niño/La Niña events. In particular, the intense and prolonged La Niña event from 2007–2008 resulted in alterations in the SST structure in the SYS. In this context, the NJNC directly influenced the *U. prolifera* source area, and its earlier northward intrusion contributed to the occurrence of *U. prolifera* events in 2007–2008. Subsequently, from 2016 onwards, the global climate exhibited a trend towards stabilization, and the temperature structure in the SYS gradually returned to the state observed before 2007. Consequently, the scale of *U. prolifera* events decreased during this period. Since the end of 2021, the global climate has been influenced by a prolonged La Niña event. Although its intensity is not as strong as that observed in 2007–2008, it is still triggered large-scale *U. prolifera* blooms in 2023, with the MCA of 998 km^2, which is the great evidence of the correlation between *U. prolifera* and global climate change.

Author Contributions: Y.P. and P.L. contributed to conception and design of the study. L.X., D.Y. and Q.F. organized the database. Y.P. and J.S. performed the statistical analysis. Y.P. wrote the first draft of the manuscript. P.L., J.S. and S.L. wrote sections of the manuscript. All authors have read and agreed to the published version of the manuscript.

Funding: This research was financially supported by the National Natural Science Foundation of China (grant numbers 42121005) and the Taishan Scholars Project (GXLI). We are also grateful to the editor and anonymous reviewers for their comments and suggestions that have helped us improve the final manuscript.

Data Availability Statement: The raw data supporting the conclusions of this article will be made available by the authors on request.

Acknowledgments: This research was supported by Ocean University of China and First Institute of Oceanography, MNR. We are also grateful to the editor and anonymous reviewers for their comments and suggestions that have helped us improve the final manuscript.

Conflicts of Interest: The authors declare no conflicts of interest.

References

1. Keesing, J.K.; Liu, D.; Fearns, P.; Garcia, R. Inter- and intra-annual patterns of Ulva prolifera green tides in the Yellow Sea during 2007–2009, their origin and relationship to the expansion of coastal seaweed aquaculture in China. *Mar. Pollut. Bull.* **2011**, *62*, 1169–1182. [CrossRef] [PubMed]
2. Li, Y.; Song, W.; Xiao, J.; Wang, Z.; Fu, M.; Zhu, M.; Li, R.; Zhang, X.; Wang, X. Tempo-spatial distribution and species diversity of green algae micro-propagules in the Yellow Sea during the large-scale green tide development. *Harmful Algae* **2014**, *39*, 40–47. [CrossRef]
3. Liu, D.; Keesing, J.K.; Xing, Q.; Shi, P. World's largest macroalgal bloom caused by expansion of seaweed aquaculture in China. *Mar. Pollut. Bull.* **2009**, *58*, 888–895. [CrossRef] [PubMed]
4. Duan, W.; Guo, L.; Sun, D.; Zhu, S.; Chen, X.; Zhu, W.; Xu, T.; Chen, C. Morphological and molecular characterization of free-floating and attached green macroalgae Ulva spp. in the Yellow Sea of China. *J. Appl. Phycol.* **2012**, *24*, 97–108. [CrossRef]
5. Fan, S.; Fu, M.; Wang, Z.; Zhang, X.; Song, W.; Li, Y.; Liu, G.; Shi, X.; Wang, X.; Zhu, M. Temporal variation of green macroalgal assemblage on *Porphyra* aquaculture rafts in the Subei Shoal, China. *Estuar. Coast. Shelf Sci.* **2015**, *163*, 23–28. [CrossRef]
6. Cui, J.; Zhang, J.; Huo, Y.; Zhou, L.; Wu, Q.; Chen, L.; Yu, K.; He, P. Adaptability of free-floating green tide algae in the Yellow Sea to variable temperature and light intensity. *Mar. Pollut. Bull.* **2015**, *101*, 660–666. [CrossRef]
7. Han, H.; Li, Y.; Ma, X.; Song, W.; Wang, Z.; Fu, M.; Zhang, X. Population differentiation in the dominant species (*Ulva prolifera*) of green tide in coastal waters of China. *Acta Oceanol. Sin.* **2022**, *41*, 108–114. [CrossRef]
8. Qi, L.; Hu, C.; Barnes, B.B.; Lapointe, B.E.; Chen, Y.; Xie, Y.; Wang, M. Climate and Anthropogenic Controls of Seaweed Expansions in the East China Sea and Yellow Sea. *Geophys. Res. Lett.* **2022**, *49*, e2022GL098185. [CrossRef]
9. Li, M.; Zhang, R.; Hong, M. Marine disaster assessment and management based on weighted Bayesian network. *Ocean Dev. Manag.* **2018**, *35*, 52–59.
10. Xiao, Y.; Zhang, J.; Cui, T. High-precision extraction of nearshore green tides using satellite remote sensing data of the Yellow Sea, China. *Int. J. Remote Sens.* **2017**, *38*, 1626–1641. [CrossRef]
11. Song, D. Temporal-Spatial Distribution and Countermeasures Study of Algae Disaster in the Bohai and Yellow Sea Based on Multi-Source Data. Ph.D. Thesis, University of Chinese Academy of Sciences (Yantai Institute of Coastal Zone Research, Chinese Academy of Sciences), Yantai, China, 2019.
12. Sun, X. Spatial and Temporal Variations and Response Mechanism of Green Tide and Chlorophyll-a Concentration. Master's Thesis, Ludong University, Yantai, China, 2018.

13. Zhang, G.; Wu, M.; Sun, X.; Zhao, D.; Xing, Q.; Liang, F. The inter-annual drift and driven force of *Ulva prolifera* bloom in the southern Yellow Sea. *Oceanol. Limnol. Sin.* **2018**, *49*, 1084–1093. [CrossRef]
14. Zhang, Z. Obsearvation and Analysis of the Coastal Current and Its Adjacent Current System in the China Offshore Waters. Ph.D. Thesis, Ocean University of China, Qingdao, China, 2016.
15. Shi, X.; Qi, M.; Tang, H.; Han, X. Spatial and temporal nutrient variations in the Yellow Sea and their effects on *Ulva prolifera* blooms. *Estuar. Coast. Shelf Sci.* **2015**, *163 Pt A*, 36–43. [CrossRef]
16. Wang, C.; Chen, C.; Su, R.; Luo, Z.; Mao, L.; Zhang, Y. Mechanism for the marked increase of *Ulva prolifera* in the south Yellow Sea: Light intensity, nitrogen, phosphorus, or co-limitations? *Mar. Ecol.-Prog. Ser.* **2021**, *671*, 97–110. [CrossRef]
17. Bai, Y.; Zhao, L.; Liu, J. The role of ecological factors in the progress of the green tide in the Yellow Sea. *Haiyang Xuebao* **2019**, *41*, 97–105. [CrossRef]
18. Hu, C.; Li, D.; Chen, C.; Ge, J.; Muller-Karger, F.E.; Liu, J.; Yu, F.; He, M.-X. On the recurrent *Ulva prolifera* blooms in the Yellow Sea and East China Sea. *J. Geophys. Res.-Ocean.* **2010**, *115*, C05017. [CrossRef]
19. Yu, Y.; Lin, J.; Jiang, J.; Hu, S.; Kang, C.-K.; Xu, N.; Li, Y. Environmental history affects the growth and photosynthesis of a green-tide macroalgae *Ulva prolifera*. *Aquac. Res.* **2022**, *53*, 2509–2517. [CrossRef]
20. Song, W.; Peng, K.; Xiao, J.; Li, Y.; Wang, Z.; Liu, X.; Fu, M.; Fan, S.; Zhu, M.; Li, R. Effects of temperature on the germination of green algae micro-propagules in coastal waters of the Subei Shoal, China. *Estuar. Coast. Shelf Sci.* **2015**, *163 Pt A*, 63–68. [CrossRef]
21. Sun, X.; Wu, M.; Xing, Q.; Song, X.; Zhao, D.; Han, Q.; Zhang, G. Spatio-temporal patterns of *Ulva prolifera* blooms and the corresponding influence on chlorophyll-a concentration in the Southern Yellow Sea, China. *Sci. Total Environ.* **2018**, *640–641*, 807–820. [CrossRef]
22. Bao, M.; Guan, W.; Yang, Y.; Cao, Z.; Chen, Q. Drifting trajectories of green algae in the western Yellow Sea during the spring and summer of 2012. *Estuar. Coast. Shelf Sci.* **2015**, *163 Pt A*, 9–16. [CrossRef]
23. Son, Y.B.; Choi, B.; Kim, Y.H.; Park, Y. Tracing floating green algae blooms in the Yellow Sea and the East China Sea using GOCI satellite data and Lagrangian transport simulations. *Remote Sens. Environ.* **2015**, *156*, 21–33. [CrossRef]
24. Hu, L.; Hu, C.; Ming-Xia, H.E. Remote estimation of biomass of *Ulva prolifera* macroalgae in the Yellow Sea. *Remote Sens. Environ.* **2017**, *192*, 217–227. [CrossRef]
25. Pan, Y.; Ding, D.; Li, G.; Liu, X.; Liang, J.; Wang, X.; Liu, S.; Shi, J. Potential Temporal and Spatial Trends of Oceanographic Conditions with the Bloom of *Ulva prolifera* in the West of the Southern Yellow Sea. *Remote Sens.* **2021**, *13*, 4406. [CrossRef]
26. Zhang, X.; Zhang, Z.; Lan, X.; Li, R. *Nan Huang Hai Qu Yu Di Zhi*; China Ocean Press: Beijing, China, 2013.
27. Li, G.; Qiao, L.; Dong, P.; Ma, Y.; Xu, J.; Liu, S.; Liu, Y.; Li, J.; Li, P.; Ding, D.; et al. Hydrodynamic condition and suspended sediment diffusion in the Yellow Sea and East China Sea. *J. Geophys. Res. Ocean.* **2016**, *121*, 6204–6222. [CrossRef]
28. Luo, Z.; Zhu, J.; Wu, H.; Li, X. Dynamics of the Sediment Plume Over the Yangtze Bank in the Yellow and East China Seas. *J. Geophys. Res.-Ocean* **2017**, *122*, 10073–10090. [CrossRef]
29. Liu, Z.; Gan, J.; Hu, J.; Wu, H.; Deng, Y. Progress on circulation dynamics in the East China Sea and southern Yellow Sea: Origination, pathways, and destinations of shelf currents. *Prog. Oceanogr.* **2021**, *193*, 102553. [CrossRef]
30. Zhu, P.; Wu, H. Origins and transports of the low-salinity coastal water in the southwestern Yellow Sea. *Acta Oceanol. Sin.* **2018**, *37*, 1–11. [CrossRef]
31. Sun, X. *Zhong Guo Jin Hai Qu Yu Hai Yang*; China Ocean Press: Beijing, China, 2006.
32. Qiao, F.; Wang, G.; Lü, X.; Dai, D. Drift characteristics of green macroalgae in the Yellow Sea in 2008 and 2010. *Chin. Sci. Bull.* **2011**, *56*, 2236–2242. [CrossRef]
33. Li, Y. *Structure and Dynamics of Ocean Circulation off the East Coast of China*; Chinese Academy of Sciences: Beijing, China, 2010.
34. Wen, S.; Song, X.; Tian, Y.; Zhang, Q.; Chen, C.; Gao, S.; Zhao, D. Technology and method for economic losses assessment of red tide disasters. *J. Catastrophol.* **2015**, *30*, 25–28. [CrossRef]
35. Xie, X.; Tao, A.; Zhang, Y.; Zeng, Y.; Zheng, J. The temporal and spatial distribution characteristics of typical marine disasters in Fujian Province. *Trans. Oceanol. Limnol.* **2018**, *4*, 21–30. [CrossRef]
36. He, E.; Ji, X.; Huang, H.; Wang, D.; Guo, M.; Gao, S.; Yang, J. The spatial and temporal distribution of *Ulva prolifera* in the Yellow Sea in recent 10 years. *Mar. Forecast.* **2021**, *38*, 1–11. [CrossRef]
37. Zhang, M.; Tang, J.; Dong, Q.; Song, Q.T.; Ding, J. Retrieval of total suspended matter concentration in the Yellow and East China Seas from MODIS imagery. *Remote Sens. Environ.* **2010**, *114*, 392–403. [CrossRef]
38. Zhang, L.; Qiao, L.; Zhong, Y.; Li, G.; Liu, Y. Alternating patterns of cross-shelf suspended sediment transport in the northern East China Sea in winter. *Mar. Geol.* **2023**, *463*, 107114. [CrossRef]
39. Zhang, Z.; Chen, Y.; Luo, F. Temporal and spatial distribution characteristics of Enteromorpha prolifera in the South Yellow Sea based on remote sensing data of 2014. *J. Huaihai Inst. Technol.* **2016**, *25*, 80–85. [CrossRef]
40. Xia, Z.; Yuan, H.; Liu, J.; Sun, Y.; Tong, Y.; Zhao, S.; Xia, J.; Li, S.; Hu, M.; Cao, J.; et al. A review of physical, chemical, and biological green tide prevention methods in the Southern Yellow Sea. *Mar. Pollut. Bull.* **2022**, *180*, 113772. [CrossRef] [PubMed]
41. Huang, Z.; Du, Y.; Wu, Y.; Xu, H. Asymmetric Response of the South China Sea SST to El Niño and La Niña. *J. Ocean Univ. China* **2013**, *12*, 272–278. [CrossRef]
42. Lin, H. Long-lead ENSO control of the boreal summer intraseasonal oscillation in the East Asian-western North Pacific region. *npj Clim. Atmos. Sci.* **2019**, *2*, 31. [CrossRef]

43. Buchmann, P.; DelSole, T. Week 3–4 Prediction of Wintertime CONUS Temperature Using Machine Learning Techniques. *Front. Clim.* **2021**, *3*, 697423. [CrossRef]
44. Hong, X.; Zhang, K.; Li, J.; Xu, Y.; Sun, M.; Jiang, J.; Xu, S.; Cai, Y.; Qiu, Y.; Chen, Z. Impacts of climate events on life history parameters of major commercial fishes in the Beibu Gulf, South China Sea in the last 15 years. *Front. Mar. Sci.* **2023**, *10*, 1234772. [CrossRef]
45. Fan, J.; Wei, S.; Liu, D.; Qin, T.; Xu, F.; Wu, C.; Liu, G.; Cheng, Y. Impact of ENSO events on meteorological drought in the Weihe River basin, China. *Front. Earth Sci.* **2023**, *11*, 1093632. [CrossRef]
46. Qi, L.; Wang, M.; Hu, C. Uncertainties in MODIS-Derived *Ulva prolifera* Amounts in the Yellow Sea: A Systematic Evaluation Using Sentinel-2/MSI Observations. *IEEE Geosci. Remote Sens. Lett.* **2023**, *20*, 1–5. [CrossRef]
47. Li, Y.; Feng, J.; Yang, X.; Zhang, S.; Chao, G.; Zhao, L.; Fu, H. Analysis of sea level variability and its contributions in the Bohai, Yellow Sea, and East China Sea. *Front. Mar. Sci.* **2024**, *11*, 1381187. [CrossRef]
48. Luo, J.; Behera, S.K.; Masumoto, Y.; Yamagata, T. Impact of Global Ocean Surface Warming on Seasonal-to-Interannual Climate Prediction. *J. Clim.* **2011**, *24*, 1626–1646. [CrossRef]

remote sensing

MDPI

Article

Behavior and Energy of the M2 Internal Tide in the Madagascar–Mascarene Region

Qian Wu [1], Jing Meng [1,*], Xu Chen [1,2] and Yulin Guo [1]

1 Frontiers Science Center for Deep Ocean Multispheres and Earth System and Key Laboratory of Physical Oceanography, Ocean University of China, Qingdao 266100, China; wq9212@stu.ouc.edu.cn (Q.W.); chenxu001@ouc.edu.cn (X.C.); guoyulin@stu.ouc.edu.cn (Y.G.)
2 Laoshan Laboratory, Qingdao 266100, China
* Correspondence: mengjing@ouc.edu.cn; Tel.: +86-135-7325-7403

Abstract: Internal tides serve as essential intermediate steps in the cascading of oceanic energy, playing a crucial role in oceanic mixing. M2 internal tides are the dominant tidal constituent in many oceanic regions, significantly influencing ocean dynamics. The Madagascar–Mascarene Region has high-energy internal tides, but due to a lack of observational studies, their propagation remains underexplored and warrants further investigation. In this study, we used satellite altimetry data to capture the sea surface manifestation of the first-mode M2 internal tides in the region. The results show that the Mascarene Plateau plays a key role in shaping the region's uneven internal tide distribution. The Mascarene Strait is the most intense generation area, with an east-west energy flux of 1.42 GW. Using the internal tidal energy concentration index, we decomposed the internal tidal beams, finding the primary beam oriented at 148°. These beams propagate outward for over 800 km, with a maximum distance exceeding 1000 km. Geostrophic currents intensify the northward refraction of westward-propagating internal tides in the Mascarene Basin, particularly between 15°S and 20°S.

Keywords: M2 internal tides; satellite altimetry; Madagascar–Mascarene; interference; refraction

1. Introduction

Internal tides are internal waves at tidal frequencies, generated by barotropic tide driving stratified seawater flowing over abrupt sloping topographies [1]. The global meridional overturning circulation represents a crucial process for mixing across densities, predominantly fueled by the movement of fluid parcels as they traverse density interfaces [2,3], playing an important role in the distribution of water nutrients and components such as dissolved oxygen [4]. With their extensive presence across the world's oceans and numerous generation locations, internal tides contribute to almost half of the energy necessary for maintaining the global meridional overturning circulation [5]. Concurrently, internal tides act as a pivotal link in the ocean's energy transfer process, supplying the kinetic energy needed for small-scale turbulence both in the deep ocean and at the ocean's surface [6,7].

M2 internal tides are the dominant tidal constituent in many oceanic regions, significantly influencing ocean dynamics [8–10]. A study utilizing satellite altimetry data demonstrated that M2 internal tides have strengthened by approximately 6% in energy over the past three decades, with notable increases observed in the Madagascar–Mascarene Region [11]. This spatially inhomogeneous strengthening indicates that M2 internal tides play a critical role in local marine dynamics and energy distribution in these waters.

The Madagascar–Mascarene Region in the southwestern Indian Ocean is one of the most intense areas for internal tides, as indicated by global-scale numerical simulations showing widespread high-energy internal tides near the Madagascar Sea with significant seasonal variability in internal tide strength [12,13]. The steep topography on the eastern side of Madagascar, adjacent to the Mascarene Basin with depths of 4000–5000 m, forms a deep-sea basin environment. Two flat-topped seamounts, about 500 m in depth, create the Mascarene Strait. This complex topography, combined with strong semidiurnal tides, is conducive to internal tide generation, making it one of the most intense tidal dissipation areas in the global ocean [14].

Numerous studies, both numerically modeled and observed through in situ measurements, have been conducted to investigate the internal tidal field within the Madagascar Region. MITgcm simulations show that the primary generation points for the semidiurnal internal tides are located in the northern and southern parts of Madagascar, with a total conversion rate reaching 19.9 GW [15]. Moored data from the eastern side of the Mascarene Strait indicate semidiurnal internal tides with wavelengths ranging between 140–150 km, propagating at an angle of 110° relative to the ridges on either side [16]. Satellite SAR images show the presence of large-amplitude internal solitary waves on both the eastern and western sides of the Mascarene Strait, with the spacing of leading waves similar to the first-mode semidiurnal internal tide peak [17,18]. Due to the lack of observational data, the internal tide phenomena in this region have not been well studied. A more detailed description of their distribution and propagation is needed. The various observed propagation patterns likely reflect different mechanisms of energy redistribution. Additionally, the complexities of refraction, reflection, and interference during the propagation of these waves require further investigation and clarification.

The sparse distribution of measurement values obtained from traditional in situ observations limits a thorough understanding of the internal tidal field. Satellite altimetry provides a practical method for large-scale observations. With an increasing number of satellites, techniques combining long-term altimetry data from multiple satellites for internal tidal fitting, such as radial basis function interpolation [19] and two-dimensional plane wave fitting [20], have been successful. Due to lower group velocities and higher shear, high-mode internal tides tend to dissipate near their generation region, while low-mode internal tides can propagate hundreds to thousands of kilometers [21,22]. Satellite altimetry is capable of capturing the phase-locked aspects of internal tides. Although the altimeter-derived signals might not fully capture the total energy of these tides, they nonetheless provide precise information on the genesis, propagation paths, and phase velocities of internal tides [23].

Satellite altimeters cover a wide area, and global maps of several major internal tides are now available. The diverse phenomena associated with the propagation of internal tides have been subjected to extensive research and analysis. The refraction and reflection of the M2 internal tides generated at the Macquarie Ridge during propagation have been investigated based on a clear distinction of propagation direction [24]. Using directional Fourier filtering, altimeter-derived data effectively capture the signature of internal tide standing waves to the west of the Luzon Strait [25]. The propagation and energy of the first-mode M2 internal tides in the Arabian Sea have been studied using satellite altimetry [26].

Limited observations have been conducted in the Mascarene Basin, capturing only a subset of the internal tide characteristics in specific directions. The advancement of satellite altimetry has enabled the systematic investigation of internal tide propagation within this region. This research focuses on the first-mode M2 internal tides in the deep waters of the Madagascar–Mascarene Region, excluding areas shallower than 500 m (Figure 1), utilizing a combination of satellite altimetry. The primary focus of our investigation is the examination of the propagation features of the M2 internal tides. The following sections are organized as follows. Section 2 presents the data and methods. Section 3 gives our results. In Section 4, discussions are carried out on some refraction, interference, and other propagation phenomena of internal tides in the study area. Section 5 summarizes the key conclusions concisely.

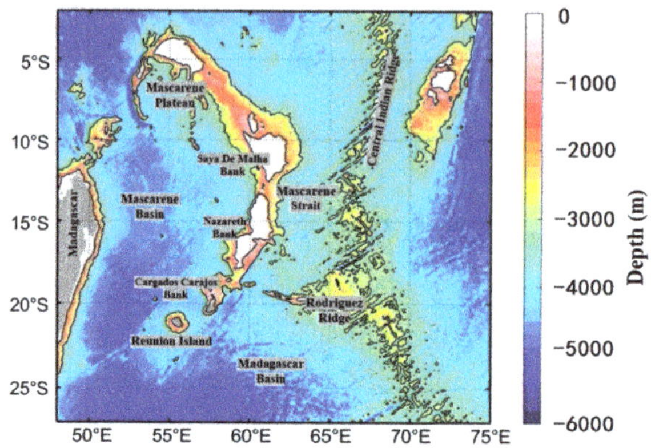

Figure 1. Bathymetric topography of the Madagascar–Mascarene Region, delineating two isobaths at 500 m and 3000 m to illustrate the underwater terrain.

2. Materials and Methods

Our study area ranges from 2°S–27°S and 48°E–75°E, and the bathymetric topography is shown in Figure 1. The two isobaths are 500 m and 3000 m, and the terrain shallower than 500 m is not our focus. The bathymetric data used for mapping in this article come from ETOPO2022 via https://doi.org/10.25921/fd45-gt74 (accessed on 7 September 2023).

The dataset utilized in this research comprises a collection of sea surface height measurements derived from both exact-repeat and non-repeat orbit satellite altimeters, spanning from 2012 to 2022. The along-track altimeter data used are available in Copernicus Marine Environment Monitoring Service (CMEMS) via https://doi.org/10.48670/moi-00146 (accessed on 14 February 2023). The specific satellite names and duration of the dataset are detailed in Figure 2. All satellite data have been processed with standard corrections for atmospheric effects, sea state bias, and geophysical effects [27].

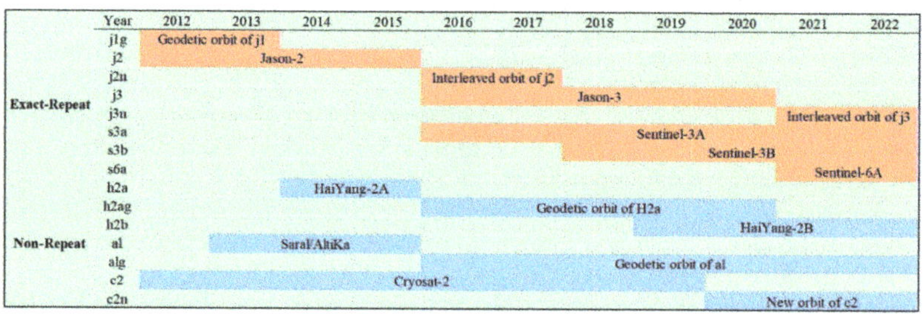

Figure 2. A diagrammatic representation of the satellite data's temporal span, which is not detailed to the monthly level.

Two-dimensional plane wave fitting [20,28,29], the method we used, assumes that the wave field within each small fitting window is uniform, an assumption that clearly does not hold at the ocean–land boundary. Therefore, this study does not include internal tide information from locations with water depths less than 500 m, nor does it utilize satellite altimetry values from shallow water areas [23]. This is also the reason why the bathymetric map exhibits blurred topography in the shallow water regions (white areas in Figure 1), which are not the focus of our study.

The harmonic analysis conducted at discrete points along the satellite's ground track is capable of capturing the internal tides that are along with the satellite's orbit, and the signal between the tracks is missing. Since internal tides in the actual ocean propagate in areas other than satellite tracks, 2D plane wave fitting has been introduced to provide a more comprehensive analysis [30].

A key feature of 2D plane wave fitting is its use of surrounding data within a square window to determine the internal tide signal at a specific location. The most suitable fitting window size for the first-mode M2 internal tides is 160 × 160 km, with the fitting location at the center of the window [31], and each window contains up to 47,000 data points in this study. The center of the window is oriented with the east as the x-axis positive direction and the north as the y-axis positive direction. The downloaded satellite data includes both sea surface height and the corresponding timestamp and location. Each datum in the window can be viewed as a function of position and time $SSH(x, y, t)$, and each set of data is fitted to a least-squares fit according to the following equation:

$$SSH(x, y, t) = \sum_{m=1}^{5} A_m \cos(kx \cos\theta_m + ky \sin\theta_m - \omega t - \phi_m) \tag{1}$$

where θ_m is the direction chosen for the fit. x, y, and t represent the position information in the horizontal direction and the time information, respectively. k is the wavenumber of the internal tides. In this paper, the theoretical wavelength is obtained by solving the Sturm–Liouville equation. A_m and ϕ_m are the amplitude and phase of the internal tides of this centroid in the direction of θ_m, respectively.

Two-dimensional plane wave fitting method applied in this paper is Zhao's original approach [31,32], which extracts five different directional wave components iteratively at each location. As illustrated in Figure 3, the method is demonstrated at a specific location (11.5°S, 56.5°E), represented by the gray square in Figure 3a, covering a 160 km square region. To obtain the internal tide information at this location, all altimetry data within the gray square are used.

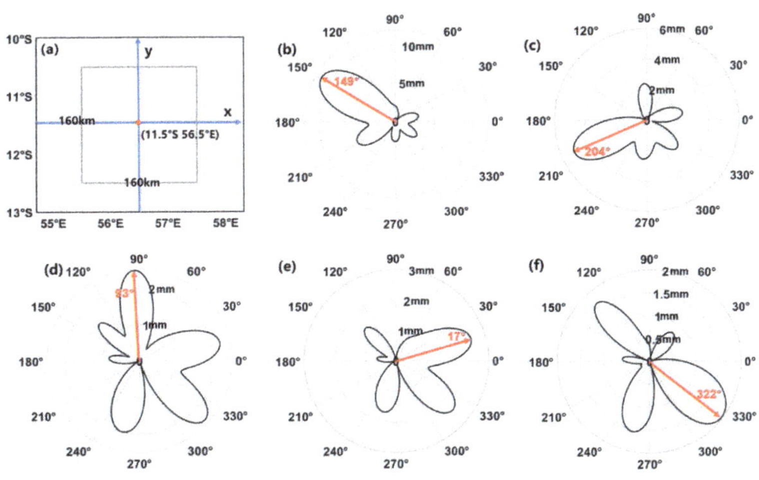

Figure 3. Iterative two-dimensional plane wave fitting process diagram. (**a**) The case study location is positioned at 11.5°S, 56.5°E, with the fitting window set to a 160 km square. (**b**–**f**) represent the amplitude and direction of the 1st to 5th wave components, respectively.

Generally speaking, the method extracts five tidal-related fluctuation components at each position through five iterations of fitting. The fitting process is performed at 1° intervals to determine the amplitude and phase of the wave in each direction. After the first iteration, the direction with the maximum amplitude (149°) is identified as the first wave component (Figure 3b). To isolate subsequent wave components, the first wave's data are subtracted before the next fitting iteration. This procedure is repeated up to the fifth wave component, with each new wave extracted after subtracting the data from the previous wave.

It has been proved that the iterative two-dimensional plane wave fitting method can effectively enhance the resolution of the propagation direction of internal tides [33].

The structure of the internal tide in the ocean is determined by the buoyancy frequency $N(z)$ and water depth. The ocean stratification data are from the World Ocean Atlas 2018 (WOA2018). The modal structure of the vertical displacements is described by the Sturm–Liouville equation [34–36] as follows:

$$\frac{d^2\Phi(z)}{dz^2} + \frac{N^2(z)}{C^2}\Phi(z) = 0 \tag{2}$$

subject to the boundary conditions $\Phi(0) = \Phi(-H) = 0$, where $\Phi(z)$ denotes the vertical eigenfunction, and C denotes the eigenspeed of the internal wave. The relationship between the horizontal structure and vertical structure is given by

$$\Pi(z) = \rho_0 C_n^2 \frac{d\Phi(z)}{dz} \tag{3}$$

where $\Pi(z)$ represents the horizontal structure, and C_n is the eigenspeed of the internal tide in the nth mode, referring to the phase velocity of the internal wave in a nonrotating fluid. Our study focuses only on the first mode, so n is 1. ρ_0 is the reference seawater density. Under the influence of the Earth's rotation, the internal wave dispersion relation is

$$\omega^2 = k^2C^2 + f^2 \tag{4}$$

where ω is the tidal frequency; f is the inertial frequency; and k is the wavenumber. Therefore; the phase speed can be expressed as

$$C_p = \frac{\omega}{k_n} = \frac{\omega}{(\omega^2 - f^2)^{1/2}}C \tag{5}$$

The vertical integral energy flux of internal tide can be expressed as follows [37]:

$$F = \frac{1}{2}A^2\left[\frac{\rho_0 g^2 \omega k_n}{\omega^2 - f^2}\int_{-H}^{0}\Pi^2 dz\right] \tag{6}$$

where A is the sea surface amplitude of the internal tide, and g is the gravitational acceleration.

3. Results

In this paper, we obtained a gridded first-mode M2 internal tide map with a resolution of 0.1° using along-track satellite altimetry data (Figure 4a). All the results presented in this study are based on the internal tide field that we have derived. To verify the accuracy of our results, we compared our map with the ZHAO model [29], depicted in Figure 4b. The ZHAO model is a globally recognized internal tide model based on multi-year satellite altimetry data, overcoming the limitations of traditional methods to generate a gridded internal tide field. The comparison reveals a similarity in the overall distribution, with high-energy internal tidal beams radiating outward on both sides of the Mascarene Strait.

Due to the interference of internal tides from different generation areas, distinct interference patterns are observed on both sides of the Mascarene Strait. The phenomenon of internal tide interference occurs when two or more internal tide waves interact under certain conditions, leading to the superposition of their oscillations and the formation of new wave patterns. During interference, the isophase lines exhibit noticeable curvature and deformation. In Figure 4a, a distinct white isophase line can be observed between the blue and red areas, showing significant undulations across most of the region. This clearly indicates the widespread presence of internal tide interference.

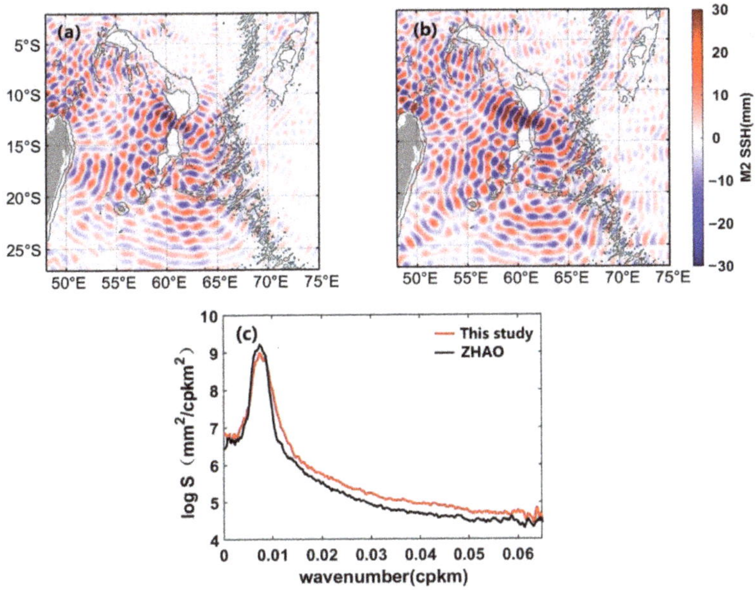

Figure 4. (**a**) Sea surface height (SSH) of the M2 internal tides in this study. (**b**) SSH of the M2 internal tides from ZHAO. (**c**) Directional wavenumber spectrums.

The wavenumber spectrum (Figure 4c) indicates that our result is also closely aligned with the ZHAO model in terms of energy. The along-track altimetry data used in this study have been filtered for mesoscale sea level anomalies [38], resulting in minor differences in the overall distribution and energy of the internal tides.

Figure 4a illustrates the SSH field of internal tides, showing a nonuniform spatial distribution. Pronounced interference patterns of internal tides are observable on the eastern and western flanks of the Mascarene Strait, within the Mascarene Basin. These phenomena indicate the presence of numerous internal tide-generation areas within this oceanic region. Internal tides originating from various source areas disperse along distinct pathways, with their interactions culminating in interference phenomena. The degree of multiwave interference R_{in} (Figure 5) is useful for further quantifying the interactions of internal tides [39] and is expressed as follows:

$$R_{in} = \left|\sum F_m\right| / \sum |F_m|, \tag{7}$$

where F_m (m = 1, 2, 3, 4, 5) represents the energy flux of the five wave components. The range of R_{in} varies from 0 to 1, with 0 indicating complete standing waves and 1 indicating complete progressive waves. The presence of a full progressive wave in the region indicates a prevailing internal tide, while the occurrence of a complete standing wave signifies the interference of two or more internal tides. The internal tides on both sides of the Mascarene Strait and the north and south sides of the Rodrigues Ridge are close to complete progressive waves. A distinct standing wave pattern, extending over 1600 km, is observed in the central Mascarene Basin (Figure 5).

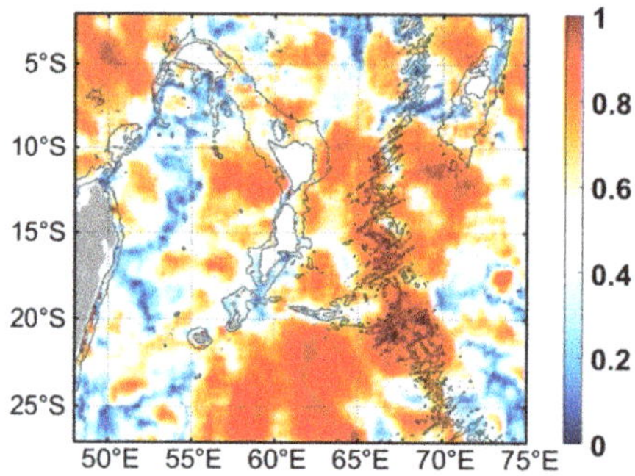

Figure 5. The degree of multiwave interference R_{in}, with 0 indicating the presence of complete standing waves and 1 denoting the presence of complete progressive wave.

To further elucidate the source regions, the barotropic tidal body force (Figure 6) is used to represent the generation source region of internal tides, expressed as [40,41]:

$$F_{body} = \frac{Q \nabla H}{\omega H^2} \int_{-H}^{0} N^2 z dz, \tag{8}$$

where H represents the water depth, ∇H is the bathymetric slope calculated based on ETOPO2022, Q is the barotropic tidal volume transport from TPXO7 [42], and ω is the M2 tidal frequency. N^2 is the square of the buoyancy frequency calculated based on the WOA2018 temperature and salinity data.

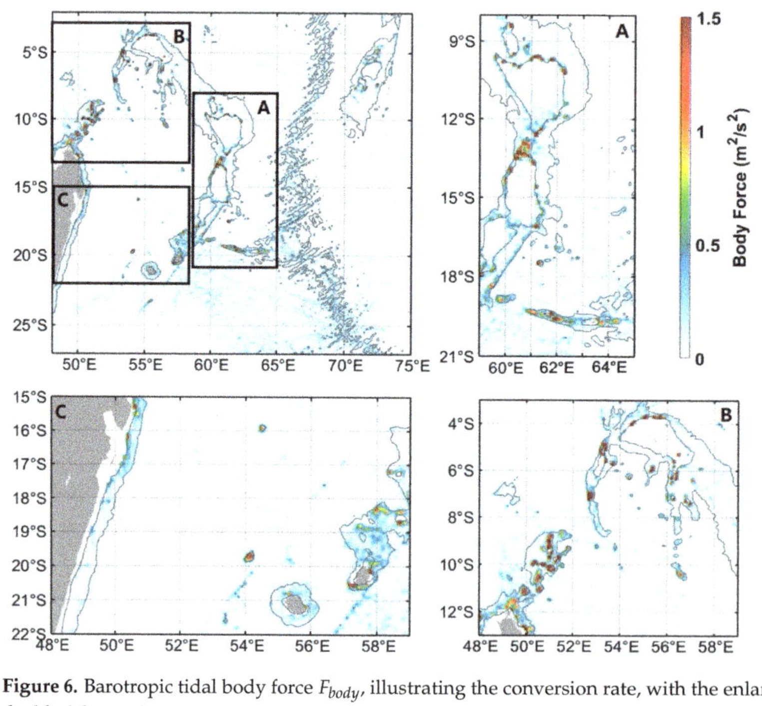

Figure 6. Barotropic tidal body force F_{body}, illustrating the conversion rate, with the enlargement of the black boxes (**A–C**).

Figure 6 shows the most significant source regions including the Mascarene Strait, northern Mascarene Plateau, Nazareth Bank, northern Madagascar, and Rodrigues Ridge.

The combination of Figures 5 and 6 indicates that near the majority of source regions, internal tides manifest as progressive waves. These waves progressively interact with those emanating from other source regions, leading to the formation of standing waves at locations removed from the generation points. The first four source regions create a semi-enclosed configuration, characterized by the prevalence of progressive waves near the generation locales and the emergence of standing waves in the middle of the basin.

Figure 7a,b illustrate the sea surface amplitude and energy flux distribution of the M2 internal tides. The maximum coherent internal tide energy is observed at the entrance of the Mascarene Strait, where peak surface amplitudes reach up to 50 mm. In the comprehensive field of internal tides with all propagation directions, the internal tidal energy radiated from the Mascarene Strait is 0.84 GW to the west and 0.58 GW to the east (integrated along the green lines in Figure 7b).

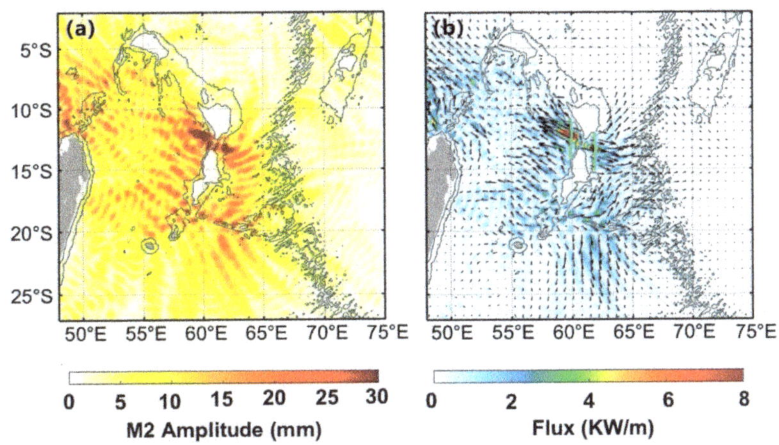

Figure 7. (**a**) M2 internal tide sea surface amplitude (**b**) M2 internal tide vertically integrated energy flux, where color indicates magnitude and arrows indicate direction. The energy is integrated along the green line.

The distribution of internal tides exhibits significant differences on either side of the Mascarene Plateau. The internal tides generated on the east side propagate westward across the basin, encountering and interfering with internal tides from the opposite side, resulting in the formation of standing wave structures characterized by numerous wave crests and troughs in the amplitude distribution. In contrast, on the west side of the Mascarene Plateau, the amplitude predominantly displays beam structures.

4. Discussion

4.1. Internal Tidal Beams from the Mascarene Strait

An individual source region can generate internal tides in multiple directions, and a single area can also contain a multitude of internal tides with varying propagation directions. Taking the Mascarene Strait as an example, to ascertain the propagation directions of the internal tides generated within the strait, it is necessary to separate the superimposed components. We introduce an internal tidal energy concentration index (Figure 8c), which can indicate the direction of concentrated energy. This index is derived from the spectral energy flux near the generation area (black box in Figure 8a, 11–14°S, 60–62°E). The amount of the same direction in the five wave components is summed and normalized by the sum energy in the small region, which is the energy share of that direction, and the formula is as follows:

$$c_\theta = \frac{\sum_{m=1}^{5} F_\mathrm{m}(direction = \theta)}{\sum_{m=1}^{5} F_m} \quad (9)$$

where F_m (m = 1, 2, 3, 4, 5) represents the flux of the mth wave component. c_θ can respond well to the direction of the internal tidal beams generated from the Mascarene Strait. Referring to Figure 8c, we identified three directional sectors: 0–60°, 300–360°, and 100–170°, which correspond to the internal tidal beams labeled as 1, 2, and 3 in Figure 8a,b,d, respectively. The internal tidal beam energy is minimal in the southwest, as shown in Figure 8c, and due to the incomplete structure, it is not presented here.

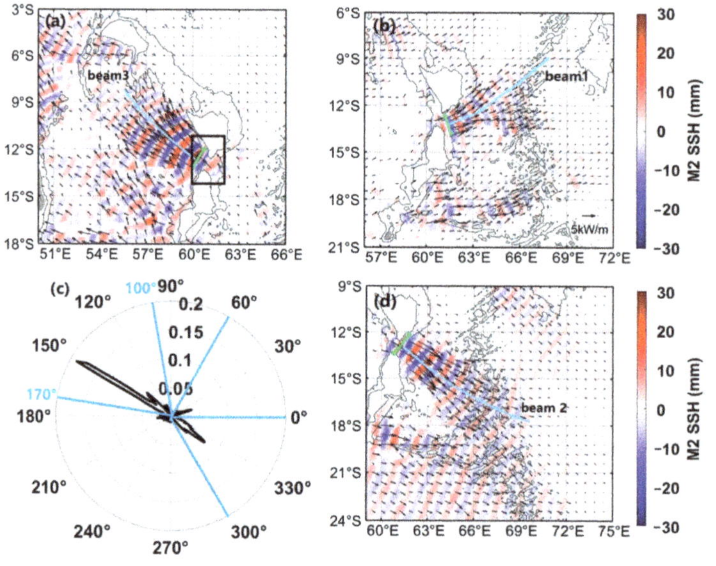

Figure 8. (**a**,**b**,**d**) The SSH of internal tides propagating in the directions of 100–170°, 0–60°, and 300–360°, with the green line representing the integral line of energy and the blue line illustrating the representative paths of tidal beams. The black line box in (**a**) is the region used to calculate the direction of the energy concentration. (**c**) Internal tidal energy concentration index, where the eastward orientation is set at 0° and rotation in a counterclockwise direction is considered positive.

The internal tides generated from the Mascarene Strait radiate eastward and westward, and the propagation directions of the internal tidal beams correspond well with the energy concentration direction. The principal axes of beams 1, 2, and 3 are 17°, 322°, and 150°, respectively. The propagation distances (up to the point where the amplitude has attenuated to 1% of its value in the generation area) and energy magnitudes of the three internal tidal beams (integrated along the green line in Figure 8) are explicitly provided in Table 1, with all three beams able to propagate more than 800 km outward. The most energetic beam is beam 3, which travels northwestward, with an energy flux of up to 0.43 GW. Next, beam 2 propagates in the opposite direction, with an energy flux of 0.19 GW. The blue lines in Figure 8a,b,d represent the representative paths based on the phase gradients. The curvatures of these blue lines indicate that the propagation directions of the three beams change with increasing propagation distance due to refraction during propagation. Beams 1 and 3 propagate toward lower latitudes, tending towards being perpendicular to the equator, while beam 2 propagates towards higher latitudes, tending toward being parallel to the equator.

Table 1. Internal tidal beams from the Mascarene Strait.

Beam	Direction (°)	Flux (GW)	Distance (km)
1	17	0.11	869
2	322	0.19	1089
3	150	0.43	823

4.2. The β Effect

For a quantitative analysis of propagation dynamics, we assessed the changes in propagation direction along the representative paths, as shown in Figure 9a. The orange line represents phase changes, and the blue line indicates the propagation direction. The periodic increments observed in the phases of beam 2 confirm the appropriateness of the selected representative path.

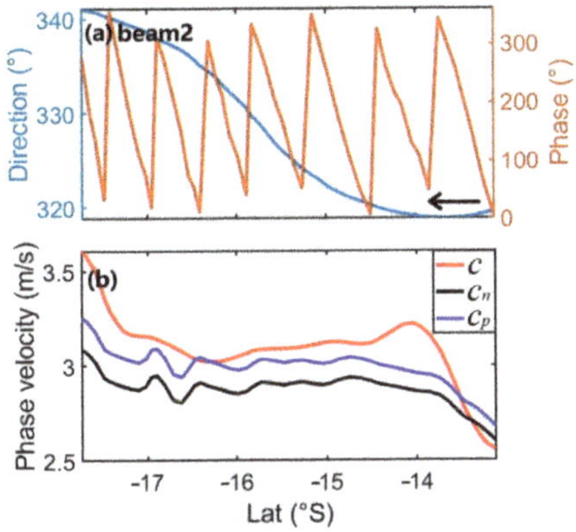

Figure 9. (**a**) The variation of the propagation direction (blue line, left y-axis) and phase (orange line, right y-axis) of beam 2, with the black arrow signifying the direction of the increasing phase. (**b**) The variation in the phase velocity of beam 2, where C indicates the observed phase velocity (gradient of the phase), C_n is the eigenspeed, and C_p is the theoretical velocity from Formula (5).

The observed velocity of the internal tide along the representative path can be derived from the gradient of the phase. Figure 9b compares the observed velocity (red line, C), eigenspeed (black line, C_n), and theoretical velocity (blue line, C_p) along the representative path. Beam 2 propagates 1089 km for 8 cycles, with an approximate 20° deviation from its intended propagation direction. If we consider only the eigenspeed C_n and disregard the influence of the Earth's rotation, the beam would be expected to cover a distance of only 1004 km over the same number of cycles, which is notably less than the observed distance. This discrepancy indicates that while variations in oceanic conditions along the path can affect the velocity of the internal tide, focusing solely on changes in eigenspeed is inadequate to account for the observed directional deviation.

The phase speed of internal tide varies with latitude, and, considering the β effect, yields a theoretical phase speed of C_p. With the β effect accounted for, the theoretical propagation distance for eight cycles is revised to approximately 1047 km, which more accurately matches the actual observed distance. The β effect enables the internal tides to achieve the refraction angle indicated by the SSH patterns.

4.3. Multiwave Interference

Two distinct internal tidal beams, originating from the east side of the Mascarene Strait, were observed to propagate into the Indian Ocean. A comprehensive characterization of their propagation dynamics is provided in the preceding section. A notable feature of the eastern of the Mascarene Strait is the multiple interference among three internal tidal beams.

In addition to the two internal tidal beams from the Mascarene Strait (Figure 10b), the Rodrigues Ridge serves as another strong source of internal tides in the region, generating two internal tidal beams to the north and south. The northward-advancing internal tides, as depicted in Figure 10c, intersect with the internal tides emanating from the Mascarene Strait at approximately 16°S.

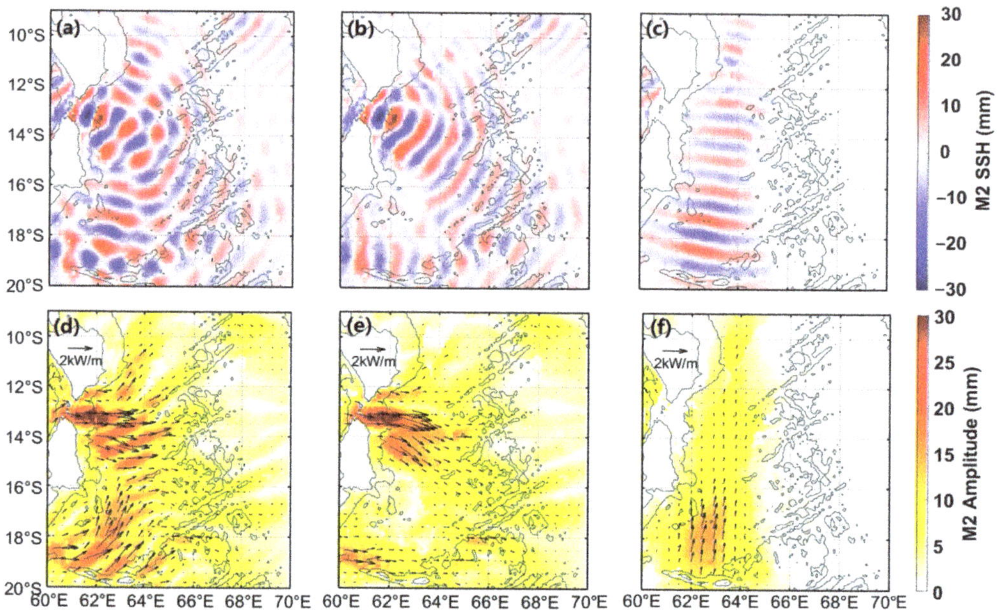

Figure 10. (**a**–**c**) The SSH of internal tides in the 300–120°, 300–60°, and 60–120° directions, respectively. (**d**–**f**) are corresponding sea surface amplitudes for (**a**), (**b**), and (**c**), respectively, where the arrows depict the direction of energy flux and the lengths of the arrows are proportional to the magnitude of the energy flux.

At the intersection, internal tides traveling in divergent directions engage in dynamic interactions, which can result in either constructive interference or destructive interference (Figure 10a). This interaction gives rise to a multiplicity of beam-like structures, as visualized in the amplitude distribution map presented in Figure 10d. The presence of beam structures within the distribution of the internal tides is indicative of the interference phenomena occurring between internal tides that originate from distinct sources [43,44].

The phenomenon of mutual interference was more distinctly observable in the Madagascar Basin. The southward propagating internal tides generated from the Rodrigues Ridge (Figure 11b) interact with the southeastward propagating internal tides from the eastern coast of Madagascar (Figure 11c) in the Madagascar Basin, where the combined internal tides with different directions are shown in Figure 11a. The amplitude map after the interference shows clear beam structures with a southeastward shift. The southeastward-propagating internal tides exert a discernible influence on the amplitude distribution, which transitions from an orientation that is nearly aligned with the due south direction to one that is modulated with a southeastward inclination. As the southeastward propagating internal tides reach the Madagascar Basin, their energy significantly decreases due to long-distance propagation, resulting in a minor impact on the overall propagation direction.

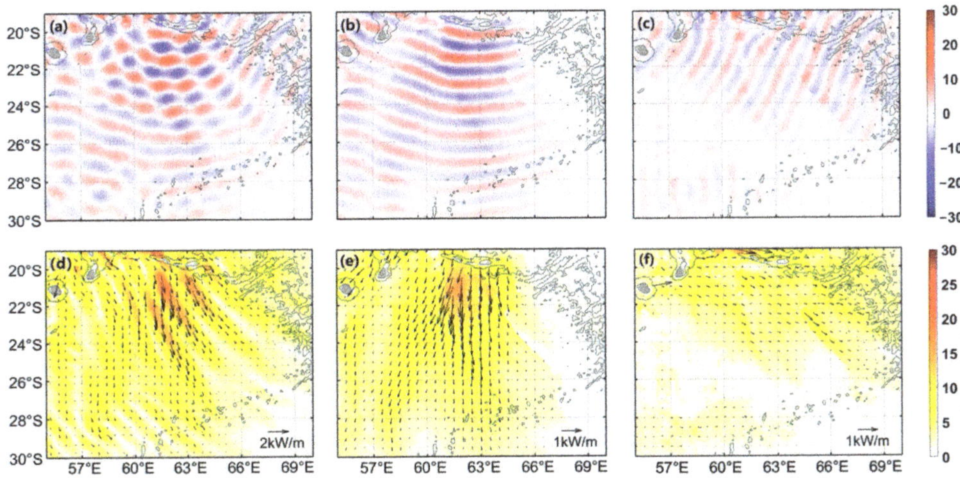

Figure 11. (**a–c**) The SSH of internal tides in the 210–60°, 300–60°, and 210–300° directions, respectively. (**d–f**) are corresponding sea surface amplitudes for (**a**), (**b**), and (**c**), respectively, where the arrows depict the direction of energy flux and the lengths of the arrows are proportional to the magnitude of the energy flux.

4.4. An Internal Tide Triangle

Two distinct internal tidal beams, originating from the east side of the Mascarene Strait, were observed to propagate into the Indian Ocean. A comprehensive characterization of their propagation dynamics is provided in the preceding section. In this section, we concentrate on the area within the Mascarene Basin bounded by 15°S and 22°S, where a distinct triangular pattern of internal tides is observed near 18°S along the eastern shelf of Madagascar (Figure 12a). To more clearly illustrate this triangular structure, we present internal tides with different propagation directions in various colors in Figure 12b. The flux directionality indicates that internal tides generated at Nazareth Bank propagate westward approximately 850 km to the east coast of Madagascar (indicated by red arrows). There are internal tides propagating from the east coast of Madagascar back towards the Mascarene Basin (indicated by blue arrows). Réunion Island also constitutes a relatively weaker generation region, yielding internal tidal beams that propagate northward (indicated by black arrows).

We can distinguish internal tides with different propagation directions in this region with the energy concentration proposed in this paper (Figure 13d). There are four wave propagation directions in this region, and we focus on internal tidal beams propagating in directions of 120–210° (Figure 13a), 60–120° (Figure 13b), and 270–60° (270–360° and 0–60°, Figure 13c). The internal tidal beam emanating westward from the Nazareth Bank is the most expansive and potent, with an energy magnitude of 0.29 GW around its generation zone and a width of approximately 400 km. Locally generated internal tides cross the Mascarene Basin to reach the east coast of Madagascar.

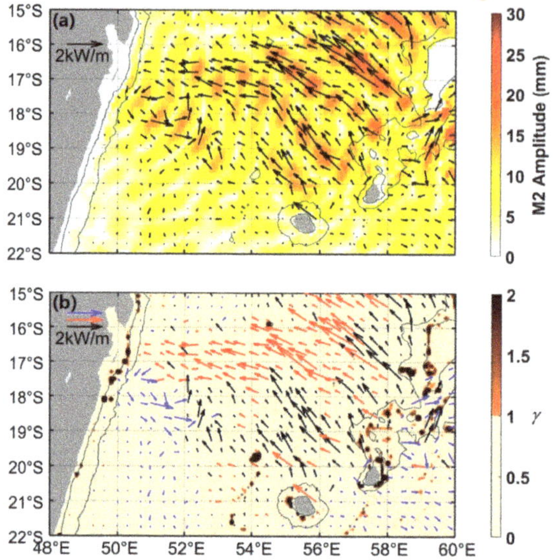

Figure 12. (**a**) Sea surface amplitude of internal tides in the 270–210° direction (excluding waves within the 210–270°), where arrows depict the energy flux directions, and the lengths of the arrows are proportional to the energy flux magnitudes. (**b**) The topographic critical parameter, with γ exceeding 1 indicating regions that are supercritical and prone to wave reflection. The arrows represent the directions of energy flux, with arrow colors distinguishing the propagation directions.

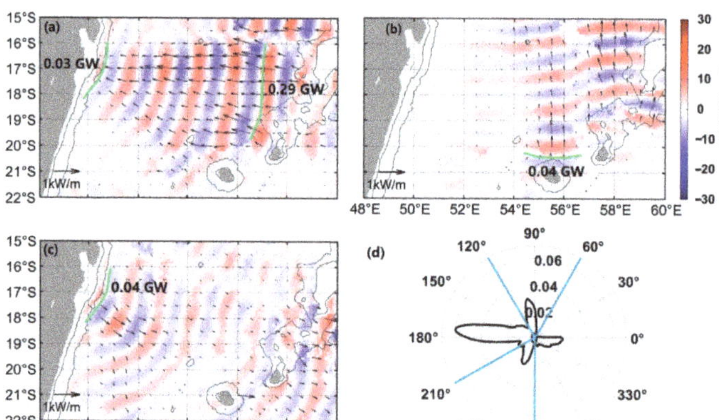

Figure 13. (**a**) The SSH of internal tides in the 120–210° direction, where arrows indicate the energy flux directions, and the lengths of the arrows are proportional to the energy flux magnitudes. Green lines signify energy integration lines. (**b**) The SSH of internal tides in the 60–120° direction. (**c**) The SSH of internal tides in the 270–60° direction. (**d**) Internal tidal energy concentration index.

Continental shelves often cause internal tide reflection. The M2 internal tides are known to reflect off the Tasmanian slope [24], and the K1 internal tides exhibit reflective behavior on the northern continental shelf of the South China Sea [45]. The topographic critical parameter of M2 internal tide helps us further confirm that internal tides could be reflected by the east coast of Madagascar. The formula is as follows [46,47]:

$$\gamma = \frac{\nabla H}{\sqrt{\frac{\omega^2 - f^2}{N^2 - \omega^2}}}, \tag{10}$$

where ∇H represents the topography gradient, ω represents the tide frequency, f represents the Coriolis parameter, and N represents the buoyancy frequency. We used the average values of the topography within ± 300 m. The γ value exceeding 1 indicates supercritical topography, which can cause internal tide reflection, while the opposite results indicate transmission.

As shown in Figure 12b, the east coast of Madagascar near 17°S is supercritical, indicating that westward propagating internal tides may reflect there. However, the energy of the westward propagating internal tides reaches the coast with an energy of only 0.03 GW, while the eastward propagating internal tides have an energy of 0.04 GW near the shelf, which suggests that the eastern coast of Madagascar presents an environment conducive to both the genesis and reflection of M2 internal tides. To distinguish the contributions of reflection at the shelf and the generation of internal wave energy, further observational data or numerical modeling studies will be required.

Figure 14a clearly illustrates the pronounced refraction of internal tides. The wave crest lines are observed to progressively deviate from a nearly meridional alignment, curving northward. To demonstrate the refraction more clearly, we selected two propagation pathways in the northern and southern parts of the region, as indicated by the blue line for the northern path and the yellow line for the southern path in Figure 14a.

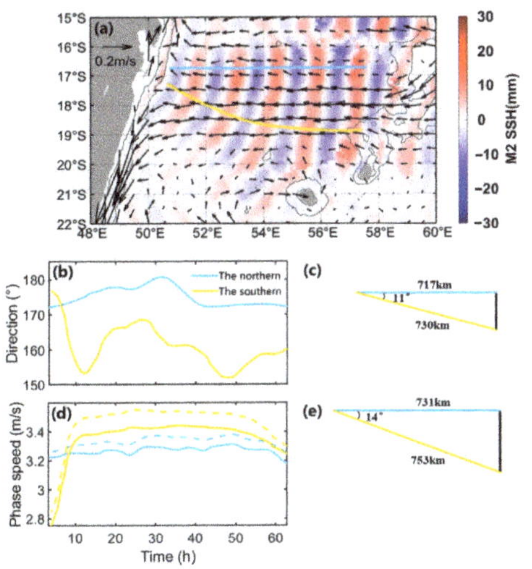

Figure 14. (**a**) The SSH of internal tides in the direction of 120–210°, where the direction of the arrows indicates the direction of the geostrophic flow, and the length of the arrows indicates the magnitude of the geostrophic flow. (**b**) The propagation directions of internal tides, with blue representing the northern part and yellow lines representing the southern part. (**d**) Phase velocities. The solid line represents the theoretical phase speed, and the dashed line represents the phase speed after geostrophic current correction, with colors corresponding to those in (**b**). (**c**,**e**) Schemas of the westward internal tidal propagation directions before and after geostrophic flow correction.

Along the representative paths, the direction in the southern part deflects 17° northward, while in the northern part, it remains relatively unchanged (Figure 14b). A comparison of the theoretical phase velocities between the two regions revealed that the phase speed of the southern part is on average 0.07 m/s greater than that of the northern part (Figure 14d). The horizontal axes of Figure 14b,d represent time, commencing from the first cycle of the selected propagation pathway. Figure 14c illustrates the internal tide propagation directions in this region as a triangle analogy, where a speed difference of 0.07 m/s over 5.5 tidal cycles can lead to a deflection of only 11°.

Figure 14a demonstrates the spatial distribution of geostrophic currents averaged over multiple years in the region, which is uneven, as shown by the arrows. The geostrophic current data are downloaded from CMEMS via https://doi.org/10.48670/moi-00148 (accessed on 19 July 2023). The geostrophic current speeds in the southern part are significantly greater than those in the northern part. To illustrate the impact of geostrophic flows on internal tide velocities, we project and superimpose the geostrophic velocity vectors onto the theoretical internal tide velocities, with the results depicted as the dashed lines in Figure 14d. Following the correction, the velocities on both the northern and southern sides increase, resulting in an average differential of 0.12 m/s. Figure 14d illustrates the corrected propagation schema, indicating that the deflection angle can extend up to 14°. The impact of currents on internal tides is primarily manifested in the nonlinearity they induce [48]. Although we have only performed a simple superposition comparison, it is still evident that within regions with stronger currents, the phase velocity of the internal tides is significantly modulated.

Near the generating area, the crest line of the internal tide is basically along the meridional direction. The phase velocity of the southern part is greater than the northern part, resulting in greater distances traveled by internal tides in the south over the same period, leading to a northward deviation. The topographical constriction between the Nazareth Bank and Cargados Carajos Bank leads to a marked increase in geostrophic current velocities within this locale. This irregular distribution of ambient flow patterns accentuates the velocity disparities between the northern and southern sectors, consequently amplifying the refraction of internal tides.

5. Conclusions

In this study, a suite of satellite altimetry data was used to investigate the propagation characteristics and energy distribution of the first-mode M2 internal tide within the Madagascar–Mascarene Region, revealing a special internal tide image for this sea area. This paper proposed an energy concentration index to assess the distribution and concentration trends of internal tidal energy, enabling clear distinction between internal tides with different propagation directions. The three main conclusions of this study are as follows:

a. The Mascarene Plateau and its surrounding regions exhibit massive standing wave structures, showing the complex interplay of internal tide. The western slope of the Mascarene Plateau radiates the M2 internal tides toward the Mascarene Basin. The combination of multiple internal tide source regions on the plateau and the source regions on the northern side and eastern coast of Madagascar form a semi-enclosed topography, resulting in a continuous standing wave structure spanning the entire basin for a distance of 1600 km. The eastern side of the Mascarene Strait experiences interference of internal tides in three directions, leading to significant beam structures due to multiwave interference effects.
b. The Mascarene Strait emerges as a potent generation area for the M2 internal tides, with a total energy flux of 1.42 GW (0.84 GW westward and 0.58 GW eastward). The internal tidal beams exhibit a cross-shaped distribution, with the northwestward and southeastward directions having the strongest energy fluxes of 0.43 GW and 0.19 GW, respectively. The internal tides propagated from the Mascarene Strait can reach distances of more than 800 km, with the southeastern internal tide capable of propagating up to 1000 km. The β effect enhances refraction during the long-distance, cross-latitude propagation of internal tides.
c. In the Mascarene Basin (15°S–20°S), a triangular configuration of internal tides arises from three generation regions. Westward-propagating internal tides from Nazareth Bank show a 17° northward refraction. The theoretical phase velocity difference (Sturm–Liouville equation) between the northern and southern parts of the internal

tides would only result in an 11° refraction. The geostrophic current speed varies significantly along the path, and considering these currents, the propagation distance difference over 5.5 cycles between the north part and the south part increases from 13 km to 22 km, with the large velocity between Nazareth Bank and Cargados Carajos Bank enhancing the refraction.

This study improves understanding of incoherent internal tide propagation in the Madagascar–Mascarene Region. It provides a solid foundation for future in situ observations and enhanced numerical simulations in this region. The supertidal kinetic energy is highest near low-latitude internal tide generation regions, such as the Bay of Bengal and Mascarene Ridge [49]. With the continued operation of the SWOT satellite and a clear understanding of steady internal tides, future studies can better assess incoherent internal tide energy and its effects on ocean mixing and climate.

Author Contributions: Conceptualization, Q.W., J.M. and X.C.; Methodology, Q.W., J.M. and X.C.; Software, Q.W.; Writing—original draft, Q.W.; Writing—review and editing, J.M., X.C. and Y.G. All authors have read and agreed to the published version of the manuscript.

Funding: This work was supported by the National Natural Science Foundation of China (grant no.41876015) and the National Key Research and Development Program of China (grant no.2021YFC3101603).

Data Availability Statement: The altimetric sea surface height measurements are downloaded from the Copernicus Marine and Environment Monitoring Service, https://doi.org/10.48670/moi-00146 (accessed on 14 February 2023). The World Ocean Atlas 2018 is available at https://www.ncei.noaa.gov/access/world-ocean-atlas-2018/ (accessed on 1 February 2023). The results in this paper are available at https://doi.org/10.6084/m9.figshare.25522873.v1 (accessed on 1 February 2023).

Acknowledgments: Figures were made with Matlab version 2022b (The MathWorks Inc., Natick, MA, USA, 2022), available at https://www.mathworks.com (accessed on 10 February 2022).

Conflicts of Interest: The authors declare no conflicts of interest. The funders had no role in the design of the study; in the collection, analyses, or interpretation of data; in the writing of the manuscript, or in the decision to publish the results.

References

1. Garrett, C.; Kunze, E. Internal Tide Generation in the Deep Ocean. *Annu. Rev. Fluid Mech.* **2007**, *39*, 57–87. [CrossRef]
2. Marshall, J.; Speer, K. Closure of the Meridional Overturning Circulation through Southern Ocean Upwelling. *Nat. Geosci.* **2012**, *5*, 171–180. [CrossRef]
3. Talley, L. Closure of the Global Overturning Circulation Through the Indian, Pacific, and Southern Oceans: Schematics and Transports. *Oceanography* **2013**, *26*, 80–97. [CrossRef]
4. MacKinnon, J.A.; Gregg, M.C. Mixing on the Late-Summer New England Shelf—Solibores, Shear, and Stratification. *J. Phys. Oceanogr.* **2003**, *33*, 1476–1492. [CrossRef]
5. Munk, W.; Wunsch, C. Abyssal Recipes II: Energetics of Tidal and Wind Mixing. *Deep Sea Res. Part I Oceanogr. Res. Pap.* **1998**, *45*, 1977–2010. [CrossRef]
6. De Lavergne, C.; Vic, C.; Madec, G.; Roquet, F.; Waterhouse, A.F.; Whalen, C.B.; Cuypers, Y.; Bouruet-Aubertot, P.; Ferron, B.; Hibiya, T. A Parameterization of Local and Remote Tidal Mixing. *J. Adv. Model. Earth Syst.* **2020**, *12*, e2020MS002065. [CrossRef]
7. Kunze, E. Internal-Wave-Driven Mixing: Global Geography and Budgets. *J. Phys. Oceanogr.* **2017**, *47*, 1325–1345. [CrossRef]
8. Kantha, L.H.; Tierney, C.C. Global Baroclinic Tides. *Progress. Oceanogr.* **1997**, *40*, 163–178. [CrossRef]
9. Vic, C.; Naveira Garabato, A.C.; Green, J.A.M.; Waterhouse, A.F.; Zhao, Z.; Melet, A.; De Lavergne, C.; Buijsman, M.C.; Stephenson, G.R. Deep-Ocean Mixing Driven by Small-Scale Internal Tides. *Nat. Commun.* **2019**, *10*, 2099. [CrossRef]
10. Niwa, Y.; Hibiya, T. Numerical Study of the Spatial Distribution of the M_2 Internal Tide in the Pacific Ocean. *J. Geophys. Res.* **2001**, *106*, 22441–22449. [CrossRef]
11. Zhao, Z. Satellite Evidence for Strengthened M_2 Internal Tides in the Past 30 Years. *Geophys. Res. Lett.* **2023**, *50*, e2023GL105764. [CrossRef]
12. Li, Z.; Von Storch, J. M_2 Internal-Tide Generation in STORMTIDE2. *J. Geophys. Res. Oceans* **2020**, *125*, e2019JC015453. [CrossRef]
13. Shriver, J.F.; Richman, J.G.; Arbic, B.K. How Stationary Are the Internal Tides in a High-Resolution Global Ocean Circulation Model? *J. Geophys. Res. Oceans* **2014**, *119*, 2769–2787. [CrossRef]
14. Niwa, Y.; Hibiya, T. Estimation of Baroclinic Tide Energy Available for Deep Ocean Mixing Based on Three-Dimensional Global Numerical Simulations. *J. Oceanogr.* **2011**, *67*, 493–502. [CrossRef]

15. Zhang, H.; Qian, Y.-K.; Wen, X.; Peng, S. Energetics of Semidiurnal Internal Tides near Madagascar in the Southwest Indian Ocean. *Progress Oceanogr.* **2023**, *211*, 102972. [CrossRef]
16. Morozov, E.G. *Oceanic Internal Tides: Observations, Analysis and Modeling*; Springer International Publishing: Cham, Switzerland, 2018. [CrossRef]
17. Da Silva, J.C.B.; New, A.L.; Magalhaes, J.M. On the Structure and Propagation of Internal Solitary Waves Generated at the Mascarene Plateau in the Indian Ocean. *Deep Sea Res. Part I Oceanogr. Res. Pap.* **2011**, *58*, 229–240. [CrossRef]
18. New, A.L.; Magalhaes, J.M.; Da Silva, J.C.B. Internal Solitary Waves on the Saya de Malha Bank of the Mascarene Plateau: SAR Observations and Interpretation. *Deep Sea Res. Part I Oceanogr. Res. Pap.* **2013**, *79*, 50–61. [CrossRef]
19. Ray, R.D.; Zaron, E.D. M2 Internal Tides and Their Observed Wavenumber Spectra from Satellite Altimetry. *J. Phys. Oceanogr.* **2016**, *46*, 3–22. [CrossRef]
20. Zhao, Z.; Alford, M.H.; Girton, J.; Johnston, T.M.S.; Carter, G. Internal Tides around the Hawaiian Ridge Estimated from Multisatellite Altimetry. *J. Geophys. Res.* **2011**, *116*, C12039. [CrossRef]
21. Mathur, M.; Carter, G.S.; Peacock, T. Topographic Scattering of the Low-Mode Internal Tide in the Deep Ocean. *J. Geophys. Res. Oceans* **2014**, *119*, 2165–2182. [CrossRef]
22. Pinkel, R.; Alford, M.; Lucas, A.; Johnston, S.; MacKinnon, J.; Waterhouse, A.; Jones, N.; Kelly, S.; Klymak, J.; Nash, J.; et al. Breaking Internal Tides Keep the Ocean in Balance. *Eos* **2015**, *96*, 1–5. [CrossRef]
23. Zhao, Z. The Global Mode-1 S_2 Internal Tide. *J. Geophys. Res. Oceans* **2017**, *122*, 8794–8812. [CrossRef]
24. Zhao, Z.; Alford, M.H.; Simmons, H.L.; Brazhnikov, D.; Pinkel, R. Satellite Investigation of the M_2 Internal Tide in the Tasman Sea. *J. Phys. Oceanogr.* **2018**, *48*, 687–703. [CrossRef]
25. Wang, M.; Zhu, X.-H.; Zheng, H.; Chen, J.; Liu, Z.-J.; Ren, Q.; Liu, Y.; Nan, F.; Yu, F.; Li, Q. Direct Evidence of Standing Internal Tide West of the Luzon Strait Observed by a Large-Scale Observation Array. *J. Phys. Oceanogr.* **2023**, *53*, 2263–2280. [CrossRef]
26. Ma, J.; Guo, D.; Zhan, P.; Hoteit, I. Seasonal M_2 Internal Tides in the Arabian Sea. *Remote Sens.* **2021**, *13*, 2823. [CrossRef]
27. Pujol, M.-I.; Taburet, G. SEALEVEL_GLO_PHY_L3_MY_008_062. 2023. Available online: https://catalogue.marine.copernicus.eu/documents/PUM/CMEMS-SL-PUM-008-032-062 (accessed on 14 February 2023).
28. Zhao, Z.; Alford, M.H.; Girton, J.B.; Rainville, L.; Simmons, H.L. Global Observations of Open-Ocean Mode-1 M_2 Internal Tides. *J. Phys. Oceanogr.* **2016**, *46*, 1657–1684. [CrossRef]
29. Zhao, Z. Mapping Internal Tides from Satellite Altimetry Without Blind Directions. *J. Geophys. Res. Oceans* **2019**, *124*, 8605–8625. [CrossRef]
30. Zhao, Z.; Alford, M.; Girton, J. Mapping Low-Mode Internal Tides from Multisatellite Altimetry. *Oceanography* **2012**, *25*, 42–51. [CrossRef]
31. Zhao, Z. Satellite Estimates of Mode-1 M_2 Internal Tides Using Nonrepeat Altimetry Missions. *J. Phys. Oceanogr.* **2022**, *52*, 3065–3076. [CrossRef]
32. Zhao, Z.; Qiu, B. Seasonal West-East Seesaw of M_2 Internal Tides from the Luzon Strait. *J. Geophys. Res. Oceans* **2023**, *128*, e2022JC019281. [CrossRef]
33. Gao, Z.; Chen, Z.; Huang, X.; Yang, H.; Wang, Y.; Ma, W.; Luo, C. Estimating the Energy Flux of Internal Tides in the Northern South China Sea Using Underwater Gliders. *J. Geophys. Res. Oceans* **2024**, *129*, e2023JC020385. [CrossRef]
34. Chelton, D.B.; DeSzoeke, R.A.; Schlax, M.G.; El Naggar, K.; Siwertz, N. Geographical Variability of the First Baroclinic Rossby Radius of Deformation. *J. Phys. Oceanogr.* **1998**, *28*, 433–460. [CrossRef]
35. Gill, A. *Atmosphere-Ocean Dynamics*; Academic Press: Cambridge, MA, USA, 1982.
36. Kelly, S.M. The Vertical Mode Decomposition of Surface and Internal Tides in the Presence of a Free Surface and Arbitrary Topography. *J. Phys. Oceanogr.* **2016**, *46*, 3777–3788. [CrossRef]
37. Zhao, Z.; Alford, M.H. New Altimetric Estimates of Mode-1 M_2 Internal Tides in the Central North Pacific Ocean. *J. Phys. Oceanogr.* **2009**, *39*, 1669–1684. [CrossRef]
38. Zaron, E.D. Baroclinic Tidal Sea Level from Exact-Repeat Mission Altimetry. *J. Phys. Oceanogr.* **2019**, *49*, 193–210. [CrossRef]
39. Zhang, P.; Xu, Z.; You, J.; Hao, Z.; Yin, B.; Hou, Y.; Robertson, R. Satellite Observed Multisource Internal Tide Radiation and Interference in the Banda Sea. *J. Geophys. Res. Oceans* **2023**, *128*, e2022JC019383. [CrossRef]
40. Li, X.; Zhao, Z.; Pichel, W.G. Internal Solitary Waves in the Northwestern South China Sea Inferred from Satellite Images. *Geophys. Res. Lett.* **2008**, *35*, GL034272. [CrossRef]
41. Niwa, Y.; Hibiya, T. Three-dimensional Numerical Simulation of M_2 Internal Tides in the East China Sea. *J. Geophys. Res. Oceans* **2004**, *109*, C04027. [CrossRef]
42. Egbert, G.D.; Erofeeva, S.Y. An Approach to Empirical Mapping of Incoherent Internal Tides with Altimetry Data. *Geophys. Res. Lett.* **2021**, *48*, e2021GL095863. [CrossRef]
43. Rainville, L.; Johnston, T.M.S.; Carter, G.S.; Merrifield, M.A.; Pinkel, R.; Worcester, P.F.; Dushaw, B.D. Interference Pattern and Propagation of the M_2 Internal Tide South of the Hawaiian Ridge. *J. Phys. Oceanogr.* **2010**, *40*, 311–325. [CrossRef]
44. Zaron, E.D.; Egbert, G.D. Time-Variable Refraction of the Internal Tide at the Hawaiian Ridge. *J. Phys. Oceanogr.* **2014**, *44*, 538–557. [CrossRef]
45. Wang, S.; Cao, A.; Li, Q.; Chen, X. Reflection of K_1 Internal Tides at the Continental Slope in the Northern South China Sea. *J. Geophys. Res. Oceans* **2021**, *126*, e2021JC017260. [CrossRef]
46. Craig, P.D. Solutions for Internal Tidal Generation over Coastal Topography. *J. Mar. Res.* **1987**, *45*, 83–105. [CrossRef]

47. Wang, S.; Chen, X.; Li, Q.; Wang, J.; Meng, J.; Zhao, M. Scattering of Low-Mode Internal Tides at Different Shaped Continental Shelves. *Cont. Shelf Res.* **2018**, *169*, 17–24. [CrossRef]
48. Alford, M.H.; Lien, R.-C.; Simmons, H.; Klymak, J.; Ramp, S.; Yang, Y.J.; Tang, D.; Chang, M.-H. Speed and Evolution of Nonlinear Internal Waves Transiting the South China Sea. *J. Phys. Oceanogr.* **2010**, *40*, 1338–1355. [CrossRef]
49. Solano, M.S.; Buijsman, M.C.; Shriver, J.F.; Magalhaes, J.; Da Silva, J.; Jackson, C.; Arbic, B.K.; Barkan, R. Nonlinear Internal Tides in a Realistically Forced Global Ocean Simulation. *J. Geophys. Res. Oceans* **2023**, *128*, e2023JC019913. [CrossRef]

remote sensing

MDPI

Article

Denoising of Photon-Counting LiDAR Bathymetry Based on Adaptive Variable OPTICS Model and Its Accuracy Assessment

Peize Li [1], Yangrui Xu [1], Yanpeng Zhao [1], Kun Liang [1,2,*] and Yuanjie Si [3]

[1] School of Electronic Information and Communications, Huazhong University of Science and Technology, Wuhan 430074, China; lipeize@hust.edu.cn (P.L.); xuyangrui@hust.edu.cn (Y.X.); zhaoyanpeng@hust.edu.cn (Y.Z.)
[2] National Key Laboratory of Multispectral Information Intelligent Processing Technology, Wuhan 430074, China
[3] School of Information Science and Technology, Shanghai Tech University, Shanghai 201210, China; siyj@shanghaitech.edu.cn
* Correspondence: liangkun@hust.edu.cn

Abstract: Spaceborne photon-counting LiDAR holds significant potential for shallow-water bathymetry. However, the received photon data often contain substantial noise, complicating the extraction of elevation information. Currently, a denoising algorithm named ordering points to identify the clustering structure (OPTICS) draws people's attention because of its strong performance under high background noise. However, this algorithm's fixed input variables can lead to inaccurate photon distribution parameters in areas near the water bottom, which results in inadequate denoising in these areas, affecting bathymetric accuracy. To address this issue, an Adaptive Variable OPTICS (AV-OPTICS) model is proposed in this paper. Unlike the traditional OPTICS model with fixed input variables, the proposed model dynamically adjusts input variables based on point cloud distribution. This adjustment ensures accurate measurement of photon distribution parameters near the water bottom, thereby enhancing denoising effects in these areas and improving bathymetric accuracy. The findings indicate that, compared to traditional OPTICS methods, AV-OPTICS achieves higher $F1$-values and lower cohesions, demonstrating better denoising performance near the water bottom. Furthermore, this method achieves an average MAE of 0.28 m and $RMSE$ of 0.31 m, indicating better bathymetric accuracy than traditional OPTICS methods. This study provides a promising solution for shallow-water bathymetry based on photon-counting LiDAR data.

Keywords: ICESat-2/ATLAS; photon-counting LiDAR; adaptive variable OPTICS; bathymetric method

1. Introduction

Shallow water refers to aquatic environments where water surface waves are significantly influenced by the water bottom topography, including inland lakes, reservoirs, coastal shallow waters, and remote islands in deep seas [1]. These shallow water resources provide essential physical environments for ecosystem sustainability and biodiversity and are crucial for supporting human necessities such as food, shelter, and transportation [2,3]. Bathymetry in shallow-water areas is fundamental to water resource exploration and is indispensable for environmental protection, sustainable development of fisheries and tourism resources, coastal safety, disaster reduction, marine engineering construction, and marine scientific research. Accurate measurements of water depth in shallow-water regions support environmental conservation measures, optimize resource utilization strategies, enhance public safety, and deepen our understanding of marine natural phenomena and geological history. Thus, bathymetry serves as a bridge connecting marine science, environmental protection, and socioeconomic development.

Currently, methods for the bathymetry of shallow water are categorized based on the carrier used, including shipborne, airborne, and spaceborne methods [4]. Shipborne bathymetry uses devices such as sounding poles, lead lines, sonar, and Doppler sounders [5,6], offering the advantages of simplicity and high accuracy. However, this method involves risks such as capsizing and grounding, inefficiency, and significant human and material resource consumption. Additionally, data updates are slow, leaving large areas unmapped and limiting long-term monitoring needs [7], particularly in offshore unmanned islands and reefs that lack systematic terrain and morphological data. Airborne bathymetry primarily employs airborne LiDAR bathymetry systems [8,9], which balance accuracy and efficiency. However, operational constraints, including climate, airspace management, and geographic location, limit its coverage capability. Unlike the aforementioned methods, satellite sensors periodically acquire optical information over large areas, providing a broader and more consistent means of data collection for shallow-water bathymetry. Various satellite remote sensing datasets have been applied to shallow-water bathymetry, including hyperspectral and multispectral images, as well as synthetic aperture radar data [10,11]. However, resulting from empirical formulas and calibration points, these techniques have limitations regarding water depth accuracy.

Spaceborne photon-counting LiDAR, a new active detection method, offers unique advantages in bathymetry [12,13]. By measuring the flight time of photons, it determines the elevation information of Earth's surface with high precision, a wide measurement range, high resolution, and real-time capabilities, making it highly promising for shallow-water bathymetry [14]. For instance, the ICESat-2 (Ice, Cloud, and Land Elevation Satellite-2) satellite, launched by NASA in September 2018, is equipped with the Advanced Topographic Laser Altimeter System (ATLAS). This system emits laser pulses with a wavelength of 532 nm to penetrate waterbodies and a pulse width of 1.5 ns at a frequency of 10 kHz. Along its orbit, ATLAS creates overlapping footprints approximately 0.7 m apart, each with a diameter of 17 m [15]. To enhance detection efficiency, the ATLAS laser transmitter divides each pulse into six beams arranged in three pairs. These beams are aligned parallel in the along-track direction, consisting of one strong and one weak signal per pair, maintaining an energy ratio of 4:1 between them. The distance along-track between pairs is approximately 3.3 km, with a separation within pairs of approximately 90 m [16,17]. Each pulse emission emits roughly 2 trillion photons, with the number of returned photons varying based on differences in Earth's surface reflectivity. Each returned photon generates a time-stamped location, resulting in a dense photon point cloud [18]. Processing of these photon point-cloud images enables the extraction of water-depth data.

In recent years, spaceborne photon-counting LiDAR detection technology has advanced significantly. However, the processing of photon point cloud data remains a focal point of research. Photon point cloud data are heavily affected by factors such as solar background, system characteristics, atmospheric scattering, water column scatter and absorption, as well as wind and wave effects, resulting in significant noise points that complicate the extraction of accurate elevation information. To address this challenge, researchers have developed various denoising algorithms based on traditional techniques to improve the extraction of elevation data. Initially, researchers transformed photon point cloud profiles into two-dimensional raster images and employed image-processing methods such as canny edge detection and median filtering for noise reduction [17,19]. Although these methods effectively reduce photon noise to a certain extent, they also lead to the loss of useful information during the conversion of photon point clouds into raster images, thereby affecting denoising accuracy [20]. Subsequently, methods based on the dense distribution of signal photons and sparse distribution of noise photons separated these two types using point-cloud density calculations and parameter thresholds, often named after the shapes of their filters, such as circular, elliptical, or rectangular filtering [21–23]. Density-based methods are straightforward and computationally efficient in low-noise conditions but struggle with high-density clustered noise scenarios. Clustering algorithms were then employed to classify photon points into signal and noise categories using density clustering techniques such as Bayesian, DBSCAN, and OPTICS [24–26]. Among these, Bayesian decision theory-based denoising algorithms excel in complex terrain but necessitate prior estimation of the signal-to-noise ratio of the original photon data [27]. DBSCAN-based algorithms generally achieve satisfactory denoising results overall but encounter challenges in complex terrain areas [28]. In contrast, the OPTICS algorithm [29–31] comprehensively considers the density and distance distribution characteristics of photon point cloud data, demonstrating good denoising performance, particularly under intense background noise and in complex terrains. As a result, it finds widespread application in processing photon point cloud data in shallow water regions [26,32]. However, this algorithm's fixed input variables can lead to inaccurate photon distribution parameters near the water bottom, influenced by water bottom photons. Consequently, this results in suboptimal denoising in areas near the water bottom, affecting bathymetric accuracy.

To address the challenge of inadequate denoising near the water bottom in OPTICS methods, an Adaptive Variable OPTICS (AV-OPTICS) model is proposed in this paper. Unlike the conventional OPTICS algorithm with fixed input variables, this method dynamically adjusts input variables based on the point cloud distribution. This adaptive approach ensures precise measurement of photon distribution parameters near the water bottom, thereby enhancing denoising effectiveness in these critical areas and improving overall bathymetric accuracy. To validate the reliability and efficacy of the AV-OPTICS model, experiments were conducted using ATL03 data from the Culebra area in Puerto Rico. Evaluation indices such as $F1$-value, cohesion, root mean square error ($RMSE$), and mean absolute error (MAE) were computed. Comparative analysis with the traditional OPTICS model was performed to highlight the advantages of the AV-OPTICS model in terms of denoising effects and bathymetric accuracy. Furthermore, this study includes a detailed discussion of the input variables specific to AV-OPTICS, the distribution of bathymetric errors, and other factors influencing bathymetry.

2. Materials and Methods

2.1. Study Area and Data

2.1.1. Study Area

The Commonwealth of Puerto Rico is situated in the eastern part of the Greater Antilles in the Caribbean Sea, encompassing the islands of Puerto Rico, Vieques, and Culebra. This study specifically focuses on the Culebra region, which spans from approximately 18.25°N to 18.36°N latitude and 65.15°W to 65.35°W longitude. Figure 1 illustrates images of Culebra and its location, labeled by a red circle.

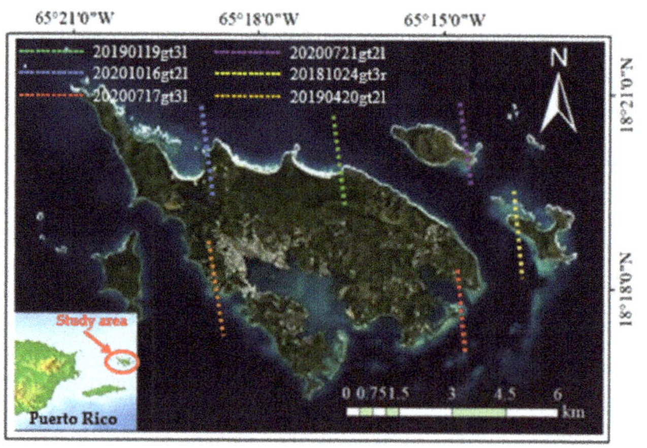

Figure 1. Study area and the detection tracks of ATLAS represented by six different colors of dashed lines.

2.1.2. ICESat-2 ATL03 Data

The study utilized ATL03 data from the study area to validate the proposed bathymetric method and assess its capability. ATL03 data consists of photon point clouds with precise geographic coordinates, capturing geospatial information from signal photons reflected by both the sea surface and seafloor, enabling the reconstruction of ocean profiles and bathymetry.

An ATL03 file comprises six groups of data corresponding to six tracks (gt1l, gt1r, gt2l, gt2r, gt3l, and gt3r, where "l" and "r" represent left and right, respectively). Due to the signal intensity and geographic coordinates, not all groups of data contain distinct signal points of the water bottom. For validation purposes, six ATL03 data with distinct signal points of water bottom were selected from the study area. as depicted in Figure 1, where dashed lines of different colors represent the various ATL03 data tracks. Table 1 presents detailed information on the six ATL03 data used in this study, including acquisition dates, tracks, geographic coordinates, and the number of LiDAR points.

Table 1. Detailed information of six ATL03 data.

Data Name	Date	Geographic Coordinates Range	Track	Number of LiDAR Points
20190119gt3l	20190119	18.2885°N, 65.2767°W~18.2936°N, 65.2773°W	gt3l	3475
20181024gt3r	20181024	18.3234°N, 65.2391°W~18.3192°N, 65.2395°W	gt3r	4442
20200717gt3l	20200717	18.2965°N, 65.2485°W~18.3001°N, 65.2488°W	gt3l	2450
20200721gt2l	20200721	18.2969°N, 65.2504°W~18.2949°N, 65.2506°W	gt2l	892
20201016gt2l	20201016	18.3290°N, 65.3147°W~18.3316°N, 65.3150°W	gt2l	2063
20190420gt2l	20190420	18.3010°N, 65.3118°W~18.3039°N, 65.3121°W	gt2l	1391

2.1.3. ALB In Situ Data

In this study, we utilized airborne LiDAR bathymetry data provided by NOAA's Office of Coastal Management as in situ data for validating the bathymetric accuracy of the ICESat-2 data in our study area. The in situ water depth data were collected on 11 July 2018 by Leading Edge Geomatics using a Riegl VQ-880-G II LiDAR system, as depicted in Figure 2.

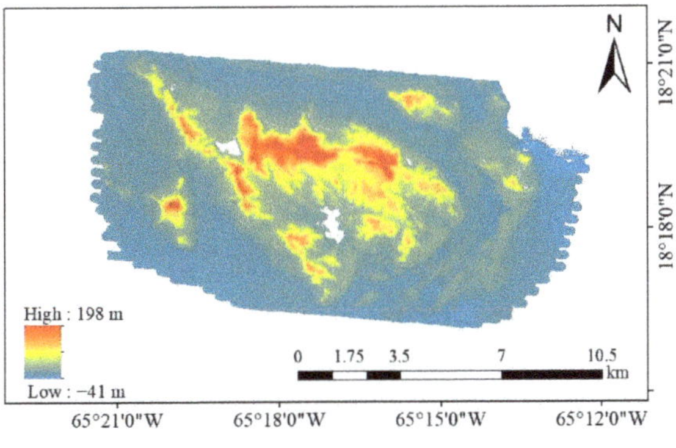

Figure 2. Water depth results of study area for ALB reference data.

2.2. Method

2.2.1. Photon-Counting Bathymetric Method

The ATL03 data obtained from Culebra contain thousands of photon points. Due to the influence of sunlight, water scattering, suspended solids, and other factors, these data include a large number of noise photons in addition to those from the sea surface and seafloor. This noise presents significant challenges for water depth extraction. Therefore, after obtaining the original ATL03 data, the AV-OPTICS model was employed to isolate pure signal photons, including those from the sea surface and seafloor. Subsequently, water depth was extracted based on the coordinates of the signal photons. During this extraction process, it is essential to consider the effects of refraction and tides on the coordinates of the signal photons. Tidal effects cause the sea level to change over time, and the refraction of seawater biases the position of underwater photons. Hence, the influence of both effects on bathymetry must be eliminated to extract the true water depth, which can then be compared with in situ data.

2.2.2. AV-OPTICS Denoising Algorithm

To handle the numerous noise photons in the ATL03 data, the AV-OPTICS model employs an elliptical filter and density threshold to define the distance parameters for each photon. It then screens out noise photons based on the size of these distance parameters. To achieve accurate and automatic denoising, the size of the elliptical filter is adjusted according to the distribution characteristics of the point cloud in different data. The proposed model is divided into three steps to realize this aim, as follows:

(a) Draw the elevation histogram, perform Gaussian curve fitting, and classify water surface photons and underwater photons based on the confidence interval.
(b) Calculate the size of the elliptical filter according to the distribution characteristics of underwater photons.
(c) Use the AV-OPTICS denoising algorithm to extract water bottom photons from underwater photons.

Finally, we integrated the water surface photons obtained in Step (a) with the water bottom photons obtained in Step (c). This enabled the denoising of different ATL03 data and the successful extraction of signal photons.

(a) Based on the elevation distribution of the original photon point cloud data, we obtained a height distribution histogram, as depicted in Figure 3. Two Gaussian peaks were identified by fitting the elevation distribution histogram to a double Gaussian distribution function. According to the characteristics of the ATL03 data, these two peaks correspond to the intervals where the water surface photons and water bottom photons are located. We preliminarily determined the elevation interval with the red profile in Figure 3 corresponding to the water surface photons based on the understanding that the elevation of the water surface is higher than that at the water bottom. We extracted the photons within the 99% confidence interval as water surface photons, while the remaining photons below the intersection of the two Gaussian functions were considered underwater photons.

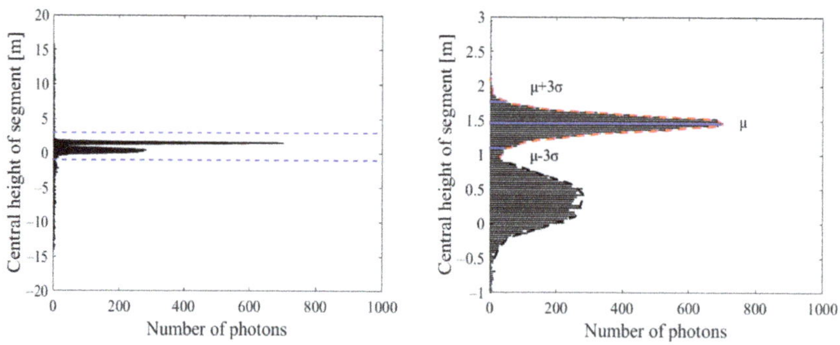

Figure 3. The elevation distribution histogram of ATL03 photon data.

(b) After obtaining the water surface photons, it is necessary to acquire the water bottom photons from the underwater photons. Owing to the absorption by the water column, signals reflected through the water bottom are much weaker compared to those from the water surface, making extraction more difficult [33]. To accurately extract water bottom photons, [32] applied the OPTICS clustering algorithm to remove noise from underwater photons. The OPTICS algorithm uses an elliptical filter of a certain size to define distribution parameters such as core distance and reach distance for each point, then calculates a reach/distance threshold based on all points in the data. By comparing the reach distance of each point with the reach/distance threshold, it determines whether the point is a signal or a noise point. A detailed introduction to the OPTICS algorithm is in Step (c). However, in the study by [32], the distribution parameters were obtained using a fixed size of the elliptical filter in the OPTICS algorithm, which does not accurately reflect the distribution characteristics of different water bottom terrains in different ATL03 data, resulting in inadequate denoising of photons near the water bottom. To address this limitation, we propose an Adaptive Variable OPTICS (AV-OPTICS) model. This model can calculate more suitable input variables based on the distribution characteristics of the water bottom photons, thereby achieving precise denoising near the bottom of the water.

The input variables of the AV-OPTICS model include the semi-minor axis b and the semi-major axis a of the elliptical filter. The value of b is related to the vertical width of the water bottom contour, whereas the value of a needs to be calculated using the K-means nearest neighbor method.

For different photon cloud data, the photon clouds were concentrated near the water bottom contour, possessing a certain width in the vertical direction, as shown in Figure 4. We measure the vertical width h of the underwater contour and take $h/2$ as the value of b. The reason for this is explained in Section 4.1. For relatively flat water bottom terrains, as shown in Figure 4a, the vertical width h of the water bottom contour can be obtained by fitting the elevation distribution histogram with a Gaussian function and then calculating the width of the confidence interval. However, for more complex water bottom terrains, as shown in Figure 4b, where the elevations of the left and right sides are not equal, the elevation distribution histogram may show a bimodal phenomenon, making it difficult to measure the vertical width h. To solve this problem, we divided the point cloud data into n segments along the horizontal direction. In this paper, n was set to 11. Each segment's data could be fitted with a Gaussian function, and the width of the 95% confidence interval was chosen as the value of h for that segment.

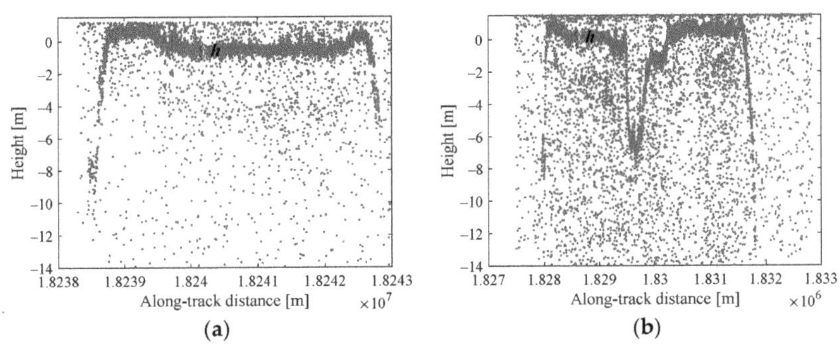

Figure 4. The contours of the water bottom terrain under different scenarios. (**a**) relatively flat water bottom terrains; (**b**) complex water bottom terrains.

To exclude outliers that may be obtained in some segments, the median of the results obtained from multiple segments was taken as the width h of the water bottom contour, and $h/2$ was taken as the value of b.

The OPTICS algorithm typically has three input variables: semi-minor axis b, semi-major axis a, and density threshold N_{MinPts}. The density threshold N_{MinPts} can be derived from the semi-minor axis b and semi-major axis a, as shown in Equation (1); therefore, it can be considered that the OPTICS algorithm has two input variables: a semi-minor axis b and a semi-major axis a.

$$\begin{cases} S_1 = \frac{\pi abN_1}{h_1 l} \\ S_2 = \frac{\pi abN_2}{h_2 l} \\ N_{MinPts} = ceil[\frac{2S_1 - S_2}{\ln(2S_1/S_2)}] \end{cases} \quad (1)$$

where S_1 denotes the total number of photons within the elliptical filter, N_1 represents the total number of photons for a given sample, h_1 and l are the maximum differences in the vertical and along-track distances in the photon cloud of the sample, respectively; S_2 denotes the number of noise photons within the elliptical filter and N_2 is the number of photons within a 5 m range upward from the deepest point and h_2 is the height of this range, set to 5 m; $ceil()$ represents the ceiling function [34].

From Equation (1), we observe that when b remains constant, the value of N_{MinPts} is a positive integer that correlates positively with a. As the value of the semi-minor axis b is already known, the optimal density threshold N_{MinPts} can be determined first, and then b and N_{MinPts} can be used to calculate the value of a via backward induction. Compared with a, the density threshold N_{MinPts} is one of the parameters that directly influences the denoising effects. An extremely small value of N_{MinPts} results in the retention of more noise points, whereas an excessively large value leads to over-denoising. Regarding the selection of N_{MinPts}, research has been conducted indicating that N_{MinPts} are typically fixed at 4 [35]. Therefore, we set the density threshold N_{MinPts} to 4 and then attempted to obtain the value of the semi-major axis a using the known values of the semi-minor axis b and N_{MinPts}.

Given the known values of b and N_{MinPts}, the K-mean nearest neighbor method [36] was employed to obtain the value of a. For data with n points, the K-mean nearest neighbor method involves three steps: (1) compute the distance distribution matrix D_{nn}, where D_{nn} is an $n \times n$ symmetric matrix, and $D_{nn}(i, j)$ represents the distance from the i th point to the j th point; (2) sort each row of the distance distribution matrix in ascending order to obtain $\overline{D}_{nn}$; (3) calculate the average values of each column of $\overline{D}_{nn}$ to obtain the list D_{es}. The list D_{es} obtained from the K-nearest neighbor method represents the sorted distances from a certain point p to other points. This study modifies Step (1) to compute a horizontal distance distribution matrix. Consequently, D_{es} obtained after these three steps represents the sorted horizontal distances from a particular point p with the other points. We take these transformed lists D_{es} as the semi-major axis parameter list D_a for a, where we choose one as the final value of a. Along with the known value of b, each value in D_a is substituted into Equation (1) to obtain a density threshold parameter list D_m for N_{MinPts}. Based on D_m and D_a, we obtained the value of a based on the value of N_{MinPts}. However, because of the rounding of N_{MinPts} values in Equation (1), D_m and D_a may include cases where multiple a values correspond to a single N_{MinPts} value. In such instances, the a value corresponding to N_{MinPts} closest to 4 was selected. In other words, the minimum a value when N_{MinPts} is equal to 4 in D_a lists is the final value chosen for the semi-major axis a.

In summary, for different underwater scenarios, this study selected half of the vertical width h of the water bottom contour as the value of the semi-minor axis of b. Then, employing the K-mean nearest neighbors method, the minimum a value when N_{MinPts} was equal to 4, was identified as the value of a, achieving automated determination of the input variables.

(c) After obtaining the input variables b and a, they were incorporated into the OPTICS algorithm to achieve denoising of underwater photons. Unlike traditional distance metrics, the OPTICS algorithm, based on an elliptical filtering kernel, defines the distance between any two points p and q, as shown in Equation (2). In this formula, x_p and x_q represent the along-track distances of points p and q; y_p and y_q denote the elevation values of points p and q.

$$d(p,q) = \sqrt{\frac{(x_p - x_q)^2}{a^2} + \frac{(y_p - y_q)^2}{b^2}} \quad (2)$$

Under this definition, the distance of the two points is quite different. As shown in Figure 5, point o is the center of the elliptical filter with a semi-minor axis b and a semi-major axis a; $w1$, $w2$, and $w3$ are three points near the point p. In these points, $w1$ is on the elliptical filter, so the distance between $w1$ and o is 1. $w2$ is inside the filter, and the distance $ow2$ is smaller than 1. Similarly, the distance $ow3$ is greater than 1. Comparing the distance of o and w with 1, we can determine whether a point w is inside the elliptical filter of o or not.

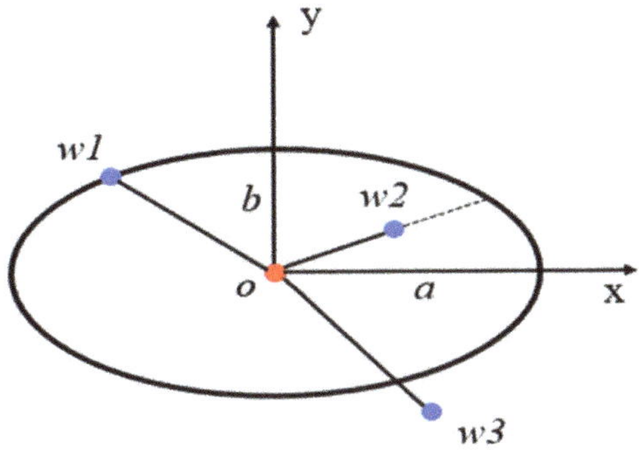

Figure 5. The distance of point o and w under the definition of OPTICS algorithm.

Based on the definition and parameters, the OPTICS algorithm defines two concepts: core distance and reach distance. Within an elliptical filter centered on photon point p, if the number of photon points is greater than or equal to the density threshold N_{MinPts}, photon point p is identified as the core point. If p is a core point, the minimum distance that makes p a core point is referred to as the core distance of p, denoted as $d_c(p)$. d_r represents the reach distance. If p is a core point, $d_r(p,q)$ is defined as the maximum distance $d(p,q)$ between sample point q and core point p or p's core distance $d_c(p)$. When p was not a core point, $d_r(p,q)$ were inexistent. The formulas for calculating the core and reach distance are shown in Equations (3) and (4), where $N_{ab}(p)$ denotes the number of points within an elliptic filter centered on p, a and b are the semi-major and semi-minor axes, respectively. x represents a point within the elliptical filter centered on p which is the N_{MinPts}th nearest to p.

$$d_c(p) = \begin{cases} undefined, N_{ab}(p) < N_{MinPts} \\ d(p,x), N_{ab}(p) \geq N_{MinPts} \end{cases} \quad (3)$$

$$d_r(p,q) = \begin{cases} undefined, N_{ab}(p) < N_{MinPts} \\ \max(d_c(p), d(p,q)), N_{ab}(p) \geq N_{MinPts} \end{cases} \quad (4)$$

For the photon cloud data, each photon point corresponds to a reach distance d_r. A larger d_r indicates a greater distance from other points, whereas a smaller d_r indicates closer proximity to other points. Based on the characteristics of the d_r list, the values of d_r less than 1 were selected and subjected to the OTSU method [37] to obtain the reach/distance threshold $d_{r,th}$. If the reach distance d_r for sample point p is smaller than $d_{r,th}$, then the sample point p is classified as a signal point; otherwise, it is classified as a noise point. This approach effectively separates the signal photon points from the noise photon points based on the reach/distance threshold $d_{r,th}$.

2.2.3. Water Depth Extraction

Based on the aforementioned method, we can obtain pure water surface and bottom signal photons, enabling the extraction of water depth based on these signal photons. However, during the propagation of water bottom photons, both their speed and direction change when they cross the air/water interface. According to Snell's law, the coordinates of the water bottom photons are shifted due to refraction, which affects the accuracy of water depth extraction. Therefore, it is necessary to perform refraction correction of the water bottom photons during water depth extraction. Simultaneously, sea level periodically changes under the influence of tidal forces from the sun and moon, known as tidal effects. To mitigate the impact of tidal effects on water depth extraction, it is essential to calculate the difference between the instantaneous water level and the reference water level and then obtain water depth values based on the reference water level.

To mitigate these two effects on water depth extraction, we first perform refraction correction to obtain the water depth under instantaneous water level. In this study, we adopted the refraction correction method outlined by [38], which considers the influence of laser incidence angles and water surface fluctuations. Unlike other refraction correction methods, they initially calculate the laser incidence angle θ using parameters ref_elev provided by the ATL03 data, as shown in Equation (5). To characterize sea surface fluctuations, a fifth Fourier series was employed to fit the water surface photons. Subsequently, the derivative of the fitted fifth Fourier series is computed to obtain the slope angle φ of each position on the sea surface, as illustrated in Equation (6). Given the known angles θ and φ, three conditions are considered, as depicted in Figure 6. By applying Snell's law and relevant trigonometric principles, the actual positions of the water bottom photons under these three conditions can be determined. Detailed methods for each case are available in the aforementioned paper.

$$\theta = \frac{\pi}{2} - ref\_elev \tag{5}$$

$$\begin{cases} f(x) = a_0 + \sum\limits_{k=1}^{5} (a_k \cos(k\omega x) + b_k \sin(k\omega x)) \\ \rho(x) = \sum\limits_{k=1}^{5} (-a_k \omega x \sin(k\omega x) + b_k \omega x \cos(k\omega x)) \\ \varphi(x) = \arctan(\rho(x)) \end{cases} \tag{6}$$

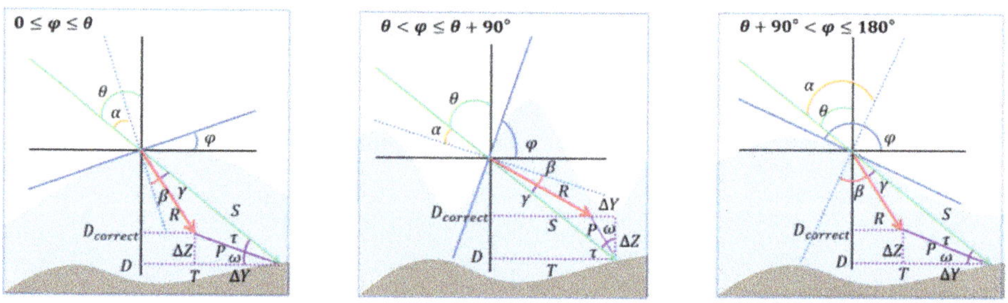

Figure 6. The spatial geometric relationships of refraction correction under different slope angles φ. The green and red vectors correspond to the original coordinate and corrected coordinate of water bottom photons, respectively [38].

After refraction correction, we obtained the water depth at the instantaneous water levels. However, owing to tidal effects, the water level varies constantly. Therefore, it was necessary to perform tidal correction to derive the water depth based on the reference water level. This study employed the S-TIDE model [39] for the tidal analysis of the study area to determine the difference between instantaneous and reference water levels. Initially, the equilibrium tide function from the S-TIDE model was used to obtain the height of the equilibrium tide at the detection time, which served as the instantaneous water level in the study area. Simultaneously, mean lower low water (MLLW) was selected as the reference water level for the study area [40]. Centered on the detection time, we utilized the model to predict the equilibrium tide height over a week and then calculated the MLLW during this time, establishing it as the reference water level. After obtaining the instantaneous and reference water levels, the difference Δh between them was calculated. Furthermore, by subtracting Δh from the water depth under instantaneous water level, the water depth based on the reference water level can be obtained. The computational process is detailed in

Equations (7)–(9), where $D_{instant}$ and D_{base} represent the water depth under instantaneous water level and reference water level, respectively; $H_{surface\_instant}$, $H_{surface_{base}}$, and $H_{seafloor}$ represent the elevation of instantaneous water surface, reference water surface, and seafloor, respectively. In ATL03 data, the water surface photons can be moved downward by Δh to characterize the reference water level.

$$D_{instant} = H_{surface\_instant} - H_{seafloor} \tag{7}$$

$$\Delta h = H_{surface\_instant} - H_{surface\_base} \tag{8}$$

$$D_{base} = H_{surface\_base} - H_{seafloor} = D_{instant} - \Delta h \tag{9}$$

After applying refraction and tide corrections, we employed a fifth-order Fourier series to model the water surface elevation from the photon data. The average value of this series was then taken as the relative elevation of the water surface. For deriving the water bottom profile, we utilized a B-spline to fit the photon data from the water bottom. Finally, we extracted water depths according to the height difference between the water surface and water bottom and validated these bathymetric values by comparing them with in situ data.

3. Results

3.1. Evaluation Methodology

To validate the accuracy of the AV-OPTICS model, six sets of point cloud data were extracted from the ATL03 data, as described in Section 2.1.2. The experiments evaluated the model's performance in terms of denoising and bathymetric accuracy. The denoising effect of the model was assessed using the $F1$ value and cohesion C_{oh}, whereas the bathymetric accuracy was evaluated using the mean absolute error (MAE) and root mean square error ($RMSE$).

To assess the denoising effectiveness of the AV-OPTICS model, we employed two metrics: the $F1$ score to gauge overall performance and cohesion C_{oh} for a detailed assessment.

The $F1$ score, widely utilized in point-cloud classification, is derived from the weighted harmonic average of Recall R and Precision P, as illustrated in Equation (10). N_{TP} denotes the number of correctly classified signal photons, N_{FN} is the number of incorrectly classified signal photons and N_{FP} is the number of incorrectly classified noise photons. A higher $F1$ value indicates better overall denoising performance of the model.

$$\begin{cases} F1 = \frac{2 \times P \times R}{P+R} \\ P = \frac{N_{TP}}{N_{TP}+N_{FP}} \\ R = \frac{N_{TP}}{N_{TP}+N_{FN}} \end{cases} \tag{10}$$

Cohesion C_{oh} reflects the extent of point-cloud dispersion within a certain region and measures the denoising effect. A smaller cohesion value indicated fewer discrete noise points within this region. The formula for cohesion is shown in Equation (11), where C_{oh} represents the value of cohesion, r denotes the number of photon points within a specific region, x_i indicates the photon points within the region, c refers to the centroid within the region, and $proximity()$ denotes the proximity formula used to compute the proximity between the two points.

$$C_{oh} = \sum_{i=1}^{r} proximity(x_i, c) \tag{11}$$

Typically, the proximity formula is based on the squared Euclidean distance. However, considering that the noise points are vertically distributed near the water bottom contour, the vertical distance is more relevant to the denoising effect. Therefore, this study adopted the squared vertical distance as the formula for proximity, with the centroid position being the average vertical distance of all points within the region.

To evaluate the bathymetric accuracy of the AV-OPTICS model, this study employed two evaluation indexes: MAE and $RMSE$. MAE is the mean absolute error between bathymetric and in situ data, providing an intuitive observation of the error. MAE can be used to evaluate the overall bathymetric accuracy of the model. $RMSE$ is the root mean square error between the bathymetric and in situ data, which is more sensitive to outliers in bathymetry data compared to MAE, making it a more effective metric for discerning subtle differences between different models. As such, we utilized $RMSE$ to compare the performance of various bathymetry models. The formulae for MAE and $RMSE$ calculations are shown in Equations (12) and (13). In the two Equations, y_i represents the value of bathymetric data obtained by the model, $\hat{y}_i$ denotes the value of in situ data, n indicates the number of data points.

$$MAE = \frac{1}{n}\sum_{i=1}^{n}|y_i - \hat{y}_i| \tag{12}$$

$$RMSE = \sqrt{\frac{1}{n}\sum_{i=1}^{n}(y_i - \hat{y}_i)^2} \tag{13}$$

3.2. Denoising Results and Comparison

The AV-OPTICS model automatically calculates input variables based on photon distribution characteristics to denoise raw photons and extract water surface and bottom photons precisely. In this study, we selected six ATL03 data from the Culebra region; among them, three data are illustrated in Figure 7a,d,g. Denoising results using the AV-OPTICS model are presented in Figure 7b,e,h, where the water surface, bottom, and noise photons are annotated with blue, red, and grey points, respectively. Figure 7b,e,h demonstrate the AV-OPTICS model's effectiveness in denoising and extracting water surface and bottom photons from the raw photons.

To demonstrate the advantages of the AV-OPTICS model, we conducted a comparative experiment using the traditional OPTICS algorithm, which employs fixed input variables determined via trial and error [32]. The denoising results of the traditional OPTICS model are shown in Figure 7c,f,i. Both methods exhibit effective denoising in regions distant from the water bottom, as depicted in Figure 7. However, near the water bottom, the AV-OPTICS model shows fewer noise photon points compared to the traditional OPTICS algorithm. To quantitatively compare the two methods, the $F1$ value and C_{oh} were computed for each method.

Based on the manually annotated signal photons, the $F1$ values for the different ATL03 data were calculated using Equation (10), and the results are listed in Table 2. Table 2 shows six sets of $F1$ values corresponding to the denoising outcomes. The average $F1$ value achieved by our AV-OPTICS algorithm is 97.53%, while the average $F1$ value obtained by the traditional OPTICS algorithm is 96.89%. This comparison indicates that the AV-OPTICS model offers an improvement in denoising accuracy compared to the traditional OPTICS model.

Table 2. $F1$-score, P, and R of our method and traditional OPTICS.

Data Name	Our Method			Traditional OPTICS		
	$F1$-Score	**P**	**R**	**$F1$-Score**	**P**	**R**
20190119gt3l	0.9704	0.9559	0.9854	0.9635	0.9310	0.9983
20181024gt3r	0.9885	0.9901	0.9870	0.9850	0.9781	0.9921
20200717gt3l	0.9631	0.9888	0.9388	0.9594	0.9203	1.0000
20200721gt2l	0.9722	0.9633	0.9813	0.9664	0.9328	1.0000
20201016gt2l	0.9740	0.9717	0.9779	0.9652	0.9476	0.9833
20190420gt2l	0.9837	0.9809	0.9865	0.9740	0.9545	0.9944
Average value	0.9753	0.9751	0.9762	0.9689	0.9441	0.9947

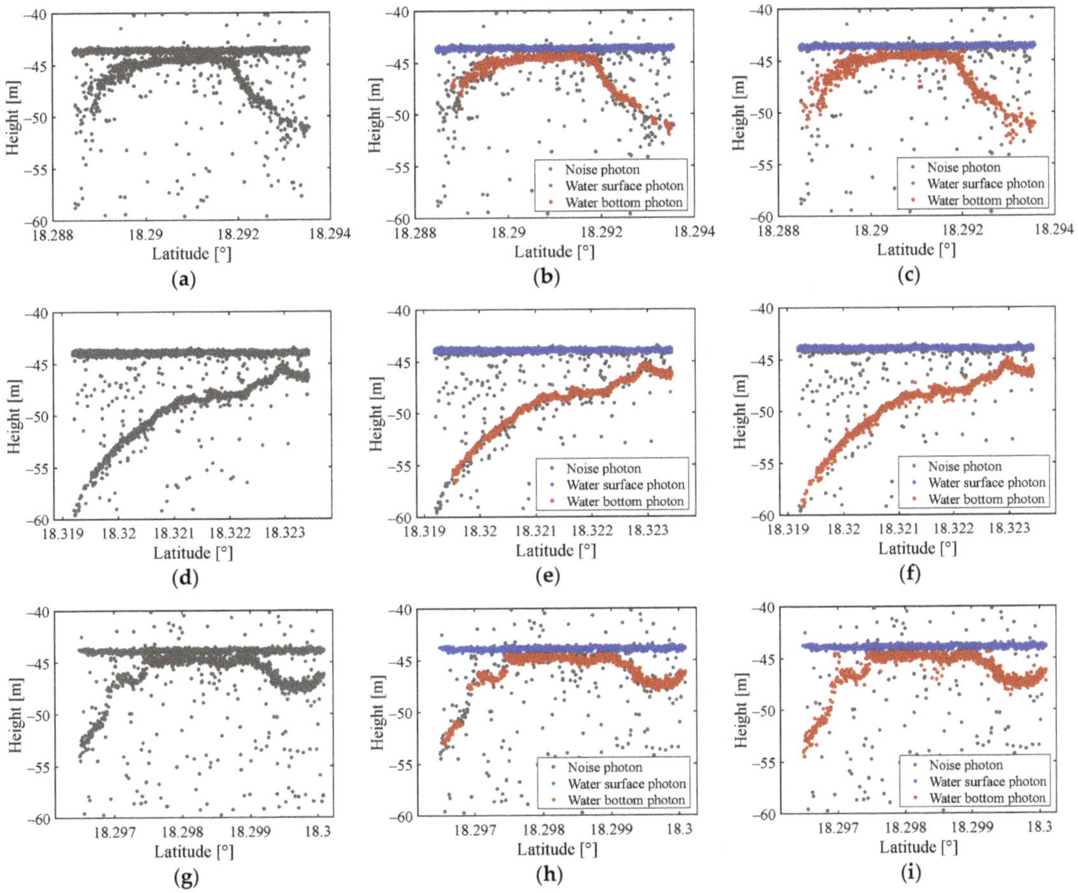

Figure 7. Denoising effects of our method and traditional OPTICS in different scenes. (**a**) 20190119gt3l—Raw data; (**b**) 20190119gt3l—Our method; (**c**) 20190119gt3l—Traditional OPTICS; (**d**) 20181024gt3r—Raw data; (**e**) 20181024gt3r—Our method; (**f**) 20181024gt3r—Traditional OPTICS; (**g**) 20200717gt3l—Raw data; (**h**) 20200717gt3l—Our method; (**i**) 20200717gt3l—Traditional OPTICS.

Combined with the observations in the previous paragraph, the AV-OPTICS model notably improves denoising accuracy, especially in areas near the water bottom.

To further emphasize the advantages of AV-OPTICS in the vicinity of water bottoms, it is crucial to magnify the contours of the water bottoms and conduct a detailed comparison between the two models. Specifically, we examined the data named 20181024gt3r and 20190420gt2l. The enlarged water bottom contours are depicted in Figure 8, highlighting that our algorithm produces fewer noise points near the water bottom in both experimental scenarios. To quantitatively reflect this observation, we applied Equation (11) to compute the cohesion of photons within the black boxes shown in Figure 8. The results are summarized in Table 3.

According to Table 3, in the vicinity of the water bottom, our method achieved an average cohesion of 24.80 m^2, whereas the traditional OPTICS algorithm recorded an average cohesion of 39.40 m^2. These findings indicate that our method achieves lower average cohesion and fewer noise points near the water bottom compared to the traditional OPTICS method.

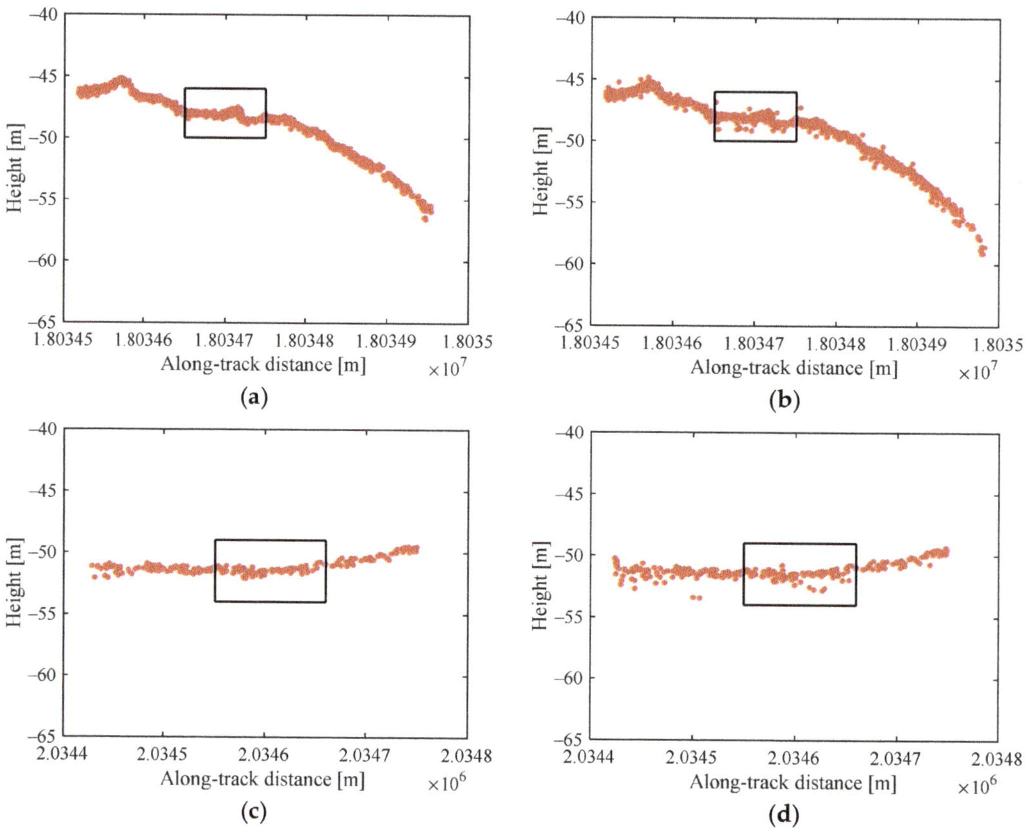

Figure 8. Comparison of denoising details of the two methods in different scenarios. (**a**) 20181024gt3r—Our method; (**b**) 20181024gt3r—Traditional OPTICS; (**c**) 20190420gt2l—Our method; (**d**) 20190420gt2l—Traditional OPTICS.

Table 3. Cohesion values of our method and traditional OPTICS within the selected region.

Data Name	**Selected Region**		**Cohesion**	
	***x* [m]**	***y* [m]**	**Our Method**	**Traditional OPTICS**
20190119gt3l	2.03308×10^6~2.03315×10^6	−50~−46	63.31	87.74
20181024gt3r	1.80346×10^7~1.80347×10^7	−50~−46	19.52	28.60
20200717gt3l	2.03345×10^6~2.03355×10^6	−49~−43	17.86	41.39
20200721gt2l	1.80376×10^7~1.80377×10^7	−48~−43	5.81	13.87
20201016gt2l	2.03760×10^6~2.03765×10^6	−52~−45	36.52	50.90
20190420gt2l	2.03455×10^6~2.03466×10^6	−54~−49	5.77	13.87
Average value			24.80	39.40

3.3. Bathymetric Accuracy and Comparison

To enhance bathymetric accuracy, it is crucial to perform coordinate correction of photons during water depth extraction to mitigate the effects of refraction and tides. Using 20190119gt3l and 20181024gt3r as examples, coordinate correction was conducted following the methodology outlined in Section 2.2.3, and the results are illustrated in Figure 9.

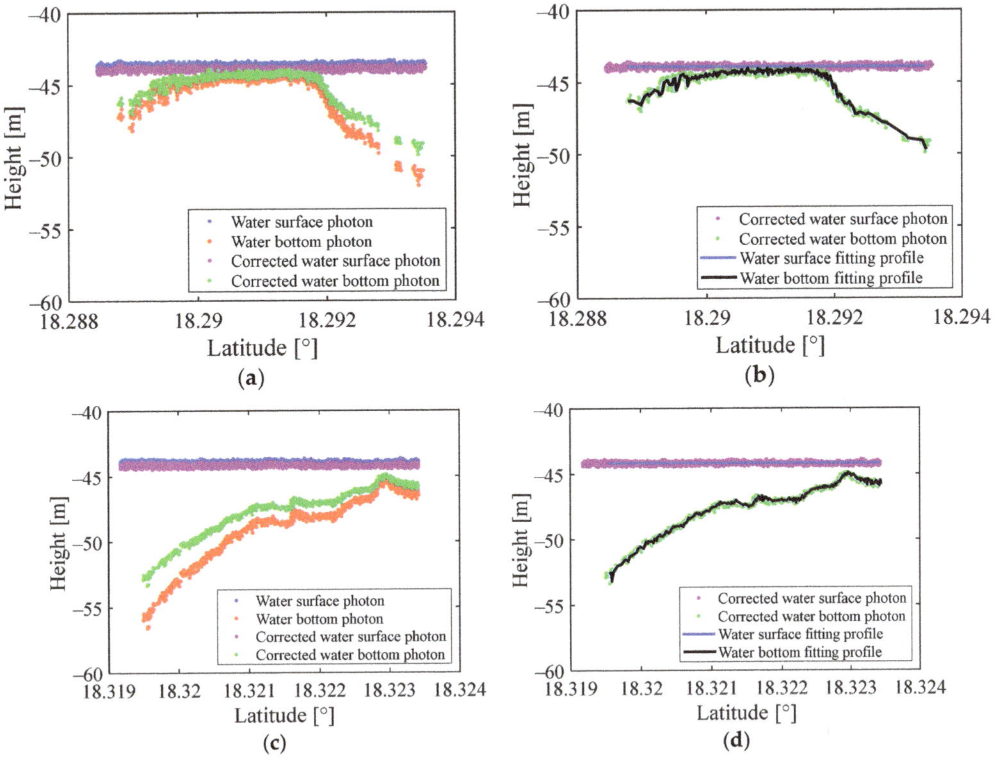

Figure 9. Coordinate correction and fitting profiles of the signal photons. (**a**) 20190119gt3l—Coordinate correction; (**b**) 20190119gt3l—Fitting profiles; (**c**) 20181024gt3r—Coordinate correction; (**d**) 20181024gt3r—Fitting profiles.

In Figure 9, the original coordinates of the surface and bottom photons are denoted by blue and red dots, respectively. Corrected coordinates after photon correction for surface and bottom photons are indicated by purple and green dots, respectively. Figure 9a,c demonstrate that the actual positions of water bottom photons are higher than their original positions due to the refraction of the seawater. Furthermore, the deeper the water, the greater the deviation between actual and original positions. Conversely, as shown in Figure 9a,c, the corrected positions of water surface photons were lower than their original positions. This adjustment aligns with the study area's reference water level, which is typically MLLW (Mean Lower Low Water); hence, instantaneous water levels are typically higher than this reference. Therefore, water surface photons were adjusted downward to the relative position of the reference water level to derive water depth data based on this standard. Following refraction and tidal corrections, a fitting procedure was employed to extract depth information from both water surface and bottom photons, as depicted in Figure 9b,d, respectively. Profiles of the water bottom and reference water surface are represented by black and blue lines, respectively, with detailed fitting methods explained in Section 2.2.3.

Using the aforementioned method, we successfully extracted bathymetric data from the ATL03 data. To validate their bathymetric accuracy, we utilized ALB data introduced in Section 2.1.3 as in situ data for verifying the bathymetric results obtained via photon-counting LiDAR. Taking data 20190119gt3l and 20190420gt2l as examples, a comparison between bathymetric data and in situ data is illustrated in Figure 10a,c, where the red and black points represent bathymetric and in situ data, respectively. From Figure 10a,c, it is evident that the water bottom profiles derived from the bathymetric data closely match the in situ data profiles. Concurrently, comparative experiments were conducted using the traditional OPTICS algorithm, as shown in Figure 10b,d. Comparing Figure 10a,c with 10b,d, it is observable that the water bottom profiles obtained using the traditional OPTICS method exhibit deviations from the in situ data in certain areas, primarily due to noise points near the water bottom. To further elucidate this observation, we quantitatively compared the bathymetric accuracy of both methods using MAE and $RMSE$. The results of this comparison are presented in Table 4.

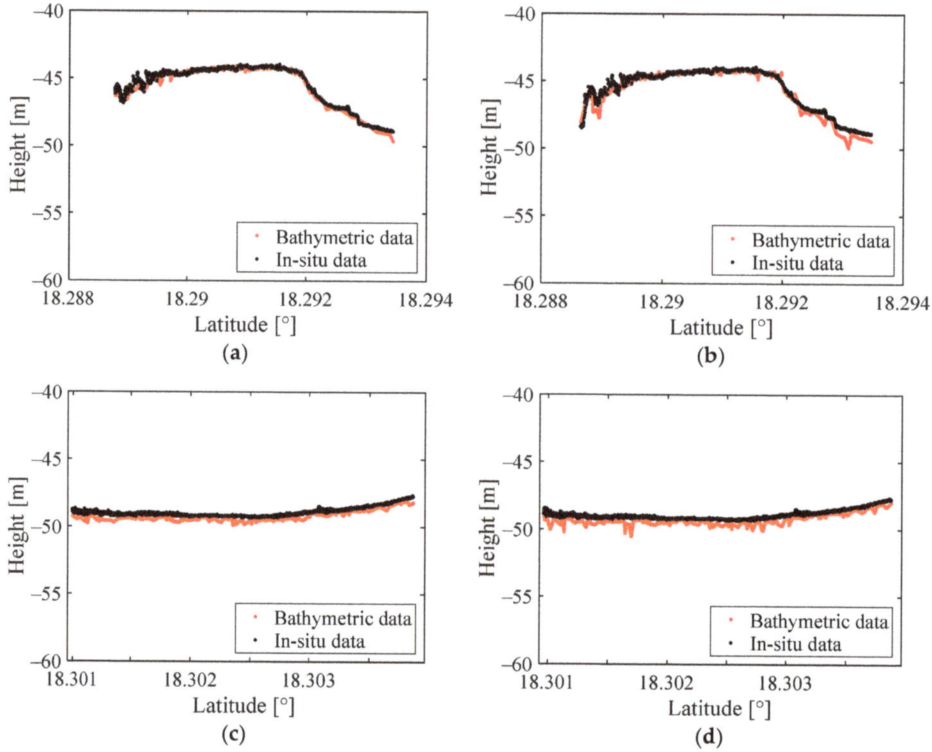

Figure 10. Bathymetric accuracy validation and comparison of our method and traditional OPTICS. (**a**,**b**) on the first row correspond to 20190119gt3l, while (**c**,**d**) on the second row correspond to 20190420gt2l. The red and black points represent the bathymetric results of the corresponding method and in situ data, respectively. (**a**) 20190119gt3l—Our method; (**b**) 20190119gt3l—Traditional OPTICS; (**c**) 20190420gt2l—Our method; (**d**) 20190420gt2l—Traditional OPTICS.

Table 4. MAE and $RMSE$ of our method and traditional OPTICS.

Data Name	Our Method		Traditional OPTICS	
	MAE	$RMSE$	MAE	$RMSE$
20190119gt3l	0.17	0.24	0.27	0.41
20181024gt3r	0.10	0.13	0.11	0.15
20200717gt3l	0.28	0.33	0.31	0.41
20200721gt2l	0.49	0.51	0.55	0.56
20201016gt2l	0.29	0.33	0.34	0.42
20190420gt2l	0.32	0.34	0.38	0.43
Average value	0.28	0.31	0.33	0.40

Using the ALB data from Section 2.1.3 as the in situ reference in Table 4, the MAE of the bathymetric data obtained by our method ranges between 0.10 m and 0.49 m, averaging 0.28 m, while the $RMSE$ ranges from 0.13 m to 0.51 m, averaging 0.31 m. From the six ATL03 data, 20181024gt3r has the lowest MAE and $RMSE$ while 20181024gt3r has the highest. That is probably related to the LiDAR point number of the data. Combining with Table 1, we can find the ATL03 data with more photon points has lower MAE and $RMSE$ in most cases, which also means the data with stronger signal intensity are more likely to have a better bathymetric result. In comparison, the MAE ranges from 0.11 m to 0.55 m, averaging 0.33 m, with $RMSE$ values of the bathymetric data obtained by traditional OPTICS model ranging from 0.15 m to 0.56 m and averaging 0.40 m. Our method achieves lower MAE and $RMSE$ values compared to the traditional OPTICS model. Specifically, our method reduces the average MAE by 0.05 m and the average $RMSE$ by 0.09 m. This demonstrates that our method provides bathymetric data with smaller errors relative to the in situ data, indicating higher accuracy compared to the traditional OPTICS model.

4. Discussion

4.1. The Parameters of AV-OPTICS

In Section 3.2, it was established that the AV-OPTICS model outperformed the traditional OPTICS model in denoising near the water bottom. This section aims to explore the underlying reasons by analyzing two key input variables: the semi-minor axis b and the semi-major axis a of the elliptical filter. Under ideal conditions, a specific contour of water bottom photons can be approximated as a rectangle. When considering the elliptical filter without horizontal dimension consideration, it resembles a vertical line segment of length $2b$, as illustrated in Figure 11.

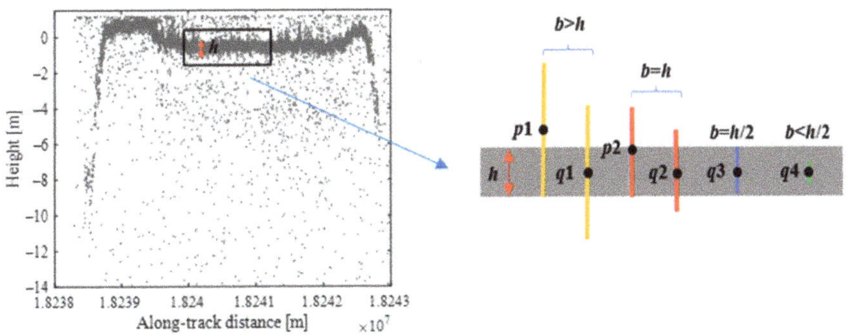

Figure 11. Idealized model of elliptic filter in vertical direction. The water bottom contour within the black block is regarded as a gray rectangle. The yellow, red, blue, and green lines are the idealized elliptical filters with different lengths of semi-minor axis, respectively.

In Figure 11, if the vertical width of the water bottom contour is denoted as h, the points outside this range are considered as noise points. Identified point $p1$ as a noise point, when the semi-minor axis $b > h$, the number of points within the idealized elliptical filter centered on $p1$ is approximately equal to the number of points within the idealized elliptical filter centered on $q1$, which is the central point of the water bottom contour. This implies that the distribution densities of points $p1$ and $q1$ are similar, making $p1$ less likely to be removed as a noise point. To effectively remove $p1$, it is necessary to reduce the value of b. When the semi-minor axis b decreases to h, the number of points within the idealized elliptical filter centered on edge point $p2$ of the water bottom contour is approximately equal to the number of points within the ideal elliptical filter centered on $q2$, which is the central point of the water bottom contour. However, given the elevation distribution of the water bottom contour, which resembles a Gaussian distribution, it is unreasonable to assume identical density distributions between the edge and central points. According to

Equation (1), when b is too small, the density threshold N_{MinPts} becomes extremely low, leaving many discrete noise points that do not satisfy the requirements. From the above analysis, it can be concluded that the value of the semi-minor axis b should be in the range of 0 to h. However, the values at the extremes of this range are not suitable. Therefore, a compromise was achieved by setting $b = h/2$.

Similar to the analysis of the semi-minor axis, considering the detailed parts of the water bottom contour as a rectangle and neglecting the vertical distance of the elliptical filter, the ellipse can be simplified as a horizontal line segment of length $2a$. The horizontal direction of the idealized model is depicted in Figure 12.

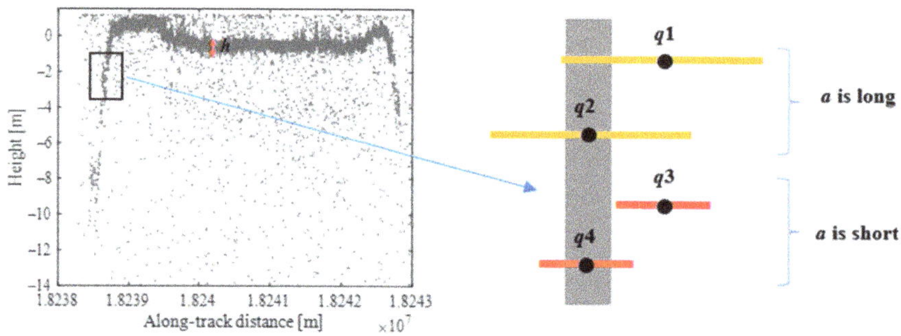

Figure 12. Idealized model of elliptic filter in horizontal direction. The water bottom contour within the black block is regarded as a gray rectangle. The yellow and red lines are the idealized elliptical filters with different lengths of semi-major axis, respectively.

Based on the above analysis, if point $q1$ is a noise point and the length of the yellow segment equals to $2a$, the distribution densities of points $q1$ and central points $q2$ are similar, and both $q1$ and the central point $q2$ are treated as noise points and removed, resulting in the disappearance of the water bottom details on both sides. Conversely, if point $q3$ is a noise point and $2a$ is shortened to the length of the red segment, the neighborhood density around noise point $q3$ and detail center $q4$ will differ, allowing $q4$ to be preserved while $q3$ is removed. However, when the value of a is too low, it causes N_{MinPts} to be extremely small, resulting in the retention of more discrete noise points. From this analysis, it is evident that maintaining the semi-minor axis b a constant and increasing the semi-major axis a leads to fewer discrete noise points but may sacrifice some water bottom details on both sides. Conversely, decreasing a preserves the water bottom details on both sides but allows more discrete noise points to persist. Therefore, the selection of a should balance the preservation of terrain details on both sides and minimize discrete noise points. Based on the theoretical basis of Section 2.2.2, the minimum value of a when $N_{MinPts} = 4$ is chosen as the semi-major axis to minimize discrete noise points while preserving water bottom details on both sides as far as possible.

Based on this analysis, we can elucidate the denoising effects of the AV-OPTICS model compared to the traditional OPTICS algorithm. In various experimental scenarios, the semi-minor b in the traditional OPTICS algorithm was approximately or more than twice the value obtained automatically using our method. As discussed, an excessively large value of b can result in noise photons near the water bottom having similar distribution densities to the water bottom signal photons at the contour's center, posing challenges for differentiation. This adversely impacts the denoising efficacy in the vicinity of the water bottom. Overall, compromised denoising near the water bottom affects precision P, resulting in decreased $F1$ values and impacting overall denoising accuracy.

4.2. Error Analysis of Bathymetry

To further demonstrate the enhanced bathymetry performance of our method, in this section, we use two data, 20190119gt3l and 20201016gt2l, as examples and assess the errors between the bathymetric data points obtained by both methods and the in situ data points by plotting error distribution histograms, as depicted in Figure 13. In these figures, the x-axis represents the error values between the bathymetric data and in situ data with an interval of 0.5 m, while the y-axis denotes the percentage of points corresponding to each error value. Figure 13a,c show the error statistics for the AV-OPTICS model, whereas Figure 13b,d indicate those for the traditional OPTICS method. From the graphs, it is evident that our method yields depth errors predominantly within the range of −1 m to 1 m, with approximately 90% of errors having an absolute value within 0.5 m. In contrast, the traditional OPTICS method shows a wider distribution range of depth errors, with fewer errors having an absolute value within 0.5 m compared to the AV-OPTICS model.

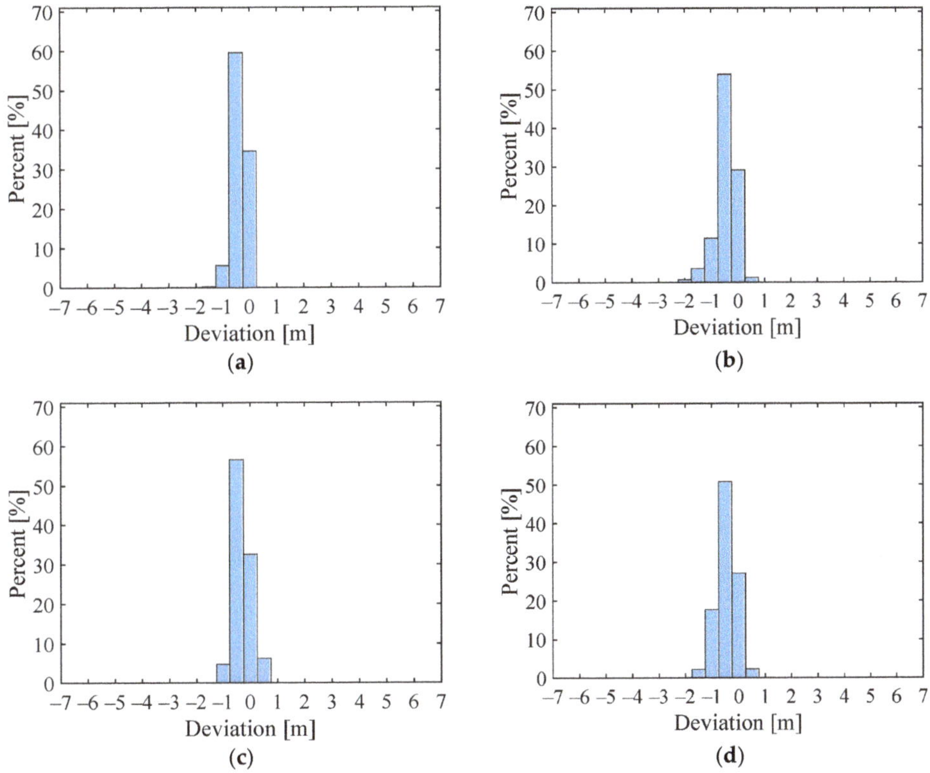

Figure 13. Deviation percentage between ICESat-2 results and ALB in situ data, with an interval of 0.5 m for each histogram column. (**a**,**b**) on the first row correspond to 20190119gt3l, while (**c**,**d**) on the second row correspond to 20201016gt2l. (**a**) 20190119gt3l—Our method; (**b**) 20190119gt3l—Traditional OPTICS; (**c**) 20201016gt2l—Our method; (**d**) 20201016gt2l—Traditional OPTICS.

To better evaluate the error distribution, we calculated the percentages of depth errors within the intervals [−1 m, 1 m] and [−0.5 m, 0.5 m] for six data under each method, summarized in Table 5. According to Table 5, our method shows that nearly all errors fall within the [−1 m, 1 m] range, with an average percentage of 99.81%. Within the narrower range of [−0.5 m, 0.5 m], the average percentage is 83.48%. In contrast, the traditional OPTICS algorithm averages 98.35% of errors within the [−1 m, 1 m] range and 76.75% within the [−0.5 m, 0.5 m] range. These percentages are lower compared to our method. This indicates that our method offers improved detection performance and bathymetric accuracy relative to the traditional OPTICS algorithm. This improvement stems from the AV-OPTICS model's effective denoising near the water bottom, which reduces the impact of nearby noise photons on the water bottom profile, thereby concentrating the error distribution and minimizing depth measurement deviations.

Table 5. Proportion of error distribution of our method and traditional OPTICS.

Data Name	Our Method		Traditional OPTICS	
	Proportion of [−1, 1]	Proportion of [−0.5, 0.5]	Proportion of [−1, 1]	Proportion of [−0.5, 0.5]
20190119gt3l	0.9974	0.9420	0.9574	0.8308
20181024gt3r	1.0000	0.9995	1.0000	0.9952
20200717gt3l	0.9923	0.8866	0.9829	0.8504
20200721gt2l	0.9991	0.4004	0.9951	0.3507
20201016gt2l	1.0000	0.8915	0.9786	0.7791
20190420gt2l	1.0000	0.8885	0.9867	0.7987
Average value	0.9981	0.8348	0.9835	0.7675

4.3. Influence of Other Factors on Bathymetry

Throughout the bathymetry process, in addition to the AV-OPTICS model, two critical processes were employed: refraction correction and tidal correction. This section investigates their impact on bathymetric accuracy using 20190119gt3l and 20181024gt3r as examples. Firstly, without refraction correction, the bathymetric data are illustrated in Figure 14b,e. These figures reveal a significant decrease in bathymetric accuracy compared to the complete process, particularly with larger deviations as water depth increases. Similarly, omitting tidal correction throughout the process yields results shown in Figure 14c,f. Here, the absence of tidal correction causes a slight downward shift in the water bottom profile compared to the complete process, leading to decreased bathymetric accuracy. The detailed evaluation indices are listed in Table 6.

Table 6. MAE and $RMSE$ of our method compared with the work without coordinate correction.

Data Name	Our Method		Our Method without Refraction Correction		Our Method without Tide Correction	
	MAE	$RMSE$	MAE	$RMSE$	MAE	$RMSE$
20190119gt3l	0.17	0.24	0.85	1.05	0.39	0.45
20181024gt3r	0.10	0.13	1.46	1.63	0.25	0.28
20200717gt3l	0.28	0.33	0.43	0.54	0.29	0.34
20200721gt2l	0.49	0.51	1.19	1.21	0.50	0.52
20201016gt2l	0.29	0.33	1.68	1.83	0.33	0.38
20190420gt2l	0.32	0.34	2.19	2.20	0.32	0.34
Average value	0.28	0.31	1.30	1.41	0.35	0.39

The observations from Figure 14 and the evaluation indices in Table 6 indicate that both refraction correction and tidal correction influence the bathymetric accuracy. Among them, refraction correction had a greater impact, with deviations increasing as the water depth increased. To determine the reasons for this, we observed the magnitudes of the depth variations induced by each process. According to calculations from the S-TIDE model, tidal correction results in depth variations of approximately 0.2 m across the six data. Conversely, refraction correction, governed by Snell's law, leads to depth variations roughly 0.25 times the original water depth, increasing proportionally with depth. For data with mean depths approximately equal to 4 m, the influence of refraction correction on MAE and $RMSE$ values of bathymetric data is on the order of 10^0 m, whereas tidal correction impacts MAE and $RMSE$ values on the order of 10^{-1} m. Therefore, the magnitude of depth variations caused by these correction processes significantly affects bathymetric accuracy.

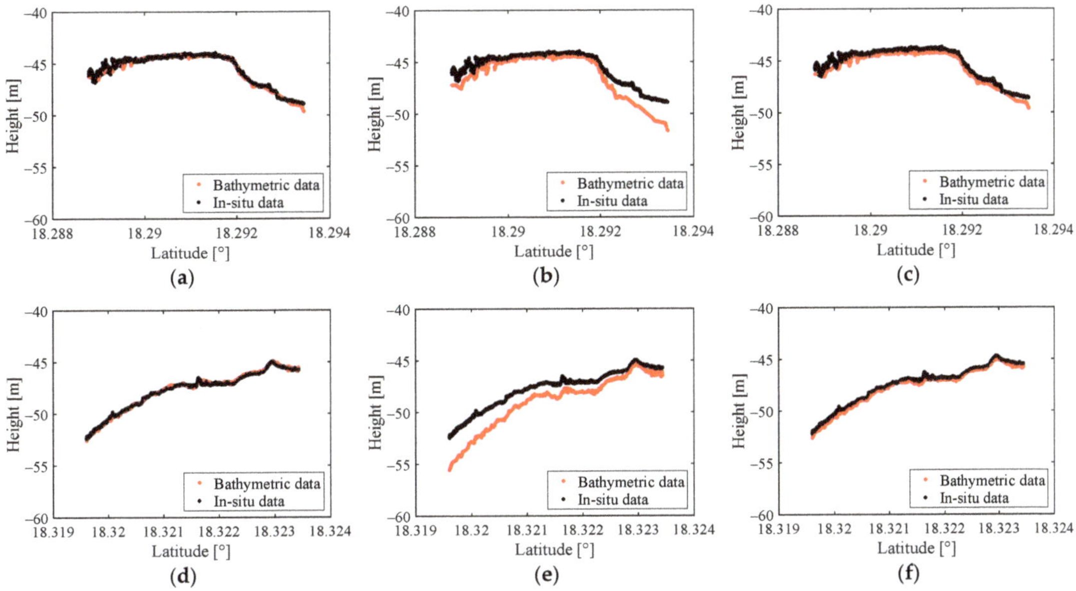

Figure 14. Bathymetric accuracy comparison of our method and that without coordinate correction. (**a–c**) on the first row correspond to 20190119gt3l, while (**d–f**) on the second row correspond to 20181024gt3r. The red and black lines represent the bathymetric results of the corresponding method and in situ data, respectively. (**a**) 20190119gt3r—Our method; (**b**) Without refraction correction; (**c**) Without tidal correction; (**d**) 20181024gt3r—Our method; (**e**) Without refraction correction; (**f**) Without tide correction.

5. Conclusions

This study presents a photon-counting LiDAR bathymetry method based on the Adaptive Variable OPTICS (AV-OPTICS) model. Unlike traditional OPTICS methods with fixed input variables, this approach automatically calculates the input variables based on point cloud distribution. This adjustment ensures precise measurement of photon distribution parameters near the water bottom, thereby enhancing denoising effects in these areas and improving bathymetric accuracy. Experimental validation using ATL03 photon data from Culebra Island, Puerto Rico, demonstrates better denoising performance compared to traditional OPTICS methods, reflected in higher $F1$-values and lower cohesion. Using ALB data as in situ reference, our method achieves an average MAE of 0.28 m and $RMSE$ of 0.31 m, indicating better bathymetric accuracy than traditional OPTICS methods. This study establishes a robust framework for shallow-water bathymetry based on photon-counting LiDAR. Future research will integrate our AV-OPTICS method with satellite remote sensing images for comprehensive 3D mapping of underwater landscapes.

Author Contributions: Conceptualization, P.L. and K.L.; methodology, P.L. and K.L.; software, P.L., Y.X. and Y.S.; validation, P.L., K.L. and Y.Z.; writing—original draft preparation, P.L.; writing—review and editing, P.L., Y.Z. and K.L. All authors have read and agreed to the published version of the manuscript.

Funding: This work was supported by the CNSA pre-research Project on Civil Aerospace Technologies, grant number D040107.

Data Availability Statement: The ATL03 photon point cloud data are available at https://search.earthdata.nasa.gov/search?q=ATLAS, accessed on 25 April 2024; The airborne LiDAR bathymetry data are available at https://www.coast.noaa.gov/dataviewer/#/LiDAR/search/, accessed on 10 May 2024.

Acknowledgments: We wish to thank NASA for providing ATL03 data and NOAA for providing Airborne LiDAR bathymetry data.

Conflicts of Interest: The authors declare no conflicts of interest.

References

1. Li, Y.; Zhou, X.; Li, G.; Guo, J.; Ma, Y.; Chen, Y. Progress and Prospect of Space-borne Photon-counting Lidar Shallow Water Bathymetry Technology. *Infrared Laser Eng.* **2022**, *51*, 107–116. Available online: https://kns.cnki.net/kcms/detail/12.1261.TN.20220310.1926.005.html (accessed on 2 February 2024).
2. Nicholls, R.J.; Cazenave, A. Sea-level rise and its impact on Coastal Zones. *Science* **2010**, *328*, 1517–1520. [CrossRef] [PubMed]
3. Wang, Y.; Zhang, J.; Zheng, Y.; Xu, Y.; Xu, J.; Jiao, J.; Su, Y.; Lv, H.; Liang, K. Brillouin scattering spectrum for liquid detection and applications in oceanography. *Opto-Electron. Adv.* **2023**, *6*, 43–53. [CrossRef]
4. Wang, B.; Ma, Y.; Zhang, J.; Zhang, H.; Zhu, H.; Leng, Z.; Zhang, X.; Cui, A. Noise Removal Algorithm Based on Adaptive Elevation Difference Thresholding for ICESat-2 Photon-counting Data. *Int. J. Appl. Earth Obs. Geoinf.* **2023**, *117*, 103207. [CrossRef]
5. Janowski, L.; Trzcinska, K.; Tegowski, J.; Kruss, A.; Rucinska-Zjadacz, M.; Pocwiardowski, P. Nearshore Benthic Habitat Mapping Based on Multi-Frequency, Multibeam Echosounder Data Using a Combined Object-Based Approach: A Case Study from the Rowy Site in the Southern Baltic Sea. *Remote Sens.* **2018**, *10*, 1983. [CrossRef]
6. Martí, A.; Portell, J.; Amblas, D.; de Cabrera, F.; Vilà, M.; Riba, J.; Mitchell, G. Compression of Multibeam Echosounders Bathymetry and Water Column Data. *Remote Sens.* **2022**, *14*, 2063. [CrossRef]
7. Casal, G.; Harris, P.; Monteys, X.; Hedley, J.; Cahalane, C.; McCarthy, T. Understanding satellite-derived bathymetry using sentinel 2 imagery and spatial prediction models. *GISci. Remote Sens.* **2020**, *57*, 271–286. [CrossRef]
8. Wright, C.W.; Kranenburg, C.; Battista, T.A.; Parrish, C. Depth calibration and validation of the Experimental Advanced Airborne Research Lidar, EAARL-B. *J. Coast. Res.* **2016**, *76*, 4–17. [CrossRef]
9. Zhao, Y.; Wang, Y.; Liang, K.; Xu, Y.; Guo, Y.; Makame, K. Underwater Temperature and Salinity Measurement by Rayleigh–Brillouin Spectroscopy Using Fizeau Interferometer and PMT Array. *Remote Sens.* **2024**, *16*, 2214. [CrossRef]
10. Renga, A.; Rufino, G.; D'Errico, M.; Moccia, A.; Boccia, V.; Graziano, M.D.; Aragno, C.; Zoffoli, S. SAR bathymetry in the Tyrrhenian Sea by COSMO-SkyMed data: A novel approach. *IEEE J. Sel. Top. Appl. Earth Obs. Remote Sens.* **2014**, *7*, 2834–2847. [CrossRef]
11. He, J.; Xu, Y.; Sun, H.; Jiang, Q.; Yang, L.; Kong, W.; Liu, Y. Sea Surface Height Wavenumber Spectrum from Airborne Interferometric Radar Altimeter. *Remote Sens.* **2024**, *16*, 1359. [CrossRef]
12. Parrish, C.E.; Magruder, L.A.; Neuenschwander, A.L.; Forfinski-Sarkozi, N.; Alonzo, M.; Jasinski, M. Validation of ICESat-2 ATLAS Bathymetry and Analysis of ATLAS's Bathymetric Mapping Performance. *Remote Sens.* **2019**, *11*, 1634. [CrossRef]
13. Kutser, T.; Hedley, J.; Giardino, C.; Roelfsema, C.; Brando, V.E. Remote sensing of shallow waters—A 50 year retrospective and future directions. *Remote Sens. Environ.* **2020**, *240*, 111619. [CrossRef]
14. Magruder, L.A.; Brunt, K.M. Performance Analysis of Airborne Photon- Counting Lidar Data in Preparation for the ICESat-2 Mission. *IEEE Trans. Geosci. Remote Sens.* **2018**, *56*, 2911–2918. [CrossRef]
15. Xie, J.; Zhong, J.; Mo, F.; Liu, R.; Li, X.; Yang, X.; Zeng, J. Denoising and Accuracy Evaluation of ICESat-2/ATLAS Photon Data for Nearshore Waters Based on Improved Local Distance Statistics. *Remote Sens.* **2023**, *15*, 2828. [CrossRef]
16. Markus, T.; Neumann, T.; Martino, A.; Abdalati, W.; Brunt, K.; Csatho, B.; Farrell, S.; Fricker, H.; Gardner, A.; Harding, D.; et al. The Ice, Cloud, and land Elevation Satellite-2 (ICESat-2): Science requirements, concept, and implementation. *Remote Sens. Environ.* **2017**, *190*, 260–273. [CrossRef]
17. Magruder, L.A.; Wharton, M.E., III; Stout, K.D.; Neuenschwander, A.L. Noise filtering techniques for photon-counting ladar data. In Proceedings of the SPIE 8379, Laser Radar Technology and Applications XVII, Baltimore, MD, USA, 24–26 April 2012; Volume 83790Q, p. 83790Q. [CrossRef]
18. Neumann, T.A.; Martino, A.J.; Markus, T.; Bae, S.; Bock, M.R.; Brenner, A.C.; Brunt, K.M.; Cavanaugh, J.; Fernandes, S.T.; Hancock, D.W.; et al. The ice, cloud, and Land Elevation Satellite—2 mission: A global geolocated photon product derived from the Advanced Topographic Laser Altimeter System. *Remote Sens. Environ.* **2019**, *233*, 111325. [CrossRef]
19. Brunt, K.M.; Neumann, T.A.; Walsh, K.M.; Markus, T. Determination of local slope on the Greenland ice sheet using a multibeam photon-counting lidar in preparation for the ICESAT-2 Mission. *IEEE Geosci. Remote Sens. Lett.* **2014**, *11*, 935–939. [CrossRef]
20. Chen, B.; Pang, Y. A denoising approach for detection of canopy and ground from ICESat-2's airborne simulator data in Maryland, USA. In Proceedings of the SPIE 9671, AOPC 2015: Advances in Laser Technology and Applications, Beijing, China, 5–7 May 2015; Volume 96711S, p. 96711S. [CrossRef]
21. Moussavi, M.S.; Abdalati, W.; Scambos, T.; Neuenschwander, A. Applicability of an automatic surface detection approach to micro-pulse photon-counting lidar altimetry data: Implications for canopy height retrieval from future ICESat-2 data. *Int. J. Remote Sens.* **2014**, *35*, 5263–5279. [CrossRef]
22. Chen, Y.; Le, Y.; Zhang, D.; Wang, Y.; Qiu, Z.; Wang, L. A photon-counting LiDAR bathymetric method based on adaptive variable ellipse filtering. *Remote Sens. Environ.* **2021**, *256*, 112326. [CrossRef]
23. Zhang, W.; Xu, N.; Ma, Y.; Yang, B.; Zhang, Z.; Wang, X.H.; Li, S. A maximum bathymetric depth model to simulate satellite photon-counting lidar performance. *ISPRS J. Photogramm. Remote Sens.* **2021**, *174*, 182–197. [CrossRef]

24. Ma, Y.; Xu, N.; Liu, Z.; Yang, B.; Yang, F.; Wang, X.H.; Li, S. Satellite-derived bathymetry using the ICESat-2 lidar and Sentinel-2 imagery datasets. *Remote Sens. Environ.* **2020**, *250*, 112047. [CrossRef]
25. Babbel, B.J.; Parrish, C.E.; Magruder, L.A. ICESat-2 Elevation Retrievals in Support of Satellite-Derived Bathymetry for Global Science Applications. *Geophys. Res. Lett.* **2021**, *48*, e2020GL090629. [CrossRef] [PubMed]
26. Zhu, X.; Nie, S.; Wang, C.; Xi, X.; Wang, J.; Li, D.; Zhou, H. A noise removal algorithm based on OPTICS for Photon-Counting LiDAR Data. *IEEE Geosci. Remote Sens. Lett.* **2021**, *18*, 1471–1475. [CrossRef]
27. Wang, X.; Pan, Z.; Glennie, C. A novel noise filtering model for photon-counting laser altimeter data. *IEEE Geosci. Remote Sens. Lett.* **2016**, *13*, 947–951. [CrossRef]
28. Zhang, J.; Kerekes, J.; Csatho, B.; Schenk, T.; Wheelwright, R. A clustering approach for detection of ground in micropulse photon-counting LiDAR altimeter data. In Proceedings of the IEEE Geoscience and Remote Sensing Symposium, Quebec City, QC, Canada, 13–18 July 2014; pp. 177–180. [CrossRef]
29. Ankerst, M.; Breunig, M.M.; Kriegel, H.P.; Sander, J. OPTICS: Ordering points to identify the clustering structure. *ACM SIGMOD Rec.* **1999**, *28*, 49–60. [CrossRef]
30. Febriana, L.N.; Sitanggang, S.I. Outlier Detection on Hotspot Data in Riau Province using OPTICS Algorithm. *IOP Conf. Ser. Earth Environ. Sci.* **2017**, *58*, 012004. [CrossRef]
31. Zhang, S.; Zhang, S.; Qiao, N.; Wang, Y.; Du, Q. Modelling and Mitigating Wind Turbine Clutter in Space–Air Bistatic Radar. *Remote Sens.* **2024**, *16*, 2674. [CrossRef]
32. Xi, X.; Wang, Z.; Wang, C. Bathymetric Extraction Method of Nearshore Based on ICESat-2/ATLAS Data. *J. Tongji Univ. (Nat. Sci.)* **2022**, *50*, 940–946. [CrossRef]
33. Jia, K.; Ma, Y.; Zhang, J.; Wang, B.; Zhang, X.; Cui, A. A Denoising Methodology for Detecting ICESat-2 Bathymetry Photons Based on Quasi Full Waveform. *IEEE Trans. Geosci. Remote Sens.* **2024**, *62*, 4207916. [CrossRef]
34. Ma, Y.; Xu, N.; Sun, J.; Wang, X.H.; Yang, F.; Li, S. Estimating water levels and volumes of lakes dated back to the 1980s using Landsat imagery and photon-counting lidar datasets. *Remote Sens. Environ.* **2019**, *232*, 111287. [CrossRef]
35. Yue, S.; Li, P.; Guo, J.; Zhou, S. A statistical information-based clustering approach in distance space. *J. Zhejiang Univ. Sci. A (Sci. Eng.)* **2005**, *01*, 72–79. [CrossRef]
36. Saputra, M.E.; Mawengkang, H.; Nababan, E.B. Determination value K in K-nearest neighbor with local mean euclidean and weight Gini Index. *IOP Conf. Ser. Mater. Sci. Eng.* **2018**, *420*, 012098. [CrossRef]
37. Khambampati, A.K.; Liu, D.; Konki, S.K.; Kim, K.Y. An automatic detection of the ROI using otsu Thresholding in Nonlinear Difference Eit Imaging. *IEEE Sens. J.* **2018**, *18*, 5133–5142. [CrossRef]
38. Chen, L.; Xing, S.; Zhang, G.; Guo, S.; Gao, M. Refraction Correction Based on ATL03 Photon Parameter Tracking for Improving ICESat-2 Bathymetry Accuracy. *Remote Sens.* **2024**, *16*, 84. [CrossRef]
39. Pan, H.; Lv, X.; Wang, Y.; Matte, P.; Chen, H.; Jin, G. Exploration of tidal-fluvial interaction in the Columbia River estuary using S_TIDE. *J. Geophys. Res. Ocean.* **2018**, *123*, 6598–6619. [CrossRef]
40. Hess, K.W. Tidal Datums and Tide Coordination. *J. Coast. Res.* **2003**, *38*, 33–43. Available online: http://www.jstor.org/stable/25736598 (accessed on 20 May 2024).

remote sensing

MDPI

Article

An Ensemble Machine Learning Approach for Sea Ice Monitoring Using CFOSAT/SCAT Data

Yanping Luo, Yang Liu *, Chuanyang Huang and Fangcheng Han

Fisheries College, Ocean University of China, Qingdao 266003, China; luoyanping@stu.ouc.edu.cn (Y.L.); huangchuanyang@stu.ouc.edu.cn (C.H.); hfc@stu.ouc.edu.cn (F.H.)
* Correspondence: yangliu315@ouc.edu.cn

Abstract: Sea ice is a crucial component of the global climate system. The China–French Ocean Satellite Scatterometer (CFOSAT/SCAT, CSCAT) employs an innovative rotating fan beam system. This study applied principal component analysis (PCA) to extract classification features and developed an ensemble machine learning approach for sea ice detection. PCA identified key features from CSCAT's backscatter information, representing outer and sweet swath observations. The ensemble model's performances (OA and Kappa) for the Northern and Southern Hemispheres were 0.930, 0.899, and 0.844, 0.747, respectively. CSCAT achieved an accuracy of over 0.9 for close ice and open water but less than 0.3 for open ice, with misclassification of open ice as closed ice. The sea ice extent discrepancy between CSCAT and the National Snow and Ice Data Center (NSIDC) was -0.06 ± 0.36 million km^2 in the Northern Hemisphere and -0.03 ± 0.48 million km^2 in the Southern Hemisphere. CSCAT's sea ice closely matched synthetic aperture radar (SAR) imagery, indicating effective sea ice and open water differentiation. CSCAT accurately distinguished sea ice from open water but struggled with open ice classification, with misclassifications in the Arctic's Greenland Sea and Hudson Bay, and the Antarctic's sea ice–water boundary.

Keywords: sea ice; CFOSAT/SCAT; PCA; ensemble machine learning

1. Introduction

Sea ice is a mixture of ice crystals, air bubbles, and brine formed in seawater. It covers approximately 7–15% of the world's oceans and is one of the most important components of the cryosphere. Sea ice influences the global ocean radiative flux balance [1], thermohaline circulation [2], biogeochemical cycles [3], and polar navigation and operations [4], as one of the global cold sources. The increase in global temperature in recent decades has led to significant changes in polar sea ice. Satellite records suggest that the extent of Arctic sea ice has repeatedly reached record lows [5], while the extent of Antarctic sea ice varies greatly, experiencing periods of positive trends and maxima of sea ice extent [6–8]. However, it has also reached record lows in recent years [9]. This change in sea ice triggers several global impacts, such as changes in marine ecosystems, atmospheric circulation, and Arctic warming [10]. Therefore, it is important to monitor the distribution of sea ice in the polar regions.

Microwave remote sensing is widely used due to its capabilities for weather-independence, all-day observation, long-term monitoring, and comprehensive coverage [11]. In particular, satellite scatterometers have attracted attention as an active microwave remote sensing system that operates without imaging capabilities. These systems are crucial for providing real-time, accurate parameters of sea ice by emitting microwave pulses towards the Earth's surface and analyzing the backscattered signals to infer surface properties. Table 1 provides a comprehensive compilation of various microwave scatterometer systems along with relevant bibliographic references, demonstrating their application in sea ice monitoring by several international agencies. It is noteworthy that these scatterometers predominantly work in the C-band (5.3 GHz) and Ku-band (13.5 GHz) frequencies. Based on their beam system, they are further classified into fixed fan beam, rotating pencil beam, and rotating fan beam scatterometers. Based on polarization, these systems are differentiated into vertical polarization (VV) and horizontal polarization (HH). The European Space Agency (ESA) is leading the advancement of C-band single polarized scatterometers with fixed fan beam, particularly through the development of instruments such as the Active Microwave Instrument-Scatterometer (AMI-SCAT) and Advanced Scatterometer (ASCAT). On the other hand, the National Aeronautics and Space Administration (NASA) focused more on Ku-band dual-polarization scatterometers. This technology has evolved significantly, moving from early fixed fan beam scatterometers (such as SeaSat-A Scatterometer System, SASS, and the NASA Scatterometer Satellite, NSCAT) to advanced rotating pencil beam scatterometers (such as SeaWinds and RapidScat). In addition to ESA and NASA, India and China have also made commendable progress in this field through the independent development of the Oceansat Scatterometer (OSCAT) and HY-2 Scatterometer (HSCAT). These systems are comparable to NASA's SeaWinds in terms of frequency, polarization, and beam system. A notable milestone was reached in 2021 with the launch of the Chinese–French Oceanography Satellite (CFOSAT), a collaborative project between China and France that carries the innovative Chinese-French Ocean Satellite Scatterometer (CSCAT)—a Ku-band dual polarization rotating fan beam scatterometer. To further expand this range of instruments, China's FY-3E satellite was deployed in 2022, equipped with Wind Radar (WindRad), which uses dual-frequency and dual-polarization. This deployment is particularly notable, as it marks the first use of horizontal C-band polarization in Earth observation [12], signifying a new chapter in the microwave remote sensing of sea ice.

Satellite scatterometers can provide daily observations of polar regions and are commonly used to map the extent of sea ice. Table 1 lists retrieval algorithms used for sea ice detection related to satellite scatterometers, which are broadly divided into two groups. The first group is called the Remund/Long-NSCAT (RL-N) algorithm, which uses linear discriminant analysis to classify ice and water by constructing features such as polarization ratio and frequency ratio [13]. The versatility of the RL-N algorithm is evident in its application to data from a variety of scatterometers, including OSCAT, HSCAT, and CSCAT, among others. These results show good consistency with sea ice concentration. However, the method has certain limitations; the accuracy of the RL-N algorithm can be significantly affected by wind-induced surface roughness and summer ice melt [14]. These effects need to be mitigated in the post-classification process by employing binary image processing techniques and sea ice growth/retreat constraint methods [15,16]. The second group refers to the Royal Netherlands Meteorological Institute (KNMI) algorithm proposed by the Royal Netherlands Meteorological Institute. This algorithm is designed for the analysis of sea ice using AMI-SCAT data and introduces geophysical model functions (GMFs) for sea ice detection. It contains Bayesian classifiers that are used to determine sea ice, as described by de Haan and Stoffelen [17] and Verspeek [18]. This approach has been successfully applied to satellite data from SeaWinds, ASCAT, HSCAT, and CSCAT. These GMFs for different sensors exhibit unique characteristics. AMI-SCAT's GMF describes sea ice probability using an ice cone line in the three-dimensional $\sigma 0$ space, assuming isotropic backscattering [17,18]. ASCAT's GMF is based on a linear relationship between the forward, mid, and aft beams, with sea ice backscatter characteristics changing with incidence angle [19]. SeaWinds' GMF represents sea ice properties through a linear relationship between vertically and horizontally polarized signals [20]. CSCAT, on the other hand, uses a look-up table to define the GMF due to the complex observation model [21]. This validation indicates that the method provides more accurate information for characterizing sea ice during the melting season. The GMFs have strong dependency on specific types of remote sensors, which can impose significant limitations when adapting to data from other sensor types or platforms.

It is worth noting that the introduction of new rotating fan beam systems such as CSCAT and WindRad introduces variability in incidence and azimuth angles, which has a significant impact on backscatter from OW and sea ice surfaces [22]. To deal with the complexity that arises from multiple angles of incidence and azimuth, the RL-N and KNMI algorithms have also been adapted for use with the CSCAT. Zhai et al. [23] addressed this problem by calculating the polarization ratio using the average of the horizontal and vertical polarization CSCAT backscatter coefficients over angles of incidence and azimuth. However, directly calculating the polarization ratio may reduce the detection accuracy because sea ice and seawater have different sensitivities to incidence and azimuth angles. At the same time, the CSCAT GMF model was first introduced in CSCAT by Liu et al. [24], which simplified the problem of mixed geometry observations by selecting backscatter observations near a 40° incidence angle to build a geophysical model of sea ice. The method of selecting a 40° incidence angle reduces the rotating fan beam system to a rotating pencil bean system. Li et al. [21] proposed a method of creating an incidence angle lookup table to construct a GMF to solve the problem of mixing incidence and azimuth observations. They observed a large standard deviation between the lowest and highest incidence angles when constructing a look-up table model, and therefore truncated the observations at both incidence angles. Moreover, the linear relationship between the incidence angle and the backscatter coefficient has also been used to correct this effect [13,25]. In essence, polarization ratio, frequency ratio, and normalization coefficients are still unable to fully mitigate the errors caused by wind-induced sea surface roughness. For CSCAT, with its multi-angle observations, the construction of GMFs is significantly more complex than traditional three-dimensional spatial distributions. It can no longer be expressed using a simple three-dimensional mathematical formula and instead requires a lookup table for implementation. Principal component analysis (PCA), traditionally viewed as a downscaling method, serves a dual role: It reduces the dimensionality of the feature space and highlights uncorrelated features [26,27]. Previous studies used PCA in sea ice retrieval by combining passive microwave radiometers and scatterometers [28] and have showed that it is effective in reducing the complexity of feature space and improving sea ice classification efficiency. However, the application of PCA to scatterometer observations for classification feature extraction has not been studied. This represents an opportunity for novel research that could potentially affect the accuracy of sea ice detection using scatterometer data.

The study aims to (1) extract classification features from CSCAT observations using PCA, (2) build an ensemble machine learning model to detect sea ice in both the Northern and Southern Hemispheres, and (3) validate the results of CSCAT sea ice detection by comparing with similar types of sea ice products and assessing the validity and feasibility of the developed model. As such, this paper not only eliminates the dependency on specific functions through the use of PCA, but also presents an automated algorithmic framework that requires no empirical parameters or manual identification, making it versatile enough to be applied across various scatterometer platforms.

Table 1. Representative microwave scatterometer system and related references for sea ice application from various agencies in different countries (modified based on Long [29]).

Sensor	Agency	Frequency	Polarization	Reference	Mission	Dates
SASS	NASA	Ku	2VV*2 2HH*2	Yueh et al. [30]	SeaSat	1978.06–1978.10
AMI-SCAT	ESA	C	3VV	Verspeek [18] [2] de Haan and Stoffelen [17] [2]	ERS-1	1991.07–1996.07
					ERS-2	1995.04–2001.01
NSCAT	NASA	Ku	3VV*2 1HH*2	Remund and Long [13] [1]	ADEOS	1996.09–1997.06
SeaWinds	NASA	Ku	HH-inner VV-outer	Belmonte Rivas and Stoffelen [20] [2]	ADEOS-2	2002.12–2003.10
					QuikSCAT	1999.06–2009.11

Table 1. *Cont.*

Sensor	Agency	Frequency	Polarization	Reference	Mission	Dates
ASCAT	ESA	C	3VV*2	Belmonte Rivas et al. [19] [2] Breivik et al. [31] [2] Lindell and Long [32] [2] Aaboe et al. [33] [2]	Metop-A	2006.10–2021.11
					Metop-B	2012.09–
					Metop-C	2018.11–
OSCAT	ISRO	Ku	HH-inner VV-outer	Hill and Long [34] [1]	OceanSat-2 ScatSAT-1	2009.09–2014.02 2016.09–2021.02
HSCAT	NSOAS	Ku	HH-inner VV-outer	Xu et al. [35] [1] Li et al. [16] [1] Zou et al. [36] [1]	HY-2A	2011.08–2022.04
					HY-2B	2018.10–
					HY-2C	2020.09–
					HY-2D	2021.05–
RapidScat	NASA	Ku	VV HH	Singh et al. [28] [1]	ISS RapidScat	2014.09–2016.08
CSCAT	CNSA	Ku	VV HH	Liu et al. [24] [2] Zhai et al. [23] [1] Li et al. [21] [2] Liu et al. [37] [2] Xu et al. [38] [1]	CFOSAT	2018.10–2023.01
WindRAD	CMA	C/Ku	VV*2 HH*2	Zhai et al. [39] [1]	FY-3E	2021.05–

[1] Algorithm belonging to RL-N. [2] Algorithm belonging to KNMI.

2. Materials and Methods

2.1. Materials

2.1.1. Input Data

The CSCAT Level 2A (L2A) products used here were maintained by the National Satellite Ocean Application Service, can be derived from https://osdds.nsoas.org.cn, accessed on 21 January 2023, and cover the period 2019–2022. CSCAT on-board CFOSAT is a real aperture radar operating in the Ku band (13.256 GHz) and collects vertical (VV) and horizontal (HH) polarization backscatter coefficient from two rotating fan beam antennas (Figure 1). The CSCAT covers a 1050 km swath divided into 25 km and 12.5 km regular grid wind vector cells (WVCs), with incidence angles ranging from 25° to 48° for fan beams. For the objectives of this study, the 25 km resolution provides sufficient detail and accuracy, while also demonstrating good quality and stability in our study area, making it the most suitable choice for this research. WVCs for cross-track orbit near ~43°S are shown in Figure 1c. According to Li et al. [40], the WVCs can be classified into three groups: outer swath WVCs (number 1–5, number 38–42); sweet swath WVCs (number 6–12, number 31–37); and nadir swath WVCs (number 13–30). CSCAT provides multiple views in the Ku band σ^0_{VV} and σ^0_{HH} per WVC, with a smaller number of views and less azimuth diversity in the outer and nadir swath and a larger number of views and more azimuth diversity in the sweet swath [41]. The incidence angle of WVCs exhibits axisymmetric properties relative to the nadir point and shows a large range of variability as the nadir point is approached. The azimuth angle shows central symmetry with respect to the nadir point. This characteristic distribution is displayed in Figure 1b and was a direct consequence of the rotating fan beam system of the CSCAT. It represented the forward- or backward-looking wind. The design of CSCAT cleverly combined the features of fixed fan beam and rotating pencil beam scatterometers. This combination not only expanded the swath, but also significantly increased the diversity of observation geometries within a specific swath.

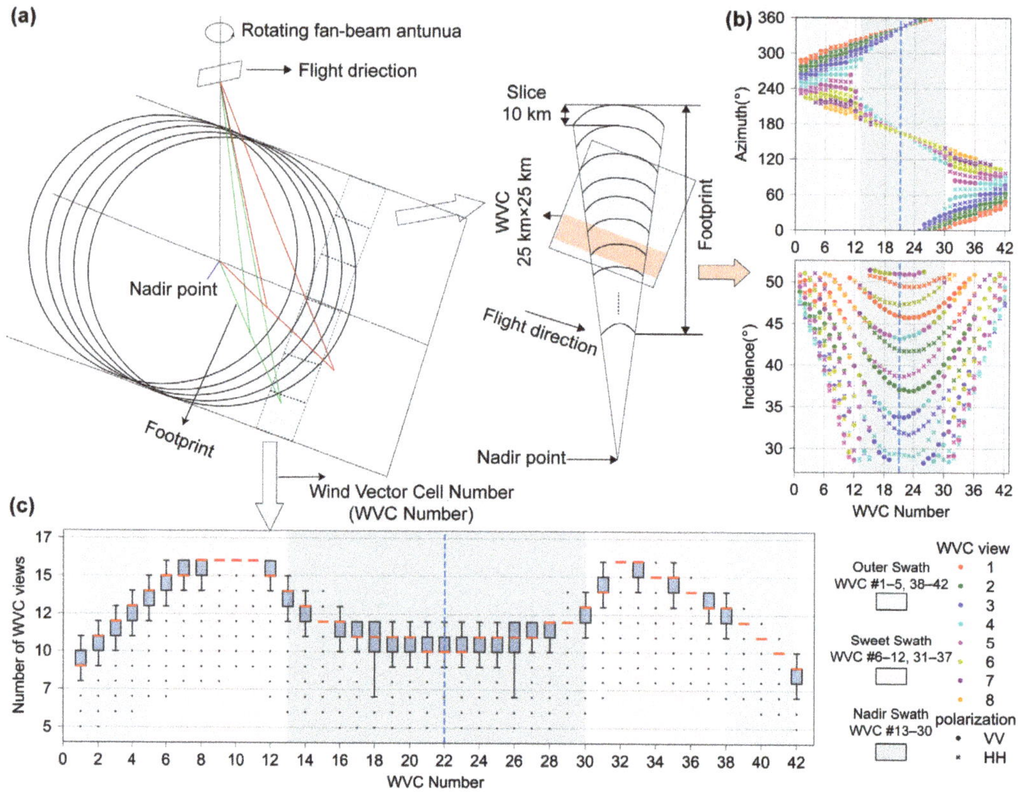

Figure 1. (**a**) Observation geometry of CSCAT adapted from Zhang et al. [42]. (**b**) Incidence and azimuth angles versus the cross-track wind vector cell (WVC) number for a row at a latitude of ~43°S from orbit observed on 1 January 2019 at 07:56:26, showcasing WVC views in color and σ^0_{VV} and σ^0_{HH} using symbolic circles and forks, respectively. (**c**) The average number of views at WVC across the swath.

2.1.2. Auxiliary Data

Auxiliary data were needed for three different purposes, as outlined in Table 2. First, it was necessary for prior reference information. Second, it was required for validating the detection results. Finally, a static dataset was needed that was relevant to the polar grid projection and controlled the quality of the detection results.

Table 2. Data used for research and model construction.

Data Set	Time Coverage	Spatial Coverage	Spatiotemporal Resolution	Data Source
Global reprocessed sea ice concentration	1978.10.25–2022.12.31	−90°–90°N −180°–180°E	Daily/25 km	https://doi.org/10.48670/moi-00136, accessed on 21 January 2023
ASCAT sea ice edge	2019.1.1–2022.12.31	−90°–90°N −180°–180°E	Daily/10 km	https://osi-saf.eumetsat.int/products/osi-402-d, accessed on 10 April 2023

Table 2. Cont.

Data Set	Time Coverage	Spatial Coverage	Spatiotemporal Resolution	Data Source
Sea Ice Concentrations from Nimbus-7 SMMR and DMSP SSM/I-SSMIS Passive Microwave Data, Version 1	2019.1.1–2022.12.31	−90°–90°N −180°–180°E	Daily/25 km	https://nsidc.org/data/nsidc-0051/versions/1, accessed on 21 January 2023
Sentinel-1 SAR	2019.6.19/2019.6.18/2019.3.8	Not Specified	Not Specified	https://vertex.daac.asf.alaska.edu/#, accessed on 1 March 2023
Polar stereographic cell area	Not Specified	−90°–90°N −180°–180°E	Not Specified/25 km	https://nsidc.org/data/nsidc-0771/versions/1, accessed on 10 February 2023
lmask_stere_100	Not Specified	−90°–90°N −180°–180°E	Not Specified/25 km	ftp://osisaf.met.no/docs/tools, accessed on 10 February 2023
Valid ice masks	1972.1.1–2007.12.31	39.5°–90°N −180°–180°E	Monthly/25 km	https://nsidc.org/data/nsidc-0622/versions/1, accessed on 20 February 2023

1. Prior reference datasets

Scatterometer-based sea ice detection is essentially a supervised classification problem. The process relies on a high-quality reference dataset to train and validate the performance of the model. According to Ivanova et al. [43] and Kern et al. [44], hybrid algorithms provided the most reliable estimates of sea ice concentration for climate monitoring purposes. In this study, we used from the CMEMS Ocean and Sea Ice Thematic Assembly Centre (OSI TAC) global reprocessed sea ice concentration product as prior reference dataset, which is a redistribution of the European Organization for the Exploitation of Meteorological Satellites (EUMETSAT) OSISAF Climate Data Record (CDR), labelled OSI-450, and the Interim Climate Data Record (ICDR), labelled OSI-430-a. This dataset was derived from Scanning Multichannel Microwave Radiometer (SMMR), Special Sensor Microwave-Imager (SSM/I), and Special Sensor Microwave-Imager/Sounder (SSMIS) data and uses a hybrid, adaptive, and self-optimizing sea ice concentration algorithm, which is an extension and improvement of the bootstrap algorithm [45] and the Bristol algorithm [46]. The algorithm utilized three channels (~19 GHz for vertical polarization and ~37 GHz for horizontal and vertical polarization) to provide optimal accuracy in both open water and consolidated ice conditions. It was characterized by quantifiability, temporal consistency, and sustainability and was therefore suitable as a prior reference dataset compared to independent estimates of sea ice concentration in regions with extremely high and low ice concentrations [47]. To match the time series of CSCAT observations, we obtained records of daily sea ice concentration for the period from 1 January 2019 to 31 December 2022. The data were projected as Equal-Area Scalable Earth 2.0 Grids (EASE-Grid 2.0) with a spatial resolution of 25 km.

2. Comparison datasets

To ensure an independent evaluation beyond OSISAF sea ice concentration, we introduced two additional datasets for third-party comparison, which allowed us to compare scatterometer and microwave radiometer sea ice cover results. Two similar sea ice products were used for spatial comparisons at a given time. As one of the comparison data sets, the study used the near real-time sea ice edge products from OSISAF of the EUMETSAT. These sea ice edge was derived from atmospherically corrected SSMIS brightness temperatures and ASCAT backscatter values through a Bayesian detection approach, which classified the data into open water (OW), open ice (OI), and close ice (CI) [31]. In addition, to validate the accuracy of the CSCAT sea ice extent detection results, sea ice concentrations from Nimbus-7 SMMR and DMSP SSM/I-SSMIS Passive Microwave Data Version 1 (hereinafter referred to as NSIDC sea ice concentration) [48] serves as an independent comparison dataset. This NSIDC dataset, generated from SSMIS brightness temperature data, used the established NASA Team (NT) algorithm developed by the Oceans and Ice Branch at NASA's Goddard Space Flight Center (GSFC) [49]. The dataset provided a 25-km daily sea ice concentration for both polar regions from 26 October 1978 to 31 May 2022.

High-resolution images from Sentinel-1, part of European Space Agency (ESA) Copernicus program, were used for local visual validation. The Sentinel-1 constellation, consisting of Sentinel-1A and 1B, carries C-band Synthetic Aperture Radar (SAR) sensors, which provide reliable all-weather, day-and-night Earth observations. SAR images, known for their high resolution, have been widely utilized to validate scatterometer-based sea ice detection [50]. The analysis used Sentinel-1A/1B SAR Level-1, extra wide (EW) swath images in ground range-detected (GRD) mode with an approximate spatial resolution of 20 m × 40 m [51], to verify the CSCAT results over main areas of the Arctic and Antarctic.

3. Static datasets

The static dataset used in this study consists of polar grid area information [52] and a land-sea mask, that provides arrays of the area in millions, as well as a land-sea mask for the 25 km grid for the Northern and Southern Hemispheres, respectively.

To improve the accuracy of sea ice detection, the retrieval results here are quality controlled using effective ice mask data. Valid sea ice is defined as the likely presence of ice where it has existed in the past based on a 35-year climatology report. Among other things, the Arctic ice mask data was obtained from the US Ice Centre [53], which is a dataset based on Arctic sea ice charts and sea ice concentration data provided by the US National Ice Centre. It contains 12 files covering the maximum sea ice extent from 1972 to 2007, with each file representing each month of the year. The Antarctic-specific study determined the maximum monthly effective sea ice extent in Antarctica by analyzing daily climatological OSI-450 sea ice concentration data from 1978 to 2020 and counting pixels with sea ice concentrations greater than 0 over a period of 42 years.

2.2. Methods

This study used PCA and ensemble machine learning model to automate the detection of sea ice using Level-2A CSCAT data from 2019 to 2022. The sea ice detected using this technique was compared with sea ice concentrations derived from passive microwave data, the sea ice edge determined by passive microwave and active scatterometer data, and SAR data for specific regions. Figure 2 illustrates the workflow for sea ice detection. The first step involved preprocessing the Level 2A CSCAT time series data. This included checking the pixel quality of the land and rain flag, reprojection, regrinding from orbit to polar stereographic grid, and calculating the polarization ratio. The result was 8 views of σ^0_{HH}, σ^0_{VV} and $\sigma^0_{VV}/\sigma^0_{HH}$ (Figure 2a). PCA was then applied to those three scatterometer parameters. The main component that most closely suggested ice and water was chosen as the distinguishing feature (Figure 2b). Regions of interest around the Arctic and Antarctic were selected, where different ice/ocean classes' characteristics are expected. Corresponding prior information on sea ice concentration data were spatially and temporally matched with CSCAT using nearest neighbor interpolation (Figure 2c). At the same time, the period length was compared and determined for statistical analysis. This was followed by sample selection, from which single and ensemble machine learning classifiers were used to distinguish sea ice, with 80% of the data trained and validated with the remaining 20%. A preliminary sea ice map was then created based on an ensemble model. In the final phase, the map was refined by applying valid ice masks to reduce misdetection noise (Figure 2d).

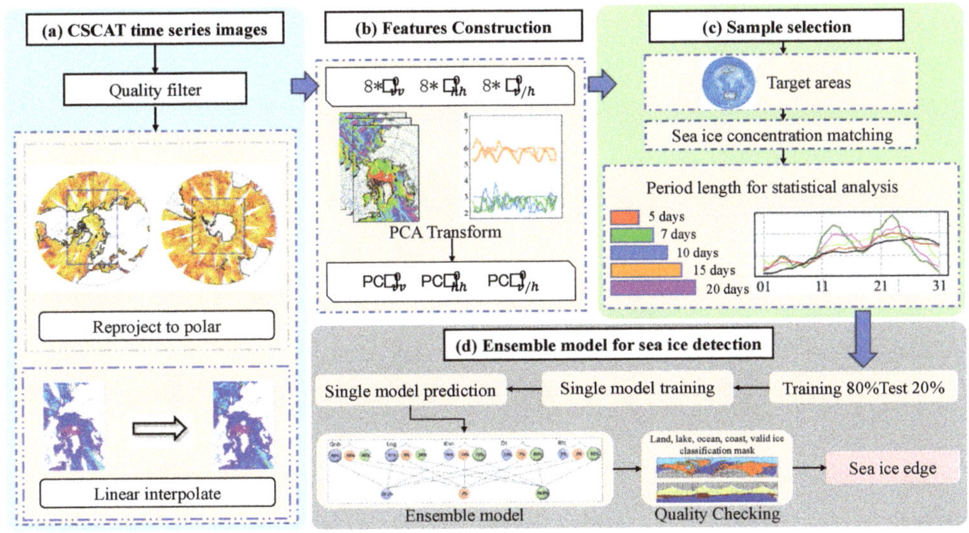

Figure 2. Workflow of this study.

2.2.1. Features Construction

The satellite orbit data were projected to the Northern and Southern Hemispheres. This was process was conducted in order to minimize deformation and keep the poles at the center of the image when studying the polar regions. For this purpose, the NSIDC projection program was used (https://nsidc.org/data/user-resources/help-center/guide-nsidcs-polar-stereographic-projection, accessed on 30 January 2023). The CSCAT L2A data we used were derived from WVC's orbit resampled data, recorded as backscattering σ^0 (dB). During the polar projection process, we filtered out pixels that were labeled as land and affected by precipitation. The precipitation information used for this filtering came from the quality indicators in the L2B wind field retrieval data, specifically the rain_fail and rain_detect flags in the wvc_quality dataset. We then divided the data into the Northern and Southern Hemispheres based on latitude and longitude and excluded pixels within which the number of WVC views was zero. The CSCAT L2A orbit data were projected onto the poles, and then the projected data was interpolated using the radial interpolation function of the SciPy library in Python to ensure consistency of the orbital centers and sides with the observed number of views in the sweet swath and to produce the projected backscattered observation dataset:

$$\sigma^0_{reproj} = \left\{\sigma^0_{VV,i}, \sigma^0_{HH,i}\right\}, i \in [1,8] \quad (1)$$

where i represents the WVC, $\sigma^0_{VV,i}$ represents the backscattering in vertical polarization for the WVC i, and $\sigma^0_{HH,i}$ represents the backscattering in horizontal polarization for the WVC i.

The polarization ratio $\sigma^0_{V/H,i}$ was calculated according to WVC with the following equation:

$$\sigma^0_{V/H,i} = \frac{\sigma^0_{VV,i}}{\sigma^0_{HH,i}}, i \in [1,8] \quad (2)$$

A polar backscatter observation dataset σ^0_{polar} was then constructed using the following equation:

$$\sigma^0_{polar} = \left\{\sigma^0_{VV,i}, \sigma^0_{HH,i}, \sigma^0_{V/H,i}\right\}, i \in [1,8] \quad (3)$$

This study selected the singular value decomposition method based on the Scikit-Learn library in Python [54]. PCA was then performed on the observations of datasets $\sigma^0_{VV,i}$, $\sigma^0_{HH,i}$, $\sigma^0_{VH,i}$, to obtain the principal components of σ^0_{PCA}:

$$\sigma^0_{PCA} = \left\{\sigma^0_{VV,PC_j}, \sigma^0_{HH,PC_j}, \sigma^0_{\frac{V}{H},PC_j}\right\}, j \in [1,8] \tag{4}$$

where j represents the number of principal components, σ^0_{VV,PC_j} refers to the principal component j of the vertical polarization, σ^0_{HH,PC_j} refers to the principal component j of the horizontal polarization, and $\sigma^0_{V/H,PC_j}$ represents the principal component j of the polarization ratio.

2.2.2. Dynamic Sampling

With the aim of obtaining representative samples of sea ice and seawater, training samples are crucial for sea ice detection. Microwave radiometric observations have revealed dynamic reference brightness temperatures of open water and sea ice (tie points) [55], demonstrating seasonal and interannual variability differences [56]. However, the seasonal and interannual variation of microwave scatter from open water and sea ice are not well understood. Currently, many methods for detecting sea ice using satellite scatterometers use a dynamic sample selection strategy or periodic model updates [34]. Liu et al. [57] collected 10 days of backscatter observations to build a lookup table for a dynamic sea ice geophysical model. The probability distribution function of open water and sea ice from OSI SAF was originally based on statistics from a fixed reference year (2007.3–2008.2) [31]. The probability density function's update frequency has gradually increased from monthly to 15 days [33]. Zhai et al. [23] also examined the effect of the length of the training sample period on the effectiveness of the model and found that the accuracy did not improve when the period of training sets exceeded 7 days. The length of the sampling period depends on the availability and accessibility of the referenced sea ice information. This information is typically obtained through various methods such as satellite remote sensing, buoy observations, and sensor measurements. Data can only be collected at specific time intervals, such as daily, weekly, or monthly. The choice of sampling period depends on the study methodology, data availability, and the need for reliable statistics. Therefore, it is necessary to analyze and determine the appropriate period length.

We used the daily CMEMS sea ice concentration climate dataset as prior reference. First, the daily CMEMS sea ice concentration climate dataset was converted to a projection plane consistent with CSCAT preprocessing in order to eliminate the impact of errors that may result from different projection systems and to ensure that the data remained consistent with the study area. For the Northern Hemisphere (Southern Hemisphere), the projections were converted from epsg6931 (epsg6932) to epsg3411 (epsg3412). According to the spatiotemporal matching principle, the eigenvalues of the principal component eigenvalues and the sea ice concentration were extracted from the regions of interest in the Northern and Southern Hemisphere, respectively (Figure 3, yellow blocks), to obtain the scattering properties and spatial distribution information of sea ice. In order to effectively detect sea ice, sea ice concentration was divided into less than 30% open water, 30%–70% open ice, or more than 70% close ice. Such a division clarified the different sea ice conditions and provided effective training examples σ^0_{Sample} for model training:

$$\sigma^0_{Sample} = \left\{\sigma^0_{VV,PC_k}, \sigma^0_{HH,PC_k}, \sigma^0_{V/H,PC_k}, SIE\right\}, k \in [1,8] \tag{5}$$

where k represented the number of principal components selected, σ^0_{VV,PC_k} was the selected principal component of vertical polarization, σ^0_{HH,PC_k} was the selected principal component of horizontal polarization, $\sigma^0_{V/H,PC_k}$ was the selected principal component of the polarization ratio, and SIE was the sample label determined by CMEMS, i.e., open water, open ice, or close ice. Microwave scattering from open water and sea ice exhibited different variation characteristics at seasonal and interannual scales, and these characteristics determined the selection of the sampling period in this study. For each specific day, we collected sea ice and open water samples from the last 5, 7, 10, 15, and 20 days (including the current day) to ensure a balanced and representative dataset. We combined five sample periods of 5, 7, 10, 15, and 20 days and analyzed the sample characteristics and model effects by constructing characteristic time series of open water, open ice, and close ice for the period 2019–2022, which were used for determination of the last sampling period.

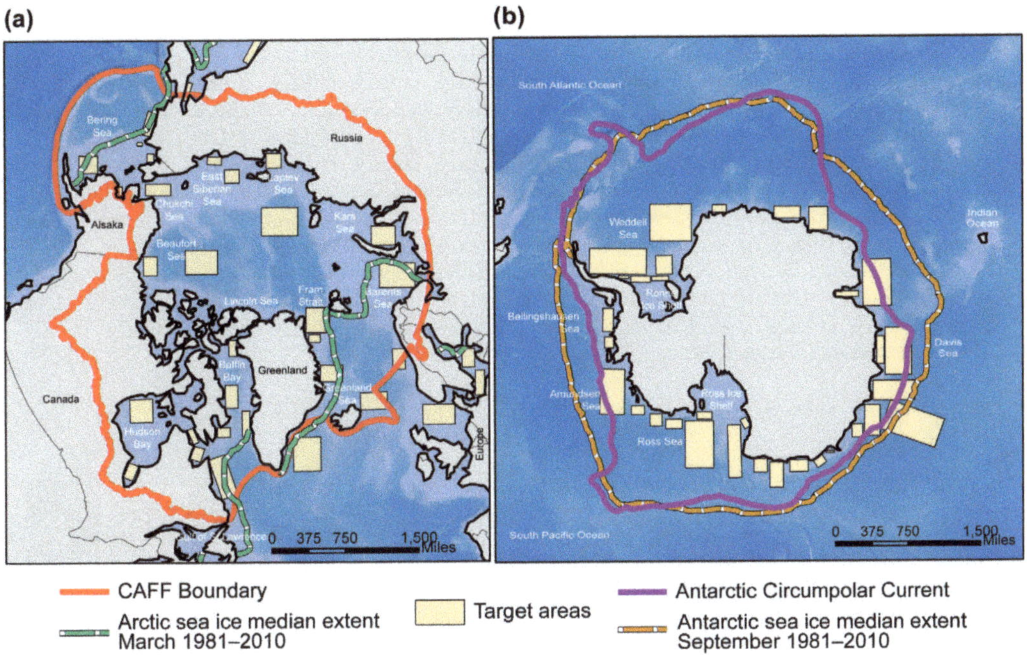

Figure 3. Location map over (**a**) the Northern Hemisphere and (**b**) the Southern Hemisphere for the regions (marked in yellow colors) used in sample selection overlaid on the CAFF Boundary [58], Antarctic Circumpolar Current (https://data.aad.gov.au/dataset/4892/download, accessed on 20 March 2023) and sea ice median extent [59].

2.2.3. Single and Ensemble Model Construction

Ensemble methods combine the predictions of several base estimators built with a given learning algorithm in order to improve generalizability/robustness over a single estimator. More generally, ensemble models can be applied to any base learner for the purposes of averaging methods such as bagging methods, model stacking, or voting, or boosting methods, such as in AdaBoost. We chose the average voting classifier as the ensemble model. The average voting ensemble model is a machine learning technique that combines multiple base estimators. Specifically, it computes the predicted probabilities for each class from each base estimator and then averages these probabilities. The final class prediction is made based on the class with the highest average probability. Such a classifier can be useful for a set of equally well-performing models in order to balance out their individual weaknesses.

In this study, we selected five base estimators: Gaussian naive Bayes (Gnb), logistic regression (Log), K-nearest neighbors (Knn), decision tree (Dt), and random forest classifier (Rfc) to build an ensemble learning model for sea ice detection in the Northern and Southern Hemispheres. The training process of the ensemble model (Figure 4) was as follows: (1)

The training and testing datasets were calculated based on the period length determined in the previous section. To ensure unbiased estimates, we divided the dataset into an 80% training set and a 20% testing set. (2) During model training, careful parameter tuning was performed to optimize the performance of each model. RandomizedSearchCV from the sklearn library was used to optimize the hyperparameters for each model. A 10-fold cross-validation approach assessed model performance, with 50 random hyperparameter combinations being sampled in order to identify the best configuration. The scoring = 'f1' metric was chosen to provide a comprehensive evaluation by considering both precision and recall. (3) After training, the probabilities predicted by each estimator were calculated for each class. That is, Gnb, Log, Knn, Dt, and Rfc generated probability maps for open water, open ice, and close ice, respectively. Specifically, the Gnb model computed the posterior probability by multiplying the conditional probability with the class prior probability. The Log model derived the class probability distribution through maximum likelihood estimation. The Knn model determined the classification probability based on the proportion of the nearest neighbors' classes. The Dt model calculated the class probability at the leaf node based on the proportion of samples in each class. Finally, the Rfc averaged the predicted probabilities across multiple decision trees to produce the final class probabilities. After obtaining the probabilities for open water, open ice, and close ice, the class with the highest probability was selected as the final classification for that pixel. (4) Then, the predicted probabilities of all classifiers were averaged to obtain the probability maps for open water, open ice, and close ice. The final prediction was based on the class with the highest probability.

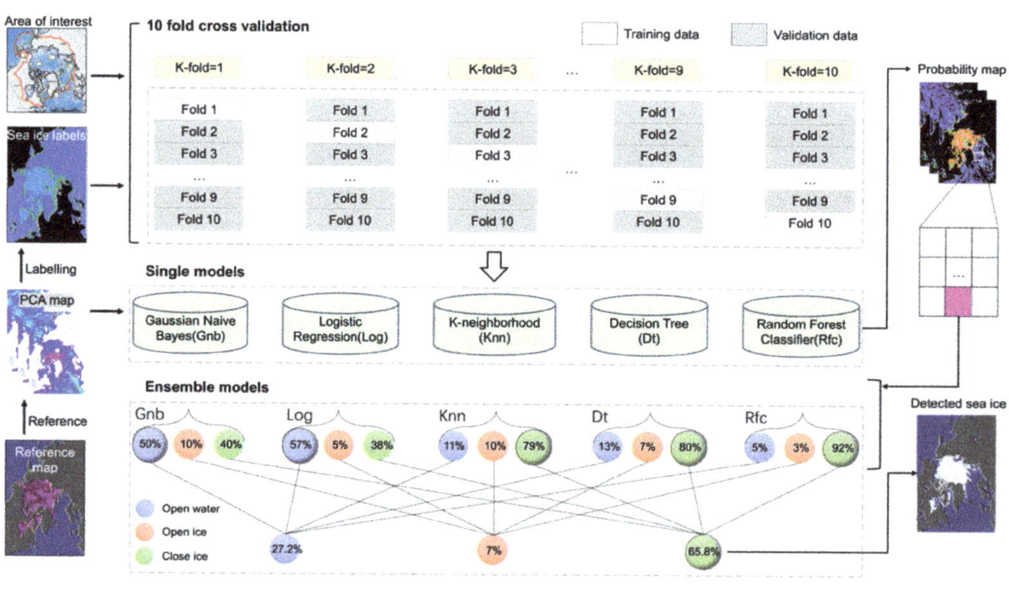

Figure 4. Model structure of the soft voting ensemble learning and training process.

2.2.4. Accuracy Assessment

To verify accuracy of models, we used 20% of the test samples to construct a confusion matrix. This yielded various validation parameters, such as user accuracy (UA), producer accuracy (PA), overall accuracy (OA), the F1 score (F1), and the kappa coefficient (Kappa).

3. Results

After reviewing the L2 orbit data and daily projected data, there were a total of 116 days where CSCAT data was unusable. Specifically, 50 days of Level 2A orbital observations were missing, and 66 days were eliminated due to quality control measures (e.g., removal of invalid values, rain-affected pixels, and land pixels). Additionally, since we needed to collect samples from the past 5 consecutive days, a total of 22 days were affected due to insufficient sample sizes. The specific relevant data and reasons for deletion are listed in the Table 3. No sea ice forecasts were made in the study for these 138 data days. The number of days used for statistical analysis was 90%.

Table 3. The list of invalid dates and reason for removal from statistical analysis from 1 January 2019 to 31 December 2022.

Invalid Date	Invalid Reason and Removed Quantity
2019/6/1, 2019/6/3, 2020/12/29, 2021/11/10, 2022/7/10 2019/7/15–2019/7/16 2019/12/21–2019/12/28 2019/12/30–2020/1/13 2022/8/10–2022/8/29	No CSCAT Level 2A data (50 days)
2019/6/4, 2019/12/6, 2019/12/20, 2019/12/29, 2020/1/14, 2020/9/3, 2021/6/16, 2022/8/30, 2022/11/9 2020/3/10–2020/3/11 2022/8/8–2022/8/9 2022/9/8–2022/9/19 2022/9/24–2022/9/26 2022/10/12–2022/10/18 2022/12/1–2022/12/31	Quality control elimination (66 days)
2019/1/1–2019/1/4 2020/1/15–2020/1/17 2021/2/20, 2021/2/24, 2022/11/13 2022/8/31–2022/9/2 2022/9/20–2022/9/23 2022/9/28–2022/9/30 2022/10/19–2022/10/20	Insufficient sample sizes (22 days)

3.1. Characteristics of CSCAT Features

The CSCAT data included eight VV polarization (σ^0_{VV}), eight HH polarization (σ^0_{HH}), and eight polarization ratios ($\sigma^0_{V/H}$). There were obvious correlations between the properties of these observations. Figure 5a,b show the Pearson's correlation for these observations in the Northern and Southern Hemisphere, respectively. As can be seen from the figures, there was not only a high correlation between adjacent σ^0_{VV} (or σ^0_{HH}), but also a correlation between the set of σ^0_{VV} and σ^0_{HH} for each WVC. We applied PCA to the CSCAT polar backscatter observation dataset σ^0_{polar} and assigned it to fewer principal components. With the bivariate plots of the principal component analysis (Figure 5c,d), we found that among the eight WVCs, the contributions of WVCs in views 3, 4, and 8 were more significant in σ^0_{VV} and σ^0_{HH}. In comparison, the results of view 8 showed a significant deviation from the results of views 3 and 4. Views 3 and 4 primarily occupied the outer swath and nadir swath, respectively. The WVCs of the nadir swath were characterized by azimuth angles of 0/360 or 180 degrees, corresponding to the forward and backward perspectives, accompanied by a wide range of incidence angles. Conversely, the WVCs in the outer swath were defined by azimuth angles of approximately 90 or 270 degrees, indicating lateral viewing angles, again with a wide range of incidence angles. View 8 was consistently located in the region designated the sweet swath, characterized by significant variability in the antenna's azimuth angle. According to Li et al. [40], the WVCs in the sweet swath were the optimal area for measuring sea surface wind. Likewise, the WVCs of the sweet swath in the principal component analysis occupied a larger amount of information. Therefore, the first two principal components of σ^0_{VV} and σ^0_{HH} effectively characterized the observed information in the outer swath, nadir swath, and sweet swath. In contrast, the bivariate plot of $\sigma^0_{V/H}$ shows that the first two principal components could only represent the outer swath, suggesting that the amount of information gathered using only the first two principal components accounted for almost 50% of the total observations. Spatial distribution analyses of the first four (out of eight) principal components conducted on 10 January 2019 for both the Northern and Southern Hemispheres (as shown in Figure 6a,b), revealed that the importance of σ^0_{VV} and σ^0_{HH} was mainly concentrated within the first two components. Meanwhile, the significance of the VH polarization was evenly distributed across all four components.

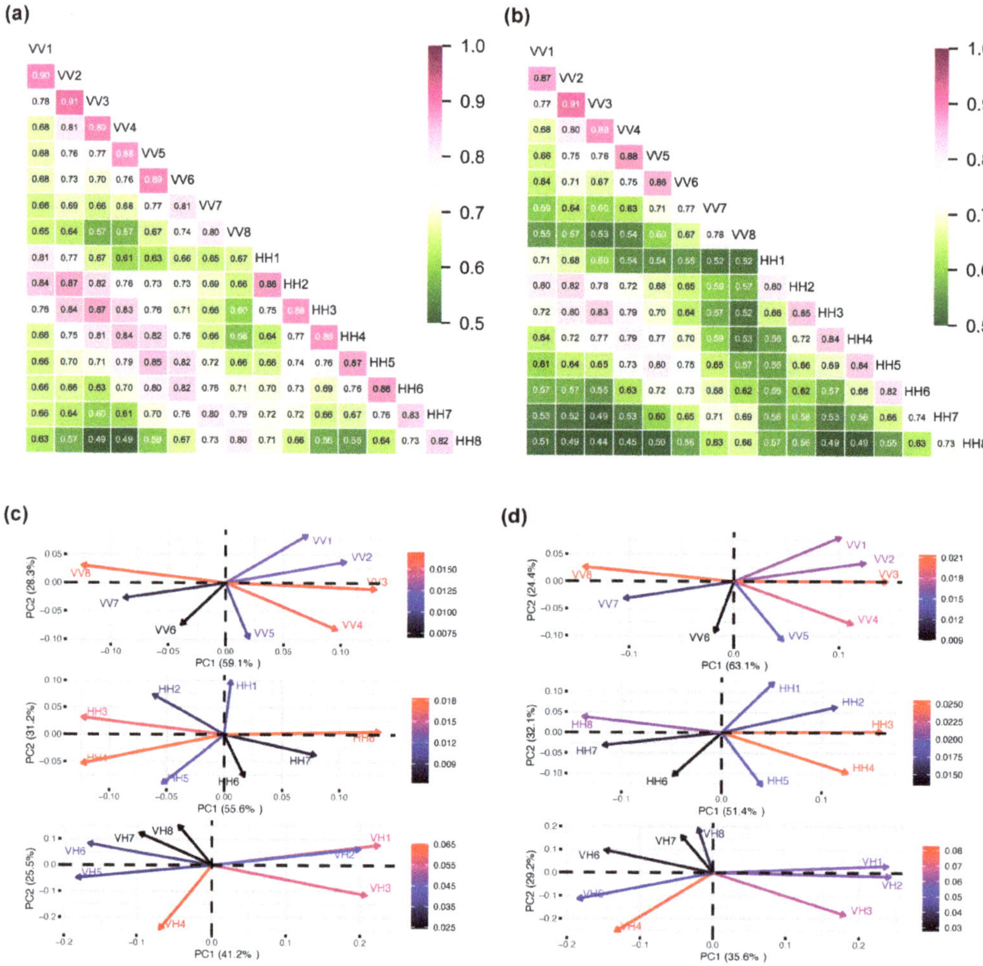

Figure 5. Pearson's correlation coefficients in the (**a**) Northern Hemisphere and (**b**) Southern Hemisphere and related principal component analysis (PCA) bioplots of CSCAT backscatter observations over the (**c**) Northern and (**d**) Southern Hemispheres on 10 January 2019.

Figure 7 shows the daily variation curves for the contributions of the first two principal components of σ^0_{VV} and σ^0_{HH} and the first four principal components of $\sigma^0_{V/H}$. These results suggested that the choice of the first two principal components for $\sigma^0_{VV}(\sigma^0_{HH})$ polarization accounted for over 80% of the total variance explained, while the choice of the first four principal components for $\sigma^0_{V/H}$ explained more than 65% of the total variance.

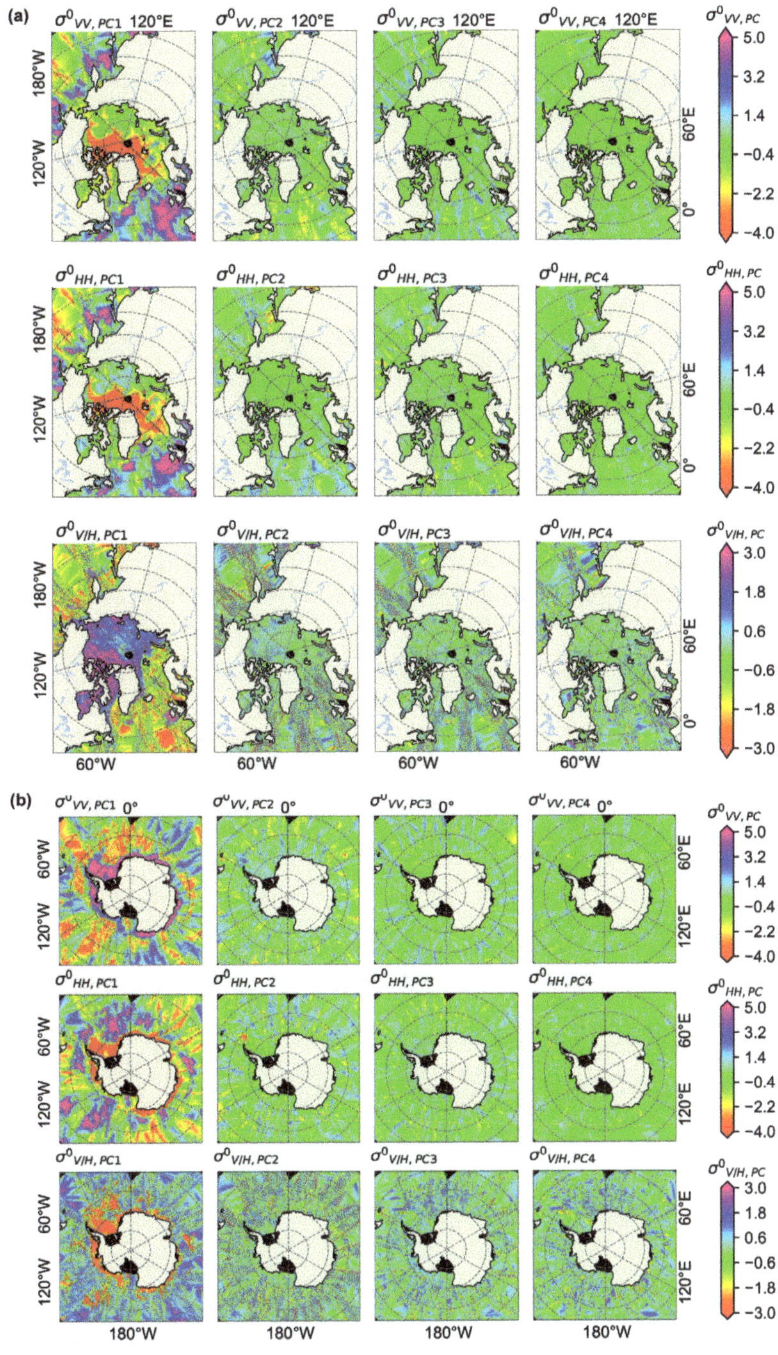

Figure 6. Spatial distribution of the first four (out of eight) principal components of σ^0_{VV}, σ^0_{HH} and σ^0_{VV} and $\sigma^0_{V/H}$ polarization in the (**a**) Northern Hemisphere and (**b**) Southern Hemisphere on 10 January 2019, respectively.

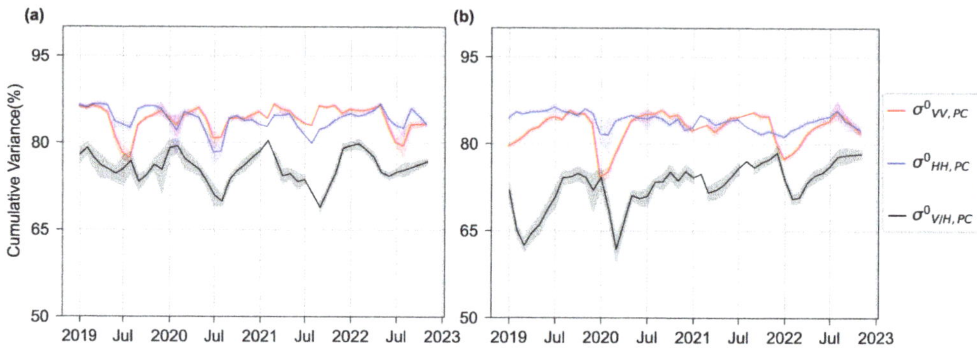

Figure 7. Time series of cumulative variance of the eigenvalues for principal components in the (**a**) Northern and (**b**) Southern Hemispheres between 2019 and 2022.

3.2. Period Choice for Dynamic Sampling

Figure 8 presents the statistical averages of the first principal component of σ^0_{VV} in both the Northern and Southern Hemispheres using different sampling periods of 5, 7, 10, 15, and 20 days. The analysis shows that shorter sample periods result in significant noise in the first principal component of the VV polarization, thereby enhancing contrast during seasonal variations. Conversely, extending the sampling period results in a smoother curve for the first principal component feature of VV polarization, suggesting a diminishing emphasis on seasonal variations. Subsequent comparisons of the detection performance of the random forest model were performed for the Northern and Southern Hemispheres in January–March and June–August, respectively, over the same range of sampling periods (5, 7, 10, 15, and 20 days). Table 4 presents the F1 scores for five different models across various sampling periods It highlights which sampling period (e.g., 5 days, 7 days, etc.) resulted in the highest F1 scores for each model. The table summarizes a total of 10 statistical scenarios, with 8 scenarios showing that the 5-day sampling period consistently achieved the highest F1 score.

Table 4. Period with the highest average F1 scores for each model (Dt, Gnb, Knn, Log, and Rfc) in the Northern and Southern Hemispheres.

Polar	Model	Accuracy	F1 Score	Period
N	Dt	0.857	0.872	5 days
	Gnb	0.874	0.875	5 days
	Knn	0.938	0.931	5 days
	Log	0.805	0.850	5 days
	Rfc	0.939	0.927	5 days
S	Dt	0.834	0.850	20 days
	Gnb	0.882	0.866	5 days
	Knn	0.950	0.941	5 days
	Log	0.810	0.831	20 days
	Rfc	0.948	0.933	5 days

After comprehensive evaluation, a sample period of 5 days was chosen for the dynamic sample statistics. Taking into account the possibility of incomplete sampling periods due to missing observations on certain dates, a forward search was performed to find available data no longer than 5 days. If the forward search exceeded 20 days, then dynamic sample data were considered missing on that date, and no modeling or sea ice detection was performed for that day.

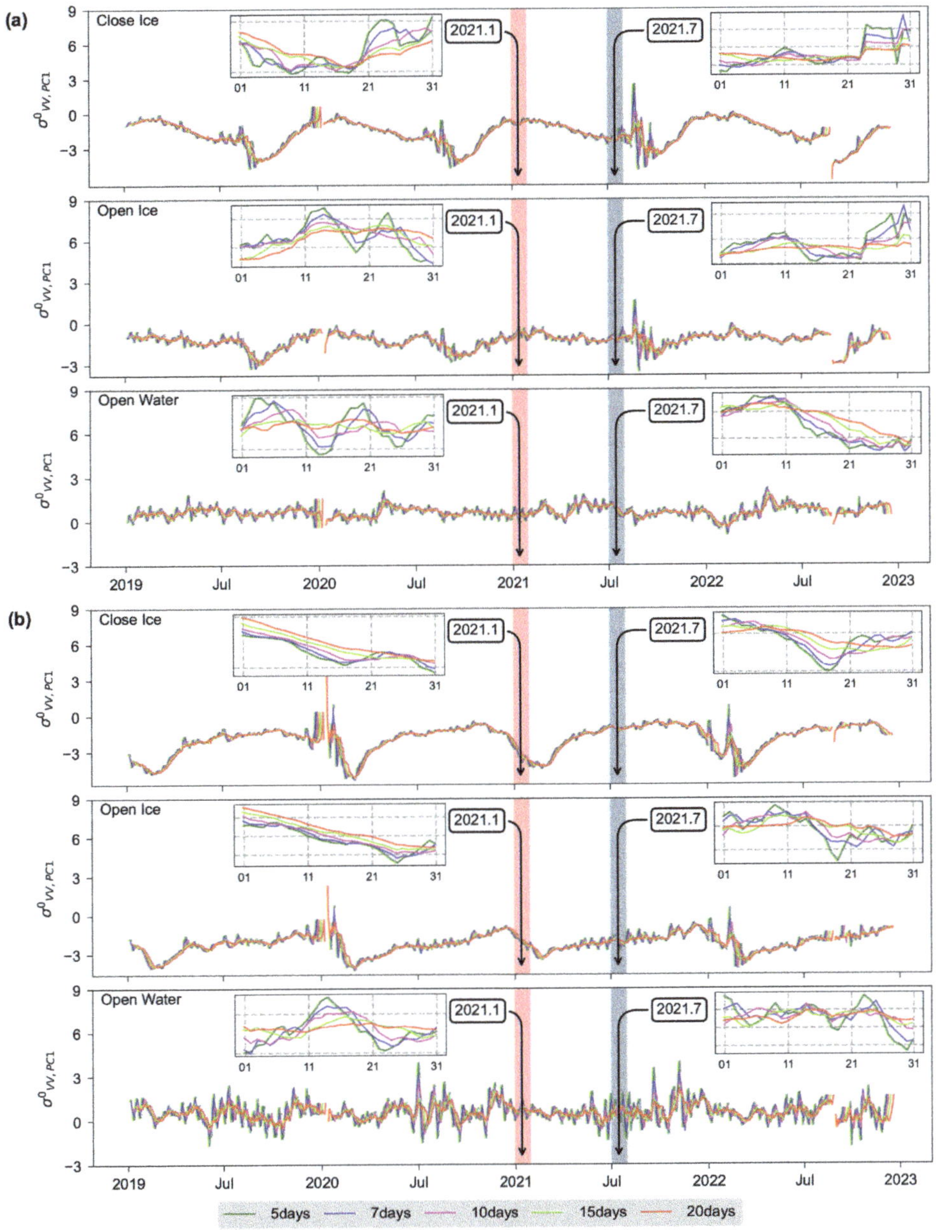

Figure 8. Time series of $\sigma^0_{VV,PC1}$ with different period lengths in the (**a**) Northern Hemisphere and (**b**) Southern Hemisphere for close ice, open ice, and open water.

3.3. *Assessment for Single and Ensemble Models*

In this study, five individual models were modelled: Gnb, Log, Knn, Dt, and Rfc. The variable importance of these models was assessed during the model building process, and cross-validation of F1 scores was performed to provide insight into the performance of the models and the relative importance of each feature. Variable importance is an important measure of how strongly features influence the predictive power of a model. We assessed feature importance using different metrics tailored to each model. For Dt and Rfc, Gini importance was used, which reflects the reduction in contamination by each feature. For Log and Gnb, feature weights from regression coefficients were used, with the magnitude of the coefficients indicating feature influence, and although Knn does not directly indicate feature importance, it can be assessed indirectly via model performance or weighted configurations.

Figure 9a shows the feature importance of these five models for the Northern and Southern Hemispheres. There are some differences in how different models rank for feature importance. Different models assigned different levels of importance to features. The most important feature was the first principal component of HH polarization. The ranking of $\sigma^0_{VV,PC1}$ and $\sigma^0_{V/H,PC1}$ varied between the models: They took second and third place in Log-ranked third and second in others The situation in the Southern Hemisphere was more complex, as there were significant differences in the order of feature importance among models. In the Dt, Knn, and Rfc models, $\sigma^0_{HH,PC1}$ was the most important feature, while in the Gnb model, $\sigma^0_{VV,PC1}$ was the most important, and in the Log model, $\sigma^0_{V/H,PC1}$ was the most important. This reflected the geographical differences between the Northern and Southern Hemispheres and the different responses to polarization features. Consequently, the models also showed variations in the ranking of radar polarization features for different regions. Figure 9b summarizes the distribution of F1 scores across the models during a 10-fold cross-validation period from 1 January 2019 to 31 December 2022. The F1 scores, which ranged from 0.660 to 0.750 for both hemispheres, indicated a commendable balance of precision and recall achieved by the models over the past four years. The Log model in the Southern Hemisphere showed the largest standard deviation (0.088), indicating greater variability in performance across different subsets compared to the other models. The standard deviations of the other models were between 0.028 and 0.069. Figure 9c shows the time series of F1 values obtained by 10-fold cross-validation for different machine learning models from 1 January 2019 to 31 December 2022. The Knn, Rfc, and ensemble models consistently showed relatively stable and high F1 values throughout both the Northern and Southern Hemispheres, indicating strong and sustained forecast performance in these regions. In contrast, Log showed larger fluctuations in F1 results and generally lower values, suggesting weaker and more unstable generalization abilities, especially in the Northern Hemisphere. For all models, there was no obvious temporal trend showing significant improvements or declines in F1 score over the analyzed period, suggesting stable model performance without noticeable deterioration or improvement. However, the different models showed differences in performance over the course of the season. Notably, in the Northern Hemisphere, the F1 score for all models increased continuously from July to September, and similarly, in the Southern Hemisphere, F1 scores were higher from January to March compared to other months. This pattern, observed continuously from 2019 to 2022, indicates that Ku-band CSCAT is particularly effective in identifying the melting status of sea ice during the summer months. On the other hand, the Log model outperformed Dt and Gnb in both hemispheres from January to March, but in the Southern Hemisphere, its F1 values were significantly lower than those of Dt and Gnb from July to September. The consistent performance of these models highlights their robustness and reliability in analyzing sea ice dynamics using radar polarization features, despite the inherent variability in environmental conditions between the Northern and Southern Hemispheres.

To evaluate the models, we summarized confusion matrices for individual and integrated models (Table 5). In both the Northern and Southern Hemispheres, the performance of the five individual models varied in different classification scenarios. For open water and close ice, Knn and Rfc achieved F1 values above 0.9, while the other three models also achieved values around 0.8. However, all five models had very low F1 values of less than 0.4 on open ice. Among them, Knn performed the best with an F1 value above 0.3, followed by Rfc in the Northern Hemisphere and Dt in the Southern Hemisphere. Among the results of five single models between 2019 and 2022, Knn and Rfc showed strong performance in the Northern Hemisphere, with F1 and OA values above 0.9 and kappa coefficients above 0.8, indicating high data consistency. In contrast, Dt, Gnb, and Log showed F1 and OA values between 0.7 and 0.9, with kappa coefficients ranging between 0.569 and 0.6887, reflecting moderate consistency. While Dt and Log had OA values around 0.7 in the Southern Hemisphere, Knn, Rfc, and Gnb reached OA values between 0.8 and 0.9. Only Knn and Rfc had kappa values above 0.7, indicating high model consistency, while Dt, Gnb, and Log had Kappa values below 0.6, indicating medium consistency. Overall, Knn and Rfc performed better than the other single models in terms of detection accuracy.

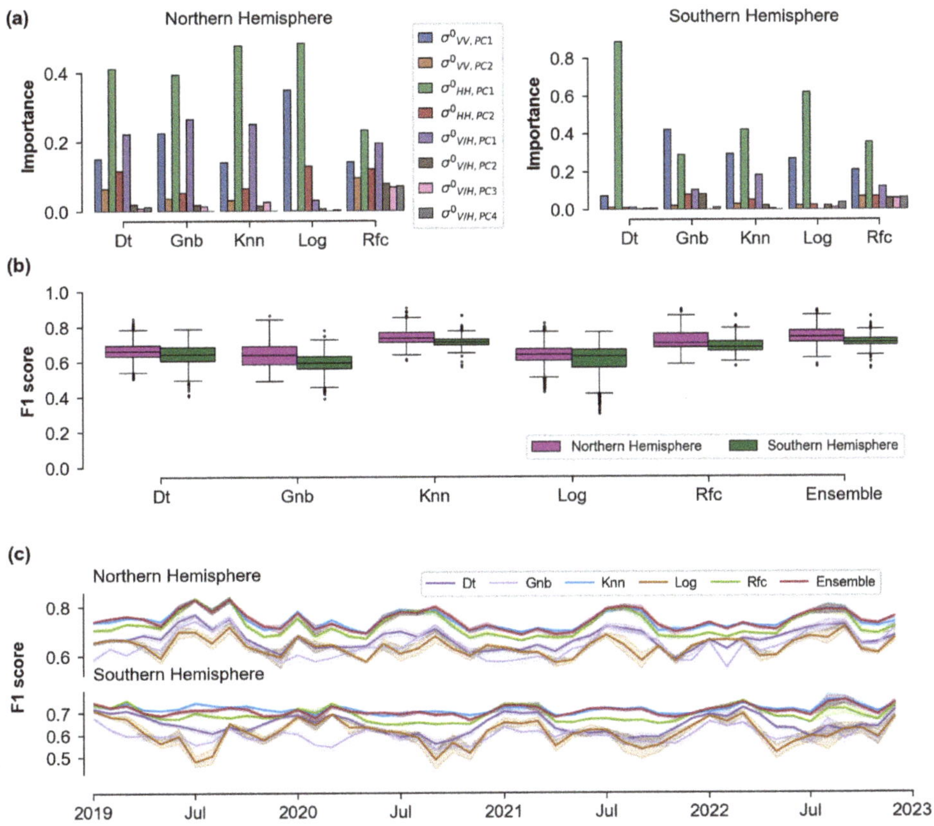

Figure 9. (**a**) Feature importance for single models on 10 January 2019 in the Northern Hemisphere (**left**) and the Southern Hemisphere (**right**). (**b**) Statistical results of 10-fold cross-validation F1 scores for different machine learning models from 1 January 2019 to 31 December 2022. (**c**) Time series of 10-fold cross-validation F1 scores for different machine learning models from 1 January 2019 to 31 December 2022.

In the Northern Hemisphere, the ensemble models slightly outperformed the best two individual models, Rfc and Knn, in terms of OA and Kappa values. Additionally, while the F1 score for open ice improved compared to Rfc, it still remained lower than that of Knn. A similar trend was observed in the Southern Hemisphere. The ensemble models showed a slight advantage over the best individual models in terms of OA and Kappa values. For open ice, the F1 score was higher than that of Rfc but still lower than Knn. Based on these observations, the ensemble models can be considered the preferred choice. Their overall performance, demonstrated by the improved OA and Kappa values, indicates their effectiveness in achieving better consistency and reliability across different scenarios. While KNN still outperforms the ensemble models in terms of F1 scores for open ice, the ensemble models' superior overall performance suggests that they offer a more balanced and reliable solution across various classification tasks.

Table 5. Summary of averaged detection accuracies through confusion matrix of sea ice edge classification in the Northern and Southern Hemisphere from 1 January 2019 to 31 December 2022.

Hemisphere	Model	Classtype	UA	PA	F1 Score	OA	Kappa
Northern	Dt	Close Ice	0.858	0.785	0.814	0.799	0.625
		Open Ice	0.209	0.618	0.295		
		Open Water	0.968	0.819	0.886		
	Gnb	Close Ice	0.783	0.896	0.833	0.861	0.702
		Open Ice	0.259	0.172	0.183		
		Open Water	0.926	0.876	0.899		
	Knn	Close Ice	0.880	0.933	0.905	0.928	0.841
		Open Ice	0.519	0.289	0.356		
		Open Water	0.962	0.966	0.964		
	Log	Close Ice	0.811	0.753	0.774	0.773	0.578
		Open Ice	0.177	0.552	0.256		
		Open Water	0.957	0.804	0.872		
	Rfc	Close Ice	0.886	0.932	0.907	0.924	0.842
		Open Ice	0.615	0.221	0.288		
		Open Water	0.954	0.975	0.964		
	Ensemble	Close Ice	0.883	0.933	0.911	0.930	0.844
		Open Ice	0.580	0.319	0.364		
		Open Water	0.965	0.956	0.965		
Southern	Dt	Close Ice	0.881	0.731	0.792	0.748	0.502
		Open Ice	0.217	0.562	0.287		
		Open Water	0.850	0.833	0.837		
	Gnb	Close Ice	0.804	0.899	0.843	0.825	0.530
		Open Ice	0.250	0.141	0.166		
		Open Water	0.822	0.713	0.756		
	Knn	Close Ice	0.872	0.936	0.902	0.897	0.750
		Open Ice	0.510	0.235	0.308		
		Open Water	0.928	0.927	0.927		
	Log	Close Ice	0.862	0.679	0.753	0.711	0.451
		Open Ice	0.197	0.532	0.268		
		Open Water	0.781	0.812	0.786		
	Rfc	Close Ice	0.865	0.947	0.902	0.892	0.734
		Open Ice	0.640	0.167	0.229		
		Open Water	0.926	0.908	0.916		
	Ensemble	Close Ice	0.868	0.952	0.907	0.899	0.747
		Open Ice	0.460	0.232	0.294		
		Open Water	0.927	0.922	0.924		

Figure 10 illustrates the time series of assessment parameters from 1 January 2019 to 31 December 2022. The overall classification metrics (OA and Kappa) had more similar daily trends to the UA, PA, and F1 values for close ice. However, these values were significantly lower from July to October in the Northern Hemisphere and from January to April in the Southern Hemisphere. UA, PA, and F1 values were generally higher in open water, with less seasonal variation and relatively low values, mainly in the Northern Hemisphere from January to April and in the Southern Hemisphere from April to October. The UA, PA, and F1 values for close ice and open water were higher than those for open ice. Figure 11 presents an error analysis based on the model classification confusion matrix, with the y-axis representing the misclassification rates for each class. Specifically, these rates were calculated by determining the number of pixels for each class (e.g., "Close Ice", "Open Ice", and "Open Water") that were incorrectly classified as other classes and then dividing these misclassified pixel counts by the total number of pixels for that class. Error analysis showed that close ice was frequently mistakenly classified as open ice from July to October in the Northern Hemisphere and from January to April in the Southern Hemisphere, resulting in decreased classification accuracy. On the other hand, open water was frequently mistakenly identified as close ice from April to October in the Southern Hemisphere and January to April in the Northern Hemisphere, which also affected the accuracy of the classification. Furthermore, a significant portion of open ice was incorrectly identified as close ice, which helps to explain why close ice classification accuracy was typically lower.

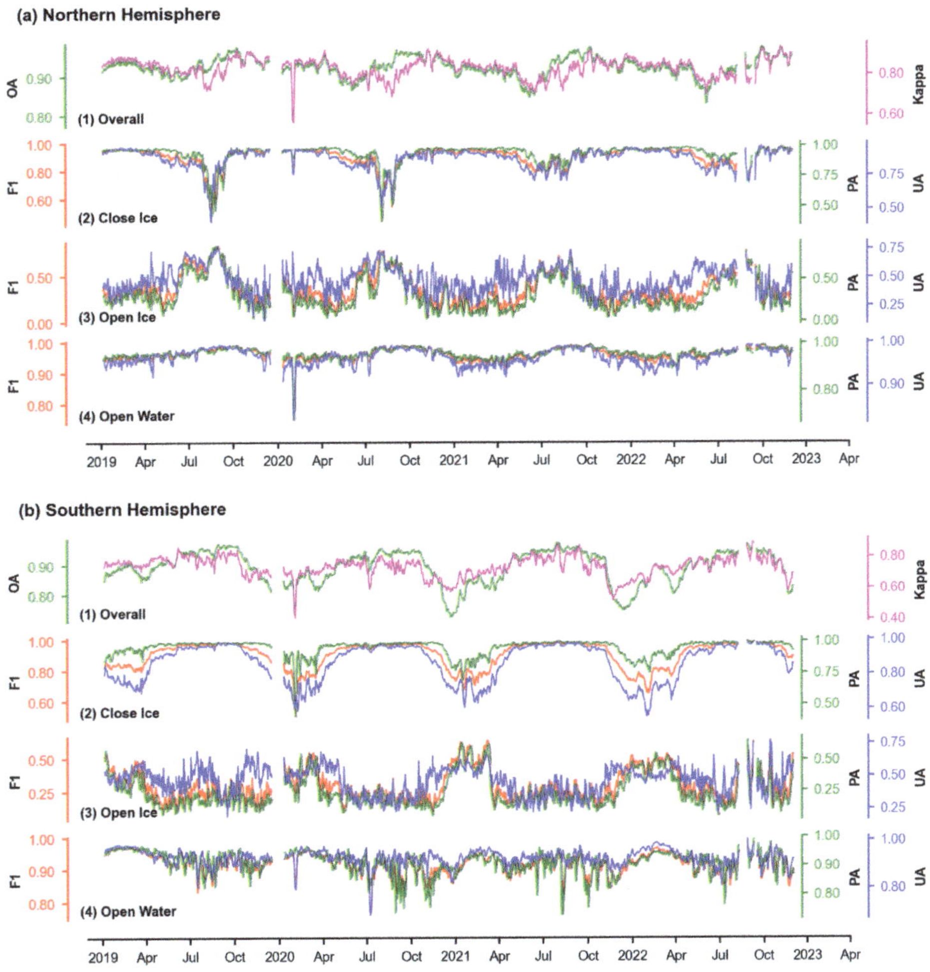

Figure 10. The time series of the evaluation parameters for (1) overall, (2) close ice, (3) open ice, and (4) open water in the sea ice monitoring ensemble training model in the (**a**) Northern Hemisphere and (**b**) Southern Hemisphere from 1 January 2019 to 31 December 2022, respectively.

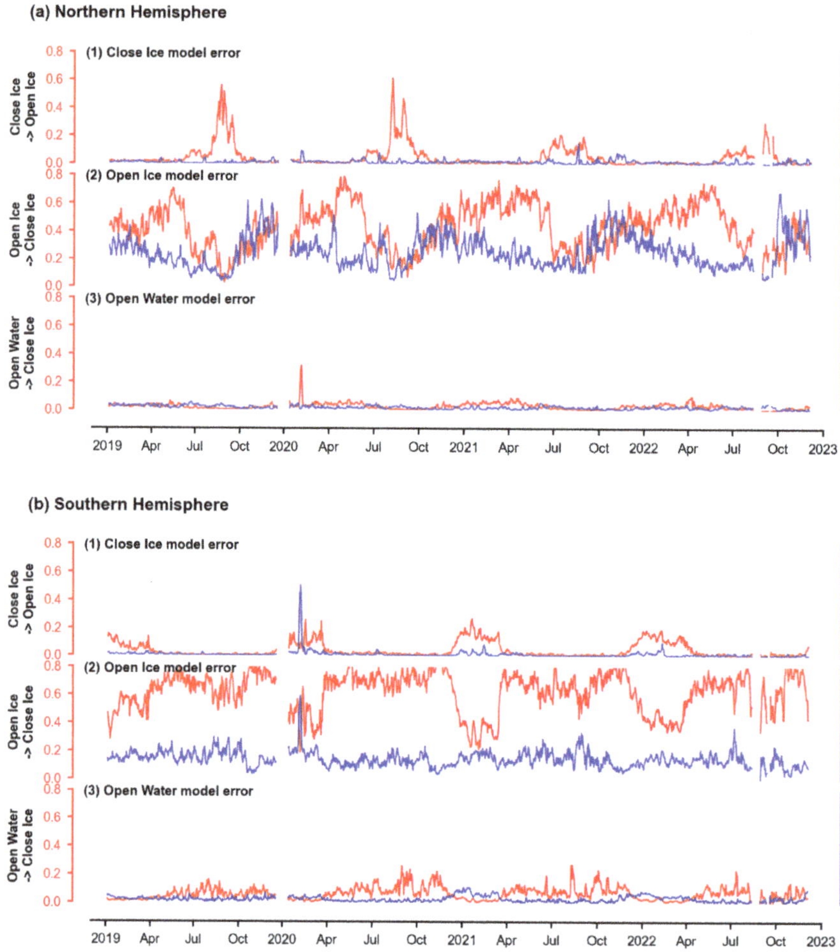

Figure 11. Daily error analysis for (1) close ice, (2) open ice, and (3) open water in the sea ice monitoring ensemble training model in the (**a**) Northern Hemisphere and (**b**) Southern Hemisphere from 1 January 2019 to 31 December 2022, respectively.

3.4. Single and Ensemble Model-Based Sea Ice Mapping

This study used five single models and an averaged ensemble model for sea ice detection. The Northern and Southern Hemisphere sea ice detection results for 10 December 2019 and 10 June 2019 are shown in Figure 12. Each column represents origin reference SIC, classified reference SIC, and different models, namely Dt, Gnb, Knn, Log, and Rfc, from left to right. The reference map provided the observed sea ice distribution, serving as a comparison baseline. It is evident that the amount of sea ice was similar in both the Northern and Southern Hemispheres. There were differences between the models in terms of sea ice detection. The Knn and Rfc models overall performed better than the other models in classifying sea ice in both the Northern and Southern Hemispheres. The Knn model detected sea ice by calculating the distance between samples, while the Rfc model classified it by constructing a decision tree integration. These models might be better suited to the capture the spatiotemporal correlation of sea ice and dealing with its nonlinear features. Compared to the original reference maps, the Knn and Rfc single models tended to misclassify open ice as close ice at the sea ice boundary, whereas Dt and Log were more prone to misclassifying close ice as open ice. These differences were likely due to variations in the algorithmic principles and feature extraction functions. The advantage of Rfc and Knn lay in their overall robustness and stability for close ice, while Dt and Log demonstrated certain advantages in handling complex boundaries, particularly in distinguishing open ice from close ice.

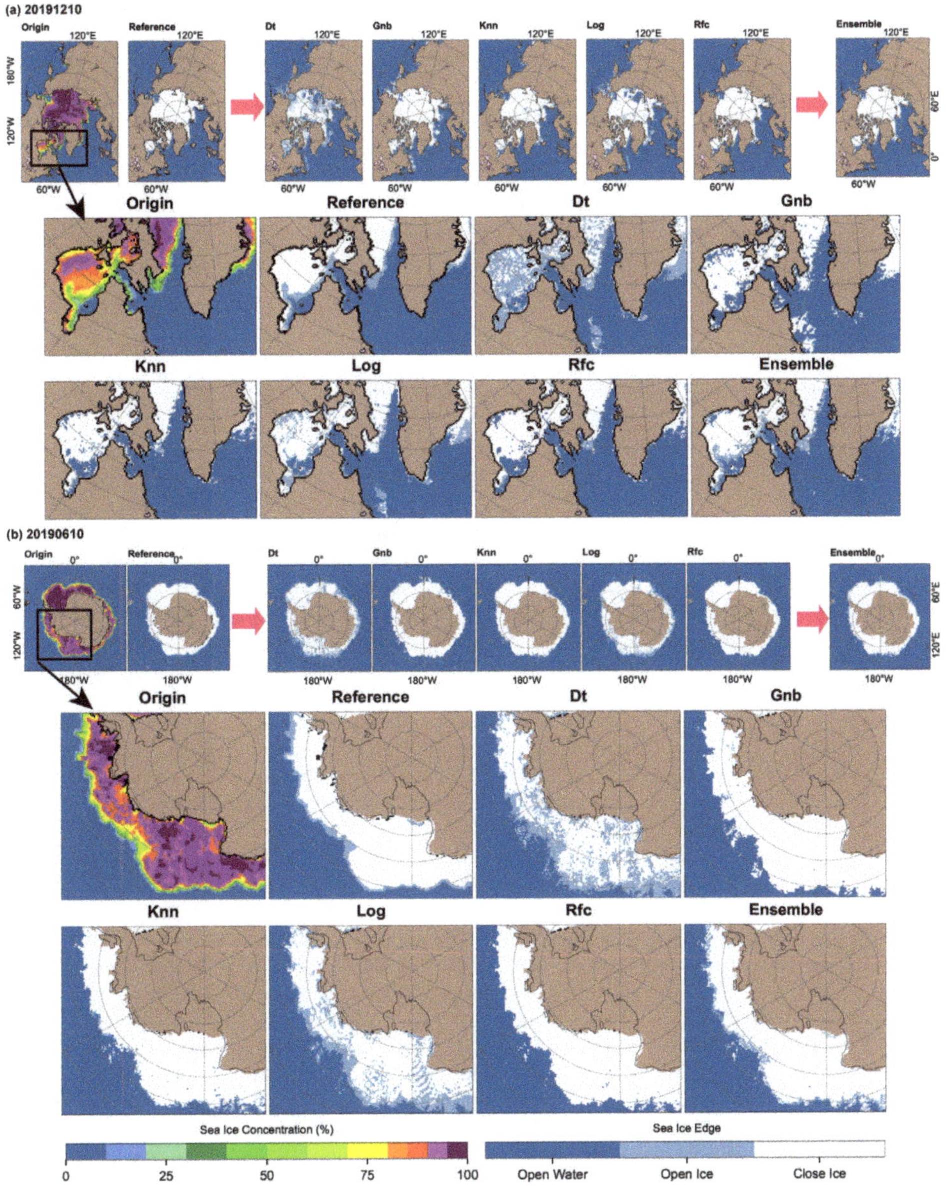

Figure 12. Sea ice detection in the (**a**) Northern Hemisphere on 10 December 2019 and (**b**) Southern Hemisphere on 10 June 2019 derived from the Dt, Gnb, Knn, Log, Rfc, and ensemble models, respectively.

By averaging the results of those five models, a sea ice map of the Northern and Southern Hemispheres was created. However, when comparing these with the ensemble model results, it was clear that the significant misclassifications in single models were effectively mitigated in the ensemble model. The ensemble model not only retained the correct classification of close ice from Knn and Rfc, but it also gained the accurate classification of open ice from the other three single models. Ensemble models reduce errors that may occur with single models by combining predictions from multiple models. Therefore, they are expected to improve the accuracy of sea ice detection to some extent.

4. Discussion

In this section, we focus on discussing the performance of the ensemble model across different time periods and spatial ranges in order to comprehensively assess its applicability and accuracy. We achieve this by using the sea ice extent derived from other sea ice concentration to validate the CSCAT sea ice classification results and by conducting detailed evaluations using metrics such as R^2 and RMSE. Additionally, we compare the CSCAT sea ice classification with similar sea ice edge and concentration products as well as with Sentinel 1 SAR images.

4.1. Comparison with Daily Sea Ice Extent

Sea ice extent (SIE) is usually defined as the sum of the area of ocean grid cells with a sea ice concentration greater than 15%. A SIC threshold of 15% is not applied regularly [60]. This parameter represents the maximum sea ice cover and is crucial for assessing climate change. However, because our model classifies sea ice using a 30%/70% SIC threshold, we adjusted our analysis to use a 30% SIC threshold to calculate sea ice extent in order to ensure a fair comparison: $SIE = \sum A_{SIC \geq 30\%}$. For CSCAT, sea ice extent was calculated by summing the area of pixels classified as close ice and open ice.

To assess the temporal accuracy of our sea ice detection, we conducted a daily comparison with the daily sea ice area data released by the NSIDC and OSISAF from 2019 to 2022. Despite significant daily fluctuations, the sea ice extent derived from CSCAT showed a high level of agreement with the officially published daily data. Figure 13b1,b2 show the time series of sea ice extent differences between CSCAT, OSISAF, and NSIDC across different hemispheres, respectively. In the Northern Hemisphere, CSCAT underestimated sea ice extent by -0.06 ± 0.36 million km^2 compared to NSIDC, while OSISAF underestimated it by -0.12 ± 0.09 million km^2. In the Southern Hemisphere, CSCAT underestimated sea ice extent by -0.03 ± 0.48 million km^2, and OSISAF underestimated it by -0.11 ± 0.09 million km^2. The comparison results showed that CSCAT generally estimated lower sea ice extent than NSIDC. Zhai et al. [23] used a random forest approach to estimate the CSCAT distribution and compared the differences to OSISAF sea ice extent. Their results suggested a lower estimate for the Northern Hemisphere, consistent with our study, but a higher estimate for the Southern Hemisphere, contradicting our results. According to the model error analysis, open ice was misclassified as close ice in most months in the Southern Hemisphere, with many open ice areas misclassified as open water from January to April (Figure 11b (2)). This likely led to the overall lower sea ice extent in the Southern Hemisphere observed in our study.

To better present the results, we have graphically displayed the daily sea ice extent on a monthly basis. The visualization showed that sea ice extent from CSCAT matched NSIDC in January, February, May, June, July, and August. However, the results were slightly overestimated in March and April and slightly underestimated from September to December. In the Southern Hemisphere, the sea ice extent in February, August, and September corresponded closely, yet we observed an underestimation from March to July and an overestimation from November to January of the following year. These variations indicated that while the model performed well under specific conditions, such as during months with relatively stable sea ice conditions, its applicability under other conditions, particularly during seasonal transitions, was challenged. The observed underestimations or overestimations were likely associated with dynamic environmental factors that significantly impacted sea ice formation and melting during these transitional periods. This is in stark contrast to the discrepancies found between QuikSCAT and ASCAT compared to AMSR-E, particularly during sea ice melt months, as reported by Belmonte Rivas et al. [19]. It is noteworthy that active scatterometers and passive microwave radiometers showed significant differences during periods of rapid sea ice change. Furthermore, linear regression analysis of the CSCAT and NSIDC sea ice extent data yielded R-squared values of 0.991 and 0.993 with corresponding RMSE values of 0.340 and 0.485 million km^2. These results confirm that our method for estimating sea ice extent provided consistent results over longer time periods compared to other accepted data sources.

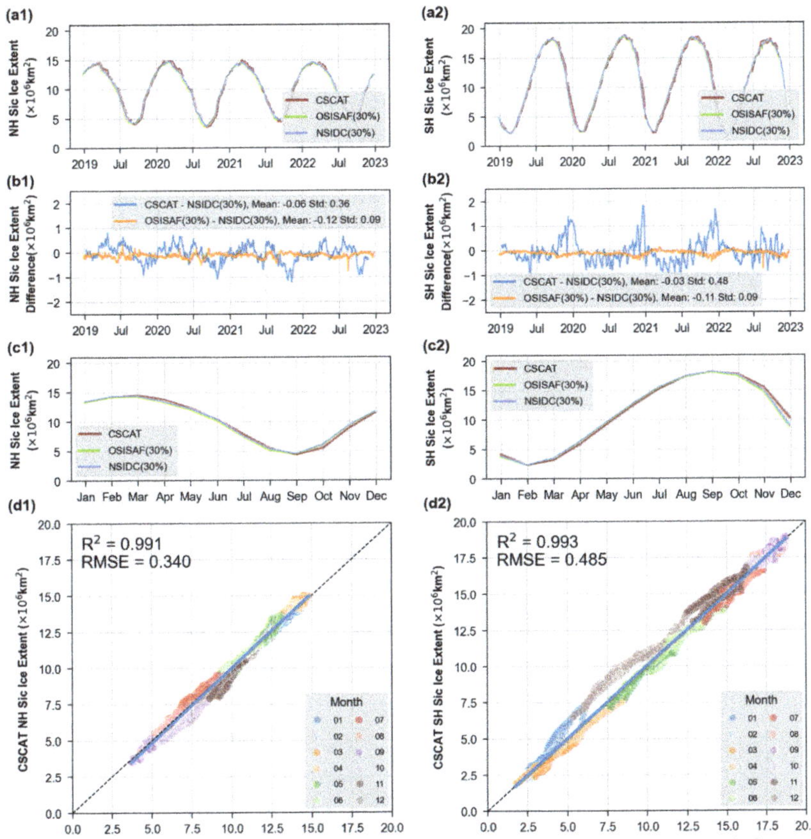

Figure 13. Daily sea ice extent from 2019 to 2022 in the (**a1**) Northern Hemisphere and (**a2**) Southern Hemisphere for CSCAT, OSISAF (30% SIC), and NSIDC (30% SIC). Daily sea ice extent difference from 2019 to 2022 in the (**b1**) Northern Hemisphere and (**b2**) Southern Hemisphere for CSCAT vs. NSIDC and OSISAF vs. NSIDC. Monthly sea ice extent from 2019 to 2022 over the (**c1**) Northern Hemisphere and (**c2**) Southern Hemisphere for CSCAT, OSISAF (30% SIC), and NSIDC (30% SIC). Scatter plot of sea ice extent between CSCAT and NSIDC over the (**d1**) Northern Hemisphere and (**d2**) Southern Hemisphere. The pairs are colored by month, and the blue line represents a trend line fitted to the data.

4.2. Comparison with Sea Ice Concentration and Sea Ice Edge Datasets

ASCAT, SSMIS, and AMSR2 are the three other sources of sea ice cover data. ASCAT is a C-band HH-polarized scatterometer that provides sea ice detection similar to CSCAT.

SSMIS and AMSR2 are the two dominant microwave radiometric instruments commonly used to determine sea ice concentration and provide more reliable sea ice distributions. To compare the spatial differences between the CSCAT sea ice edge and other sea ice cover products, we selected the sea ice edge from ASCAT and sea ice concentration from NSIDC. The comparison involved contrasting the CSCAT and ASCAT sea ice edges with the SSMIS sea ice concentration. This comparison employed a statistical method similar to a confusion matrix (Equation (6)). First, the NSIDC sea ice concentration was divided into three classes based on thresholds: less than 30% was open water (OW), 30–70% was open ice (OI), and more than 70% was close ice (CI). The CSCAT and ASCAT sea ice edge results also included these three types. Monthly mode statistics were calculated for each pixel compared to the NSIDC sea ice edge, resulting in the monthly average differences for sea ice edge between CSCAT, ASCAT, and NSIDC. The consistency of CI OI, and OW can be calculated using Equation (7).

$$\text{confusion matrix} = \begin{bmatrix} CI\_CI & CI\_OI & CI\_OW \\ OI\_CI & OI\_OI & OI\_OW \\ OW\_CI & OW\_OI & OW\_OW \end{bmatrix} \tag{6}$$

$$\text{Consisitency} = \frac{\sum Pixel_x^x}{\sum Pixel_{CI}^x + Pixel_{OI}^x + Pixel_{OW}^x}, x \in (CI, OI, OW) \tag{7}$$

For this study, we compared sea ice detection in the Northern Hemisphere on 10 January 2019 and in the Southern Hemisphere on 10 June 2019 from threedifferent sources: CSCAT, ASCAT, and NSIDC (Figure 14). The results of each data source were represented by a color-coded scale indicating the sea ice concentration and sea ice edge results. For NSIDC SIC, the scale ranges from dark red (0%, open water) to deep purple (100%, close ice). For sea ice edge products, the scale ranges from dark blue (open water) to light blue (open ice) to white (close ice). The comparison results indicate that the sea ice edge obtained by CSCAT was consistent with the sea ice acquired by ASCAT and NSIDC, but there were also some deviations. In the Northern Hemisphere, CSCAT detected open ice in the Greenland Sea, Baffin Bay, and Hudson Bay, while in the Southern Hemisphere, open ice was detected at the ice edge regions of the Ross Sea. Compared to ASCAT, this open ice tended to be less, which can also be observed in the sea ice distribution detected by Li et al. [21] using GMFs.

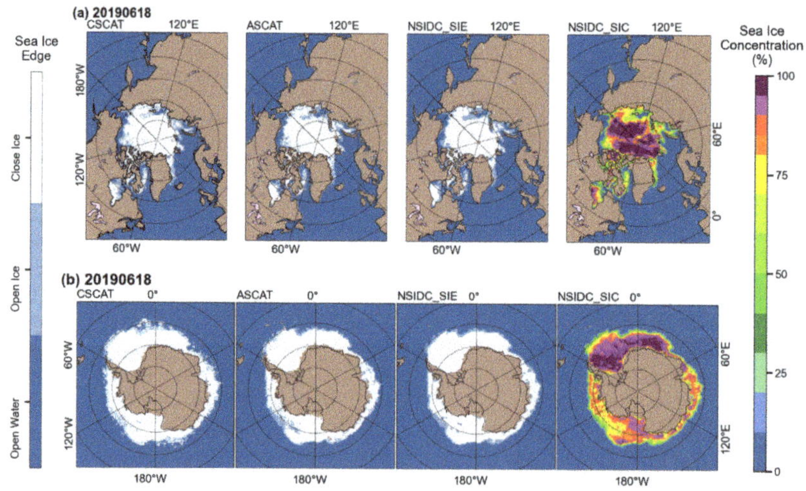

Figure 14. Sea ice mapping in the (**a**) Northern Hemisphere on 18 June 2019 and (**b**) Southern Hemisphere on 18 June 2019 derived from CSCAT, ASCAT, NSIDC sea ice edge (SIE), and NSIDC sea ice concentration (SIC), respectively.

From the consistency time series for sea ice edge (Figure 15), it was observed that, in both the Southern and Northern Hemispheres, the consistency of close ice observed by CSCAT with that observed by NSIDC (0.903, 0.931) was slightly better than that of ASCAT (0.898, 0.922), while ASCAT (0.998, 0.997) showed slightly better performance for open water compared to CSCAT (0.988, 0.988). For open ice, CSCAT performed worse overall than ASCAT, although with slight differences in the different hemispheres. In the Northern Hemisphere, CSCAT showed lower consistency with NSIDC compared to ASCAT in most months. In the Southern Hemisphere, the differences between ASCAT and CSCAT in terms of open ice exhibited a seasonal symmetry: in the months when the consistency of ASCAT was poor (January to April), CSCAT performed well, and in the months when the consistency of CSCAT was poor (May to December), ASCAT performed well. This seasonal symmetry for the consistency of open ice in the Southern Hemisphere was also observed in Zhai et al. [23]. This suggested that CSCAT had a better ability to identify open ice compared to ASCAT from January to April. These differences were mainly due to the different frequencies of the two sensors. CSCAT operates in the Ku band, which has a shorter wavelength than the C-band ASCAT. The shorter wavelength is more sensitive to the rough surface of multiyear ice but has a weaker response to thin ice edges.

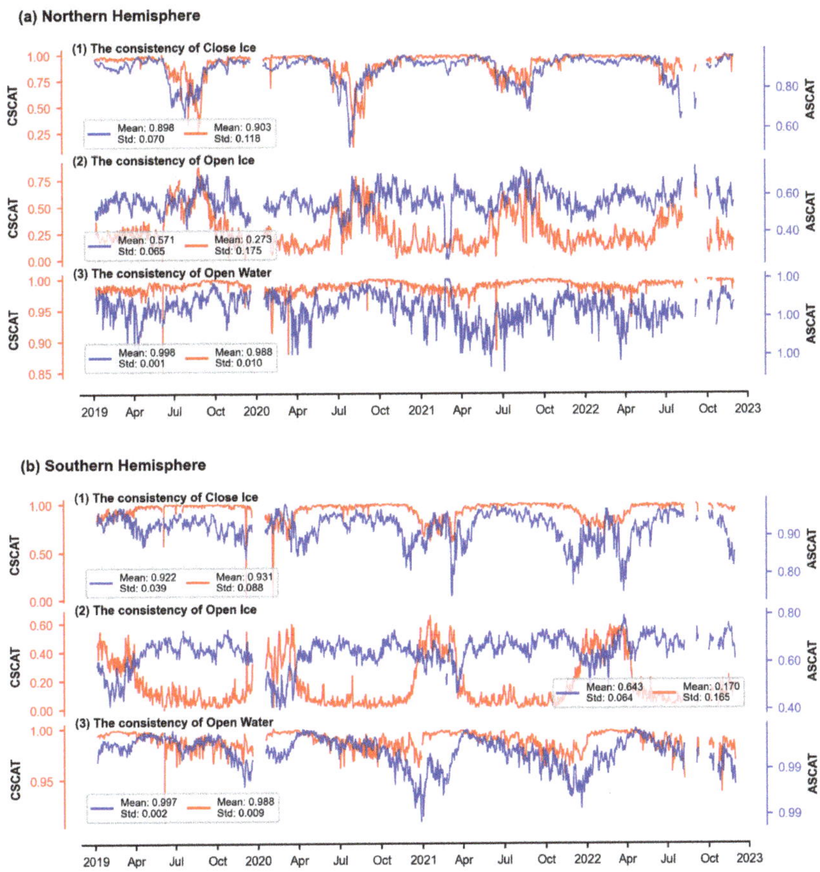

Figure 15. Daily consistency compared to NSIDC for (1) close ice, (2) open ice, and (3) open water over the (**a**) Northern Hemisphere and (**b**) Southern Hemisphere from 1 January 2019 to 31 December 2022, respectively.

In the Northern Hemisphere, CSCAT showed lower consistency for open ice from January to April and lower consistency for close ice from May to August (Figure 15a (1),(2)). The spatial distribution of sea ice cover differences between CSCAT, ASCAT, and NSIDC in February and May revealed that the misclassification of sea ice by CSCAT in the Northern Hemisphere was mainly concentrated along the ice edges in the Barents Sea and the Greenland Sea, where close ice was incorrectly classified as open ice, which was less common in the spatial differences observed with ASCAT (Figure 16a,c). In August, CSCAT misclassified open ice as close ice, especially in the northern sea areas of Canada, while ASCAT showed less misclassification in these areas. In the Southern Hemisphere, CSCAT showed lower consistency for open ice from April to December but higher consistency from January to March. Misclassification still occurred at the ice edge, with open ice being mostly classified as close ice in August and November (Figure 16b,d). In March, mutual misclassification between close and open ice occurred at the sea ice edge of East Antarctica.

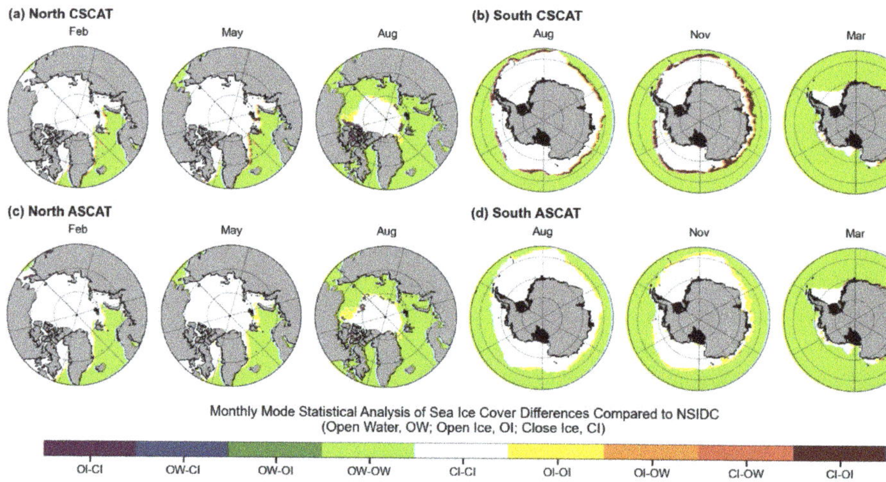

Figure 16. Monthly mode statistics for CSCAT over the (**a**) Northern Hemisphere and (**b**) Southern Hemisphere and for ASCAT over the (**c**) Northern Hemisphere and (**d**) Southern Hemisphere, showing sea ice cover differences compared to NSIDC.

The comparison between CSCAT and ASCAT revealed that CSCAT had slightly better consistency with NSIDC for close ice in both hemispheres, while ASCAT performed better for open water. CSCAT generally underperformed compared to ASCAT for open ice, especially in the Northern Hemisphere. However, in the Southern Hemisphere, CSCAT outperformed ASCAT from January to April, whereas ASCAT performed better from May to December. Misclassifications were primarily observed along the ice edges in the Barents Sea and Greenland Sea, where CSCAT often misclassified close ice as open ice, unlike ASCAT. In August, CSCAT also misclassified open ice as close ice in central sea areas, while ASCAT's misclassifications occurred mainly at the boundary between open and close ice. In the Southern Hemisphere, CSCAT showed lower consistency in detecting open ice from April to December but higher consistency from January to March, with misclassifications mostly occurring along the ice edges.

4.3. Comparison with High Resolution SAR Imagery

To confirm the regional detection of sea ice, a comparison with Sentinel-1 SAR data was performed. For the validation process, a section of the Northern Hemisphere on 19 June and 8 March 2019 and a section of the Southern Hemisphere on 19 June 2019 were selected and verified using the map data. During the process, we performed various data processing procedures using the SNAP 8.0.0 software, namely orbital corrections, radiometric calibrations, dB conversions, and latitude/longitude projections. We then created a geo-mosaic using ArcGIS 10.7 software, which allowed us to create SAR images covering many parts of the polar regions. In addition, we used ArcGIS software to extract the peripheral sea ice boundary from the sea ice results and overlay it on the corresponding SAR image.

Figure 17a depicts the expanse of open ice in the Greenland Sea as captured by the Sentinel-1 SAR imagery. The open ice appears as a lighter shade of gray in contrast to the deep gray of the seawater. The features observed were consistent with the sea ice boundaries extracted by CSCAT. Moreover, the Sentinel-1 SAR image dated 8 March 2019, facilitated the clear identification of sea ice and open water in Baffin Bay. This identification closely aligned with our extracted sea ice boundary positions, as illustrated in Figure 17b. Similar outcomes were demonstrated over two regions in the Southern Hemisphere above the Indian Ocean. The sea ice boundaries extracted by CSCAT showed superior performance in delineating sea ice from open waters (Figure 17c). Near the Antarctic Peninsula, CSCAT's detection of sea ice boundaries differed from the SAR images and incorrectly marked some open water as sea ice. Despite this, CSCAT had accurately identified a distinct, isolated region of sea ice. Our detection results showed similarity in the shape and position of sea ice compared to Sentinel-1 SAR imagery while also capturing the variations and structural features at the sea ice edge.

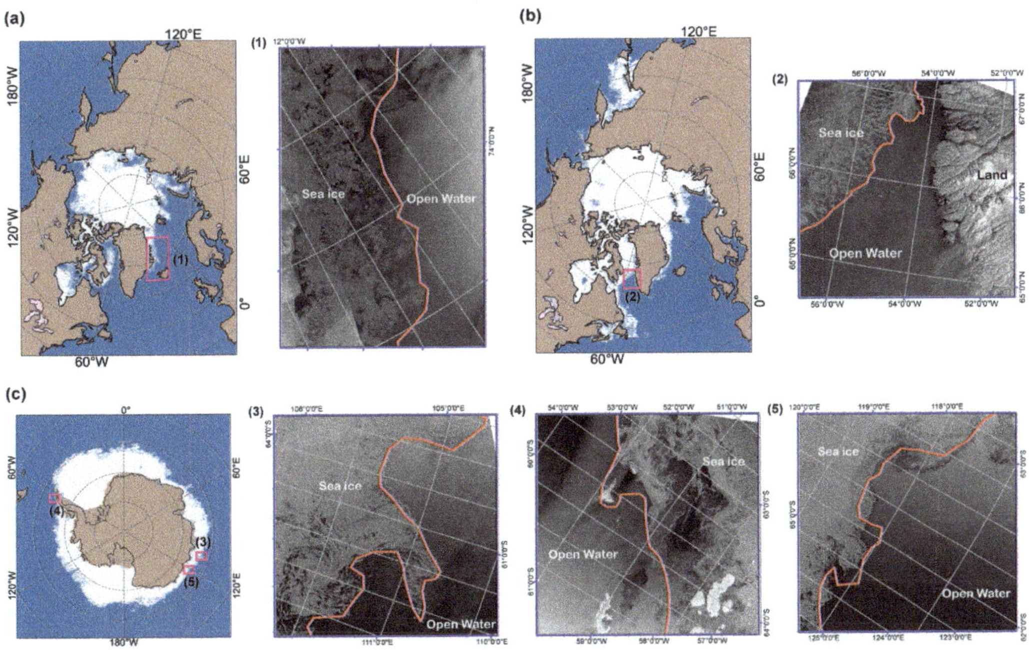

Figure 17. Comparative analysis of sea ice detection and high-resolution synthetic aperture radar (SAR) images. Comparison between CSCAT-derived sea ice detection results and Sentinel-1 SAR images in the Northern Hemisphere taken on (**a**) 18 June 2019 and (**b**) 8 March 2019 and in the Southern Hemisphere on (**c**) 19 June 2019. The thick red line represents the CSCAT-derived sea ice detection results.

5. Conclusions

The aim of this study was to automatically extract feature information from CSCAT using PCA to retrieve sea ice data in the Northern and Southern Hemispheres and to use an ensemble machine learning algorithm to obtain reliable daily sea ice distributions from 2019 to 2022. PCA effectively extracted principal component features representing outer swath, zones, and close ice. We trained ensemble models based on Knn, Log, Dt, Gnb, and Rfc. Rfc/Knn exhibited high error rates in detecting open ice, often misclassifying it as close ice, particularly at the sea ice boundary. In contrast, Dt/Log/Gnb performed more effectively in identifying open ice at the sea ice boundary. By combining these models, we improved overall classification accuracy both for open ice and close ice. The sea ice edge detected by CSCAT was also independently validated against NSIDC's sea ice concentration and ASCAT's sea ice edge, showing high correlation in sea ice extent and temporal–spatial consistency as well as good alignment with SAR imagery at the sea ice–water boundary. The PCA extraction method significantly enhances the feature extraction capabilities of scatterometers with fan-beam and rotating-beam configurations. It complements traditional sea ice detection approaches and allows for the precise and reliable classification of sea ice and open water. Although CSCAT performed well in distinguishing between sea ice and open water, it was prone to confusion between different types of sea ice, with especially limited capability to identify open ice. In the Arctic, such misclassifications were most notable in the Greenland Sea in February and May and in parts of the central region in August; in the Antarctic, they were primarily observed across the entire Antarctic sea ice–water boundary in August and November and along the Antarctic coastline and sea water boundary in May. Open ice is typically associated with the marginal ice zone (MIZ), which is more important than the central region (close ice) for shipping route development, fishery resources, ecosystems, and climate responses. So, good performance on open ice is more important. Future work should focus on further optimizing the algorithm to improve open ice recognition and extending its application to other radar systems, incorporating additional data from ship-based observations, optical imagery, and SAR to enhance detection performance.

Author Contributions: Y.L. (Yanping Luo): investigation, conceptualization, methodology, data curation, writing—original draft, writing—review and editing. Y.L. (Yang Liu): conceptualization, formal analysis, writing—review and editing. C.H.: methodology, data curation. F.H.: methodology, data curation. All authors contributed to the manuscript and approved the submitted version. All authors have read and agreed to the published version of the manuscript.

Funding: This work was supported by the research program "Impact and Response of Antarctic Seas to Climate Change, IRASCC2020-2022" (Grant No. IRASCC 01-02-05C) from the Chinese Arctic and Antarctic Administration (CAA), Ministry of Natural Resources of the People's Republic of China.

Data Availability Statement: The data underlying this article will be shared upon reasonable request to the corresponding authors.

Conflicts of Interest: The authors declare no conflicts of interest.

References

1. Turner, J.; Orr, A.; Gudmundsson, G.H.; Jenkins, A.; Bingham, R.G.; Hillenbrand, C.-D.; Bracegirdle, T.J. Atmosphere-ocean-ice interactions in the Amundsen Sea Embayment, West Antarctica. *Rev. Geophys.* **2017**, *55*, 235–276. [CrossRef]
2. Boers, N. Observation-based early-warning signals for a collapse of the Atlantic Meridional Overturning Circulation. *Nat. Clim. Chang.* **2021**, *11*, 680–688. [CrossRef]
3. Lin, Y.; Moreno, C.; Marchetti, A.; Ducklow, H.; Schofield, O.; Delage, E.; Meredith, M.; Li, Z.; Eveillard, D.; Chaffron, S.; et al. Decline in plankton diversity and carbon flux with reduced sea ice extent along the Western Antarctic Peninsula. *Nat. Commun.* **2021**, *12*, 4948. [CrossRef] [PubMed]
4. Gui, D.; Pang, X.; Lei, R.; Zhao, X.; Wang, J. Changes in sea ice kinematics in the Arctic outflow region and their associations with Arctic Northeast Passage accessibility. *Acta Oceanol. Sin.* **2019**, *38*, 101–110. [CrossRef]
5. Comiso, J.C.; Meier, W.N.; Gersten, R. Variability and trends in the Arctic Sea ice cover: Results from different techniques. *J. Geophys. Res. Ocean.* **2017**, *122*, 6883–6900. [CrossRef]
6. Comiso, J.C.; Gersten, R.A.; Stock, L.V.; Turner, J.; Perez, G.J.; Cho, K. Positive Trend in the Antarctic Sea Ice Cover and Associated Changes in Surface Temperature. *J. Clim.* **2017**, *30*, 2251–2267. [CrossRef]
7. Turner, J.; Hosking, J.S.; Phillips, T.; Marshall, G.J. Temporal and spatial evolution of the Antarctic sea ice prior to the September 2012 record maximum extent. *Geophys. Res. Lett.* **2013**, *40*, 5894–5898. [CrossRef]

8. Lieser, J.; Massom, R.; Reid, P.; Scambos, T.; Stammerjohn, S. The record 2013 Southern Hemisphere sea-ice extent maximum. *Ann. Glaciol.* **2015**, *56*, 99–106. [CrossRef]
9. Turner, J.; Phillips, T.; Marshall, G.J.; Hosking, J.S.; Pope, J.O.; Bracegirdle, T.J.; Deb, P. Unprecedented springtime retreat of Antarctic sea ice in 2016. *Geophys. Res. Lett.* **2017**, *44*, 6868–6875. [CrossRef]
10. Cohen, J.; Screen, J.A.; Furtado, J.C.; Barlow, M.; Whittleston, D.; Coumou, D.; Francis, J.; Dethloff, K.; Entekhabi, D.; Overland, J.; et al. Recent Arctic amplification and extreme mid-latitude weather. *Nat. Geosci.* **2014**, *7*, 627–637. [CrossRef]
11. Sandven, S.; Johannessen, O.M. The use of microwave remote sensing for sea ice studies in the Barents Sea. *ISPRS J. Photogramm. Remote Sens.* **1993**, *48*, 2–18. [CrossRef]
12. Zhang, P.; Hu, X.; Lu, Q.; Zhu, A.; Lin, M.; Sun, L.; Chen, L.; Xu, N. FY-3E: The First Operational Meteorological Satellite Mission in an Early Morning Orbit. *Adv. Atmos. Sci.* **2022**, *39*, 1–8. [CrossRef]
13. Remund, Q.P.; Long, D.G. Sea ice extent mapping using Ku band scatterometer data. *J. Geophys. Res. Ocean.* **1999**, *104*, 11515–11527. [CrossRef]
14. Remund, Q.P.; Long, D.G. A Decade of QuikSCAT Scatterometer Sea Ice Extent Data. *IEEE Trans. Geosci. Remote Sens.* **2014**, *52*, 4281–4290. [CrossRef]
15. Abreu, R.D.; Wilson, K.; Arkett, M.; Langlois, D. Evaluating the use of QuikSCAT data for operational sea ice monitoring. In Proceedings of the IEEE International Geoscience and Remote Sensing Symposium, Toronto, ON, Canada, 24–28 June 2002; Volume 3035, pp. 3032–3033.
16. Li, M.; Zhao, C.; Zhao, Y.; Wang, Z.; Shi, L. Polar Sea Ice Monitoring Using HY-2A Scatterometer Measurements. *Remote Sens.* **2016**, *8*, 688. [CrossRef]
17. de Haan, S.; Stoffelen, A. *Ice Discrimination Using ERS Scatterometer*. 2001. Available online: https://www.knmi.nl/research/publications/ice-discrimination-using-ers-scatterometer (accessed on 23 August 2024).
18. Verspeek, J.A. *Sea Ice Classification Using Bayesian Statistics*; KNMI, 2006. Available online: https://www.knmi.nl/research/publications/sea-ice-classification-using-bayesian-statistics (accessed on 23 August 2024).
19. Belmonte Rivas, M.; Verspeek, J.; Verhoef, A.; Stoffelen, A. Bayesian Sea Ice Detection With the Advanced Scatterometer ASCAT. *IEEE Trans. Geosci. Remote Sens.* **2012**, *50*, 2649–2657. [CrossRef]
20. Belmonte Rivas, M.; Stoffelen, A. New Bayesian Algorithm for Sea Ice Detection With QuikSCAT. *IEEE Trans. Geosci. Remote Sens.* **2011**, *49*, 1894–1901. [CrossRef]
21. Li, Z.; Verhoef, A.; Stoffelen, A. Bayesian Sea Ice Detection Algorithm for CFOSAT. *Remote Sens.* **2022**, *14*, 3569. [CrossRef]
22. Hersbach, H.; Stoffelen, A.; de Haan, S. An improved C-band scatterometer ocean geophysical model function: CMOD5. *J. Geophys. Res. Ocean.* **2007**, *112*, C03006. [CrossRef]
23. Zhai, X.; Wang, Z.; Zheng, Z.; Xu, R.; Dou, F.; Xu, N.; Zhang, X. Sea Ice Monitoring with CFOSAT Scatterometer Measurements Using Random Forest Classifier. *Remote Sens.* **2021**, *13*, 4686. [CrossRef]
24. Liu, L.; Dong, X.; Lin, W.; Lang, S.; Wang, L. Polar Sea Ice Detection with the CFOSAT Scatterometer. In Proceedings of the 2021 IEEE International Geoscience and Remote Sensing Symposium IGARSS, Brussels, Belgium, 11–16 July 2021; pp. 5645–5648. [CrossRef]
25. Xu, R.; Zhao, C.; Zhai, X.; Chen, G. Arctic Sea Ice Type Classification by Combining CFOSCAT and AMSR-2 Data. *Earth Space Sci.* **2022**, *9*, e2021EA002052. [CrossRef]
26. Hotelling, H. Analysis of a complex of statistical variables into principal components. *J. Educ. Psychol.* **1933**, *24*, 498–520. [CrossRef]
27. Zabalza, J.; Ren, J.; Yang, M.; Zhang, Y.; Wang, J.; Marshall, S.; Han, J. Novel Folded-PCA for improved feature extraction and data reduction with hyperspectral imaging and SAR in remote sensing. *ISPRS J. Photogramm. Remote Sens.* **2014**, *93*, 112–122. [CrossRef]
28. Singh, R.K.; Singh, K.N.; Maisnam, M.; P., J.; Maity, S. Antarctic Sea Ice Extent from ISRO's SCATSAT-1 Using PCA and An Unsupervised Classification. *Proceedings* **2018**, *2*, 340. [CrossRef]
29. Long, D.G. Polar Applications of Spaceborne Scatterometers. *IEEE J. Sel. Top. Appl. Earth Obs. Remote Sens.* **2017**, *10*, 2307–2320. [CrossRef]
30. Yueh, S.H.; Kwok, R.; Lou, S.H.; Tsai, W.Y. Sea ice identification using dual-polarized Ku-band scatterometer data. *IEEE Trans. Geosci. Remote Sens.* **1997**, *35*, 560–569. [CrossRef]
31. Breivik, L.A.; Eastwood, S.; Lavergne, T. Use of C-Band Scatterometer for Sea Ice Edge Identification. *IEEE Trans. Geosci. Remote Sens.* **2012**, *50*, 2669–2677. [CrossRef]
32. Lindell, D.; Long, D. Multiyear Arctic Ice Classification Using ASCAT and SSMIS. *Remote Sens.* **2016**, *8*, 294. [CrossRef]
33. Aaboe, S.; Down, E.J.; Eastwood, S. EUMETSAT Ocean and Sea Ice Satellite Application Facility, Global Seaice Edge Near-Real-Time Product-Multimission (2020), OSI-402-d, (Data Extracted from OSI SAF FTP Server/EUMETSAT Data Center: Accessed 10-01-2023. 2020. Available online: https://osi-saf.eumetsat.int/products/osi-402-d (accessed on 23 August 2024).
34. Hill, J.C.; Long, D.G. Extension of the QuikSCAT Sea Ice Extent Data Set with OSCAT Data. *IEEE Geosci. Remote Sens. Lett.* **2017**, *14*, 92–96. [CrossRef]
35. Xu, R.; Zhao, C.; Zhai, X.; Zhao, K.; Shen, J.; Chen, G. Polar Sea Ice Identification and Classification Based on HY-2A/SCAT Data. *J. Ocean Univ. China* **2022**, *21*, 331–346. [CrossRef]

36. Zou, J.; Zeng, T.; Guo, M.; Cui, S. The study on an Antarctic sea ice identification algorithm of the HY-2A microwave scatterometer data. *Acta Oceanol. Sin.* **2016**, *35*, 74–79. [CrossRef]
37. Liu, L.; Zhai, H.; Dong, X.; Zhao, F. SEA ICE Extent Retrieval with Ku-Band Rotating Fan Beam Scatterometer Data. In Proceedings of the IGARSS 2022-2022 IEEE International Geoscience and Remote Sensing Symposium, Kuala Lumpur, Malaysia, 17–22 July 2022; pp. 3810–3813. [CrossRef]
38. Xu, C.; Wang, Z.; Zhai, X.; Lin, W.; He, Y. SVM-Based Sea Ice Extent Retrieval Using Multisource Scatterometer Measurements. *Remote Sens.* **2023**, *15*, 1630. [CrossRef]
39. Zhai, X.; Tian, S.; Ye, Y.; Cao, G.; Chen, L.; Xu, N.; Zheng, Z. First Results of Antarctic Sea Ice Classification Using Spaceborne Dual-Frequency Scatterometer FY-3E WindRAD. *IEEE Geosci. Remote Sens. Lett.* **2024**, *21*, 2000105. [CrossRef]
40. Li, Z.; Stoffelen, A.; Verhoef, A.; Verspeek, J. Numerical Weather Prediction Ocean Calibration for the Chinese-French Oceanography Satellite Wind Scatterometer and Wind Retrieval Evaluation. *Earth Space Sci.* **2021**, *8*, e2020EA001606. [CrossRef]
41. Li, Z.; Stoffelen, A.; Verhoef, A.; Verspeek, J. NWP Ocean Calibration for the CFOSAT Wind Scatterometer. In Proceedings of the 2021 IEEE International Geoscience and Remote Sensing Symposium IGARSS, Brussels, Belgium, 11–16 July 2021; pp. 443–446. [CrossRef]
42. Zhang, K.; Dong, X.; Zhu, D.; Yun, R.; Wang, B.; Yu, M. An Improved Method of Noise Subtraction for the CFOSAT Scatterometer. *IEEE J. Sel. Top. Appl. Earth Obs. Remote Sens.* **2021**, *14*, 7506–7515. [CrossRef]
43. Ivanova, N.; Pedersen, L.T.; Tonboe, R.T.; Kern, S.; Heygster, G.; Lavergne, T.; Sørensen, A.; Saldo, R.; Dybkjær, G.; Brucker, L.; et al. Inter-comparison and evaluation of sea ice algorithms: Towards further identification of challenges and optimal approach using passive microwave observations. *Cryosphere* **2015**, *9*, 1797–1817. [CrossRef]
44. Kern, S.; Lavergne, T.; Notz, D.; Pedersen, L.T.; Tonboe, R.T.; Saldo, R.; Sørensen, A.M. Satellite passive microwave sea-ice concentration data set intercomparison: Closed ice and ship-based observations. *Cryosphere* **2019**, *13*, 3261–3307. [CrossRef]
45. Comiso, J.C. Characteristics of Arctic winter sea ice from satellite multispectral microwave observations. *J. Geophys. Res. Ocean.* **1986**, *91*, 975–994. [CrossRef]
46. Smith, D.M. Extraction of winter total sea-ice concentration in the Greenland and Barents Seas from SSM/I data. *Int. J. Remote Sens.* **1996**, *17*, 2625–2646. [CrossRef]
47. Lavergne, T.; Sørensen, A.M.; Kern, S.; Tonboe, R.; Notz, D.; Aaboe, S.; Bell, L.; Dybkjær, G.; Eastwood, S.; Gabarro, C.; et al. Version 2 of the EUMETSAT OSI SAF and ESA CCI sea-ice concentration climate data records. *Cryosphere* **2019**, *13*, 49–78. [CrossRef]
48. Cavalieri, D.J.; Parkinson, C.; Gloersen, P.; Zwally, H.J. Sea Ice Concentrations from Nimbus-7 SMMR and DMSP SSM/I-SSMIS Passive Microwave Data. 1996. Available online: https://nsidc.org/data/nsidc-0051/versions/1 (accessed on 21 January 2023).
49. Cavalieri, D.J.; Gloersen, P.; Campbell, W.J. Determination of sea ice parameters with the Nimbus 7 SMMR. *J. Geophys. Res. Atmos.* **1984**, *89*, 5355–5369. [CrossRef]
50. Dabboor, M.; Shokr, M. A new Likelihood Ratio for supervised classification of fully polarimetric SAR data: An application for sea ice type mapping. *ISPRS J. Photogramm. Remote Sens.* **2013**, *84*, 1–11. [CrossRef]
51. Torres, R.; Snoeij, P.; Geudtner, D.; Bibby, D.; Davidson, M.; Attema, E.; Potin, P.; Rommen, B.; Floury, N.; Brown, M.; et al. GMES Sentinel-1 mission. *Remote Sens. Environ.* **2012**, *120*, 9–24. [CrossRef]
52. Stewart, J.S.; Meier, W.N.; Scott, D.J. *Polar Stereographic Ancillary Grid Information, Version 1*; National Snow and Ice Data Center: Boulder, CO, USA, 2022. [CrossRef]
53. Meier, W.N.; Stroeve, J.; Fetterer, F.; Wilcox, H. *Polar Stereographic Valid Ice Masks Derived from National Ice Center Monthly Sea Ice Climatologies, Version 1*; National Snow and Ice Data Center: Boulder, CO, USA, 2015. [CrossRef]
54. Halko, N.; Martinsson, P.-G.; Tropp, J.A. Finding structure with randomness: Stochastic algorithms for constructing approximate matrix decompositions. *arXiv* **2009**, arXiv:0909.4061. [CrossRef]
55. Sandven, S.; Spreen, G.; Heygster, G.; Girard-Ardhuin, F.; Farrell, S.L.; Dierking, W.; Allard, R.A. Sea Ice Remote Sensing—Recent Developments in Methods and Climate Data Sets. *Surv. Geophys.* **2023**, *44*, 1653–1689. [CrossRef]
56. Hao, H.; Su, J.; Shi, Q.; Li, L. Arctic sea ice concentration retrieval using the DT-ASI algorithm based on FY-3B/MWRI data. *Acta Oceanol. Sin.* **2021**, *40*, 176–188. [CrossRef]
57. Liu, J.; Liu, S.; Lin, W.; Lang, S.; He, Y. Sea ice identification based on CFOSAT scatterometer. *Haiyang Xuebao* **2023**, *45*, 134–140. [CrossRef]
58. CAFF. Boundary for Conservation of Arctic Flora and Fauna (CAFF) Working Group of the Arctic Council. 2017. Available online: http://geo.abds.is/geonetwork/srv/eng/catalog.search#/metadata/2ad7a7cb-2ad7-4517-a26e-7878ef134239 (accessed on 1 December 2023).
59. Fetterer, F.; Fetterer, F.; Knowles, K.; Meier, W.N.; Savoie, M.; Windnagel, A.K. *Sea Ice Index, Version 3 [Data Set]*; National Snow and Ice Data Center: Boulder, CO, USA, 2017; Available online: https://doi.org/10.7265/N5K072F8 (accessed on 10 January 2023).
60. Kern, S.; Lavergne, T.; Notz, D.; Pedersen, L.T.; Tonboe, R. Satellite passive microwave sea-ice concentration data set intercomparison for Arctic summer conditions. *Cryosphere* **2020**, *14*, 2469–2493. [CrossRef]

remote sensing MDPI

Technical Note

Wavelength Cut-Off Error of Spectral Density from MTF3 of SWIM Instrument Onboard CFOSAT: An Investigation from Buoy Data

Yuexin Luo [1,2,3], Ying Xu [2], Hao Qin [3,4] and Haoyu Jiang [1,2,4,*]

[1] College of Life Sciences and Oceanography, Shenzhen University, Shenzhen 518000, China
[2] Key Laboratory of Space Ocean Remote Sensing and Application, Ministry of Natural Resources, Beijing 100081, China
[3] College of Marine Science and Technology, China University of Geosciences, Wuhan 430074, China
[4] Shenzhen Research Institute, China University of Geosciences, Shenzhen 518000, China
* Correspondence: haoyujiang@szu.edu.cn

Abstract: The Surface Waves Investigation and Monitoring instrument (SWIM) provides the directional wave spectrum within the wavelength range of 23–500 m, corresponding to a frequency range of 0.056–0.26 Hz in deep water. This frequency range is narrower than the 0.02–0.485 Hz frequency range of buoys used to validate the SWIM nadir Significant Wave Height (SWH). The modulation transfer function used in the current version of the SWIM data product normalizes the energy of the wave spectrum using the nadir SWH. A discrepancy in the cut-off frequency/wavelength ranges between the nadir and off-nadir beams can lead to an overestimation of off-nadir cut-off SWHs and, consequently, the spectral densities of SWIM wave spectra. This study investigates such errors in SWHs due to the wavelength cut-off effect using buoy data. Results show that this wavelength cut-off error of SWH is small in general thanks to the high-frequency extension of the resolved frequency range. The corresponding high-frequency cut-off errors are systematic errors amenable to statistical correction, and the low-frequency cut-off error can be significant under swell-dominated conditions. By leveraging the properties of these errors, we successfully corrected the high-frequency cut-off SWH error using an artificial neural network and mitigated the low-frequency cut-off SWH error with the help of a numerical wave hindcast. These corrections significantly reduced the error in the estimated cut-off SWH, improving the bias, root-mean-square error, and correlation coefficient from 0.086 m, 0.111 m, and 0.9976 to 0 m, 0.039 m, and 0.9994, respectively.

Keywords: surface waves investigation and monitoring; China–France oceanography satellite; wave spectrum; significant wave height; modulation transfer function

1. Introduction

Wind-generated sea surface gravity waves (hereafter, referred to simply as waves) are one of the most common dynamic phenomena on the ocean surface. Observation of the waves plays a very important role across various domains including ocean engineering, marine disaster prevention and mitigation, and offshore structural design. Currently, operational wave observation relies primarily on two data sources—wave buoys and satellite remote sensing. For wave remote sensing, the two most commonly used types of remote sensors are satellite altimeters and Synthetic Aperture Radars (SARs), each with its own advantages and disadvantages. Altimeters can provide accurate observations of Significant Wave Height (SWH) [1–3] but are unable to capture wave spectra; SARs can obtain parts of the directional wave spectrum, but the so-called "azimuth cut-off" effect caused by the nonlinear imaging mechanism leads to the loss of high-frequency wave information in the wave spectrum [4]. The launching of the

China–France Oceanography SATellite (CFOSAT) in October 2018 opened new prospects for the observation of ocean waves.

The Surface Waves Investigation and Monitoring instrument (SWIM) onboard the CFOSAT is the world's first space-borne 'wave spectrometer'. SWIM is a six-beam rotating radar with small incidence angles (0–10°) operating at 13.575 GHz [5]. Its main task is to measure directional wave spectra in global open oceans. SWIM not only obtains wind speed and SWH information from the nadir beam like an altimeter but also obtains wave spectrum information from off-nadir beams with incidence angles of 6°–10°. At these low incidence angles, the normalized radar cross-section is sensitive to local slopes associated with long-wave tilts and is insensitive to wind-induced small-scale roughness as well as hydrodynamic modulations caused by the interaction between short and long waves. As a result, SWIM avoids the "azimuth cut-off" associated with SARs during wave spectrum measurements. It is capable of resolving waves with wavelengths of 70–500 m in the field-of-view direction [5]; however, according to the actual data files, the range of resolvable wavelengths is about 23–500 m. By combining unidirectional wave spectra from different directions through sensor rotation, SWIM produces a complete wave directional spectrum.

For retrieving wave spectral densities, the Modulation Transfer Function (MTF) currently used in the official SWIM data product is MTF3. This method directly normalizes the energy of the wave spectrum by using the energy derived from the nadir SWH as the total energy of the wave spectrum. Although MTF3 is more accurate in estimating the total energy of the wave spectrum compared to its predecessor, MTF1 [5], it introduces a potential wavelength cut-off error in the obtained wave spectrum. The wave spectrum measured by CFOSAT-SWIM has a wavelength range of 23–500 m, corresponding to a frequency range of 0.056–0.26 Hz in deep water. In contrast, the nadir SWH is an integral part of the spectral densities over a wider frequency range. The nadir SWH is often validated by in situ observations, such as the buoy data from the National Data Buoy Center (NDBC, e.g., [1–3]), and most NDBC buoys have a frequency range of 0.02–0.485 Hz. Therefore, the SWH corresponding to the SWIM spectrum should be considered as a cut-off SWH, and normalizing the energy of the cut-off wave spectrum with the SWH without the cut-off may result in an overestimation of the cut-off SWH and hence the spectral densities of the SWIM wave spectrum. This study aims to investigate the impact of SWIM's wavelength cut-off on wave spectral density using NDBC buoy data and to mitigate this impact to some extent.

2. Materials and Methods

2.1. NDBC Buoy Data

NDBC has a coastal-marine automated network comprising more than 100 buoys that measure wave spectra. Although these buoys can provide the first five Fourier coefficients of waves using their translational and pitch–roll data, only the energy spectra (i.e., the first Fourier coefficients) are used in this study because directional information is not considered. The data are available from the NDBC website (https://www.ndbc.noaa.gov/, accessed on 19 July 2024). We selected the data during the period from April 2019 to March 2023, applying the following criteria for buoy selection: (1) The buoys should have the data of the energy spectra covering the frequency range of 0.02–0.485 Hz; (2) The buoys should be more than 150 km from the coastlines, and the corresponding water depth should be more than 200 m so that the SWIM cut-off wavelengths can be easily converted to cut-off frequencies (0.056–0.26 Hz) using deep water dispersion relation. Following this screening process, 35 buoys meeting the criteria were retained, as depicted in Figure 1.

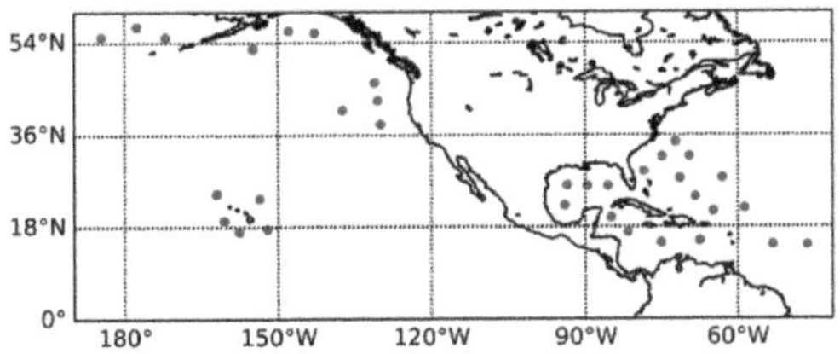

Figure 1. The location of the NDBC buoys used in this study.

2.2. Numerical Wave Model Hindcast

The wave spectrum data from the Integrated Ocean Waves for Geophysical and other Applications (IOWAGA) dataset are used as the axillary data to study the correction of the wavelength cut-off error in SWH. IOWAGA is a global hindcast wave field dataset generated using the WAVEWATCH III® model (version 6.07) [6] based on the source term package of ST4 [7] and forced by the global 10 m wind from the ERA5 dataset, surface currents from the CMEMS-Globcurrent, and ice concentrations from the IFREMER SSMI-derived daily product. Although the numerical wave model does not assimilate any observations, IOWAGA data show good agreement with both the in situ and altimeter observations [8]. The directional wave spectrum in this wave model is uniformly spaced across 24 directions and exponentially spaced across 36 frequencies ranging from 0.034 to 0.95 Hz with a 1.1 increment factor. The IOWAGA hindcast is run at a resolution of $0.5^\circ \times 3$ h, but a nested grid with a higher spatial resolution of $1/6^\circ$ is used in coastal regions. Full directional wave spectra and frequency spectra are available at more than 10,000 points along coastlines worldwide and at the locations of many moored buoys including those shown in Figure 1. More details of the dataset and the numerical wave model to generate this dataset can be found in [8].

2.3. Cut-Off SWH

Two types of SWHs are computed from the buoy wave spectra, the total SWH and the cut-off SWH. They are both computed using the following equations:

$$SWH = 4\sqrt{m_0} \tag{1}$$

$$m_n = \int_{f_{dn}}^{f_{up}} f^n E(f)\,\mathrm{d}f \tag{2}$$

where f is the frequency, m_n is the nth moment of the spectral density function $E(f)$, and f_{up} and f_{dn} are the high-frequency and low-frequency cut-offs of the measuring devices, respectively. For the total SWH, f_{up} and f_{dn} are set to 0.02 and 0.485 Hz, respectively, which are the cut-off frequencies for buoys. For the cut-off SWH, f_{up} and f_{dn} are set to 0.056 and 0.26 Hz, respectively, which are the cut-off frequencies for SWIM. The difference between the total and cut-off SWHs is then the wavelength cut-off error of SWH, which is equivalent to the error of spectral density due to wavelength cut-off when the spectral density is normalized by the total SWH. Hereafter, this wavelength/frequency cut-off error of SWH will be referred to simply as the cut-off SWH error, or cut-off error.

To test whether the cut-off error comes from the high-frequency or low-frequency part of the wave spectrum, we also compute the corresponding high-frequency and low-frequency cut-off SWHs where the integral frequency ranges are set to 0.02–0.26 Hz and 0.056–0.485 Hz, respectively. The high-frequency and low-frequency cut-off SWH errors are defined as the differences between the total SWH and the high-frequency and low-frequency cut-off SWHs, respectively.

2.4. Artificial Neural Network

An Artificial Neural Network (ANN) is used to correct the high-frequency cut-off SWH errors. An ANN can be thought of as a network of "neurons" organized into layers, including an input layer, one or more hidden layers, and an output layer. Upon receiving input and output data, the ANN establishes weighted connections between the input and output nodes through interconnected computing elements known as "neurons" situated in the hidden layers. The main advantage of an ANN is its theoretical ability to approximate any function with any number of input parameters, provided there are enough neurons and layers, making it a powerful non-linear fitting tool.

In Section 3, we demonstrate the nonlinear correlations of high-frequency cut-off errors with both wind speeds and SWHs, suggesting that the ANN might be well-suited for predicting the high-frequency cut-off error using wind speed and SWH as inputs. The buoy data are randomly divided into two groups, one for the training (10%) and the other for the validation (90%). Using only 10% of the data for training ensures the robustness of the statistical correction model. The ANN used here has an input layer with two neurons (wind speed and SWH), two hidden layers with 32 neurons each, and an output layer with one neuron. The activation function between different layers is the rectified linear unit (ReLU). The ANN is trained to minimize the Mean-Square Error (MSE) between the outputs and the targets (high-frequency cut-off SWH errors) using the Adam optimizer with a batch size of 512. The learning rate (initially set to 0.0002) decreased by 50% if the MSE of the training set did not decrease for two epochs, and the training process stopped when the MSE of the validation set did not decrease for ten epochs. It is also shown that having more hidden layers and hidden neurons does not significantly impact the model's performance.

3. Results

3.1. Evaluation of SWH Cut-Off Errors

Comparisons between the total and the three different cut-off SWHs, integrating over 0.056–0.26 Hz, 0.02–0.26 Hz, and 0.056–0.485 Hz, at all the buoys used in this study are shown as scatter plots in Figure 2a, Figure 2b, and Figure 2c, respectively. To evaluate the impact of the wavelength cut-off effect on the estimation of SWH, three error metrics are employed—bias, root-mean-square error (RMSE), and correlation coefficient (CC)—as follows:

$$Bias = \frac{1}{n}\sum_{i=1}^{n}(y_i - x_i) \tag{3}$$

$$RMSE = \sqrt{\frac{1}{n}\sum_{i=1}^{n}(y_i - x_i)^2} \tag{4}$$

$$R = \sum_{i=1}^{n}(y_i - \overline{y})(x_i - \overline{x}) / \left[\sqrt{\sum_{i=1}^{n}(y_i - \overline{y})^2}\sqrt{\sum_{i=1}^{n}(x_i - \overline{x})^2}\right] \tag{5}$$

where x and y denote the cut-off (including the high-frequency and low-frequency cut-off) SWHs and total SWHs, respectively, and the bars over them denote their mean values.

From Figure 2a, the first conclusion that can be drawn is that the wavelength cut-off does not significantly impact SWH estimation. The majority of the data closely align with the y = x line, with the bias, RMSE, and CC values of 0.086 m, 0.111 m, and 0.9976, respectively. Such a low value of RMSE is comparable with the SWH random noise observed by the NDBC buoys and altimeters [2], suggesting that the frequency/wavelength cut-off effect is not a major error source in SWIM's retrieval of wave spectra. This is probably one of the most important conclusions of this study, as it indicates that the cut-off effect will not significantly impact most applications of SWIM's wave spectral data. However, it is also noted that a large portion of the error comes from the systematic bias that can be corrected statistically, and there are still a few cases with significant cut-off errors, indicating that this issue warrants further investigation.

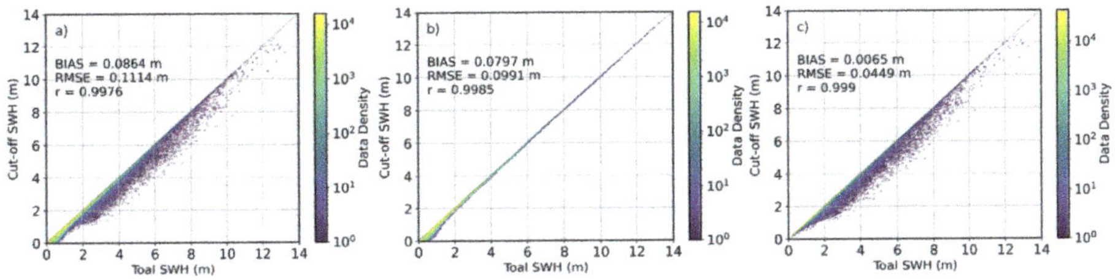

Figure 2. The scatter plot between the SWHs integrated from full buoy wave spectra (0.02–0.485 Hz) and from cut-off buoy wave spectra. The integral frequency range for the cut-off SWH is (**a**) 0.056–0.26 Hz, (**b**) 0.02–0.26 Hz (high-frequency cut-off SWH), and (**c**) 0.056–0.485 Hz.

A closer examination of the high-frequency and low-frequency cut-off SWHs reveals an interesting pattern, where most of the errors stem from the cut-off of the high-frequency tails of the spectra (Figure 2b), while the few cases with large errors are due to the low-frequency cut-off (Figure 2c). Although both the bias and RMSE in Figure 2b are significantly larger than those in Figure 2c, and the CC is lower in Figure 2b than in Figure 2c, the scatter plot in Figure 2b appears to show better agreement than Figure 2c, as it is much less scattered. This suggests that a large proportion of the errors in Figure 2b originates from the correctable systematic bias between the total and cut-off SWHs. Despite the more scattered appearance of Figure 2c, it is important to highlight that these "outliers" represent a very small fraction of the dataset and have only a minimal impact on the overall error metrics, where only 1.2% of the data has residual >0.1 m in Figure 2c, and this value decreases to 0.5% for residual exceeding 0.25 m.

To further investigate the dependence of these errors on other variables, the cut-off errors of SWH as functions of total SWH and wind speed are shown in Figure 3. A weak yet statistically significant correlation (CC $\approx$ 0.22) is found between the low-frequency cut-off errors and SWHs, but no clear pattern is found in Figure 3a. Similarly, in Figure 3b, the correlation between the low-frequency cut-off errors and wind speeds is not statistically significant. In contrast, the high-frequency cut-off errors exhibit stronger correlations with both wind speeds and SWHs, with CCs of 0.33 and -0.33, respectively. Although the CCs of ± 0.33 are still low (but statistically significant and higher than the CC in Figure 3a), clear patterns can be observed in Figure 3c,d, where high-frequency cut-off errors increase with the increase in wind speed and with the decrease in SWH. These dependencies might be helpful for the correction of high-frequency cut-off errors.

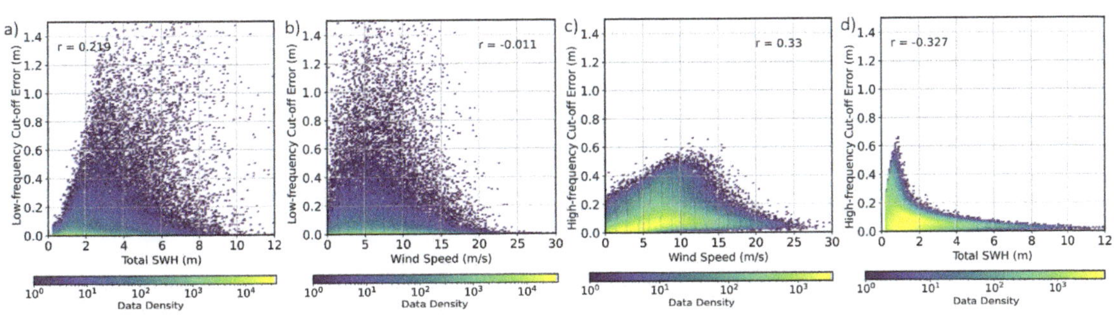

Figure 3. Wavelength cut-off error of SWH as functions of total SWH and wind speed. (**a**,**b**) Low-frequency cut-off errors as a function of SWH and wind speed, respectively. (**c**,**d**) The same as subplots a and b but for high-frequency cut-off errors.

3.2. Correction of High-Frequency Cut-Off Errors of SWH

The high-frequency cut-off errors appear to be systematic, with their magnitudes significantly correlated to both wind speeds and SWHs, as shown in Figure 3c,d. From a physical point of view, previous studies have shown that these high-frequency tails of the wave spectrum, known as the "equilibrium range", respond rapidly to local wind, especially in the open ocean [9]. Additionally, it has been shown that the shape of the "equilibrium range" can be utilized to estimate wind speed [10,11]. These correlations suggest that the correction of high-frequency cut-off errors is promising.

Since wind speed is available, the high-frequency tail can be computed using the f^{-4} relation (the impact of the saturation range f^{-5} on the SWH is small):

$$E(f) = \frac{4\beta I u_* g}{(2\pi)^3} f^{-4} \tag{6}$$

where β is an empirical constant, and I was introduced to allow for enhanced tail energy levels for different directional spread. If one assumes that I is constant and the wind profile is known (e.g., the log profile), then the cut-off energy can be estimated using Equation (6). However, this semi-analytical method might involve relatively large errors, as shown by previous attempts to estimate wind speed using wave spectra [10,11].

Given the availability of wind speed and total SWH data from SWIM, a model was developed to correct high-frequency cut-off SWH errors, employing wind speed and SWH as inputs. As depicted in Figure 3c,d, the correlations of high-frequency cut-off errors with both wind speed and SWHs are nonlinear. Therefore, we try to use an ANN to conduct this correction simply because ANN can handle the problem of multivariate nonlinear fitting well and is widely used in the statistical correction of systematic errors in wave remote sensing (e.g., [2,12]).

After applying the ANN-based correction, on which more information is available in Section 2.4, the comparison between the corrected high-frequency cut-off SWHs and the buoy-derived high-frequency cut-off SWHs (for the validation set) is shown in Figure 4a. Compared to the results in Figure 2b, the bias is eliminated, the RMSE decreases from 0.099 m to 0.035 m, and the CC increases from 0.9985 to 0.9994, showing that this statistical method can effectively correct the high-frequency cut-off error. After the correction, the most significant errors (although they are small) are observed under conditions of low SWHs, primarily because even a very small error in spectral density can result in a relatively large error in SWH when SWHs are low.

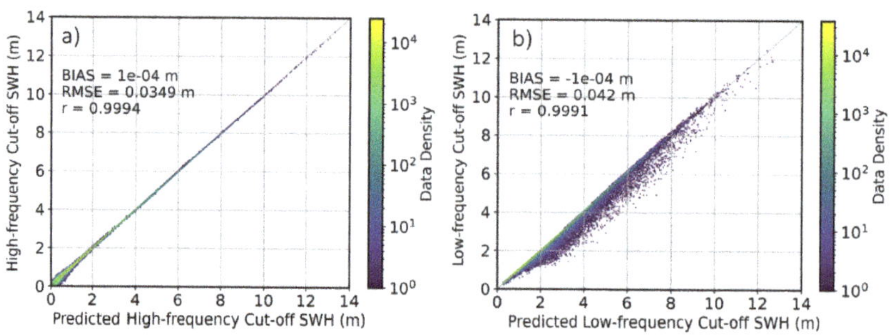

Figure 4. (**a**) Comparison between the high-frequency cut-off SWHs (0.02–0.26 Hz) from ANN correction and from the original buoy data. (**b**) The same as (**a**) but for low-frequency cut-off SWHs (0.056–0.485 Hz).

3.3. Correction of Low-Frequency Cut-Off Errors of SWH

Compared to the high-frequency errors that are correlated to local winds and SWHs, the correction of low-frequency errors is more challenging. As previously discussed, the low-frequency cut-off error of SWH comes from long swells generated by the westerlies and propagated into the buoy locations. These long swells are not directly correlated to any local variables from a physical standpoint because they are simply remote signals passing through the given area. Consequently, correcting the low-frequency cut-off error of SWH through statistical methods is unlikely to be effective. We attempted to apply the aforementioned ANN for this correction, but the results in Figure 4b aligned with our expectations—the statistical correction was ineffective for correcting the low-frequency cut-off error. The ANN marginally reduced the RMSE from 0.045 m to 0.042 m and increased the CC from 0.9990 to 0.9991, but the scattering of data in Figure 4b remained almost unchanged compared to Figure 2c.

Given the ineffectiveness of statistical methods in correcting low-frequency cut-off errors, such an error might only be corrected using the dynamic way, i.e., with the help of a numerical wave model forecast or hindcast.

For spectral densities at frequencies lower than 0.056 Hz, the difference between the results from wave forecasts and hindcasts is small because the energy at such low frequencies is dependent only on the historical and remote wind field [13]. In this study, we tried to use the omnidirectional wave spectra from the IOWAGA dataset to correct the low-frequency cut-off errors. Although the lowest frequency resolved by the IOWAGA dataset is 0.034 Hz, which is higher than that which has been resolved by the buoy (0.02 Hz), it can be seen from Figure 2b that there is almost no wave energy for frequencies lower than 0.034 Hz in the buoy data, even for swell-dominated cases. Therefore, the cut-off frequency of the model is not an issue for the energy correction. Although contemporary numerical models still face challenges in accurately modeling swells that propagate over large distances [4,14,15], we believe that they can at least reflect a part of the cut-off low-frequency error.

Figure 5a shows the comparison of the low-frequency cut-off errors computed from the buoy data and the corresponding model data. Although the agreement between these two sets of data might not seem satisfactory at first glance, it is noted that the CC between the two error values exceeds 0.8, which can be regarded as a high value. This means that the low-frequency cut-off error can be, at least, partly reflected and corrected using a numerical wave model. The error correction process is carried out as follows: First, the modeled proportions of cut-off energy (energy below 0.056 Hz) are linearly regressed against the buoy-observed proportions of cut-off energy to reduce the systematic error of energy between the model and buoy in low frequencies. Then, the low-frequency cut-off error δ_{LF} is estimated as follows:

$$\delta_{LF} = \sqrt{SWH^2 \times p} \tag{7}$$

where p is the modeled proportions of cut-off energy after the correction in the first step. This correction method yields slightly better results than a direct correction using the model-estimated low-frequency cut-off error (also adjusted by linear regression). Figure 5b shows the comparison between the high-frequency cut-off SWHs (0.056–0.45 Hz) after this correction and from the original buoy data. Compared to the results in Figure 4b, most outliers with large errors are effectively eliminated, the RMSE decreases from 0.042 m to 0.020 m, and the CC increases from 0.9991 to 0.9998. Finally, combining the corrections for both high-frequency and low-frequency cut-off errors, the comparison between the predicted and buoy-measured cut-off SWH is shown in Figure 5c. It is evident that the potential wavelength cut-off error of SWH is significantly reduced, and it will be more reasonable for the MTF of SWIM to use the estimated cut-off SWH to normalize the wave spectra.

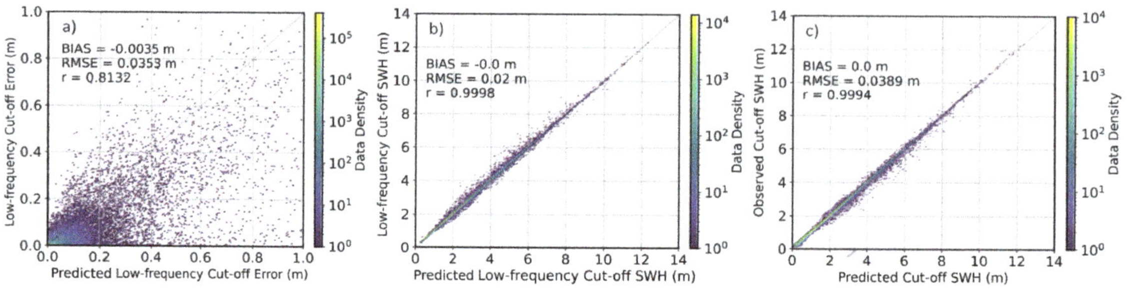

Figure 5. (**a**) The scatter plot between the low-frequency cut-off errors of SWH derived from buoys and those from IOWAGA. (**b**) The scatter plot between the low-frequency cut-off SWHs (0.056–0.485 Hz) after correction using the numerical wave model and the original buoy data. (**c**) The scatter plot between the cut-off SWHs (0.056–0.26 Hz) from direct buoy observations and those from the correction method presented in this study.

4. Discussion

To better understand the cut-off error of SWH, Figure 6a shows four mean buoy spectra for different SWHs. The figure shows that a small portion of wave energy is cut off at both high and low frequencies. When SWHs are high (e.g., the red curve), the cut-off energy is only a very small proportion of the total energy, especially in the high-frequency range, resulting in a small error in SWH estimation. However, when the SWHs are low (e.g., the orange curve), the proportion of high-frequency cut-off energy becomes large, resulting in SWHs which are in agreement with Figures 2b and 3d, where the high-frequency cut-off errors are negatively correlated to SWH. This can also be regarded as the physical basis for using the total SWH as the input of ANN for correcting high-frequency cut-off errors.

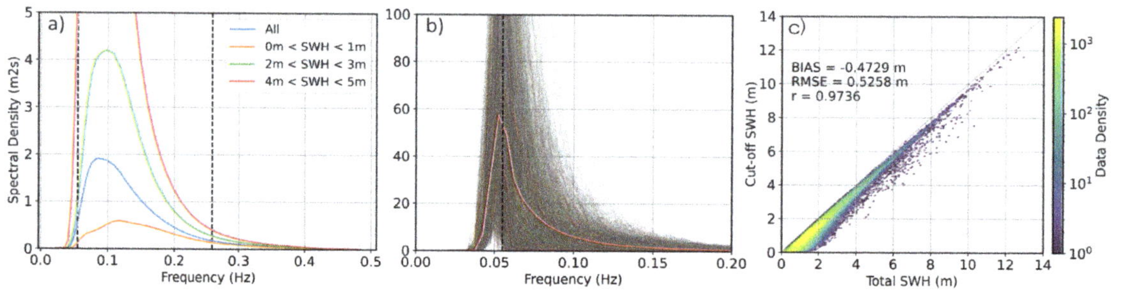

Figure 6. (**a**) Mean buoy spectra for the entire sample (blue) and for different SWH ranges (other colors) with vertical dashed line indicates the cut-off frequencies. (**b**) Buoy spectra for cases with cut-off errors exceed 0.5 m, where the red curve indicates their mean spectrum. (**c**) The same as Figure 2a, but using the nominal cut-off range of SWIM, 70 to 500 m.

For SWH < 3 m, the low-frequency cut-off energy remains lower than the high-frequency cut-off energy, and its contribution to the total energy diminishes as SWH increases. This explains why the low-frequency cut-off effect is negligible in most cases, regardless of whether the SWH is low or high.

Another noteworthy feature in Figure 6a is that the wavelength cut-off error of SWH, particularly the high-frequency cut-off error, would be significant if the resolved wavelength ranged only from 70 to 500 m (0.0565 to 0.15 Hz). This is because spectral densities at frequencies of above 0.15 Hz contribute significantly to the total energy in cases with SWHs at below 3 m. The comparison between the total and cut-off SWHs using the nominal cut-off range of SWIM (70 to 500 m), as shown in Figure 6c, supports the above inference. Both the bias and RMSE are about four times greater than those in Figure 2a, with errors being particularly significant in low-SWH cases. However, thanks to the extension of the resolved frequency range, the impact of high-frequency cut-off error is greatly reduced.

Figure 6b shows the mean spectrum for instances with cut-off errors exceeding 0.5 m (this threshold is arbitrarily selected, yet it was checked that varying it from 0.2 to 0.8 m has negligible impact on the result). For the cases of large cut-off error, the mean spectrum exhibits a distinct profile, characterized by the peak frequency being lower than the cut-off frequency, indicating that all significant cut-off errors originate from the cut-off in low frequencies. In these instances, the spectra are dominated by long swells that probably cannot be fully resolved by SWIM, and this will lead to a significant overestimation of the SWIM spectral density by MTF3.

Further examination of cases with low-frequency cut-off errors in excess of 0.2 m revealed that more than 95% of them have a peak frequency lower than 0.07 Hz (equivalent to a peak period of ~14 s). Most waves with such low peak frequencies are swells generated by remote wind (especially extra-tropical storms in the westerlies). Although cases of relatively large low-frequency cut-off errors in SWHs make up only a small fraction of the buoy dataset, this cut-off error has a large impact on the observation and study of long ocean swells propagating across the ocean. When using SWIM data to track very long swells, it is important to account for this low-frequency cut-off effect, as the spectral densities in the current version of the data product might be significantly overestimated for such long swell events.

5. Conclusions

The SWIM is capable of resolving the directional wave spectrum within a wavelength range of 23–500 m, which corresponds to a frequency range of 0.056–0.26 Hz in deep water. Meanwhile, the SWHs obtained by the nadir beam of SWIM, which are used to normalize the off-nadir SWIM wave spectra in MTF3, are validated using the buoy-observed SWHs calculated from the spectral energy within a broader frequency range of 0.02 Hz to 0.485 Hz. This mismatch in frequency ranges can potentially lead to an overestimation of the SWIM spectral densities. This study investigated this frequency/wavelength cut-off error using the wave spectra from buoys. The comparison of SWHs before and after the frequency cut-off revealed that the wavelength cut-off error is generally small. However, the high-frequency cut-off can cause a systematic bias in SWH of nearly 0.1 m, while the low-frequency cut-off can lead to significant errors under conditions of long swells. These cut-off errors in wave energy will be proportionally distributed across the SWIM wave spectra, leading to a proportional overestimation of the energy in each wave spectrum.

Understanding the nature of these errors, it was found that the high-frequency cut-off error can be statistically corrected using wind speed and SWH information. This is because the high-frequency cut-off error arises from the tails of the wave spectra that respond rapidly to local wind, and a proportion of the tail energy within the entire spectrum decreases with increasing SWH. An ANN, using wind speed and SWH as inputs, was employed and successfully corrected the high-frequency cut-off error.

In contrast, the low-frequency cut-off error, which comes from swells generated elsewhere, could not be effectively corrected using statistical methods. Fortunately, contemporary numerical wave models can provide some information on these long swells. Thus, the low-frequency cut-off error of SWH is partially corrected with the help of the output of a numerical wave model. Although it is understood that access to modeled wave spectra is not universally available across the global ocean, we failed to figure out a better way to correct low-frequency cut-off errors. Thus, it is also recommended that the future SWIM data product incorporates the along-track modeled wave spectra.

Following these corrections, the error in the estimated cut-off SWH was substantially reduced. The bias, RMSE, and CC improved from 0.086 m, 0.111 m, and 0.9976 to 0 m, 0.039 m, and 0.9994, respectively. Although this wavelength/frequency cut-off error may be considered second-order compared to other known error sources such as speckle noise and parasitic peaks, its significance should not be overlooked, particularly in cases involving low SWHs and long swells. Given that this error can be substantially corrected using our method, we hope that future versions of the SWIM data product will incorporate this correction or simply further extend the frequency range of the spectrum.

Author Contributions: Conceptualization, H.J.; methodology, H.J. and Y.L.; software, H.J. and Y.L.; validation, H.J.; formal analysis, H.J. and Y.L.; investigation, H.J. and Y.L.; resources, H.J. and Y.X.; data curation, Y.L. and H.J.; writing—original draft preparation, Y.L. and H.J.; writing—review and editing, H.Q. and Y.X.; visualization, H.J.; supervision, H.J.; project administration, Y.X.; funding acquisition, H.J. All authors have read and agreed to the published version of the manuscript.

Funding: This research was funded partly by the National Key Research and Development Program of China, grant number 2022YFC3104900-2022YFC3104905, partly by the National Natural Science Foundation of China, grant number 42376172, partly by the Key Laboratory of Space Ocean Remote Sensing and Application, MNR, grant number 2023CFO008, and partly by the Guangdong Basic and Applied Basic Research Foundation, grant number, 2022A1515240069/2024A1515012032/2023A1515240047.

Data Availability Statement: The NDBC buoy data are available from NDBC website (ASCII, https://www.ndbc.noaa.gov/, accessed on 19 July 2024). The IOWAGA dataset is available from (Alday et al., 2021) and can be downloaded from the official website of LOPS, IFREMER (https://www.umr-lops.fr/Donnees/Vagues, accessed on 19 July 2024).

Conflicts of Interest: The authors declare no conflicts of interest.

References

1. Dodet, G.; Piolle, J.-F.; Quilfen, Y.; Abdalla, S.; Accensi, M.; Ardhuin, F.; Ash, E.; Bidlot, J.-R.; Gommenginger, C.; Marechal, G.; et al. The Sea State CCI dataset v1: Towards a sea state climate data record based on satellite observations. *Earth Syst. Sci. Data* **2020**, *12*, 1929–1951. [CrossRef]
2. Jiang, H. Random, Environmental, and representativeness errors in ocean remote sensing versus in situ data: An example of wave heights from altimeters. *IEEE Trans. Geosci. Remote Sens.* **2023**, *61*, 4205613. [CrossRef]
3. Ribal, A.; Young, I.R. 33 years of globally calibrated wave height and wind speed data based on altimeter observations. *Sci. Data* **2019**, *6*, 77. [CrossRef] [PubMed]
4. Ardhuin, F.; Chapron, B.; Collard, F. Observation of swell dissipation across oceans. *Geophys. Res. Lett.* **2009**, *36*, L06607. [CrossRef]
5. Hauser, D.; Tourain, C.; Hermozo, L.; Alraddawi, D.; Aouf, L.; Chapron, B.; Dalphinet, A.; Delaye, L.; Dalila, M.; Dormy, E.; et al. New Observations from the SWIM Radar On-Board CFOSAT: Instrument Validation and Ocean Wave Measurement Assessment. *IEEE Trans. Geosci. Remote Sens.* **2021**, *59*, 5–26. [CrossRef]
6. The WAVEWATCH III® Development Group. User Manual and System Documentation of WAVEWATCH III® Version 5.16. Tech. Note, NOAA/NWS/NCEP/MMAB, College Park. MD, USA. 2016. Available online: https://polar.ncep.noaa.gov/waves/wavewatch/manual.v5.16.pdf (accessed on 19 July 2024).
7. Ardhuin, F.; Rogers, E.; Babanin, A.V.; Filipot, J.-F.; Magne, R.; Roland, A.; van der Westhuysen, A.; Queffeulou, P.; Lefevre, J.-M.; Aouf, L.; et al. Semiempirical Dissipation Source Functions for Ocean Waves. Part I: Definition, Calibration, and Validation. *J. Phys. Oceanogr.* **2010**, *40*, 1917–1941. [CrossRef]
8. Alday, M.; Accensi, M.; Ardhuin, F.; Dodet, G. A global wave parameter database for geophysical applications. Part 3: Improved forcing and spectral resolution. *Ocean Model.* **2021**, *166*, 101848. [CrossRef]
9. Phillips, O.M. Spectral and statistical properties of the equilibrium range in wind-generated gravity waves. *J. Fluid Mech.* **1985**, *156*, 505–531. [CrossRef]
10. Voermans, J.J.; Smit, P.B.; Janssen, T.T.; Babanin, A. V Estimating Wind Speed and Direction Using Wave Spectra. *J. Geophys. Res. Ocean.* **2020**, *125*, e2019JC015717. [CrossRef]
11. Jiang, H. Wind speed and direction estimation from wave spectra using deep learning. *Atmos. Meas. Tech.* **2022**, *15*, 1–9. [CrossRef]
12. Jiang, H.; Song, Y.; Mironov, A.; Yang, Z.; Xu, Y.; Liu, J. Accurate mean wave period from SWIM instrument on-board CFOSAT. *Remote Sens. Environ.* **2022**, *280*, 113149. [CrossRef]
13. Jiang, H.; Mu, L. Wave climate from spectra and its connections with local and remote wind climate. *J. Phys. Oceanogr.* **2019**, *49*, 543–559. [CrossRef]

14. Jiang, H.; Stopa, J.E.; Wang, H.; Husson, R.; Mouche, A.; Chapron, B.; Chen, G. Tracking the attenuation and nonbreaking dissipation of swells using altimeters. *J. Geophys. Res. Ocean.* **2016**, *121*, 1446–1458. [CrossRef]
15. Jiang, H.; Babanin, A.V.; Chen, G. Event-based validation of swell arrival time. *J. Phys. Oceanogr.* **2016**, *46*, 3563–3569. [CrossRef]

MDPI AG
Grosspeteranlage 5
4052 Basel
Switzerland
Tel.: +41 61 683 77 34

Remote Sensing Editorial Office
E-mail: remotesensing@mdpi.com
www.mdpi.com/journal/remotesensing

www.ingramcontent.com/pod-product-compliance
Lightning Source LLC
Chambersburg PA
CBHW061813210926
51850CB00006B/688